# Applied Mathematical Sciences
Volume 42

# Applied Mathematical Sciences

*(continued following index)*

John Guckenheimer
Philip Holmes

# Nonlinear Oscillations, Dynamical Systems, and Bifurcations of Vector Fields

With 206 Illustrations

Springer-Verlag
New York Berlin Heidelberg London Paris
Tokyo Hong Kong Barcelona Budapest

John Guckenheimer
Department of Mathematics
Cornell University
Ithaca, NY 14853
U.S.A.

Philip Holmes
Department of Theoretical and
  Applied Mechanics and
Center for Applied Mathematics
Cornell University
Ithaca, NY 14853
U.S.A.

*Editors*

F. John
Courant Institute of
  Mathematical Sciences
New York University
New York, NY 10012
U.S.A.

J. E. Marsden
Department of
  Mathematics
University of California
Berkeley, CA 94720
U.S.A.

L. Sirovich
Division of
  Applied Mathematics
Brown University
Providence, RI 02912
U.S.A.

Mathematics Subject Classification (1991): 34A34, 34C15, 34C35, 58FXX, 70KXX

Library of Congress Cataloging in Publication Data
Guckenheimer, John.
   Nonlinear oscillations, dynamical systems and
bifurcations of vector fields.
   (Applied mathematical sciences; v. 42)
   Bibliography: p.
   Includes index.
   1. Nonlinear oscillations.   2. Differentiable
dynamical systems.   3. Bifurcation theory.   4. Vector
fields.   I. Holmes, Philip.   II. Title.   III. Series:
Applied mathematical sciences (Springer-Verlag New York
Inc.); v. 42.
QA1.A647  vol. 42  [QA867.5]  510s  [531'.322]  82-19641

Typeset by Composition House Ltd., Salisbury, England.
Printed and bound by R. R. Donnelley & Sons, Harrisonburg, Virginia.
Printed in the United States of America.

9  8  7  6  5  4  (fourth printing, 1993)

ISBN 0-387-90819-6 Springer-Verlag New York Berlin Heidelberg Tokyo
ISBN 3-540-90819-6 Springer-Verlag Berlin Heidelberg New York Tokyo

To
G. Duffing, E. N. Lorenz and B. van der Pol,
pioneers in a chaotic land

# Correspondances

La Nature est un temple où de vivants piliers
Laissent parfois sortir de confuses paroles ;
L'homme y passe à travers des forêts de symboles
Qui l'observent avec des regards familiers.

Comme de longs échos qui de loin se confondent
Dans une ténébreuse et profonde unité,
Vaste comme la nuit et comme la clarté,
Les parfums, les couleurs et les sons se répondent.

Il est des parfums frais comme des chairs d'enfants,
Doux comme les hautbois, verts comme les prairies,
— Et d'autres, corrompus, riches et triomphants,

Ayant l'expansion des choses infinies,
Comme l'ambre, le musc, le benjoin et l'encens,
Qui chantent les transports de l'esprit et des sens.

CHARLES BAUDELAIRE
*Les Fleurs du Mal*, 1857

# Preface

## Introductory Remarks

Problems in dynamics have fascinated physical scientists (and mankind in general) for thousands of years. Notable among such problems are those of celestial mechanics, especially the study of the motions of the bodies in the solar system. Newton's attempts to understand and model their observed motions incorporated Kepler's laws and led to his development of the calculus. With this the study of models of dynamical problems as differential equations began.

In spite of the great elegance and simplicity of such equations, the solution of specific problems proved remarkably difficult and engaged the minds of many of the greatest mechanicians and mathematicians of the eighteenth and nineteenth centuries. While a relatively complete theory was developed for linear ordinary differential equations, nonlinear systems remained largely inaccessible, apart from successful applications of perturbation methods to weakly nonlinear problems. Once more, the most famous and impressive applications came in celestial mechanics.

Analysis remained the favored tool for the study of dynamical problems until Poincaré's work in the late-nineteenth century showed that perturbation methods might not yield correct results in all cases, because the series used in such calculations diverged. Poincaré then went on to marry analysis and geometry in his development of a qualitative approach to the study of differential equations.

The modern methods of qualitative analysis of differential equations have their origins in this work (Poincaré [1880, 1890, 1899]) and in the work of Birkhoff [1927], Liapunov [1949], and others of the Russian School: Andronov and co-workers [1937, 1966, 1971, 1973] and Arnold [1973, 1978,

1982]. In the past 20 years there has been an explosion of research. Smale, in a classic paper [1967], outlined a number of outstanding problems and stimulated much of this work. However, until the mid-1970s the new tools were largely in the hands of pure mathematicians, although a number of potential applications had been sketched, notably by Ruelle and Takens [1971], who suggested the importance of "strange attractors" in the study of turbulence.

Over the past few years applications in solid and structural mechanics as well as fluid mechanics have appeared, and there is now widespread interest in the engineering and applied science communities in strange attractors, chaos, and dynamical systems theory. We have written this book primarily for the members of this community, who do not generally have the necessary mathematical background to go directly to the research literature. We see the book primarily as a "user's guide" to a rapidly growing field of knowledge. Consequently we have selected for discussion only those results which we feel are applicable to physical problems, and have generally excluded proofs of theorems which we do not feel to be illustrative of the applicability. Nor have we given the sharpest or best results in all cases, hoping rather to provide a background on which readers may build by direct reference to the research literature.

This is far from a complete treatise on dynamical systems. While it may irritate some specialists in this field, it will, we hope, lead them in the direction of important applications, while at the same time leading engineers and physical scientists in the direction of exciting and useful "abstract" results. In writing for a mixed audience, we have tried to maintain a balance in our statement of results between mathematical pedantry and readability for those without formal mathematical training. This is perhaps most noticable in the way we define terms. While major new terms are defined in the traditional mathematical fashion, i.e., in a separate paragraph signalled by the word **Definition**, we have defined many of the more familiar terms as they occur in the body of the text, identifying them by *italics*. Thus we formally define *structural stability* on p. 39, while we define *asymptotic stability* (of a fixed point) on p. 3. For the reader's convenience, the index contains references to the terms defined in both manners.

The approach to dynamical systems which we adopt is a geometric one. A quick glance will reveal that this book is liberally sprinkled with illustrations—around 200 of them! Throughout we stress the geometrical and topological properties of solutions of differential equations and iterated maps. However, since we also wish to convey the important analytical underpinning of these illustrations, we feel that the numerous exercises, many of which require nontrivial algebraic manipulations and even computer work, are an essential part of the book. Especially in Chapter 2, the direct experience of watching graphical displays of numerical solutions to the systems of differential equations introduced there is extraordinarily valuable in developing an intuitive feeling for their properties. To help the reader along, we have tried to indicate which exercises are fairly routine applications of theory and which require more substantial effort. However, we warn the reader that, towards the end of the book, and especially in Chapter 7, some of our exercises become reasonable material for Ph.D. theses.

We have chosen to concentrate on applications in nonlinear oscillations for three reasons:

(1) There are many important and interesting problems in that field.
(2) It is a fairly mature subject with many texts available on the classical methods for analysis of such problems: the books of Stoker [1950], Minorsky [1962], Hale [1962], Hayashi [1964], or Nayfeh and Mook [1979] are good representatives. The geometrical analysis of two-dimensional systems (free oscillations) is also well established in the books of Lefschetz [1957] and Andronov and co-workers [1966, 1971, 1973].
(3) Most abstract mathematical examples known in dynamical systems theory occur "naturally" in nonlinear oscillator problems.

In this context, the present book should be seen as an attempt to extend the work of Andronov *et al.* [1966] by one dimension. This aim is not as modest as it might seem: as we shall see, the apparently innocent addition of a (small) periodic forcing term $f(t) = f(t + T)$ to a single degree of freedom nonlinear oscillator,

$$\ddot{x} + g(x, \dot{x}) = 0,$$

to yield the three-dimensional system

$$\ddot{x} + g(x, \dot{x}) = f(t),$$

or

$$\dot{x} = y,$$
$$\dot{y} = -g(x, y) + f(\theta),$$
$$\dot{\theta} = 1,$$

can introduce an uncountably infinite set of new phenomena, in addition to the fixed points and limit cycles familiar from the planar theory of non-linear oscillations. This book is devoted to a partial description and under-standing of these phenomena.

A somewhat glib observation, which, however, contains some truth, is that the pure mathematician tends to think of some nice (or nasty) property and then construct a dynamical system whose solutions exhibit it. In con-trast, the traditional rôle of the applied mathematician or engineer is to take a given system (perhaps a model that he or she has constructed) and find out what its properties are. We mainly adopt the second viewpoint, but our exposition may sometimes seem schizophrenic, since we are applying ideas of the former group to the problems of the latter group. Moreover, we feel strongly that the properties of specific systems cannot be discovered unless one knows what the possibilities are, and these are often revealed only by the general abstract theory. Practice and theory progress best hand-in-hand.

## The Contents of This Book

This book is concerned with the application of methods from dynamical systems and bifurcation theories to the study of nonlinear oscillations. The mathematical models we consider are (fairly small) sets of ordinary differential equations and mappings. Many of the results discussed in this book can be extended to infinite-dimensional evolution systems arising from partial differential equations. However, the main ideas are most easily seen in the finite-dimensional context, and it is here that we shall remain. Almost all the methods we describe also generalize to dynamical systems whose phase spaces are differentiable manifolds, but once more, so as not to burden the reader with technicalities, we restrict our exposition to systems with Euclidean phase spaces. However, in the final section of the last chapter we add a few remarks on partial differential equations.

In Chapter 1 we provide a *review* of basic results in the theory of dynamical systems, covering both ordinary differential equations (flows) and discrete mappings. (We concentrate on diffeomorphisms: smooth invertible maps.) We discuss the connection between diffeomorphisms and flows obtained by their Poincaré maps and end with a review of the relatively complete theory of two-dimensional differential equations. Our discussion moves quickly and is quite cursory in places. However, the bulk of this material has been treated in greater detail from the dynamical systems viewpoint in the books of Hirsch and Smale [1974], Irwin [1980], and Palis and de Melo [1982], and from the oscillations viewpoint in the books of Andronov and his co-workers [1966, 1971, 1973] and we refer the reader to these texts for more details. Here the situation is fairly straightforward and solutions generally behave nicely.

Chapter 2 presents four examples from nonlinear oscillations: the famous oscillators of van der Pol [1927] and Duffing [1918], the Lorenz equations [1963], and a bouncing ball problem. We show that the solutions of these problems can be markedly chaotic and that they seem to possess strange attractors: attracting motions which are neither periodic nor even quasiperiodic. The development of this chapter is not systematic, but it provides a preview of the theory developed in the remainder of the book. We recommend that either the reader skim this chapter to gain a general impression before going on to our systematic development of the theory in later chapters, or read it with a microcomputer at hand, so that he can simulate solutions of the model problems we discuss.

We then retreat from the chaos of these examples to muster our forces. Chapter 3 contains a discussion of the methods of local bifurcation theory for flows and maps, including center manifolds and normal forms. Rather different, less geometrical, and more analytical discussions of local bifurcations can be found in the recent books by Iooss and Joseph [1981] and Chow and Hale [1982].

In Chapter 4 we develop the analytical methods of averaging and perturbation theory for the study of periodically forced nonlinear oscillators, and show that they can yield surprising global results. We end this chapter with a brief discussion of chaos and nonintegrability in Hamiltonian systems and the Kolmogorov–Arnold–Moser theory. More complete introductions to this area can be found in Arnold [1978], Lichtenberg and Lieberman [1982], or, for the more mathematically inclined, Abraham and Marsden [1978].

In Chapter 5 we return to chaos, or rather to the close analysis of geometrically defined two-dimensional maps with complicated invariant sets. The famous horseshoe map of Smale is discussed at length, and the method of symbolic dynamics is described and illustrated. A section on one-dimensional (noninvertible) maps is included, and we return to the specific examples of Chapter 2 to provide additional information and illustrate the analytical methods. We end this chapter with a brief discussion of Liapunov exponents and invariant measures for strange attractors.

In Chapter 6 we discuss global homoclinic and heteroclinic bifurcations, bifurcations of one-dimensional maps, and once more illustrate our results with the examples of Chapter 2. Finally, in our discussion of global bifurcations of two-dimensional maps and wild hyperbolic sets, we arrive squarely at one of the present frontiers of the field. We argue that, while the one-dimensional theory is relatively complete (cf. Collet and Eckmann [1980]), the behavior of two-dimensional diffeomorphisms appears to be considerably more complex and is still incompletely understood. We are consequently unable to complete our analysis of the nonlinear oscillators of van der Pol and Duffing, but we are able to give a clear account of much of their behavior and to show precisely what presently obstructs further analysis.

In the final chapter we show how the global bifurcations, discussed previously, reappear in degenerate local bifurcations, and we end with several more models of physical problems which display these rich and beautiful behaviors.

Throughout the book we continually return to specific examples, and we have tried to illustrate even the most abstract results. In our Appendix we give suggestions for further reading. We make no claims for the completeness of our bibliography. We have, however, tried to include references to the bulk of the papers, monographs, lecture notes, and books which have proved useful to ourselves and our colleagues, but we recognize that our biases probably make this a rather eclectic selection.

We have included a glossary of the more important terminology for the convenience of those readers lacking a formal mathematical training.

Finally, we would especially like to acknowledge the encouragement, advice, and gentle criticisms of Bill Langford, Clark Robinson and David Rod, whose careful readings of the manuscript enabled us to make many corrections and improvements.

Nessen MacGiolla Mhuiris, Xuehai Li, Lloyd Sakazata, Rakesh, Kumarswamy Hebbale, and Pat Hollis suffered through the preparation of this manuscript as students in TAM 776 at Cornell, and pointed out many errors almost as quickly as they were made. Edgar Knobloch, Steve Shaw, and David Whitley also read and commented on the manuscript. The comments of these and many other people have helped us to improve this book, and it only remains for each of us to lay the blame for any remaining errors and omissions squarely on the shoulders of the other.

Barbara Boettcher prepared the illustrations from our rough notes and Dolores Pendell deserves more thanks than we can give for her patient typing and retyping of our almost illegible manuscripts.

Finally, we thank our wives and children for their understanding and patience during the production of this addition to our families.

JOHN GUCKENHEIMER                                    PHILIP HOLMES
*Santa Cruz, Spring 1983*                              *Ithaca, Spring 1983*

## Preface to the Second Printing

The reprinting of this book some $2\frac{1}{2}$ years after its publication has provided us with the opportunity of correcting many minor typographical errors and a few errors of substance. In particular, errors in Section 6.5 in the study of the Šilnikov return map have been corrected, and we have rewritten parts of Sections 7.4 and 7.5 fairly extensively in the light of recent work by Carr, Chow, Cushman, Hale, Sanders, Zholondek, and others on the number of limit cycles and bifurcations in these unfoldings. In the former case the main result is unaffected, but in the latter case some of our intuitions (as well as the incorrect calculations with which we supported them) have proved wrong. We take some comfort in the fact that our naive assertions stimulated some of the work which disproved them.

Although progress in some areas of applied dynamical systems has been rapid, and significant new developments have occurred since the first printing, we have not seen fit to undertake major revisions of the book at this stage, although we have briefly noted some of the developments which bear directly on topics discussed in the book. These comments appear at the end of the book, directly after the Appendix. A complete revision will perhaps be appropriate 5 or 10 years from now. (Anyone wishing to perform it, please contact us!) In the same spirit, we have not attempted to bring the bibliography up to date, although we have added about 75 references, including those mentioned above. References that were in preprint form at the first printing have been updated in cases where the journal of publication is known. In cases in which the publication date of the journal differs from that of the preprint, the journal date is given at the end of the reference. We note that a useful bibliography due to Shiraiwa [1981] has recently been updated (Shiraiwa [1985]); it contains over 4,400 items.

In preparing the revisions we have benefited from the advice and corrections supplied by many readers, including Marty Golubitsky, Kevin Hockett, Fuhua Ling, Wei-Min Liu, Clark Robinson, Jan Sanders, Steven Shaw, Ed Zehnder, and Zhaoxuan Zhu. Professor Ling, of the Shanghai Jiao Tong University, with the help of his students and of Professor Zhu, of Peking University, has prepared a Chinese translation of this book.

JOHN GUCKENHEIMER<br>
PHILIP HOLMES<br>
*Ithaca, Fall 1985*

# Contents

# CHAPTER 1
# Introduction: Differential Equations and Dynamical Systems

In this introductory chapter we review some basic topics in the theory of ordinary differential equations from the viewpoint of the global geometrical approach which we develop in this book. After recalling the basic existence and uniqueness theorems, we consider the linear, homogeneous, constant coefficient system and then introduce nonlinear and time-dependent systems and concepts such as the Poincaré map and structural stability. We then review some of the better-known results on two-dimensional autonomous systems and end with a statement and sketch of the proof of Peixoto's theorem, an important result which summarizes much of our knowledge of two-dimensional flows.

In the first two sections our review of basic theory and the linear system $\dot{x} = Ax$ is rapid. We assume that the reader is fairly familiar with this material and with the fundamental notions from analysis used in its derivation. Most standard courses in ordinary differential equations deal with these topics, and the material covered in these sections is treated in detail in the books of Hirsch and Smale [1974] and Arnold [1973], for example. We especially recommend the former text as one of the few elementary introductions to the geometric theory of ordinary differential equations. However, most books on differential equations contain versions of the main results.

## 1.0. Existence and Uniqueness of Solutions

For the purposes of this book, it is generally sufficient to regard a differential equation as a system

$$\frac{dx}{dt} \overset{\text{def}}{=} \dot{x} = f(x), \qquad (1.0.1)$$

where $x = x(t) \in \mathbb{R}^n$ is a vector valued function of an independent variable (usually time) and $f : U \to \mathbb{R}^n$ is a smooth function defined on some subset $U \subseteq \mathbb{R}^n$. We say that the *vector field* $f$ generates a *flow* $\phi_t : U \to \mathbb{R}^n$, where $\phi_t(x) = \phi(x, t)$ is a smooth function defined for all $x$ in $U$ and $t$ in some interval $I = (a, b) \subseteq \mathbb{R}$, and $\phi$ satisfies (1.0.1) in the sense that

$$\frac{d}{dt}(\phi(x, t))|_{t=\tau} = f(\phi(x, \tau)) \tag{1.0.2}$$

for all $x \in U$ and $\tau \in I$. We note that (in its domain of definition) $\phi_t$ satisfies the group properties (i) $\phi_0 = $ id, and (ii) $\phi_{t+s} = \phi_t \circ \phi_s$. Systems of the form (1.0.1), in which the vector field does not contain time explicitly, are called *autonomous*.

Often we are given an initial condition

$$x(0) = x_0 \in U, \tag{1.0.3}$$

in which case we seek a solution $\phi(x_0, t)$ such that

$$\phi(x_0, 0) = x_0. \tag{1.0.4}$$

(We will also sometimes write such a solution as $x(x_0, t)$, or simply $x(t)$.)

In this case $\phi(x_0, \cdot) : I \to \mathbb{R}^n$ defines a *solution curve, trajectory,* or *orbit* of the differential equation (1.0.1) *based at* $x_0$. Since the vector field of the autonomous system (1.0.1) is invariant with respect to translations in time, solutions based at times $t_0 \neq 0$ can always be translated to $t_0 = 0$.

In classical texts on ordinary differential equations, such as Coddington and Levinson [1955], the stress is on individual solution curves and their properties. Here we shall be more concerned with families of such curves, and hence with the global behavior of the flow $\phi_t : U \to \mathbb{R}^n$ defined for (all) points $x \in U$; see Figure 1.0.1. In particular, the concepts of smooth *invariant manifolds* composed of solution curves, discussed in the books of Hartman [1964] and Hale [1969] will be of importance. We will introduce these ideas in the context of linear systems in the next section.

We will not usually need the more general concept of a dynamical system as a flow on a differentiable manifold $M$ arising from a vector field, regarded as a map

$$f : M \to TM,$$

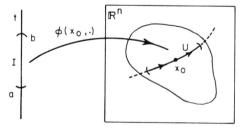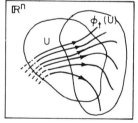

Figure 1.0.1. A solution curve and the flow. (a) The solution curve $\phi_t(x_0)$; (b) the flow $\phi_t$.

where $TM$ is the tangent bundle of $M$. We will therefore not need much from the theory of differential topology. For those interested, Chillingworth's [1976] book provides a good introduction; also see Arnold [1973]. In almost all cases in which we work explicitly with phase spaces which are manifolds, we will have a global coordinate system (a single chart) and we can essentially work in the covering space: i.e., in $\mathbb{R}^n$ modulo some suitable identification, as in the cases of the torus $T^2 = \mathbb{R}^2/\mathbb{Z}^2$ and the cylinder $S^1 \times \mathbb{R} = \mathbb{R}^2/\mathbb{Z}$. Such systems typically arise when the vector field $f$ is periodic in (some of) its components. We meet the first such systems in Sections 1.4 and 1.5.

In discussing submanifolds of solutions such as the stable, unstable, and center manifolds, we shall be able to work with copies of real Euclidean spaces defined locally by graphs.

We now state, without proof, the basic local existence and uniqueness theorem (cf. Coddington and Levinson [1955], Hirsch and Smale [1974]):

**Theorem 1.0.1.** *Let $U \subset \mathbb{R}^n$ be an open subset of real Euclidean space (or of a differentiable manifold $M$), let $f : U \to \mathbb{R}^n$ be a continuously differentiable $(C^1)$ map and let $x_0 \in U$. Then there is some constant $c > 0$ and a unique solution $\phi(x_0, \cdot) : (-c, c) \to U$ satisfying the differential equation $\dot{x} = f(x)$ with initial condition $x(0) = x_0$.*

In fact $f$ need only be (locally) Lipschitz, i.e., $|f(y) - f(x)| \leq K|x - y|$ for some $K < \infty$, where $K$ is called the *Lipschitz constant* for $f$. Thus we can deal with piecewise linear functions, such as one gets in "stick-slip" friction problems and in the clock problem (cf. Andronov *et al.* [1966], pp. 186ff.).

Intuitively, any solution may leave $U$ after sufficient time. We therefore say that the theorem is only *local*. We can easily construct vector fields $f : U \to \mathbb{R}^n$ such that $x(t)$ leaves any subset $U \subset \mathbb{R}^n$ in a finite time, for example,

$$\dot{x} = 1 + x^2, \tag{1.0.5}$$

which has the general solution $x(t) = \tan(t + c)$. Thus, although there are many equations on non-compact phase spaces (such as $\mathbb{R}^n$) for which solutions do exist globally in time, we cannot assert this in specific cases without further investigation.

*Fixed points*, also called an *equilibria* or *zeroes*, are an important class of solutions of a differential equation. Fixed points are defined by the vanishing of the vector field $f(x)$: $f(\bar{x}) = 0$. A fixed point $\bar{x}$ is said to be *stable* if a solution $x(t)$ based nearby remains close to $\bar{x}$ for all time, i.e., if for every neighborhood $V$ of $\bar{x}$ in $U$ there is a neighborhood $V_1 \subset V$ such that every solution $x(x_0, t)$ with $x_0 \in V_1$ is defined and lies in $V$ for all $t > 0$. If, in addition, $V_1$ can be chosen so that $x(t) \to \bar{x}$ as $t \to \infty$ then $\bar{x}$ is said to be *asymptotically stable*. See Figure 1.0.2.

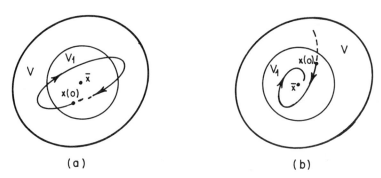

Figure 1.0.2. (a) Stability; (b) asymptotic stability.

EXERCISE 1.0.1. Show that the fixed points of the linear systems

(a) $\dot{x} = y, \dot{y} = -x$;
(b) $\dot{x} = y, \dot{y} = -x - y$,

are both stable. Which one is asymptotically stable? (Routine.)

The type of stability illustrated in Figure 1.0.2(a) is sometimes called *neutral* and is typified by fixed points such as *centers*. Asymptotically stable fixed points are called *sinks*. A fixed point is called *unstable* if it is not stable: *saddle points* and *sources* provide examples of such equilibria. Hirsch and Smale [1974, Chapter 9] give a detailed discussion of stability of fixed points.

The notions of stability defined above are *local* in nature: they concern only the behavior of solutions near the fixed point $\bar{x}$. Even if such solutions remain bounded for all time, other solutions may not exist globally.

EXERCISE 1.0.2. Find the fixed points for the equation $\dot{x} = -x + x^2$ and discuss their stability. Show that this equation has solutions which exist for all time as well as solutions which become unbounded in finite time. (Solving this equation is straightforward, but this interpretation of the behavior of solutions may be new to you.)

Often a Liapunov function approach suffices to show that an energy-like quantity decreases for $|x|$ sufficiently large, so that $x(t)$ remains bounded for all $t$ and all (bounded) initial conditions $x(0)$. Since it is so useful, we outline the method here for completeness. For more details, see Hirsch and Smale [1974, §9.3] or LaSalle and Lefschetz [1961]. The method relies on finding a positive definite function $V: U \rightarrow \mathbb{R}$, called the *Liapunov function*, which decreases along solution curves of the differential equation:

**Theorem 1.0.2** (Hirsch and Smale [1974], pp. 192ff.). *Let $\bar{x}$ be a fixed point for* (1.0.1) *and $V: W \rightarrow \mathbb{R}$ be a differentiable function defined on some neighborhood $W \subseteq U$ of $\bar{x}$ such that:*

(i) $V(\bar{x}) = 0$ and $V(x) > 0$ if $x \neq \bar{x}$; and
(ii) $\dot{V}(x) \leq 0$ in $W - \{\bar{x}\}$.

*Then $\bar{x}$ is stable. Moreover, if*

(iii) $\dot{V}(x) < 0$ *in* $W - \{\bar{x}\}$;

*then $\bar{x}$ is asymptotically stable.*

Here

$$\dot{V} = \sum_{j=1}^{n} \frac{\partial V}{\partial x_j} \dot{x}_j = \sum_{j=1}^{n} \frac{\partial V}{\partial x_j} f_j(x)$$

is the derivative of $V$ along solution curves of (1.0.1).

If we can choose $W (= U) = \mathbb{R}^n$ in case (iii), then $x$ is said to be *globally asymptotically stable*, and we can conclude that all solutions remain bounded and in fact approach $\bar{x}$ as $t \to \infty$. Thus the stability of equilibria and boundedness of solutions can be tested without actually solving the differential equation. There are, however, no general methods for finding suitable Liapunov functions, although in mechanical problems the energy is often a good candidate.

EXAMPLE. Consider the motion of a particle of mass $m$ attached to a spring of stiffness $k(x + x^3)$, $k > 0$, where $x$ is displacement. The differential equation governing the system is

$$m\ddot{x} + k(x + x^3) = 0, \tag{1.0.6}$$

or, letting $\dot{x} = y$

$$\dot{x} = y,$$

$$\dot{y} = -\frac{k}{m}(x + x^3). \tag{1.0.7}$$

The associated total energy of the system is

$$E(x, y) = \frac{my^2}{2} + k\left(\frac{x^2}{2} + \frac{x^4}{4}\right). \tag{1.0.8}$$

We note that $E(x, y)$ provides a Liapunov function for (1.0.7), since $E(0, 0) = 0$ at the (unique) equilibrium $(x, y) = (0, 0)$ and $E(x, y) > 0$ for $(x, y) \neq (0, 0)$. Moreover, we have

$$\dot{E} = my\dot{y} + k(x + x^3)\dot{x}$$
$$= -ky(x + x^3) + k(x + x^3)y \equiv 0; \tag{1.0.9}$$

thus $(x, y) = (0, 0)$ is (neutrally) stable. If we add some damping $\alpha > 0$, to the system, so that the equation of motion becomes

$$\dot{x} = y,$$

$$\dot{y} = -\frac{k}{m}(x + x^3) - \alpha y, \tag{1.0.10}$$

then the same Liapunov function yields

$$\dot{E} = -\alpha m y^2, \qquad (1.0.11)$$

which is negative for all $(x, y) \neq (0, 0)$ except on the $x$-axis. We therefore modify the Liapunov function slightly to

$$V(x, y) = \frac{my^2}{2} + k\left(\frac{x^2}{2} + \frac{x^4}{4}\right) + \beta\left(xy + \frac{\alpha x^2}{2}\right), \qquad (1.0.12)$$

so that

$$\dot{V} = my\dot{y} + k(x + x^3)\dot{x} + \beta(\dot{x}y + x\dot{y} + \alpha x\dot{x})$$

$$= (my + \beta x)\left(-\frac{k}{m}(x + x^3) - \alpha y\right) + k(x + x^3)y + \beta y^2 + \alpha\beta xy$$

$$= -\beta\frac{k}{m}(x^2 + x^4) - (\alpha m - \beta)y^2. \qquad (1.0.13)$$

If we choose $\beta$ sufficiently small, $V$ remains positive definite and $\dot{V}$ is strictly negative for *all* $(x, y) \neq (0, 0)$. Thus $(0, 0)$ is globally asymptotically stable for $\alpha > 0$.

In differentiating $V$ along solution curves we are trying to verify that all solutions cross the level curves of $V$ "inwards." A sketch of the level curves of $E$ and the modified function $V$ for this example show that those of $V$ are slightly tilted, so that the vector field is nowhere tangent to them, whereas, even with damping present, the vector field is tangent to $E = \text{constant}$ on $y = 0$ (Figure 1.0.3).

EXERCISE 1.0.3. Using the Liapunov function $V = \frac{1}{2}(x^2 + \sigma y^2 + \sigma z^2)$, obtain conditions on $\sigma$, $\rho$, and $\beta$ sufficient for global asymptotic stability of the origin $(x, y, z) = (0, 0, 0)$ in the Lorenz equations

$$\dot{x} = \sigma(y - x); \qquad \dot{y} = \rho x - y - xz; \qquad \dot{z} = -\beta z + xy; \qquad \sigma, \beta > 0.$$

Are your conditions also necessary?

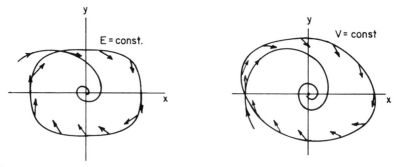

Figure 1.0.3. Level curves of the Liapunov functions $E$ and $V$ and the vector field of equations (1.0.10).

For problems with multiple equilibria, local Liapunov functions can be sought, or one can attempt to find a compact hypersurface $S \subset \mathbb{R}^n$ such that the vector field is directed everywhere inward on $S$. If such a surface exists, then any solution starting on or inside $S$ can never leave the interior of $S$ and thus must remain bounded for all time. We use this approach in several examples later in this book.

The local existence theorem (Theorem 1.0.1) becomes global in all cases when we work on *compact* manifolds $M$ instead of open spaces like $\mathbb{R}^n$:

**Theorem 1.0.3** (Chillingworth [1976], pp. 187–188). *The differential equation* $\dot{x} = f(x)$, $x \in M$, *with* $M$ *compact, and* $f \in C^1$, *has solution curves defined for all* $t \in \mathbb{R}$.

Thus flows on spheres and tori are globally defined, since there is no way in which solutions can escape from such manifolds.

The local theorem can be extended to show that solutions depend in a "nice" way on initial conditions (cf. Coddington and Levinson [1955], Hirsch and Smale [1974]):

**Theorem 1.0.4.** *Let* $U \subseteq \mathbb{R}^n$ *be open and suppose* $f : U \to \mathbb{R}^n$ *has a Lipschitz constant* $K$. *Let* $y(t), z(t)$ *be solutions to* $\dot{x} = f(x)$ *on the closed interval* $[t_0, t_1]$. *Then, for all* $t \in [t_0, t_1]$,

$$|y(t) - z(t)| \le |y(t_0) - z(t_0)|e^{K(t - t_0)}.$$

We note that this continuous dependence does not preclude the exponentially fast separation of solutions typical of the chaotic flows to be encountered in subsequent chapters, cf. Figure 1.0.4.

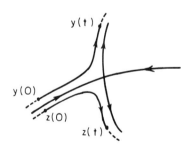

Figure 1.0.4. Exponential separation of neighboring solutions near a saddle point.

EXERCISE 1.0.4. Which of the following systems give rise to globally defined flows?

(a) $\dot{x} = x, x \in \mathbb{R}$;
(b) $\dot{x} = x^2, x \in \mathbb{R}$;
(c) $\dot{x} = 2 + \cos x, x \in \mathbb{R}$;
(d) $\dot{x} = \cos^2 x, x \in S^1$;
(e) $\dot{x} = -x^3, x \in \mathbb{R}$;
(f) $\dot{x} = Ax, x \in \mathbb{R}^n$, where $A$ is an $n \times n$ constant matrix.

(You can integrate all of these directly, but will need linear algebra, reviewed in the next section, for the last one.)

EXERCISE 1.0.5. Show that $\dot{x} = x^{2/3}$ does not have unique solutions for all initial points $x(0)$. Under what conditions are solutions unique? (This example is an old favorite in classical texts on differential equations.)

## 1.1. The Linear System $\dot{x} = Ax$

We first review some features of the linear system

$$\frac{dx}{dt} \stackrel{\text{def}}{=} \dot{x} = Ax, \qquad x \in \mathbb{R}^n, \tag{1.1.1}$$

where $A$ is an $n \times n$ matrix with constant coefficients. For more information and background see a standard introductory text on differential equations such as Braun [1978]; for a more detailed review of the linear algebra from the viewpoint of dynamical systems theory, Hirsch and Smale [1974] or Arnold [1973] are recommended.

By a solution of (1.1.1) we mean a vector valued function $x(x_0, t)$ depending on time $t$ and the initial condition

$$x(0) = x_0; \tag{1.1.2}$$

$x(x_0, t)$ is thus a solution of the initial value problem (1.1.1)–(1.1.2). In terms of the flow $\phi_t$, we have $x(x_0, t) \equiv \phi_t(x_0)$. Theorem 1.0.4 guarantees that the solution $x(x_0, t)$ of the linear system is defined for all $t \in \mathbb{R}$ and $x_0 \in \mathbb{R}^n$. Note that such *global* existence in time does not generally hold for nonlinear systems, as we have already seen. However, no such problems occur for (1.1.1), the solution of which is given by

$$x(x_0, t) = e^{tA}x_0, \tag{1.1.3}$$

where $e^{tA}$ is the $n \times n$ matrix obtained by exponentiating $A$. We will see how $e^{tA}$ can be calculated most conveniently in a moment, but first note that it is defined by the convergent series

$$e^{tA} = \left[ I + tA + \frac{t^2}{2!}A^2 + \cdots + \frac{t^n}{n!}A^n + \cdots \right]. \tag{1.1.4}$$

We leave it to the reader to make use of (1.1.4) to prove that (1.1.3) does indeed solve (1.1.1)–(1.1.2).

A *general* solution to (1.1.1) can be obtained by linear superposition of $n$ linearly independent solutions $\{x^1(t), \ldots, x^n(t)\}$:

$$x(t) = \sum_{j=1}^{n} c_j x^j(t), \tag{1.1.5}$$

where the $n$ unknown constants $c_j$ are to be determined by initial conditions.

If $A$ has $n$ linearly independent eigenvectors $v^j$, $j = 1, \ldots, n$, then we may take as a basis for the space of solutions the vector valued functions

$$x^j(t) = e^{\lambda_j t} v^j, \tag{1.1.6}$$

where $\lambda_j$ is the eigenvalue associated with $v^j$. For complex eigenvalues without multiplicity, $\lambda_j$, $\bar{\lambda}_j = \alpha_j \pm i\beta_j$, having eigenvectors $v^R \pm iv^I$, we may take

$$x^j = e^{\alpha_j t}(v^R \cos \beta t - v^I \sin \beta t),$$
$$x^{j+1} = e^{\alpha_j t}(v^R \sin \beta t + v^I \cos \beta t), \tag{1.1.7}$$

as the associated pair of (real) linearly independent solutions. When there are repeated eigenvalues and less than $n$ eigenvectors, then one generates the *generalized* eigenvectors as described by Braun [1978], for example. Again one obtains a set of $n$ linearly independent solutions. We denote the *fundamental solution matrix* having these $n$ solutions for its columns as

$$X(t) = [x^1(t), \ldots, x^n(t)]. \tag{1.1.8}$$

The columns $x^j(t)$, $j = 1, \ldots, n$ of $X(t)$ form a basis for the space of solutions of (1.1.1). It is easy to show that

$$e^{tA} = X(t)X(0)^{-1}; \tag{1.1.9}$$

we again leave the proof as an exercise.

EXERCISE 1.1.1. Find $e^{tA}$ for

$$A = \begin{bmatrix} 2 & 1 & 3 \\ 0 & 2 & 0 \\ 1 & 0 & 0 \end{bmatrix}.$$

Then solve $\dot{x} = Ax$ for initial conditions

$$x_0 = \begin{pmatrix} 1 \\ 1 \\ 1 \end{pmatrix}, \quad \begin{pmatrix} -2 \\ 0 \\ 2 \end{pmatrix}, \quad \text{and} \quad \begin{pmatrix} 5 \\ -3 \\ 2 \end{pmatrix}.$$

What do you notice about the last two solutions? Look carefully at the geometry of the solutions and eigenspaces.

Equation (1.1.1) may also be solved by first finding an invertible transformation $T$ which diagonalizes $A$ or at least puts it into Jordan normal form (if there are repeated eigenvalues). Equation (1.1.1) becomes

$$\dot{y} = Jy, \tag{1.1.10}$$

where $J = T^{-1}AT$ and $x = Ty$. Equation (1.1.10) is easy to work with, but since the columns of $T$ are the (generalized) eigenvectors of $A$, just as much work is required as in the former method. The exponential $e^{tA}$ may be computed as

$$e^{tA} = Te^{tJ}T^{-1} \tag{1.1.11}$$

(cf. Hirsch and Smale [1974], pp. 84–87), where exponentials are evaluated for the three $2 \times 2$ Jordan form matrices:

$$A = \begin{bmatrix} \lambda_1 & 0 \\ 0 & \lambda_2 \end{bmatrix}, \qquad e^{tA} = \begin{bmatrix} e^{\lambda_1 t} & 0 \\ 0 & e^{\lambda_2 t} \end{bmatrix};$$

$$A = \begin{bmatrix} \alpha & -\beta \\ \beta & \alpha \end{bmatrix}, \qquad e^{tA} = e^{\alpha t} \begin{bmatrix} \cos \beta t & -\sin \beta t \\ \sin \beta t & \cos \beta t \end{bmatrix}; \qquad (1.1.12)$$

$$A = \begin{bmatrix} \lambda & 0 \\ 1 & \lambda \end{bmatrix}, \qquad e^{tA} = e^{\lambda t} \begin{bmatrix} 1 & 0 \\ t & 1 \end{bmatrix}.$$

We also note that if $v^j$ is an eigenvector belonging to a real eigenvalue $\lambda_j$ of $A$, then $v^j$ is also an eigenvector belonging to the eigenvalue $e^{\lambda_j}$ of $e^A$. Moreover, if span$\{\text{Re}(v^j), \text{Im}(v^j)\}$ is an eigenspace belonging to a complex conjugate pair $\lambda_j$, $\bar{\lambda}_j$ of eigenvalues, then it is also an eigenspace belonging to $e^{\lambda_j}$, $e^{\bar{\lambda}_j}$.

## 1.2. Flows and Invariant Subspaces

The matrix $e^{tA}$ can be regarded as a mapping from $\mathbb{R}^n$ to $\mathbb{R}^n$: given any point $x_0$ in $\mathbb{R}^n$, $x(x_0, t) = e^{tA} x_0$ is the point at which the solution based at $x_0$ lies after time $t$. The operator $e^{tA}$ hence contains *global* information on the set of *all* solutions of (1.1.1), since the formula (1.1.3) holds for all points $x_0 \in \mathbb{R}^n$. As in the general case, described in Section 1.0, we say that $e^{tA}$ defines a *flow* on $\mathbb{R}^n$ and that this flow (or "phase flow") is *generated by* the vector field $Ax$ defined on $\mathbb{R}^n$: $e^{tA}$ is our first specific example of a flow $\phi_t$.

The flow $e^{tA}: \mathbb{R}^n \to \mathbb{R}^n$ can be thought of as the set of all solutions to (1.1.1). In this set certain solutions play a special role; those which lie in the linear subspaces spanned by the eigenvectors. These subspaces are *invariant* under $e^{tA}$, in particular, if $v^j$ is a (real) eigenvector of $A$, and hence of $e^{tA}$, then a solution based at a point $c_j v^j \in \mathbb{R}^n$ remains on span$\{v^j\}$ for all time; in fact

$$x(cv^j, t) = cv^j e^{\lambda_j t}. \qquad (1.2.1)$$

Similarly, the (two-dimensional) subspace spanned by $\text{Re}\{v^j\}$, $\text{Im}\{v^j\}$, when $v^j$ is a complex eigenvector, is invariant under $e^{tA}$. In short, the eigenspaces of $A$ are invariant subspaces for the flow. It is worth returning to Exercise 1.1.1 in the light of this discussion.

We divide the subspaces spanned by the eigenvectors into three classes:

the *stable subspace*, $E^s = \text{span}\{v^1, \ldots, v^{n_s}\}$,

the *unstable subspace*, $E^u = \text{span}\{u^1, \ldots, u^{n_u}\}$,

the *center subspace*, $E^c = \text{span}\{w^1, \ldots, w^{n_c}\}$,

where $v^1, \ldots, v^{n_s}$ are the $n_s$ (generalized) eigenvectors whose eigenvalues have negative real parts, $u^1, \ldots, u^{n_u}$ are the $n_u$ (generalized) eigenvectors whose eigenvalues have positive real parts and $w^1, \ldots, w^{n_c}$ are those whose eigenvalues have zero real parts. Of course, $n_s + n_c + n_u = n$. The names reflect the facts that solutions lying on $E^s$ are characterized by exponential decay (either monotonic or oscillatory), those lying in $E^u$ by exponential growth, and those lying in $E^c$ by neither. In the absence of multiple eigenvalues, these latter either oscillate at constant amplitude (if $\lambda, \bar{\lambda} = \pm i\beta$) or remain constant (if $\lambda = 0$). A schematic picture appears in Figure 1.2.1, with two specific examples.

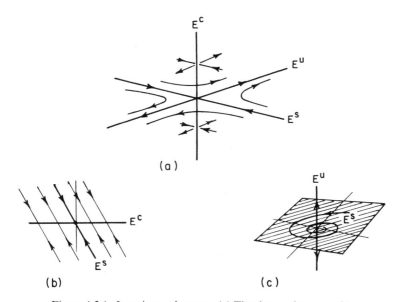

Figure 1.2.1. Invariant subspaces. (a) The three subspaces; (b)

$$A = \begin{bmatrix} 0 & 1 \\ 0 & -4 \end{bmatrix},$$

$(E^s = \text{span}(1, -4), E^c = \text{span}(1, 0), E^u = \varnothing)$; (c)

$$A = \begin{bmatrix} -1 & -1 & 0 \\ 1 & -1 & 0 \\ 0 & 0 & 2 \end{bmatrix}$$

$(E^s = \text{span}\{(1, 0, 0), (1, 1, 0)\}, E^c = \varnothing, E^u = (0, 0, 1))$.

When there are multiple eigenvalues for which algebraic and geometric multiplicities differ, then one may have growth of solutions in $E^c$, as the following exercise demonstrates:

EXERCISE 1.2.1. Find general solutions for the linear system $\dot{x} = Ax$, $x \in \mathbb{R}^2$ with

(a) $A = \begin{bmatrix} 0 & 0 \\ 0 & 0 \end{bmatrix}$;    (b) $A = \begin{bmatrix} 0 & 0 \\ 1 & 0 \end{bmatrix}$.

For more information on the flow $e^{tA}$, and for a complete classification of two- and three-dimensional systems, the reader is referred to Hirsch and Smale [1974] or Arnold [1973].

## 1.3. The Nonlinear System $\dot{x} = f(x)$

We must start by admitting that almost nothing beyond general statements can be made about most nonlinear systems. In the remainder of this book we will meet some of the delights and horrors of such systems, but the reader must bear in mind that the line of attack we develop in this text is only one, and that any other tool in the workshop of applied mathematics, including numerical integration, perturbation methods, and asymptotic analysis, can and should be brought to bear on a specific problem.

We recall that the basic existence–uniqueness theorem for ordinary differential equations, given in Section 1.0, implies that, for smooth functions* $f(x)$, the solution to the initial value problem

$$\dot{x} = f(x); \qquad x \in \mathbb{R}^n, \qquad x(0) = x_0 \qquad (1.3.1)$$

is defined at least in some neighborhood $t \in (-c, c)$ of $t = 0$. Thus a *local* flow $\phi_t : \mathbb{R}^n \to \mathbb{R}^n$ is defined by $\phi_t(x_0) = x(t, x_0)$ in a manner analogous to that in the linear case, although of course we cannot give a general formula like $e^{tA}$.

A good place to start the study of the nonlinear system $\dot{x} = f(x)$ is by finding the *zeros* of $f$ or the *fixed points* of (1.3.1). These are also referred to as *zeros, equilibria,* or *stationary solutions.* Even this may be a formidable task, although in most of our examples it will not be. Suppose then that we have a fixed point $\bar{x}$, so that $f(\bar{x}) = 0$, and we wish to characterize the behavior of solutions near $\bar{x}$. We do this by linearizing (1.3.1) at $\bar{x}$, that is, by studying the linear system

$$\dot{\xi} = Df(\bar{x})\xi, \qquad \xi \in \mathbb{R}^n, \qquad (1.3.2)$$

where $Df = [\partial f_i / \partial x_j]$ is the Jacobian matrix of first partial derivatives of the function $f = (f_1(x_1, \ldots, x_n), f_2(x_1, \ldots, x_n), \ldots, f_n(x_1, \ldots, x_n))^T$ ($T$ denotes transpose), and $x = \bar{x} + \xi, |\xi| \ll 1$. Since (1.3.2) is just a linear system of the form (1.1.1), we can do this easily. In particular, the linearized flow map $D\phi_t(\bar{x})\xi$ arising from (1.3.1) at a fixed point $\bar{x}$ is obtained from (1.3.2) by integration:

$$D\phi_t(\bar{x})\xi = e^{tDf(\bar{x})}\xi. \qquad (1.3.3)$$

The important question is, what can we say about the solutions of (1.3.1) based on our knowledge of (1.3.2)? The answer is provided by two fundamental results of dynamical systems theory which we give below, and may be

---

* Throughout this book by *smooth* we generally mean $C^\infty$, unless stated otherwise. We note that we do not always concentrate upon obtaining optimal smoothness in our results.

summed up by saying that local behavior (for $|\xi|$ small) does carry over in certain "nice" cases.

**Theorem 1.3.1** (Hartman-Grobman). *If $Df(\bar{x})$ has no zero or purely imaginary eigenvalues then there is a homeomorphism $h$ defined on some neighborhood $U$ of $\bar{x}$ in $\mathbb{R}^n$ locally taking orbits of the nonlinear flow $\phi_t$ of (1.3.1), to those of the linear flow $e^{tDf(\bar{x})}$ of (1.3.2). The homeomorphism preserves the sense of orbits and can also be chosen to preserve parametrization by time.*

A more delicate situation in which the nonlinear and linear flows are related via *diffeomorphisms* (Sternberg's theorem) requires certain non-resonance conditions among the eigenvalues of $Df(\bar{x})$. We shall not consider this here, but see the discussion of normal forms in Chapter 3.

When $Df(\bar{x})$ has no eigenvalues with zero real part, $\bar{x}$ is called a *hyperbolic* or *nondegenerate* fixed point and the asymptotic behavior of solutions near it (and hence its stability type) is determined by the linearization. If any one of the eigenvalues has zero real part, then stability cannot be determined by linearization, as the example

$$\ddot{x} + \varepsilon x^2 \dot{x} + x = 0 \tag{1.3.4}$$

shows. Rewritten as a system (with $x_1 = x$, $x_2 = \dot{x}$),

$$\begin{pmatrix} \dot{x}_1 \\ \dot{x}_2 \end{pmatrix} = \begin{pmatrix} 0 & 1 \\ -1 & 0 \end{pmatrix} \begin{pmatrix} x_1 \\ x_2 \end{pmatrix} - \varepsilon \begin{pmatrix} 0 \\ x_1^2 x_2 \end{pmatrix}, \tag{1.3.5}$$

we find eigenvalues $\lambda, \bar{\lambda} = \pm i$. However, unless $\varepsilon = 0$, the fixed point $(x_1, x_2) = (0, 0)$ is not a center, as in the linear system, but a nonhyperbolic or weak attracting spiral sink if $\varepsilon > 0$, and a repelling source if $\varepsilon < 0$.

EXERCISE 1.3.1. Verify that $(x_1, x_2) = (0, 0)$ is globally asymptotically stable for (1.3.5) when $\varepsilon > 0$. (Use a Liapunov function approach, cf. equation (1.0.10).)

Before the next result we need a couple of definitions. We define the *local stable and unstable manifolds* of $\bar{x}$, $W^s_{\text{loc}}(\bar{x})$, $W^u_{\text{loc}}(\bar{x})$ as follows

$$W^s_{\text{loc}}(\bar{x}) = \{x \in U \mid \phi_t(x) \to \bar{x} \text{ as } t \to \infty, \text{ and } \phi_t(x) \in U \text{ for all } t \geq 0\},$$
$$\tag{1.3.6}$$
$$W^u_{\text{loc}}(\bar{x}) = \{x \in U \mid \phi_t(x) \to \bar{x} \text{ as } t \to -\infty, \text{ and } \phi_t(x) \in U \text{ for all } t \leq 0\},$$

where $U \subset \mathbb{R}^n$ is a neighborhood of the fixed point $\bar{x}$. The invariant manifolds $W^s_{\text{loc}}$ and $W^u_{\text{loc}}$ provide nonlinear analogues of the flat stable and unstable eigenspaces $E^s$, $E^u$ of the linear problem (1.3.2). The next result tells us that $W^s_{\text{loc}}$ and $W^u_{\text{loc}}$ are in fact tangent to $E^s$, $E^u$ at $\bar{x}$.

**Theorem 1.3.2** (Stable Manifold Theorem for a Fixed Point). *Suppose that $\dot{x} = f(x)$ has a hyperbolic fixed point $\bar{x}$. Then there exist local stable and unstable manifolds $W^s_{\text{loc}}(\bar{x})$, $W^u_{\text{loc}}(\bar{x})$, of the same dimensions $n_s$, $n_u$ as those of* *the eigenspaces $E^s$, $E^u$ of the linearized system (1.3.2), and tangent to $E^s$, $E^u$ at $\bar{x}$. $W^s_{loc}(\bar{x})$, $W^u_{loc}(\bar{x})$ are as smooth as the function $f$.*

For proofs of these two theorems see, for example, Hartman [1964] and Carr [1981], or, for a more modern treatment, Nitecki [1971], Shub [1978], or Irwin [1980]. Hirsch *et al.* [1977] contains a more general result. The two results may be illustrated as in Figure 1.3.1.

Note that we have not yet said anything about a center manifold, tangent to $E^c$ at $\bar{x}$, and have, in fact, confined ourselves to hyperbolic cases in which $E^c$ does not exist. We shall consider nonhyperbolic cases later when we deal with bifurcation theory in Chapter 3.

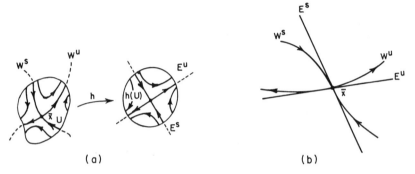

Figure 1.3.1. Linearization and invariant subspaces. (a) Hartman's theorem; (b) local stable and unstable manifolds.

The local invariant manifolds $W^s_{loc}$, $W^u_{loc}$ have global analogues $W^s$, $W^u$, obtained by letting points in $W^s_{loc}$ flow backwards in time and those in $W^u_{loc}$ flow forwards:

$$W^s(\bar{x}) = \bigcup_{t \leq 0} \phi_t(W^s_{loc}(\bar{x})),$$

$$W^u(x) = \bigcup_{t \geq 0} \phi_t(W^u_{loc}(\bar{x})). \qquad (1.3.7)$$

Existence and uniqueness of solutions of (1.3.1) ensure that two stable (or unstable) manifolds of distinct fixed points $\bar{x}^1$, $\bar{x}^2$ cannot intersect, nor can $W^s(\bar{x})$ (or $W^u(\bar{x})$) intersect itself. However, intersections of stable and unstable manifolds of distinct fixed points or the same fixed point can occur and, in fact, are a source of much of the complex behavior found in dynamical systems. The global stable and unstable manifolds need not be embedded submanifolds of $\mathbb{R}^n$ since they may wind around in a complex manner, approaching themselves arbitrarily closely. We give an example of a map possessing such a structure in the next section.

To illustrate the ideas of this section, we consider a simple system on the plane:

$$\dot{x} = x,$$
$$\dot{y} = -y + x^2, \qquad (1.3.8)$$

which has a unique fixed point at the origin. For the linearized system we have the following invariant subspaces:

$$E^s = \{(x, y) \in \mathbb{R}^2 \,|\, x = 0\},$$
$$E^u = \{(x, y) \in \mathbb{R}^2 \,|\, y = 0\}. \tag{1.3.9}$$

In this case we can integrate the nonlinear system exactly. Rather than obtaining a solution in the form $(x(t), y(t))$, we rewrite (1.3.8) as a (linear) first-order system by eliminating time:

$$\frac{dy}{dx} = \frac{-y}{x} + x. \tag{1.3.10}$$

This can be integrated directly to obtain the family of solution curves

$$y(x) = \frac{x^2}{3} + \frac{c}{x}, \tag{1.3.11}$$

where $c$ is a constant determined by initial conditions. Now Theorem 1.3.1, together with (1.3.9), implies that $W^u_{\text{loc}}(0, 0)$ can be represented as a graph $y = h(x)$ with $h(0) = h'(0) = 0$, since $W^u_{\text{loc}}$ is tangent to $E^u$ at $(0, 0)$. Thus $c = 0$ in (1.3.10) and we have

$$W^u(0, 0) = \left\{ (x, y) \in \mathbb{R}^2 \,\Big|\, y = \frac{x^2}{3} \right\}. \tag{1.3.12}$$

Finally, noting that if $x(0) = 0$, then $\dot{x} \equiv 0$, and hence $x(t) \equiv 0$, we see that $W^s(0, 0) \equiv E^s$. Note that, for this example, we have found the global manifolds; see Figure 1.3.2.

EXERCISE 1.3.2. Find and classify the fixed points of the following systems by linearizing about the fixed points (i.e., find eigenvalues and eigenvectors and sketch the local flows). Start by rewriting the second-order equations as first-order systems:

(a) $\ddot{x} + \varepsilon\dot{x} - x + x^3 = 0$;
(b) $\ddot{x} + \sin x = 0$;
(c) $\ddot{x} + \varepsilon\dot{x}^2 + \sin x = 0$;
(d) $\dot{x} = -x + x^2, \dot{y} = x + y$;
(e) $\ddot{x} + \varepsilon(x^2 - 1)\dot{x} + x = 0$;

(where $\varepsilon$ appears let $\varepsilon < 0, \varepsilon = 0, \varepsilon > 0$). Can you calculate (or guess) the global structure of stable and unstable manifolds in any of these cases? (This last part is quite hard if you are not familiar with the tricks outlined later in this chapter.)

It is well known that nonlinear systems possess limit sets other than fixed points; for example, *closed or periodic orbits* frequently occur. A periodic solution is one for which there exists $0 < T < \infty$ such that $x(t) = x(t + T)$ for all $t$. We consider the stability of such orbits in Section 1.5, but note here that they have stable and unstable manifolds just as do fixed points.

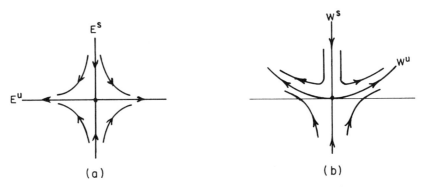

Figure 1.3.2. Stable and unstable manifolds for equation (1.3.8). (a) the linear system; (b) the nonlinear system.

Let $\gamma$ denote the closed orbit and $U$ be some neighborhood of $\gamma$; then we define

$$W^s_{\mathrm{loc}}(\gamma) = \{x \in U \mid |\phi_t(x) - \gamma| \to 0 \text{ as } t \to \infty, \text{ and } \phi_t(x) \in U \text{ for } t \geq 0\},$$

$$W^u_{\mathrm{loc}}(\gamma) = \{x \in U \mid |\phi_t(x) - \gamma| \to 0 \text{ as } t \to -\infty, \text{ and } \phi_t(x) \in U \text{ for } t \leq 0\}.$$

Examples will follow in the sections below.

## 1.4. Linear and Nonlinear Maps

We have seen how the linear system (1.1.1) gives rise to flow map $e^{tA}: \mathbb{R}^n \to \mathbb{R}^n$, when $e^{tA}$ is an $n \times n$ matrix. For fixed $t = \tau$ let $e^{\tau A} = B$, then $B$ is a constant coefficient matrix and the difference equation

$$x_{n+1} = Bx_n \quad \text{or} \quad x \mapsto Bx, \tag{1.4.1}$$

is a discrete dynamical system obtained from the flow of (1.1.1). Similarly, a nonlinear system and its flow $\phi_t$ give rise to a nonlinear map

$$x_{n+1} = G(x_n) \quad \text{or} \quad x \mapsto G(x), \tag{1.4.2}$$

where $G = \phi_\tau$ is a nonlinear vector valued function. If the flow $\phi_t$ is smooth (say $r$-times continuously differentiable), then $G$ is a smooth map with a smooth inverse: i.e., a *diffeomorphism*. This is one example of the way in which a continuous flow gives rise to a discrete map; a more important one, the Poincaré map, will be considered in Section 1.5.

Diffeomorphisms or discrete dynamical systems can also be studied in their own right and more generally we might also consider noninvertible maps such as

$$x \mapsto x - x^2. \tag{1.4.3}$$

EXERCISE 1.4.1. Show that the map $(x, y) \mapsto (y, bx + dy - y^3)$ is a diffeomorphism for $b \neq 0$ and calculate its inverse. (This map was suggested as an approximation to the Poincaré map of the Duffing equation, cf. Section 2.2 below, and Holmes [1979a].)

An orbit of a linear map $x \to Bx$ is a sequence of points $\{x_i\}_{i=-\infty}^{\infty}$ defined by $x_{i+1} = Bx_i$. Any initial point generates a unique orbit provided that $B$ has no zero eigenvalues.

We define stable, unstable, and center subspaces in a manner analogous to that for linear vector fields:

$E^s = \text{span}\{n_s \text{ (generalized) eigenvectors}$

$\qquad\qquad\qquad\qquad\qquad\qquad \text{whose eigenvalues have modulus} < 1\},$

$E^u = \text{span}\{n_u \text{ (generalized) eigenvectors}$

$\qquad\qquad\qquad\qquad\qquad\qquad \text{whose eigenvalues have modulus} > 1\},$

$E^c = \text{span}\{n_c \text{ (generalized) eigenvectors}$

$\qquad\qquad\qquad\qquad\qquad\qquad \text{whose eigenvalues have modulus} = 1\},$

where the orbits in $E^s$ and $E^u$ are characterized by contraction and expansion, respectively. If there are no multiple eigenvalues, then the contraction and expansion are bounded by geometric series: i.e., there exist constants $c > 0$, $\alpha < 1$ such that, for $n \geq 0$,

$$|x_n| \leq c\alpha^n |x_0| \quad \text{if } x_0 \in E^s,$$
$$|x_{-n}| \leq c\alpha^n |x_0| \quad \text{if } x_0 \in E^u. \tag{1.4.4}$$

If multiple eigenvalues occur, then much as in the case of flows, the contraction (or expansion) need not be exponential, as the following exercise illustrates. However, an exponential bound can still be found if $|\lambda_j| < 1$ for all eigenvalues.

EXERCISE 1.4.2. Compute orbits for

$$x \to \begin{bmatrix} \frac{1}{2} & 1 \\ 0 & \frac{1}{2} \end{bmatrix} x \quad \text{and} \quad x \to \begin{bmatrix} 1 & 0 \\ 1 & 1 \end{bmatrix} x$$

and sketch them on the plane. Show that $(0, 0)$ is asymptotically stable in the first case, while it is unstable in the second case, even though $|\lambda| = 1$ (cf. Exercise 1.2.1).

In spite of problems caused by multiplicities, if $B$ has no eigenvalues of unit modulus, the eigenvalues alone serve to determine stability. In this case $x = 0$ is called a *hyperbolic* fixed point and, in general, if $\bar{x}$ is a fixed point for $G$ ($G(\bar{x}) = \bar{x}$) and $DG(\bar{x})$ has no eigenvalues of unit modulus, then $\bar{x}$ is called a hyperbolic fixed point.

There is a theory for diffeomorphisms parallel to that for flows, and in particular the linearization theorem of Hartman–Grobman and the invariant manifold results apply to maps just as the flows (Hartman [1964], Nitecki [1971], Shub [1978]):

**Theorem 1.4.1** (Hartman–Grobman). *Let $G: \mathbb{R}^n \to \mathbb{R}^n$ be a $(C^1)$ diffeomorphism with a hyperbolic fixed point $\bar{x}$. Then there exists a homeomorphism $h$ defined on some neighborhood $U$ on $\bar{x}$ such that $h(G(\xi)) = DG(\bar{x})h(\xi)$ for all $\xi \in U$.*

**Theorem 1.4.2** (Stable Manifold Theorem for a Fixed Point). *Let $G: \mathbb{R}^n \to \mathbb{R}^n$ be a $(C^1)$ diffeomorphism with a hyperbolic fixed point $\bar{x}$. Then there are local stable and unstable manifolds $W^s_{\text{loc}}(\bar{x})$, $W^u_{\text{loc}}(\bar{x})$, tangent to the eigenspaces $E^s_{\bar{x}}$, $E^u_{\bar{x}}$ of $DG(\bar{x})$ at $\bar{x}$ and of corresponding dimensions. $W^s_{\text{loc}}(\bar{x})$, $W^u_{\text{loc}}(\bar{x})$ are as smooth as the map $G$.*

Global stable and unstable manifolds are defined as for flows, by taking unions of backward and forward iterates of the local manifolds. We have

$$W^s_{\text{loc}}(\bar{x}) = \{x \in U \,|\, G^n(x) \to \bar{x} \text{ as } n \to +\infty, \text{ and } G^n(x) \in U, \forall n \geq 0\},$$

$$W^u_{\text{loc}}(\bar{x}) = \{x \in U \,|\, G^{-n}(x) \to \bar{x} \text{ as } n \to +\infty, \text{ and } G^{-n}(x) \in U, \forall n \geq 0\},$$

and

$$W^s(\bar{x}) = \bigcup_{n \geq 0} G^{-n}(W^s_{\text{loc}}(\bar{x})),$$

$$W^u(\bar{x}) = \bigcup_{n \geq 0} G^{n}(W^u_{\text{loc}}(\bar{x})).$$

The reader should bear in mind, however, that flows and maps differ crucially in that, while the *orbit* or *trajectory* $\phi_t(p)$ of a flow is a *curve* in $\mathbb{R}^n$, the orbit $\{G^n(p)\}$ of a map is a *sequence of points*. Thus, while the invariant manifolds of flows are composed of the unions of solution curves, those of maps are unions of discrete orbit points; see Figure 1.4.1. This distinction will be important later, in the discussion of global behavior.

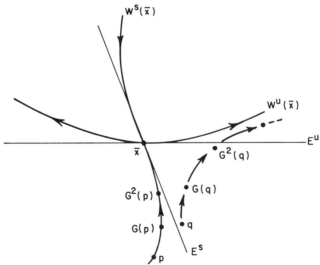

Figure 1.4.1. Invariant manifolds and orbits for a map $G: \mathbb{R}^2 \to \mathbb{R}^2$.

We note that, when we write $G^2(p)$, we mean $G(G(p))$ and, similarly, that $G^n(p)$ means the $n$th iterate of $p$ under $G$. Thus, if there is a cycle of $k$ distinct points $p_j = G^j(p_0)$, $j = 0, \ldots, k - 1$, and $G^k(p_0) = p_0$, we have a *periodic orbit* of period $k$. The stability of such an orbit is determined by the linearized map $DG^k(p_0)$, or, equivalently $DG^k(p_j)$ for any $j$. By the chain rule, we have $DG^k(p_0) = DG(G^{k-1}(p_0)) \cdots DG(G(p_0)) \cdot DG(p_0)$.

Much as for flows, the behavior of the linear map (1.4.1) is governed by the eigenvalues and eigenvectors of $B$. Since maps are rarely dealt with in texts on differential equations or nonlinear oscillations, we include some details here. For a one-dimensional map, where $B = b$ is a scalar and the orbit of a point $\{p_j\}_{j=0}^{\infty}$ is simply given by the geometric sequence $p_j = b^j p_0$, there are four "common" cases and three "unusual" ones listed below in Table 1.4.1. We shall see precisely what we mean by "common" and "unusual" later in this book.

Table 1.4.1. Behavior of the Linear Map $x \to bx$.

| Case | Description | Sketch |
|------|-------------|--------|
| 1. $b < -1$ | Orientation reversing source | $p_1 \quad 0 \quad p_0 \quad p_2$ |
| 2. $b \in (-1, 0)$ | Orientation reversing sink | $p_1 \quad 0 \quad p_2 \quad p_0$ |
| 3. $b \in (0, 1)$ | Orientation preserving sink | $0 \quad p_2 \; p_1 \quad p_0$ |
| 4. $b > 1$ | Orientation preserving source | $0 \quad p_0 \quad p_1 \quad p_2$ |
| 5. $b = -1$ | Orientation reversing, all points of period 2 | $p_1 \quad 0 \quad p_0 = p_2$ |
| 6. $b = 0$ | All points go to 0 on first iterate (noninvertible) | $0 = p_j, j \geq 1 \quad p_0$ |
| 7. $b = +1$ | Orientation preserving, all points fixed | $0 \quad p_0 = p_j, \forall j$ |

In general, the stability type of the fixed point $x = 0$ is determined by the magnitude of the eigenvalues of $B$. If $|\lambda_j| < 1$ for all eigenvalues, then we have a sink; if $|\lambda_j| > 1$ for some eigenvalues and $|\lambda_i| < 1$ for the others: a saddle point, and if $|\lambda_j| > 1$ for all eigenvalues: a source. If $|\lambda_j| = 1$ for any eigenvalues then a norm is preserved in the directions $v^j$ associated with those eigenvalues (unless they are multiple with nontrivial Jordan blocks).

EXERCISE 1.4.3. Develop a classification scheme similar to that of Table 1.4.1 for the two-dimensional map $x \mapsto Bx$;

$$B = \begin{pmatrix} b_{11} & b_{12} \\ b_{21} & b_{22} \end{pmatrix}.$$

Work in terms of the eigenvalues of $B$. (Hint: For help see Hsu [1977], or Bernoussou [1977].)

If an even number of eigenvalues have negative real parts, then the map $x \mapsto Bx$ is *orientation preserving*, while if an odd number have negative real parts it reverses orientation. We give some two-dimensional examples (part of the answer to Exercise 1.4.3) in Figure 1.4.2.

To get a feel for the rich and complex behavior possible for nonlinear maps the reader may like to experiment with the following two examples. Solutions may be conveniently obtained on a programmable pocket calculator or a minicomputer:

EXERCISE 1.4.4. How many fixed and periodic points can you find for the following one-dimensional map and two-dimensional diffeomorphism? Discuss their stability. Let the parameter $\mu$ vary over the ranges indicated. Can you find "bifurcation" values of $\mu$ at which new periodic points appear?

(a) $x \mapsto \mu x(1 - x)$; $\mu \in [0, 4]$,
(b) $(x, y) \mapsto (y, -\frac{1}{2}x + \mu y - y^3)$; $\mu \in [2, 4]$

(This problem is much harder than it looks. For instance, there are infinitely many periodic points for (a) if $3.7 < \mu \le 4$. We only expect you to find a few low period ones in each case.)

As a final example of a two-dimensional map with rather rich behavior, consider the simple linear map

$$\begin{pmatrix} x \\ y \end{pmatrix} \mapsto \begin{pmatrix} 1 & 1 \\ 1 & 2 \end{pmatrix} \begin{pmatrix} x \\ y \end{pmatrix}, \qquad (x, y) \in T^2 = \mathbb{R}^2/\mathbb{Z}^2, \qquad (1.4.5)$$

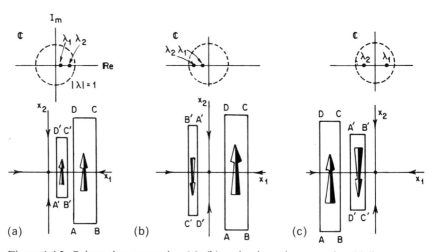

Figure 1.4.2. Orientation preserving (a), (b) and orientation reversing (c) linear maps

$$x \mapsto \begin{pmatrix} \lambda_1 & 0 \\ 0 & \lambda_2 \end{pmatrix} x.$$

Position of eigenvalues with respect to unit circle in complex plane shown above orbit structures. The oriented rectangle $ABCD$ is mapped to $A'B'C'D'$ in each case.

where the phase space is the two-dimensional torus. On the plane (the covering space) we simply have a saddle point, with eigenvectors $v^{1,2} = (1, (1 \pm \sqrt{5})/2)^T$ belonging to the eigenvalues $\lambda_{1,2} = (3 \pm \sqrt{5})/2$. Since the map is linear, $W^s(0) = E^s$, $W^u(0) = E^u$ and thus span$\{(1, (1 + \sqrt{5})/2)^T\}$ is the unstable manifold and span$\{1, (1 - \sqrt{5})/2)^T\}$ the stable manifold. However, our phase space is the torus, $T^2$, obtained by identifying points whose coordinates differ by integers. The map is well defined on $T^2$ since it preserves the periodic lattice. Any point of the unit square $[0, 1) \times [0, 1)$ mapped into another square is translated back into the original square; for example, if $(x, y) = (-1.4, +1.2)$, we set $(x, y) = (0.6, 0.2)$. See Figure 1.4.3. Thus the unstable manifold "runs off the square" at $(2/(1 + \sqrt{5}), 1)$ and reappears, with the same slope, at $(2/(1 + \sqrt{5}), 0)$, to run off at $(1, (\sqrt{5} - 1)/2)$, etc. Since the slopes of $W^s$ and $W^u$ are irrational $((1 \pm \sqrt{5})/2)$ these manifolds are dense in the unit square (or wind densely around the torus). Thus each manifold approaches itself arbitrarily closely, and hence is not an embedded submanifold of $T^2$.

EXERCISE 1.4.5. Show that the map of equation (1.4.5) has a countable infinity of periodic points and that the set of such points is dense in $T^2$. (First show that a point $\bar{x}$ is periodic if and only if both components of $\bar{x}$ are rational numbers with the same denominator.)

EXERCISE 1.4.6. Describe the set $\Lambda = W^s(0) \cap W^u(0)$ of intersections of the invariant manifolds for the linear map on the torus. What do you think that this implies for the structure of "typical" orbits? (Hint: See Chillingworth [1976], pp. 235–237.)

Arnold and Avez [1968, pp. 5–7] have nice illustrations of the torus map. Also, see Chapter 5 for more information on the invariant sets of such maps.

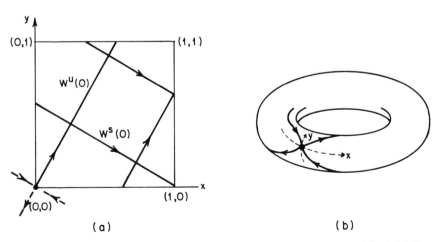

Figure 1.4.3. The linear map on the torus (the hyperbolic toral automorphism). (a) On $\mathbb{R}^2$, the covering space; (b) on $T^2$.

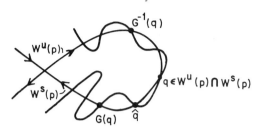

Figure 1.4.4. Homoclinic orbits.

This example might seem rather artificial, but as we shall see, many physically interesting systems have similar properties. In subsequent chapters we shall see that a Poincaré map associated with the forced Duffing equation with negative linear stiffness

$$\ddot{x} + \alpha \dot{x} - x + x^3 = \beta \cos \omega t, \tag{1.4.6}$$

which is a nonlinear diffeomorphism of the plane, possesses a hyperbolic saddle point $p$ whose stable and unstable manifolds intersect transversely somewhat as in the torus map above (cf. Figure 1.4.4). It is fairly easy to see that, if there is one point $q \in W^u(p) \cap W^s(p)$, with $q \neq p$, then, since $G^n(q) \to p$ as $n \to \pm\infty$, and the approach is governed by the linear system for $|q - p|$ small, there must be an infinite set of such *homoclinic* points. Moreover, if the map is orientation preserving (as our Poincaré maps are), then the two homoclinic points $q$, $G(q)$ must be separated by at least one further point in $W^u(p) \cap W^s(p)$ (marked $\hat{q}$ in Figure 1.4.4). The orbit $\{G^n(q)\}$ of $q$ is called a *homoclinic orbit* and plays an important rôle in the global dynamics of the map $G$. In particular, the violent winding of the global manifolds $W^u(p)$ and $W^s(p)$ in the neighborhood of $p$ leads to a sensitive dependence of orbits $\{G^n(x_0)\}$ on the initial condition $x_0$, so that the presence of homoclinic orbits tends to promote erratic behavior. This underlies the chaotic behavior exhibited by the examples of Chapter 2 and in the subject of much of Chapters 5 and 6. If the stable and unstable manifolds $W^s(p_1)$, $W^u(p_2)$ of two distinct fixed points intersect then the resulting orbit is called *heteroclinic*.

EXERCISE 1.4.7. Show that the stable manifold of a saddle point of a two-dimensional map cannot intersect itself.

## 1.5. Closed Orbits, Poincaré Maps, and Forced Oscillations

In classical texts on differential equations the stability of closed orbits or periodic solutions is discussed in terms of the characteristic or Floquet multipliers. Here we wish to introduce a more geometrical view which is in essence equivalent: the Poincaré map. Since the ideas are so important, we devote a considerable amount of space to familiar examples from forced oscillations.

Let $\gamma$ be a periodic orbit of some flow $\phi_t$ in $\mathbb{R}^n$ arising from a nonlinear vector field $f(x)$. We first take a *local cross section* $\Sigma \subset \mathbb{R}^n$, of dimension $n - 1$. The hypersurface $\Sigma$ need not be planar, but must be chosen so that the flow is everywhere *transverse* to it. This is achieved if $f(x) \cdot n(x) \neq 0$ for all $x \in \Sigma$, where $n(x)$ is the unit normal to $\Sigma$ at $x$. Denote the (unique) point where $\gamma$ intersects $\Sigma$ by $p$, and let $U \subseteq \Sigma$ be some neighborhood of $p$. (If $\gamma$ has multiple intersections with $\Sigma$, then shrink $\Sigma$ until there is only one inter-section.) Then the *first return* or *Poincaré map* $P: U \rightarrow \Sigma$ is defined for a point $q \in U$ by

$$P(q) = \phi_\tau(q), \tag{1.5.1}$$

where $\tau = \tau(q)$ is the time taken for the orbit $\phi_t(q)$ based at $q$ to first return to $\Sigma$. Note that $\tau$ generally depends upon $q$ and need not be equal to $T = T(p)$, the period of $\gamma$, However, $\tau \rightarrow T$ as $q \rightarrow p$.

Clearly $p$ is a fixed point for the map $P$, and it is not difficult to see that the stability of $p$ for $P$ reflects the stability of $\gamma$ for the flow $\phi_t$. In particular, if $p$ is hyperbolic, and $DP(p)$, the linearized map, has $n_s$ eigenvalues with modulus less than one and $n_u$ with modulus greater than one $(n_s + n_u = n - 1)$, then dim $W^s(p) = n_s$, and dim $W^u(p) = n_u$ for the map. Since the orbits of $P$ lying in $W^s$ and $W^u$ are formed by intersections of orbits (solution curves) of $\phi_t$ with $\Sigma$, the dimensions of $W^s(\gamma)$ and $W^u(\gamma)$ are each one greater than those for the map. This is most easily seen in the sketches of Figure 1.5.1.

As an example, consider the planar system

$$\dot{x} = x - y - x(x^2 + y^2),$$
$$\dot{y} = x + y - y(x^2 + y^2), \tag{1.5.2}$$

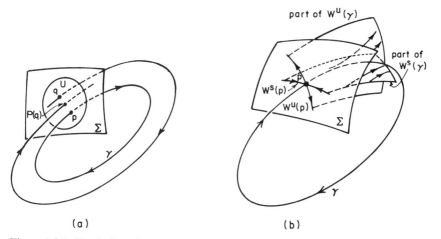

Figure 1.5.1. The Poincaré map. (a) The cross section and the map; (b) a closed orbit.

and take as our cross section

$$\Sigma = \{(x, y) \in \mathbb{R}^2 \,|\, x > 0, y = 0\}.$$

Transforming (1.5.2) to polar coordinates $r = (x^2 + y^2)^{1/2}$, $\theta = \arctan(y/x)$, we obtain

$$\begin{aligned} \dot{r} &= r(1 - r^2), \\ \dot{\theta} &= 1, \end{aligned} \tag{1.5.3}$$

and the section becomes

$$\Sigma = \{(r, \theta) \in \mathbb{R}^+ \times S^1 \,|\, r > 0, \theta = 0\}.$$

It is easy to solve (1.5.3) to obtain the global flow

$$\phi_t(r_0, \theta_0) = \left( \left(1 + \left(\frac{1}{r_0^2} - 1\right)e^{-2t}\right)^{-1/2}, t + \theta_0 \right).$$

The time of flight $\tau$ for any point $q \in \Sigma$ is simply $\tau = 2\pi$, and thus the Poincaré map is given by

$$P(r_0) = \left(1 + \left(\frac{1}{r_0^2} - 1\right)e^{-4\pi}\right)^{-1/2}. \tag{1.5.4}$$

Clearly, $P$ has a fixed point at $r_0 = 1$, reflecting the circular closed orbit of radius 1 of (1.5.3). Here $P$ is a one-dimensional map and its linearization is given by

$$DP(1) = \frac{dP}{dr_0}\bigg|_{r_0 = 1} = -\frac{1}{2}\left(1 + \left(\frac{1}{r_0^2} - 1\right)e^{-4\pi}\right)^{-3/2} \cdot \left(-\frac{2e^{-4\pi}}{r_0^3}\right)\bigg|_{r_0 = 1}$$

$$= e^{-4\pi} < 1. \tag{1.5.5}$$

Thus $p = 1$ is a stable fixed point and $\gamma$ is a stable or attracting closed orbit.

We note that we could have computed $DP(1)$ a little more simply by considering the flow of the vector field (1.5.3) linearized near the closed orbit $r = 1$. Since $(d/dr)(r - r^3) = 1 - 3r^2$, this is

$$\begin{aligned} \dot{\xi} &= -2\xi, \\ \dot{\theta} &= 1, \end{aligned} \tag{1.5.6}$$

with flow

$$D\phi_t(\xi_0, \theta_0) = (e^{-2t}\xi_0, t + \theta_0). \tag{1.5.7}$$

Hence $DP(1) = e^{-2(2\pi)} = e^{-4\pi}$, as above.

To demonstrate the general relationship between Poincaré maps and linearized flows we must review a little Floquet theory (Hartman [1964], §§IV.6, IX.10). Let $\bar{x}(t) = \bar{x}(t + T)$ be a solution lying on the closed orbit $\gamma$, based at $x(0) = p \in \Sigma$. Linearizing the differential equation about $\gamma$, we obtain the system

$$\dot{\xi} = Df(\bar{x}(t))\xi, \tag{1.5.8}$$

where $Df(\bar{x}(t))$ is an $n \times n$, $T$-periodic matrix. It can be shown that any fundamental solution matrix of such a $T$-periodic system can be written in the form

$$X(t) = Z(t)e^{tR}; \qquad Z(t) = Z(t + T), \tag{1.5.9}$$

where $X$, $Z$, and $R$ are $n \times n$ matrices (cf. Hartman [1964], p. 60). In particular, we can choose $X(0) = Z(0) = I$, so that

$$X(T) = Z(T)e^{TR} = Z(0)e^{TR} = e^{TR}. \tag{1.5.10}$$

It then follows that the behavior of solutions in the neighborhood of $\gamma$ is determined by the eigenvalues of the constant matrix $e^{TR}$. These eigenvalues, $\lambda_1, \ldots, \lambda_n$, are called the *characteristic (Floquet) multipliers or roots* and the eigenvalues $\mu_1, \ldots, \mu_n$ of $R$ are the *characteristic exponents* of the closed orbit $\gamma$. The multiplier associated with perturbations along $\gamma$ is always unity; let this be $\lambda_n$. The moduli of the remaining $n - 1$, if none are unity, determine the stability of $\gamma$.

Choosing the basis appropriately, so that the last column of $e^{TR}$ is $(0, \ldots, 0, 1)^T$, the matrix $DP(p)$ of the linearized Poincaré map is simply the $(n - 1) \times (n - 1)$ matrix obtained by deleting the $n$th row and column of $e^{TR}$. Then the first $n - 1$ multipliers $\lambda_1, \ldots, \lambda_{n-1}$ are the eigenvalues of the Poincaré map.

Although the matrix $R$ in (1.5.9) is not determined uniquely by the solutions of (1.5.8) (Hartman [1964], p. 60), the eigenvalues of $e^{TR}$ are uniquely determined ($e^{TR}$ can be replaced by any similar matrix $C^{-1}e^{TR}C$). However, to compute these eigenvalues we still need a representation of $e^{TR}$, and this can only be obtained by actually generating a set of $n$ linearly independent solutions to form $X(t)$. Except in special cases, like the simple example above, this is generally difficult and requires perturbation methods or the use of special functions.

EXERCISE 1.5.1. Repeat the analysis above for the three-dimensional systems obtained by adding the components $\dot{z} = \mu z$ and then $\dot{z} = \mu - z^2$ to (1.5.3): consider $\mu < 0$, $\mu = 0$, and $\mu > 0$. Sketch the stable and unstable manifolds of the periodic orbits in each case. (This is fairly simple.)

EXERCISE 1.5.2. Find the closed orbits of the following system for different values of $\mu_1$ and $\mu_2$: $\dot{r} = r(\mu_1 + \mu_2 r^2 - r^4)$, $\dot{\theta} = 1 - r^2$. Discuss their stability in terms of the Poincaré map. (While the analysis is simple here, since the $r$ and $\theta$ equations uncouple, this is a nontrivial example which will reappear in Chapter 7.)

We have seen how a vector field $f(x)$ on $\mathbb{R}^n$ gives rise to a flow map $\phi_t$ on $\mathbb{R}^n$ and, in the neighborhood of a closed orbit, to a (local) Poincaré map $P$

on a transversal hypersurface $\Sigma$. Another important way in which a flow gives rise to a map is in non-autonomous, periodically forced oscillations. Consider a system

$$\dot{x} = f(x, t); \qquad (x, t) \in \mathbb{R}^n \times \mathbb{R}, \qquad (1.5.11)$$

where $f(\cdot, t) = f(\cdot, t + T)$ is periodic of period $T$ in $t$. System (1.5.11) may be rewritten as an autonomous system at the expense of an increase in dimension by one, if time is included as an explicit state variable:

$$\begin{aligned} \dot{x} &= f(x, \theta), \\ \dot{\theta} &= 1; \qquad (x, \theta) \in \mathbb{R}^n \times S^1. \end{aligned} \qquad (1.5.12)$$

The phase space is the manifold $\mathbb{R}^n \times S^1$, where the circular component $S^1 = \mathbb{R} \,(\text{mod } T)$ reflects the periodicity of the vector field $f$ in $\theta$. For this problem we can define a *global* cross section

$$\Sigma = \{(x, \theta) \in \mathbb{R}^n \times S^1 | \theta = \theta_0\}, \qquad (1.5.13)$$

since all solutions cross $\Sigma$ transversely, in view of the component $\dot{\theta} = 1$ of (1.5.12). The Poincaré map $P: \Sigma \to \Sigma$, if it is defined globally, is given by

$$P(x_0) = \pi \cdot \phi_T(x_0, \theta_0), \qquad (1.5.14)$$

where $\phi_t: \mathbb{R}^n \times S^1 \to \mathbb{R}^n \times S^1$ is the flow of (1.5.12) and $\pi$ denotes projection onto the first factor. Note that here the time of flight $T$ is the same for all points $x \in \Sigma$. Alternatively, $P(x_0) = x(x_0, T + \theta_0)$, where $x(x_0, t)$ is the solution of (1.5.12) based at $x(x_0, \theta_0) = x_0$.

The Poincaré map can also be derived as a discrete dynamical system arising from the flow $\psi(x, t)$ of the time-dependent vector field of (1.5.11). Since $f$ is $T$-periodic, we have $\psi(x, nT) \equiv \psi^n(x, T) \overset{\text{def}}{=} \psi_T^n(x)$. The map $P(x_0) = \psi_T(x_0)$ is in this sense another example of a discrete dynamical system of the type considered at the beginning of Section 1.4.

The system

$$\begin{aligned} \dot{x} &= x^2, \\ \dot{\theta} &= 1, \end{aligned} \qquad (1.5.15)$$

with solution

$$\phi_t(x_0, \theta_0) = \left( \left(\frac{1}{x_0} - t\right)^{-1}, t + \theta_0 \right),$$

and the Poincaré map

$$P(x_0) = \left(\frac{1}{x_0} - 2\pi\right)^{-1}, \qquad x_0 \in (-\infty, 1/2\pi)$$

on $\Sigma = \{(x, \theta) | \theta = 0\}$ shows that $P$ may not be globally defined. Here, trajectories of $\phi_t$ based at $x_0 \geq 1/2\pi$ approach $\infty$ at a time $t \leq 2\pi$. However, $P: U \to \Sigma$ is usually defined for some subset $U \subset \Sigma$.

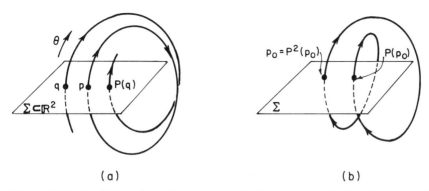

(a)                                                    (b)

Figure 1.5.2. The Poincaré map for forced oscillations. (a) A periodic orbit of period $T$ and the fixed point $p = P(p)$; (b) a subharmonic of period $2T$.

We illustrate the Poincaré map for forced oscillations in Figure 1.5.2. As in the previous case, it is easy to see that a fixed point $p$ of $P$ corresponds to a periodic orbit of period $T$ for the flow. In addition, a periodic point of period $k > 1$ ($P^k(p) = p$ but $P^j(p) \neq p$ for $1 \leq j \leq k - 1$) corresponds to a subharmonic of period $kT$. Here $P^k$ means $P$ iterated $k$ times, thus $P^2(p_0) = P(P(p_0))$; etc. This, of course, also applies for the autonomous case discussed earlier. Such periodic points must always come in sets of $k$: $p_0, \ldots, p_{k-1}$ such that $P(p_i) = p_{i+1}, 0 \leq i \leq k - 2$ and $p_0 = P(p_{k-1})$.

EXERCISE 1.5.3. (a) Show that the periodic orbits of (1.5.12) can only have periods $kT$ for integers $k$.
(b) Show that periodic orbits can only have period $T$ if $n = 1$.
(c) Show that the Poincaré map for forced oscillations is orientation preserving.

(Hint: Use uniqueness of solutions in $\mathbb{R}^n \times S^1$.)

Since the definition of the Poincaré map relies on knowledge of the flow of the differential equation, Poincaré maps cannot be computed unless general solutions of these equations are available. However, as we shall see in Chapter 4, perturbation and averaging methods can be used to approximate the map in appropriate cases and valuable information can thus be obtained from the marriage of conventional methods with the geometric approach of dynamical systems theory.

We now consider two examples from the theory of oscillations.

## Forced Linear Oscillations

We start with a problem for which a general solution can be found and the Poincaré map computed explicitly. Consider the system

$$\ddot{x} + 2\beta\dot{x} + x = \gamma \cos \omega t; \quad 0 \leq \beta < 1, \tag{1.5.16}$$

or

$$\begin{pmatrix} \dot{x}_1 \\ \dot{x}_2 \end{pmatrix} = \begin{bmatrix} 0 & 1 \\ -1 & -2\beta \end{bmatrix} \begin{pmatrix} x_1 \\ x_2 \end{pmatrix} + \begin{pmatrix} 0 \\ \gamma \cos \omega\theta \end{pmatrix}, \tag{1.5.17}$$

$$\dot{\theta} = 1.$$

Here the forcing is of period $T = 2\pi/\omega$. Since the system is linear, its solution is easily obtained by conventional methods (cf. Braun [1978]):

$$x(t) = e^{-\beta t}(c_1 \cos \omega_d t + c_2 \sin \omega_d t) + A \cos \omega t + B \sin \omega t, \tag{1.5.18}$$

where $\omega_d = \sqrt{1 - \beta^2}$ is the damped natural frequency and $A$ and $B$, the coefficients in the particular solution, are given as

$$A = \frac{(1 - \omega^2)\gamma}{[(1 - \omega^2)^2 + 4\beta^2\omega^2]}; \qquad B = \frac{2\beta\omega\gamma}{[(1 - \omega^2)^2 + 4\beta^2\omega^2]}. \tag{1.5.19}$$

The constants $c_1$, $c_2$ are determined by the initial conditions. Letting $x = x_1 = x_{10}$ and $\dot{x} = x_2 = x_{20}$, at $t = 0$, we have

$$\left.\begin{array}{l} x(0) = x_{10} = c_1 + A \\ \dot{x}(0) = x_{20} = -\beta c_1 + \omega_d c_2 + \omega B \end{array}\right\} \Rightarrow \left.\begin{array}{l} c_1 = x_{10} - A, \\ c_2 = (x_{20} + \beta(x_{10} - A) - \omega B)/\omega_d \end{array}\right\}.$$

$$(1.5.20)$$

Thus, since $\phi_t(x_{10}, x_{20}, 0)$ is given by (1.5.18) and

$$\begin{aligned} x_2(t) = \dot{x}_1(t) = e^{-\beta t}\{&-\beta(c_1 \cos \omega_d t + c_2 \sin \omega_d t) \\ &+ \omega_d(-c_1 \sin \omega_d t + c_2 \cos \omega_d t)\} \\ &- \omega(A \sin \omega t - B \cos \omega t), \end{aligned}$$

we can compute the Poincaré map explicitly as $\pi \cdot \phi_{2\pi/\omega}(x_{10}, x_{20}, 0)$. In the case of *resonance*, $\omega = \omega_d = \sqrt{1 - \beta^2}$, we obtain

$$P(x_{10}, x_{20}) = ((x_{10} - A)e^{-2\pi\beta/\omega} + A, (x_{20} - \omega B)e^{-2\pi\beta/\omega} + \omega B). \tag{1.5.21}$$

As expected, the map has an attracting fixed point given by $(x_1, x_2) = (A, \omega B)$ or $c_1 = c_2 = 0$. The map is, of course, linear and since the matrix

$$\begin{bmatrix} \dfrac{\partial P_1}{\partial x_{10}} & \dfrac{\partial P_1}{\partial x_{20}} \\ \dfrac{\partial P_2}{\partial x_{10}} & \dfrac{\partial P_2}{\partial x_{20}} \end{bmatrix} = \begin{bmatrix} e^{-2\pi\beta/\omega} & 0 \\ 0 & e^{-2\pi\beta/\omega} \end{bmatrix} \tag{1.5.22}$$

is diagonal with equal eigenvalues, the orbits of $P$ approach $(A, \omega B)$ radially, cf. Figure 1.5.3.

EXERCISE 1.5.4. Compute the Poincaré map for the linear oscillator in the case when $\omega \neq \sqrt{1 - \beta^2} = \omega_d$. What happens when $\beta = 0$ and $\omega = 1$?

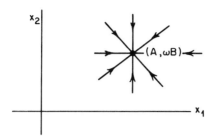

Figure 1.5.3. The Poincaré map of the linear oscillator equation (1.5.16).

EXERCISE 1.5.5. Consider the forced "negative stiffness" Duffing equation $\ddot{x} + \alpha \dot{x} - x + x^3 = \beta \cos t$, $\alpha > 0$, $\beta \geq 0$. Show that: (a) solutions remain bounded for all time ($\phi_t$ is globally defined); (b) for $1 \gg \alpha \gg \beta > 0$ there are precisely three periodic orbits of period $2\pi$, one a saddle and the other two attractors. Discuss the structure of stable and unstable manifolds of these periodic orbits by considering the structure of the associated manifolds of the fixed points of the Poincaré map. (This is quite difficult. For (a) you must find a closed curve on which all solutions are directed inward. For (b) you can perturb from the case $\beta = 0$, which is quite simple to analyze.)

The Duffing problem will be taken up in more detail in Chapter 2. As a second example we take a nonlinear system which we linearize about two equilibria.

## The Periodically Perturbed Pendulum

The equation of motion of a pendulum with a periodically excited support may be written as a nonlinear Mathieu equation:

$$\ddot{\phi} + (\alpha^2 + \beta \cos t) \sin \phi = 0; \qquad \beta \geq 0, \qquad (1.5.23a)$$

or

$$\left.\begin{aligned} \dot{\phi} &= v, \\ \dot{v} &= -(\alpha^2 + \beta \cos \theta) \sin \phi, \\ \dot{\theta} &= 1. \end{aligned}\right\} ; \qquad (\phi, v, \theta) \in S^1 \times \mathbb{R} \times S^1. (= \mathbb{R} \times T^2)$$

$$(1.5.23b)$$

Note that the equilibrium positions $(\phi, v) = (0, 0)$ and $(\pi, 0)$ of the unforced problem ($\beta = 0$) still yield $\dot{\phi} = \dot{v} = 0$ when $\beta \neq 0$. Thus for all $\beta$ we have periodic orbits given by $(0, 0; \theta(t)), (\pi, 0; \theta(t))$ with $\theta(t) = t + t_0$. Linearizing the vector field about these orbits we obtain the linear Mathieu equations

$$\begin{aligned} \dot{\phi} &= v, & \dot{\phi} &= v, \\ \dot{v} &= -(\alpha^2 + \beta \cos \theta)\phi, & \text{and} \quad \dot{v} &= (\alpha^2 + \beta \cos \theta)\phi, \\ \dot{\theta} &= 1, & \dot{\theta} &= 1, \end{aligned}$$

$$(1.5.24a, b)$$

respectively, where we have retained the same variables $(\phi, v)$ as for the original nonlinear problem.

We now investigate the stability of these periodic orbits. When $\beta \neq 0$, equations (1.5.24) represent problems in Floquet theory. When $\beta = 0$ we have the simple pendulum and solutions near $\phi = 0$ and $\phi = \pi$ are given, respectively, by the general solutions of the linear oscillator $\ddot{\phi} \pm \alpha^2 \phi = 0$:

$$\begin{pmatrix} \phi \\ v \end{pmatrix} = c_1^0 \begin{pmatrix} \cos \alpha t \\ -\alpha \sin \alpha t \end{pmatrix} + c_2^0 \begin{pmatrix} \sin \alpha t \\ \alpha \cos \alpha t \end{pmatrix}$$

$$\text{and} \quad \begin{pmatrix} \phi \\ v \end{pmatrix} = c_1^\pi \begin{pmatrix} e^{\alpha t} \\ \alpha e^{\alpha t} \end{pmatrix} + c_2^\pi \begin{pmatrix} e^{-\alpha t} \\ -\alpha e^{-\alpha t} \end{pmatrix} \qquad \text{(1.5.25a, b)}$$

Letting $(\phi(0), v(0)) = (\phi_0, v_0)$, we find that $c_1^0 = \phi_0$, $c_2^0 = v_0/\alpha$ and

$$c_1^\pi = \frac{\phi_0 + v_0/\alpha}{2}, \qquad c_2^\pi = \frac{\phi_0 - v_0/\alpha}{2}.$$

Integrating these solutions for one period $T = 2\pi$ of the forcing perturbation, we obtain the linearized Poincaré maps

$$DP_0(0, 0)\begin{pmatrix} \phi_0 \\ v_0 \end{pmatrix} = \begin{pmatrix} \phi_0 \cos(2\pi\alpha) + \dfrac{v_0}{\alpha} \sin(2\pi\alpha) \\[2mm] -\alpha\phi_0 \sin(2\pi\alpha) + v_0 \cos(2\pi\alpha) \end{pmatrix}, \qquad \text{(1.5.26a)}$$

and

$$DP_0(\pi, 0)\begin{pmatrix} \phi_0 \\ v_0 \end{pmatrix} = \frac{1}{2}\begin{pmatrix} \left(\phi_0 + \dfrac{v_0}{\alpha}\right)e^{2\pi\alpha} + \left(\phi_0 - \dfrac{v_0}{\alpha}\right)e^{-2\pi\alpha} \\[2mm] (\alpha\phi_0 + v_0)e^{2\pi\alpha} - (\alpha\phi_0 - v_0)e^{-2\pi\alpha} \end{pmatrix} \qquad \text{(1.5.26b)}$$

Thus the linearized operators are

$$DP_0(0, 0) = \begin{bmatrix} \cos(2\pi\alpha) & \dfrac{1}{\alpha} \sin(2\pi\alpha) \\[3mm] -\alpha \sin(2\pi\alpha) & \cos(2\pi\alpha) \end{bmatrix}$$

$$\text{and} \quad DP_0(\pi, 0) = \begin{bmatrix} \cosh(2\pi\alpha) & \dfrac{1}{\alpha} \sinh(2\pi\alpha) \\[3mm] \alpha \sinh(2\pi\alpha) & \cosh(2\pi\alpha) \end{bmatrix}. \qquad \text{(1.5.27a, b)}$$

The eigenvalues of these matrices are

$$\begin{aligned} \lambda_{1,2}^0 &= \cos(2\pi\alpha) \pm i \sin(2\pi\alpha) \quad \text{and} \quad \lambda_{1,2}^\pi = \cosh(2\pi\alpha) \pm \sinh(2\pi\alpha) \\ &= e^{i2\pi\alpha}, e^{-i2\pi\alpha}, \qquad\qquad\qquad\qquad\quad = e^{2\pi\alpha}, e^{-2\pi\alpha}. \end{aligned}$$

$$\text{(1.5.28a, b)}$$

We conclude that the orbit $(0, 0, \theta(t))$ is neutrally stable, with eigenvalues on the unit circle, and that at $(\pi, 0, \theta(t))$ is of saddle type with one eigenvalue within and one outside the unit circle. Note, however, that when $\alpha = n/2$; $n = 0, 1, 2, \ldots$ the eigenvalues of the neutrally stable orbit are $+1$ (resp. $-1$) with multiplicity two. We shall return to this below.

We now turn to the more interesting case when $\beta \neq 0$. As is well known, the general solution of (1.5.24) can be written as

$$\begin{pmatrix} \phi(t) \\ v(t) \end{pmatrix} = c_1 x^1(t) + c_2 x^2(t), \tag{1.5.29}$$

where $x^1(t)$ and $x^2(t)$ are two linearly independent solutions. Thus $X(t) = [x^1(t), x^2(t)]$ is a fundamental solution matrix. The linearized Poincaré map can be obtained as

$$DP_\beta = X(2\pi)X^{-1}(0), \tag{1.5.30}$$

since, using (1.5.29) we have

$$\begin{pmatrix} \phi(2\pi) \\ v(2\pi) \end{pmatrix} = X(2\pi)\begin{pmatrix} c_1 \\ c_2 \end{pmatrix} \quad \text{and} \quad \begin{pmatrix} c_1 \\ c_2 \end{pmatrix} = X^{-1}(0)\begin{pmatrix} \phi(0) \\ v(0) \end{pmatrix}. \tag{1.5.31}$$

Our problem now becomes one of calculating a pair of linearly independent solutions, a problem solved in many classical textbooks by special functions (Mathieu functions) arising from series solutions, or by perturbation methods (cf. Nayfeh and Mook [1979]). Rather than repeating such analyses, we shall derive an interesting property of the eigenvalues of $DP_\beta$ and use this to discuss the stability of solutions for $\beta \neq 0$, small. We choose an independent pair of solutions $x^1(t)$, $x^2(t)$ such that

$$x^1(0) = \begin{pmatrix} 1 \\ 0 \end{pmatrix}, \quad x^2(0) = \begin{pmatrix} 0 \\ 1 \end{pmatrix} \quad \text{and} \quad X(0) = I = X^{-1}(0).$$

Then we have

$$DP_\beta = X(2\pi) = \begin{bmatrix} \phi^1(2\pi) & \phi^2(2\pi) \\ v^1(2\pi) & v^2(2\pi) \end{bmatrix}, \tag{1.5.32}$$

where $X(t) = D\phi_t$, the linearized flow. We claim that the determinant of $DP_\beta$ (the Wronskian of the solutions $x^1$, $x^2$) is unity for our system. To see this, consider the determinant of the linearized flow $D\phi_t$:

$$\Delta = \det(D\phi_t) = \phi^1 v^2 - \phi^2 v^1,$$

$$\frac{d\Delta}{dt} = \dot{\phi}^1 v^2 + \phi^1 \dot{v}^2 - \dot{\phi}^2 v^1 - \phi^2 \dot{v}^1 = \phi^1 \dot{v}^2 - \phi^2 \dot{v}^1$$

$$= \phi^1[\pm(\alpha^2 + \beta \cos t)\phi^2] - \phi^2[\pm(\alpha^2 + \beta \cos t)\phi^1] \equiv 0. \tag{1.5.33}$$

Thus $\Delta$ maintains its value. But setting $t = 0$ and using

$$x_1(0) = \begin{pmatrix} \phi^1(0) \\ v^1(0) \end{pmatrix} = \begin{pmatrix} 1 \\ 0 \end{pmatrix}, \quad x_2(0) = \begin{pmatrix} \phi^2(0) \\ v^2(0) \end{pmatrix} = \begin{pmatrix} 0 \\ 1 \end{pmatrix},$$

we have $\Delta = 1$. Therefore $\det DP_\beta = \det D\phi_{2\pi} = 1$ also, and the (linearized) Poincaré map is area preserving. The eigenvalues of (1.5.32) are

$$\lambda_{1,2} = a \pm \sqrt{a^2 - 1}, \qquad a = \tfrac{1}{2}(\phi^1(2\pi) + v^2(2\pi)), \qquad (1.5.34)$$

and we have

$$\lambda_1 \lambda_2 = 1. \qquad (1.5.35)$$

Thus, when $\beta \neq 0$ as well as when $\beta = 0$, the eigenvalues are either complex conjugates with nonzero imaginary parts, real and reciprocal, or multiple and equal to $+1$ or $-1$.

Now as $\beta$ increases from zero, the eigenvalues of $DP_\beta$ vary continuously, starting as those of $DP_0$, and we therefore have the following result.

**Proposition 1.5.1.** *The periodic orbit $(0, 0, \theta(t))$ is neutrally stable for $\beta \neq 0$ and sufficiently small, provided that $\alpha \neq n/2, n \in \mathbb{Z}$. The periodic orbit $(\pi, 0, \theta(t))$ is of saddle type for $\beta \neq 0$, sufficiently small and all $\alpha \neq 0$.*

PROOF. The second assertion is easily proved, since if $\alpha < 0$ then the eigenvalues of $DP_0(\pi, 0)$ are given by $e^{-2\pi\alpha} < 1 < e^{2\pi\alpha}$ and since those of $DP_\beta(\pi, 0)$ vary continuously with $\beta$, we can choose a $\beta_0 > 0$ such that for all $0 < \beta < \beta_0$, $DP_\beta(\pi, 0)$ has eigenvalues $1/\lambda_\beta < 1 < \lambda_\beta$. Proof of the first assertion proceeds similarly, except that we must exclude the critical values $\alpha = 0, \tfrac{1}{2}, 1, \tfrac{3}{2}, \ldots$, since for those values the eigenvalues of $DP_0(0, 0)$ are $\pm 1$ with multiplicity two, and we cannot know how they will split as $\beta$ increases from zero. These critical values are, of course, the resonance conditions familiar from texts on linear parametric excitation (cf. Nayfeh and Mook [1979]).  $\square$

Note that when $\alpha = \tfrac{1}{2}, \tfrac{3}{2}, \ldots$ and $\lambda_{1,2} = -1$ the eigenvalues can split and take the form $-\lambda_\beta < -1 < -1/\lambda_\beta$ ($\lambda_\beta > 0$) when $\beta \neq 0$. As we shall see in Chapter 3, this bifurcation of the Poincaré map typically involves the appearance of an orbit twice the original period. Here, for example, the instability corresponding to the second subharmonic ($T = 4\pi$) appears for $\beta > 0$ and $\alpha = \tfrac{1}{2}$.

Arnold [1978, §25] has a nice treatment of parametric resonance in periodically perturbed Hamiltonian systems. He also works in terms of the Poincaré map.

EXERCISE 1.5.6. Consider the system

$$\dot\phi = v,$$

$$\dot v = -(\alpha^2 + \beta \cos t)\phi - \gamma v,$$

where the damping parameter $\gamma > 0$ is fixed. Using arguments similar to those above, show that the solution $(\phi, v) = (0, 0)$ is asymptotically stable for *all* $\alpha$ and sufficiently small $\beta$. (Hint: First show that $\det DP_\beta = e^{-2\pi\gamma}$ in this case.)

## 1.6. Asymptotic Behavior

Before we can get down to some specific examples of flows and maps, we need a little more technical apparatus. In this section we define various limit sets which represent asymptotic behavior of certain classes of solutions, and in the next section we discuss equivalence relations. While our definitions are fairly general, we concentrate on two-dimensional flows and maps for most of our examples. In Section 8 we shall give a more complete review of two-dimensional flows.

We first define an *invariant set* $S$ for a flow $\phi_t$ or map $G$ on $\mathbb{R}^n$ as a subset $S \subset \mathbb{R}^n$ such that

$$\phi_t(x) \in S \text{ (or } G(x) \in S) \quad \text{for } x \in S \quad \text{for all } t \in \mathbb{R}. \tag{1.6.1}$$

The stable and unstable manifolds of a fixed point or periodic orbit provide examples of invariant sets. However, the *nonwandering* set is perhaps more important to the study of long-term behavior. We have already seen that fixed points and closed orbits are important in the study of dynamical systems, since they represent stationary or repeatable behavior. A generalization of these sets is the nonwandering set, $\Omega$. A point $p$ is called *nonwandering for the flow* $\phi_t$ (resp. the map $G$) if, for any neighborhood $U$ of $p$, there exists arbitrarily large $t$ (resp. $n > 0$) such that $\phi_t(U) \cap U \neq \varnothing$ (resp. $G^n(U) \cap U \neq \varnothing$). $\Omega$ is the set of all such points $p$. Thus a nonwandering point lies on or near orbits which come back within a specified distance of themselves. Fixed points and periodic orbits are clearly nonwandering. For the damped harmonic oscillator

$$\ddot{x} + \alpha\dot{x} + x = 0, \tag{1.6.2}$$

$(x, \dot{x}) = (0, 0)$ is the only nonwandering point; but for the undamped oscillator

$$\ddot{x} + x = 0, \tag{1.6.3}$$

all points $p \in \mathbb{R}^2$ are nonwandering, since the $(x, \dot{x})$ phase plane is filled with a continuous family of periodic orbits.

EXERCISE 1.6.1. Find the nonwandering sets for the following flows and maps:

(a) $\ddot{x} + \varepsilon(x^2 - 1)\dot{x} + x = 0$ (van der Pol's equation, $\varepsilon > 0$).
(b) $\dot{\theta} = \mu - \sin\theta$, $\theta \in S^1$ (take $\mu < 1$, $\mu = 1$, and $\mu > 1$).
(c) $\ddot{\theta} + \sin\theta = \frac{1}{2}$, $\theta \in S^1$.

(d) $x \to \begin{pmatrix} 1 & 1 \\ 1 & 2 \end{pmatrix} x$, $x \in T^2$ (see §1.4).

((d) is difficult: try to find a dense orbit; recall that points with rational coordinates are periodic; cf. Exercises 1.4.5–1.4.6.)

Not all invariant sets consist of nonwandering points. For example, the linear map

$$x \mapsto \begin{bmatrix} 1 & 1 \\ 0 & \frac{1}{2} \end{bmatrix} x; \qquad x \in \mathbb{R}^2 \tag{1.6.4}$$

has invariant subspaces (eigenspaces) given by

$$E^s = (2, -1)^T,$$
$$E^c = (1, 0)^T,$$

but points $p \in E^s$ are all wandering (except $(0, 0)$). The points $q \in E^c$ are, however, nonwandering, since they are all fixed.

Since the set of wandering points is open, $\Omega$ is closed, and it must contain the closure of the set of fixed points and periodic orbits. Wandering points correspond to transient behavior, while "long-term" or asymptotic behavior corresponds to orbits of nonwandering points. In particular, the attracting set and attractors will be important. However, before defining an attractor we need another two ideas.

A point $p$ is an $\omega$-*limit point* of $x$ if there are points $\phi_{t_1}(x), \phi_{t_2}(x), \ldots$ on the orbit of $x$ such that $\phi_{t_i}(x) \to p$ and $t_i \to \infty$. A point $q$ is an $\alpha$-*limit point* if such a sequence exists with $\phi_{t_i}(x) \to q$ and $t_i \to -\infty$. For maps $G$ the $t_i$ are integers. The $\alpha$- (resp. $\omega$-) limit sets $\alpha(x)$, $\omega(x)$ are the sets of $\alpha$ and $\omega$ limit points of $x$. See Figure 1.6.1.

A closed invariant set $A \subset \mathbb{R}^n$ is called an *attracting set* if there is some neighborhood $U$ of $A$ such that $\phi_t(x) \in U$ for $t \geq 0$ and $\phi_t(x) \to A$ as $t \to \infty$, for all $x \in U$.* The set $\bigcup_{t \leq 0} \phi_t(U)$ is the *domain of attraction* of $A$ (it is, of course, the stable manifold of $A$). An attracting set ultimately captures all orbits starting in its domain of attraction. A *repelling set* is defined analogously, replacing $t$ by $-t$. Domains of attraction of disjoint attracting sets are necessarily nonintersecting and separated by the stable manifolds of nonattracting sets. See Figure 1.6.2.

In many problems we are able to find a "trapping region," a closed connected set $D \subset \mathbb{R}^n$ such that $\phi_t(D) \subset D$ for all $t > 0$. For this, it is sufficient to show that the vector field is directed everywhere inward on the boundary of $D$. In this case we can define the associated attracting set as

$$A = \bigcap_{t \geq 0} \phi_t(D).$$

For maps, a closed set $A$ is an attracting set if it has some neighborhood $U$ such that $G^n(U) \to A$ as $n \to \infty$. As in the case of flows, if $D$ is a trapping region ($G(U) \subset U$), then the associated attracting set is

$$A = \bigcap_{n \geq 0} G^n(D).$$

In Chapter 2 we shall use this idea in studies of several problems.

---

* In Chapter 5 we shall relax this definition somewhat.

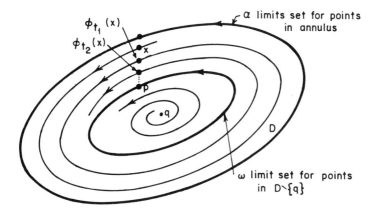

Figure 1.6.1. Examples of $\alpha$ and $\omega$ limit sets. $D$ is the open disc bounded by the outer periodic orbit.

EXERCISE 1.6.2. Show that there is a trapping region for the flow of the system

$$\dot{x} = \mu_1 x - x(x^2 + y^2) - xy^2,$$
$$\dot{y} = \mu_2 y - y(x^2 + y^2) - yx^2; \qquad (x, y) \in \mathbb{R}^2,$$

for all finite values of $\mu_1, \mu_2$. Find the fixed points and discuss their stability. Show that, for $\mu_1 = \mu_2 > 0$, the line $x = y$ separates two distinct domains of attraction. (Hint: Let $D$ be the closed disc with boundary $x^2 + y^2 = c$ for large $c$.)

EXERCISE 1.6.3. Show that there is a solid ellipsoid $E$ given by $\rho x^2 + \sigma y^2 + \sigma(z - 2\rho)^2 \leq c < \infty$ such that, for suitable choices of $\sigma$, $\beta$, $\rho \geq 0$, all solutions of the Lorenz equations

$$\dot{x} = \sigma(y - x); \qquad \dot{y} = \rho x - y - xz; \qquad \dot{z} = -\beta z + xy;$$

enter $E$ within finite time and thereafter remain in $E$ (cf. Sparrow [1982], Appendix C).

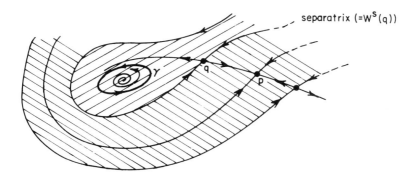

Figure 1.6.2. Domains of attraction: of the closed orbit $\gamma$ ▨▨▨ and of the fixed point $p$ ▧▧▧ .

EXERCISE 1.6.4. Consider the system

$$\dot{x} = x - x^3,$$
$$\dot{y} = -y; \qquad (x, y) \in \mathbb{R}^2.$$

Find the nonwandering set, $\Omega$, and the $\alpha$ and $\omega$ limit points of typical points $x \in \mathbb{R}^2$. Show that the closed interval $[-1, 1]$ of the $x$-axis is an attracting set, although most points in it are wandering. Where do you think most orbits will end up?

The last problem should help motivate our working definition of an *attractor* as an *attracting set which contains a dense orbit*. A repellor is defined analogously. Thus, in Exercise 1.6.4 there are two distinct attractors: the points $(\pm 1, 0)$. As we shall see in Chapter 5, it is very difficult to show in examples that a dense orbit exists, and in fact many of the numerically observed "strange attractors" may not be true attractors but merely attracting sets, since they may contain stable periodic orbits. We shall meet the first such examples in Chapter 2.

An example due to Ruelle [1981] shows that, even in one-dimensional flows, attracting sets can be quite complicated. Consider the system

$$\dot{x} = -x^4 \sin\left(\frac{\pi}{x}\right), \qquad (1.6.5)$$

which has a countable set of fixed points at $x = 0$ and $\pm 1/n$, $n = 1, 2, \ldots$. The interval $[-1, 1]$ is an attracting set, but it contains a countable set of repelling fixed points at $\pm 1/2n$, $n = 1, 2, \ldots$ and attracting fixed points at $\pm 1/(2n - 1)$, $n = 1, 2, \ldots$, as the reader can check by considering the linearized vector field

$$\left(-4x^3 \sin\left(\frac{\pi}{x}\right) + \pi x^2 \cos\left(\frac{\pi}{x}\right)\right)\Bigg|_{x = \pm 1/n} = \frac{\pi}{n^2} \cos n\pi. \qquad (1.6.6)$$

However, the fixed point $x = 0$ is itself neither a repellor nor an attractor. Conley [1978] defined "quasiattractors" earlier to cover this type of example.

EXERCISE 1.6.5. Describe the set of fixed points for the map

$$f : x \to |x|^\alpha \cos\left(\ln\left(\frac{1}{|x|}\right)\right), \qquad x \in [-1, 1]$$

for $\alpha < 0$, $\alpha = 0$, and $\alpha > 0$.

A further example may help to illustrate some of the ideas of this section. In the analysis of the weakly forced van der Pol equation, which we shall outline in Section 1 of Chapter 2, the phase portrait shown in Figure 1.6.3 occurs. Clearly, the closed curve $\gamma \cup \{p\}$, including the fixed point $p$, is an

Figure 1.6.3. A planar phase portrait in the averaged van der Pol equation.

attracting set, but the point $p$ is neither an attractor nor a repellor, being simultaneously the $\alpha$ and $\omega$ limit point for all points $x \in \gamma$. In fact $\gamma$ is filled with wandering points and the fixed points $p$ and $q$ are the only components of the nonwandering set. Since there is a dense orbit in $\gamma \cup \{p\}$, our attracting set is in fact an attractor, but it is clear that, in the absence of perturbations, all solutions except those based at $q$ tend towards $p$ from the left as $t \to +\infty$. This example, among others, should warn us that our definition of an attractor may not be the most appropriate for physical applications, and we shall therefore modify it in Chapter 5 in the light of examples arising from physical problems.

This example also illustrates why we include the requirement that $\phi_t(x) \in U$ for all $t \geq 0$, $x \in U$, since there are orbits starting to the right of $p$ which leave a neighborhood of $p$ only to eventually return as $t \to \infty$. The reader should compare this requirement with our definitions of local stable and unstable manifolds in Section 1.3.

EXERCISE 1.6.6. Show that the circle $r = 1$ is an attracting set for the flow arising from the vector field

$$\dot{r} = r - r^3, \qquad \dot{\theta} = 1 - \cos 2\theta.$$

Which of the equilibrium points are attractors and which repellors? Describe the $\alpha$ and $\omega$ limit sets for typical points inside and outside the circle $r = 1$ and in the upper and lower half planes.

EXERCISE 1.6.7. Construct an example of a two-dimensional flow with an attractor which contains no fixed points or closed orbits. (Hint: Consider linear translation on the torus $T^2 = \mathbb{R}^2/Z^2$ given by the vector field $\dot{\theta} = \alpha$, $\dot{\phi} = \beta$.)

We note that we have not specified that an attractor should be persistent with respect to small perturbations of the vector field or map. While this has been a requirement in many previous definitions, many of the examples which we consider in this book almost certainly do not have such structurally stable attractors. Nonetheless, the idea of structural stability plays an important rôle in dynamical systems theory, and it is to this that we now turn.

## 1.7. Equivalence Relations and Structural Stability

The idea of a "robust" or "coarse" system—one that retains its qualitative properties under small perturbations or changes to the functions involved in its definition—originated in the work of Andronov and Pontryagin [1937]. A very readable introduction to the special case of planar vector fields may be found in the text on nonlinear oscillations by Andronov *et al.* [1966]. In Chapter 5 we shall question the conventional wisdom that robustness or structural stability is an essential property for models of physical systems, but since the concept has played such a large rôle in the development of dynamical systems theory we discuss it briefly here. We first discuss the idea of perturbations of maps and vector fields.

Given a map $F \in C^r(\mathbb{R}^n)$, we want to specify what is meant by a perturbation $G$ of $F$. Intuitively, $G$ should be "close to" $F$, but there are technical issues involved in making a workable definition. We refer the reader to Hirsch [1976] for a full discussion of function spaces and their topologies. Since we have avoided the use of function spaces in this book, we make the following definition which suffices for our discussion of structural stability.

**Definition 1.7.1.** If $F \in C^r(\mathbb{R}^n)$, $r$, $k \in \mathbb{Z}^+$, $k \leq r$, and $\varepsilon > 0$, then $G$ is a $C^k$ *perturbation of size* $\varepsilon$ if there is a compact set $K \subset \mathbb{R}^n$ such that $F = G$ on the set $\mathbb{R}^n - K$ and for all $(i_1, \ldots, i_n)$ with $i_1 + \cdots i_n = i \leq k$ we have $|(\partial^i/\partial x_1^{i_1} \cdots \partial x_n^{i_n})(F - G)| < \varepsilon$.

We remark that in this definition the functions $F$ and $G$ might be vector fields or maps.

Now that we can discuss the "closeness" of maps or vector fields, we can consider the questions of topological equivalence and structural stability:

**Definition 1.7.2.** Two $C^r$ maps $F$, $G$ are $C^k$ *equivalent* or $C^k$ *conjugate* $(k \leq r)$ if there exists a $C^k$ homeomorphism $h$ such that $h \circ F = G \circ h$. $C^0$ equivalence is called *topological equivalence*.

This definition implies that $h$ takes an orbit $\{F^n(x)\}$ to an orbit $\{G^n(x)\}$. The notion of orbit-equivalence is also what we need in the case of vector fields:

**Definition 1.7.3.** Two $C^r$ vector fields, $f$, $g$ are said to be $C^k$ *equivalent* $(k \leq r)$ if there exists a $C^k$ diffeomorphism $h$ which takes orbits $\phi_t^f(x)$ of $f$ to orbits $\phi_t^g(x)$ of $g$, preserving senses but not necessarily parametrization by time. If $h$ does preserve parametrization by time, then it is called a *conjugacy*.

The definition of equivalence implies that for any $x$ and $t_1$, there is a $t_2$ such that

$$h(\phi_{t_1}^f(x)) = \phi_{t_2}^g(h(x)). \qquad (1.7.1)$$

One reason that parametrization by time cannot, in general, be preserved is that the periods of closed orbits in flows can differ.

We now come to the major definition:

**Definition 1.7.4.** A map $F \in C^r(\mathbb{R}^n)$ (resp. a $C^r$ vector field $f$) is *structurally stable* if there is an $\varepsilon > 0$ such that all $C^1$, $\varepsilon$ perturbations of $F$ (resp. of $f$) are topologically equivalent to $F$ (resp. $f$).

At first sight the use of $C^0$ equivalence might seem crude and we might be tempted to use $C^k$ equivalence with $k > 0$. This is too strict, however, because it implies that if $f$ and $g$ have fixed points $p$ and $q = h(p)$, then the eigenvalues of the linearized systems $\dot\xi = Df(p)\xi$ and $\dot\eta = Dg(q)\eta$ must be in the same ratios (we prove this in the appendix to this section). For example, the linear systems

$$\begin{pmatrix} \dot x \\ \dot y \end{pmatrix} = \begin{bmatrix} 1 & 0 \\ 0 & 1 \end{bmatrix}\begin{pmatrix} x \\ y \end{pmatrix} \tag{1.7.2a}$$

and

$$\begin{pmatrix} \dot x \\ \dot y \end{pmatrix} = \begin{bmatrix} 1 & 0 \\ 0 & 1 + \varepsilon \end{bmatrix}\begin{pmatrix} x \\ y \end{pmatrix} \tag{1.7.2b}$$

are not $C^k$ orbit equivalent for any $\varepsilon \neq 0$ and $k \geq 1$. In this example the lack of differentiable equivalence is clear, since in the first case solution curves are given by graphs of the form $y = C_1 x$, and in the second by $y = C_2|x|^{1+\varepsilon}$. Any such pair of curves with $C_1, C_2 \neq 0$ are not diffeomorphic at the origin.

Note that homeomorphic equivalence does not distinguish among nodes, improper nodes, and foci: for example, the two-dimensional linear vector fields with matrices

$$\begin{bmatrix} -1 & 0 \\ 0 & -2 \end{bmatrix}, \quad \begin{bmatrix} -1 & 1 \\ 0 & -1 \end{bmatrix} \quad \text{and} \quad \begin{bmatrix} -3 & -1 \\ 1 & -3 \end{bmatrix}$$

all have flows which are $C^0$ equivalent to that of the node with matrix

$$\begin{bmatrix} -1 & 0 \\ 0 & -1 \end{bmatrix}. \tag{1.7.3}$$

However, the $C^0$ equivalence relation clearly *does* distinguish between sinks, saddles, and sources.

As a further illustration of structural stability of both flows and maps, consider the two-dimensional linear differential equation

$$\dot x = Ax, \quad x \in \mathbb{R}^2, \tag{1.7.4}$$

and the map

$$x \mapsto Bx, \quad x \in \mathbb{R}^2. \tag{1.7.5}$$

Suppose in the first case that $A$ has no eigenvalues with zero real part, and in the second that $B$ has no eigenvalues of unit modulus. We claim that, if these conditions hold, then both systems are structurally stable.

Consider a small perturbation of (1.7.4):

$$\dot{x} = Ax + \varepsilon f(x), \tag{1.7.6}$$

where $f$ has support in some compact set. Since $A$ is invertible, by the implicit function theorem, the equation

$$Ax + \varepsilon f(x) = 0 \tag{1.7.7}$$

continues to have a unique solution $\bar{x} = 0 + \mathcal{O}(\varepsilon)$ near $x = 0$, for sufficiently small $\varepsilon$. Moreover, since the matrix of the linearized system

$$\dot{\xi} = [A + \varepsilon Df(\bar{x})]\xi$$

has eigenvalues which depend continuously on $\varepsilon$, no eigenvalues can cross the imaginary axis if $\varepsilon$ remains small with respect to the magnitude of the real parts of the eigenvalues of $A$. Thus the perturbed system (1.7.7) has a unique fixed point with eigenspaces and invariant manifolds of the same dimensions as those of the unperturbed system, and which are $\varepsilon$-close locally in position and slope to the unperturbed manifolds. Similar observations apply to the discrete system (1.7.5) and a corresponding small perturbation

$$x \mapsto Bx + \varepsilon g(x). \tag{1.7.8}$$

In both cases the problem is that of finding a homeomorphism which takes orbits of the linear system to those of the perturbed, nonlinear system. Specifically, for the discrete systems, we must prove that there is a homeomorphism $h$ such that the following diagram commutes:

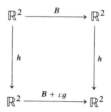

For the flow we replace $B$ by $e^{t_1 A}$ and $B + \varepsilon g$ by the flow $\phi_{t_1}$ generated by the vector field (1.7.7) (they are conjugate in this case).

The proof is now essentially the same as that of Hartman's theorem and the reader is referred to Pugh [1969] or Hartman [1964, Chapter 9].

It should be clear that a vector field (or map) possessing a non-hyperbolic fixed point cannot be structurally stable, since a small perturbation can remove it, if the linearized matrix is noninvertible, having a zero eigenvalue, or turn it into a hyperbolic sink, a saddle, or a source, if the matrix has purely imaginary eigenvalues. Similar observations apply to periodic orbits and we are therefore in a position to state an important requirement for structural stability of flows or maps: *all fixed points and closed orbits must be hyperbolic*. However, as we shall see, this condition alone is not enough to guarantee structural stability, since more subtle, global effects also come into play.

Structurally stable systems have rather "nice" properties in the sense that, if a system is structurally stable, then any sufficiently close system has the same qualitative behavior. However, as we shall see, structurally stable behavior can be extremely complex for flows of dimension $\geq 3$ or diffeomorphisms of dimension $\geq 2$. Also, it will turn out that structural stability is not even a *generic property*—that is, we can find structurally unstable (and complicated) systems which *remain* unstable under small perturbations, and which, in fact continually change their topological equivalence class as we perturb them. We shall meet our first examples of such systems in Chapter 2.

We have not defined or discussed the notion of generic properties in this section because their definition is formulated in terms of function spaces. Interested readers should consult Chillingworth [1976] or Hirsch and Smale [1974] for introductions to the subject.

Before closing this section we wish to stress that the definition of structural stability is relative to the class of systems we deal with. In our main definition we have allowed all $C^1$, $\varepsilon$ perturbations by $C^r$ vector fields on $\mathbb{R}^n$. If we restrict ourselves to some subset, say all $C^r$ *Hamiltonian* vector fields on $\mathbb{R}^2$, then things are different and we find that the linear system

$$\begin{aligned} \dot{x} &= y, \\ \dot{y} &= -\omega^2 x, \end{aligned} \qquad \omega \neq 0, \qquad (1.7.9)$$

possessing an elliptic center at $(x, y) = (0, 0)$ surrounded by a continuous family of non-hyperbolic closed orbits, *is* stable to small perturbations within this subset. However, the system

$$\begin{aligned} \dot{x} &= y, \\ \dot{y} &= 0, \end{aligned} \qquad (1.7.10)$$

possessing a degenerate line of fixed points on the x-axis, is not structurally stable, since we can find a Hamiltonian perturbation which yields an isolated fixed point near $(0, 0)$ which is either a center or a hyperbolic saddle point. Of course, both systems are structurally unstable with respect to perturbations by general $C^r$ vector fields. In this book we concentrate on dissipative systems and pay little attention to the special properties of Hamiltonian systems (but see Section 4.8).

EXERCISE 1.7.1. Show that the vector fields $\dot{x} = -x$, $\dot{x} = -4x$, and $\dot{x} = -x^3$ are all $C^0$ equivalent. Which ones are structurally stable? Find explicitly the homeomorphisms relating their orbits.

EXERCISE 1.7.2. Which of the following systems are structurally stable in the set of all one- (or two-) dimensional systems?

(a) $\dot{x} = x$;                                  (b) $\dot{x} = x^2$;
(c) $\dot{x} = \sin x$;                              (d) $\ddot{x} = \sin x$;
(e) $\ddot{x} + 2\dot{x} + x = 0$;                   (f) $\ddot{x} + \dot{x}^2 + x = 0$;
(g) $\ddot{x} + \dot{x} + x^3 = 0$;                  (h) $\ddot{x} + (x^2 - 1)\dot{x} + x = 0$;
(i) $\dot{\theta} = 1, \dot{\phi} = 2; (\theta, \phi) \in T^2$;    (j) $\dot{\theta} = 1, \dot{\phi} = \pi; (\theta, \phi) \in T^2$;
(k) $\dot{\theta} = 2 - \sin \theta, \dot{\phi} = 1; (\theta, \phi) \in T^2$;    (l) $\dot{\theta} = 1 - 2\sin \theta, \dot{\phi} = 1; (\theta, \phi) \in T^2$.

(The criteria for determining structural stability of one and two dimensional flows are discussed in the next two sections.)

## Appendix to Section 1.7: On $C^k$ Equivalence

Here we show that $C^k$ equivalence, $k \geq 1$, implies that two systems must have the same ratios of eigenvalues when linearized at corresponding fixed points.

If $X$ and $Y$ are $C^k$ orbit equivalent then there is a $C^k$ diffeomorphism $h: \mathbb{R}^n \to \mathbb{R}^n$ such that $h$ "converts" the system $\dot{x} = X(x)$ into $\dot{y} = Y(y)$, i.e., since $y = h(x)$ we have

$$Dh(x)X(x) = \tau(h(x))Y(h(x)), \qquad (1.7.11)$$

where $\tau: \mathbb{R}^n \to \mathbb{R}$ is a positive scalar function which allows reparametrization of time.

We now suppose $x = p$ is a fixed point for the flow of $X$, so that $y = q = h(p)$ is a fixed point for $Y$. Differentiating (1.7.11) and setting $x = p$, $y = q$, we obtain

$$D^2h(x)X(x) + Dh(x)DX(x) = D\tau(h(x))Dh(x)Y(h(x))$$
$$+ \tau(h(x))DY(h(x))Dh(x);$$

since $X(p) = Y(q) = 0$, this gives                                (1.7.12)

$$Dh(p)DX(p) = \tau(q)DY(q)Dh(p),$$

or

$$DX(p) = \tau(q)Dh(p)^{-1}DY(q)Dh(p). \qquad (1.7.13)$$

Thus the two matrices $DX(p)$ and $DY(q)$ are similar, up to a uniform scaling by the constant $\tau(q)$. Hence ratios between eigenvalues are preserved. (If we do not allow reparametrization of time, then $DX(p)$ and $DY(q)$ are strictly similar.)

## 1.8. Two-Dimensional Flows

In this section we provide a review of some of the theory of two dimensional flows. The Jordan curve theorem, and the fact that solution curves are one dimensional, make the range of solution types on the plane rather limited. The planar system is therefore fairly well understood. However, the reader should realize that we provide only a sampling of the many results here. Andronov and co-workers [1966, 1971, 1973] have well over a thousand pages on the subject and Lefschetz [1957] should also be consulted for more details. Many examples of two-dimensional systems arising in engineering and physics are given by Andronov *et al.* [1966] and in the books of Minorsky [1962], Hayashi [1964], and Nayfeh and Mook [1979]. Here we concentrate on some special classes of systems and give a number of examples, some classical and others less familiar, which will prepare us for the material to follow.

Systems on two manifolds other than $\mathbb{R}^2$ are more complicated and can display surprisingly subtle behavior. In the remainder of this chapter we therefore concentrate on the planar system, although we end with some examples of systems on cylinders and tori.

Suppose that we are given a differential equation

$$
\begin{aligned}
\dot{x} &= f(x, y), \\
\dot{y} &= g(x, y),
\end{aligned}
\qquad (x, y) \in U \subseteq \mathbb{R}^2,
\qquad (1.8.1)
$$

where $f$ and $g$ are (sufficiently smooth) functions specified by some physical model. In approaching equation (1.8.1) we normally first seek fixed points, at which $f(x, y) = g(x, y) = 0$. Linearizing (1.8.1) at such a point $(\bar{x}, \bar{y})$, we obtain

$$
\begin{pmatrix} \dot{\xi}_1 \\ \dot{\xi}_2 \end{pmatrix} =
\begin{bmatrix}
\dfrac{\partial f}{\partial x}(\bar{x}, \bar{y}) & \dfrac{\partial f}{\partial y}(\bar{x}, \bar{y}) \\[2ex]
\dfrac{\partial g}{\partial x}(\bar{x}, \bar{y}) & \dfrac{\partial g}{\partial y}(\bar{x}, \bar{y})
\end{bmatrix}
\begin{pmatrix} \xi_1 \\ \xi_2 \end{pmatrix}
\quad \text{or} \quad \dot{\xi} = Df(\bar{x}, \bar{y})\xi. \qquad (1.8.2)
$$

If the eigenvalues of the matrix $Df(\bar{x}, \bar{y})$ have nonzero real parts, then the solution $\xi(t) = e^{tDf(\bar{x}, \bar{y})}\xi(0)$ of (1.8.2) not only yields local asymptotic behavior, but, by Hartman's theorem and the stable manifold theorem, also provides the local topological structure of the phase portrait. The following exercise shows that the insistence of nonzero eigenvalues is necessary:

EXERCISE 1.8.1. Sketch the phase portraits of the following two nonlinear oscillators and of their linearizations. (You may need to read on to review analytical methods for two-dimensional systems.)

(a) $\ddot{x} + \varepsilon|\dot{x}|\dot{x} + x = 0; \varepsilon < 0, \varepsilon = 0, \varepsilon > 0;$
(b) $\ddot{x} + \dot{x} + \varepsilon x^2 = 0; \varepsilon < 0, \varepsilon = 0, \varepsilon > 0.$

After locating the fixed points and studying their stability (perhaps using (local) Liapunov functions in the case of nonhyperbolic points), we next wish to ascertain whether (1.8.1) has any periodic orbits. Here the following two results are useful:

**Theorem 1.8.1** (The Poincaré–Bendixson Theorem). *A nonempty compact ω- or α-limit set of a planar flow, which contains no fixed points, is a closed orbit.*

For a proof, see Hirsch and Smale [1974, p. 248] or Andronov *et al.* [1966, p. 361].

This result can be used to establish the existence of closed orbits, as in the exercise below:

EXERCISE 1.8.2. Use the Poincaré–Bendixson theorem to prove that the van der Pol oscillator $\ddot{x} + \varepsilon(x^2 - 1)\dot{x} + x = 0$ has at least one closed orbit. (First find two nested closed curves, $C_1$, $C_2$ such that the flow crosses $C_1$ inward and crosses $C_2$ outward.)

Proving uniqueness in the above example is considerably harder unless $\varepsilon \ll 1$; see Hirsch and Smale [1974, Chapter 10], for example. However, if the vector field is such that $C_1$ and $C_2$ can be chosen to bound a narrow annulus, $R$, then it may be possible to prove uniqueness by showing that $\partial f/\partial x + \partial g/\partial y$ is everywhere negative (or positive) in $R$. We give such an example in Section 1 of Chapter 2.

The next result enables us to rule out the occurrence of closed orbits in some cases:

**Theorem 1.8.2** (Bendixson's Criterion). *If on a simply connected region $D \subseteq \mathbb{R}^2$ the expression $\partial f/\partial x + \partial g/\partial y$ is not identically zero and does not change sign, then equation (1.8.1) has no closed orbits lying entirely in D.*

PROOF. This result is a simple consequence of Green's theorem, for on any solution curve of (1.8.1) we have $dy/dx = g/f$ or, in particular

$$\int_\gamma (f(x, y)\, dy - g(x, y)\, dx) = 0$$

on any closed orbit $\gamma$. This implies, via Green's theorem, that

$$\iint_S \left(\frac{\partial f}{\partial x} + \frac{\partial g}{\partial y}\right) dx\, dy = 0, \tag{1.8.3}$$

where $S$ is the interior of $\gamma$. But if $\partial f/\partial x + \partial g/\partial y > 0$ (or $<0$) on $D$, then we cannot find a region $S \subseteq D$ such that (1.8.3) holds. Hence there can be no closed orbits entirely in $D$. ☐

EXERCISE 1.8.3. Find sufficient conditions for the system $\ddot{x} + \alpha\dot{x} + \beta x + x^2\dot{x} + x^3 = 0$ to have no closed orbits. Are they also necessary?

For a generalization of Bendixson's criterion, Dulac's criterion, see Andronov *et al.* [1966, p. 305].

In addition to fixed points and closed orbits, we have already met examples of other limit sets of two-dimensional flows in Section 1.6. In fact for planar flows, all the possible nonwandering sets fall into three classes (Andronov *et al.* [1966], §VI.2).

(i) Fixed points;
(ii) closed orbits; and
(iii) the unions of fixed points and the trajectories connecting them.

The latter are referred to as *heteroclinic orbits* when they connect distinct points and *homoclinic orbits* when they connect a point to itself. Closed paths formed of heteroclinic orbits are called *homoclinic cycles*. We note that the fixed points contained in such cycles must all be saddle points (if they are hyperbolic), since sinks and sources necessarily have wandering points in their neighborhoods. Some examples of such limit sets are shown in Figure 1.8.1. We will meet specific systems which display almost all of these behaviors later in this book.

EXERCISE 1.8.4. All the examples of planar flows in Figure 1.8.1 are structurally unstable. Why? (Hint: In (a)–(c), try adding a small perturbation near the homoclinic cycles.)

In flows on nonplanar two-dimensional manifolds, such as the torus, limit sets which are neither closed orbits, fixed points, nor homoclinic cycles can arise. In particular, irrational linear flow such as that generated by the vector field

$$\dot{\theta} = 1, \qquad (\theta, \phi) \in T^2; \qquad (1.8.4)$$
$$\dot{\phi} = \pi,$$

has a dense orbit and thus *every* point on $T^2$ is nonwandering. In spite of its apparent artificiality, we shall subsequently see that this example arises naturally in the study of coupled oscillators.

Thus, in two-dimensional flows the global structures of solution curves are generally far richer than those of one-dimensional systems, in which periodic orbits cannot occur and the fixed points are ordered and necessarily connected to their immediate neighbors and only to them. Whether or not such heteroclinic connections are likely to exist in higher-dimensional systems depends upon the relative dimensions of stable and unstable manifolds of neighboring fixed points, but in any case they are generally very difficult to find, unless the system possesses special symmetries or other properties. Our first main example illustrates this point, as well as introducing two important special classes of systems: Hamiltonian and gradient flows. In each type of system, the level curves of a real valued function determine the global structure of the flow.

We consider the example

$$\dot{x} = -\zeta x - \lambda y + xy,$$
$$\dot{y} = \lambda x - \zeta y + \tfrac{1}{2}(x^2 - y^2), \qquad (1.8.5)$$

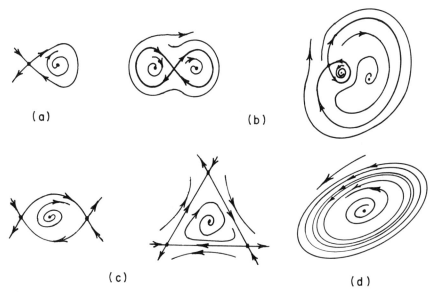

(a)

(b)

(c)

(d)

Figure 1.8.1. Some limit sets for flows on the plane. (a) A homoclinic orbit or saddle-loop; (b) double saddle-loops; (c) homoclinic cycles formed of heteroclinic orbits; (d) bands of periodic orbits.

which arose as an averaged system (cf. Chapter 4) in wind induced oscillation studies (Holmes [1979b]). Here $0 \le \zeta \ll 1$ is a damping factor and $\lambda\ (|\lambda| \ll 1)$ is a detuning parameter. When $\zeta = 0$, (1.8.5) becomes a *Hamiltonian* system (Goldstein [1980]):

$$\dot{x} = \frac{\partial H}{\partial y}, \qquad \dot{y} = -\frac{\partial H}{\partial x}, \qquad (1.8.6)$$

for which the Hamiltonian (energy) function

$$H(x, y) = -\frac{\lambda}{2}(x^2 + y^2) + \frac{1}{2}\left(xy^2 - \frac{x^3}{3}\right) \qquad (1.8.7)$$

is a map $H: \mathbb{R}^2 \to \mathbb{R}$. The critical points of $H$ correspond to the fixed points of the flow of Hamilton's equations (1.8.6). Moreover, since

$$\frac{dH}{dt} = \frac{\partial H}{\partial x}\dot{x} + \frac{\partial H}{\partial y}\dot{y} = \frac{\partial H}{\partial x}\frac{\partial H}{\partial y} - \frac{\partial H}{\partial y}\frac{\partial H}{\partial x} \equiv 0, \qquad (1.8.8)$$

the level curves $H(x, y) =$ constant are solution curves for (1.8.6). Thus for our example the phase portrait may easily be drawn as in Figure 1.8.2. Such a system is said to be integrable, since the solutions, or integral curves, lie along level curves of a smooth function.

Notice that the three saddle-points at $p_3 = (-2\lambda, 0)$ and $p_{2,1} = (\lambda, \pm\sqrt{3}\lambda)$, are connected. The connecting curves $\Gamma_{ij} = W^u(p_i) \cap W^s(p_j)$ are examples of saddle connections or heteroclinic orbits (if such a curve connects a saddle

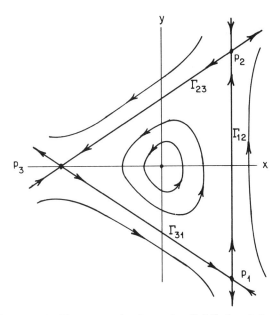

Figure 1.8.2. Phase portrait of equation (1.8.5); $\zeta = 0, \lambda > 0$.

to itself it is referred to as a homoclinic orbit). Here these orbits occur as a result of the Hamiltonian integral constraint, although saddle connections can occur in non-Hamiltonian systems. The reader should check that when $\zeta > 0$ all three connections are broken and the unstable manifolds $W^u(p_i)$ now have components which approach the fixed point at $(x, y) = (0, 0)$ as $t \to +\infty$. This point is then a sink, with eigenvalues $-\zeta \pm i\lambda$. Realizing that the $\Gamma_{ij}$ are each intersections of two one-dimensional curves, $W^u(p_i)$ and $W^s(p_j)$, in the plane, we would indeed expect such intersections to occur only under special circumstances, and, if they do occur, that they would be broken by arbitrarily small perturbations. Such connections are thus *structurally unstable* in the space of all vector fields on $\mathbb{R}^2$. We shall return to this point later, in sketching the proof of Peixoto's theorem. Note that intersection of $W^u(p_i)$ and $W^s(p_j)$ implies that portions of the two curves are, in fact, *identified*: they cannot merely intersect, as shown in Figure 1.8.3(b), or the solution based at the intersection point $q$ would have two possible futures and pasts, thus violating uniqueness of solutions.

Upon linearizing a planar Hamiltonian system at a fixed point, one finds that trace($Df$) $\equiv 0$ and thus all fixed points are either saddles or centers; no sinks or sources can exist. This reflects the more general fact that Hamiltonian flows preserve volume (or, in the two-dimensional case, area); a result known as Liouville's theorem. For more information on this and other results applicable to higher-dimensional Hamiltonian systems, the reader should refer to classical mechanics texts such as Goldstein [1980] or Arnold [1978].

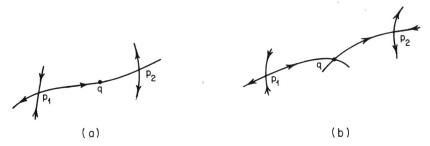

( a )                                        ( b )

Figure 1.8.3. Heteroclinic points $q \in W^u(p_1) \cap W^s(p_2)$ for flows in $\mathbb{R}^2$. (a) Admissible; (b) not admissible.

EXERCISE 1.8.5. Show that a differential equation of the form $\dot{x} = f(y)$, $\dot{y} = g(x)$ always possesses a first integral $F(y) + G(x)$, the level curves of which are solution curves. Use this fact to study the global solution structure of the system $\dot{x} = -y + y^3$, $\dot{y} = x - x^3$.

A special subset of conservative nonlinear oscillator problems with which we shall be concerned in this book takes the form

$$\ddot{x} + f(x) = 0, \tag{1.8.9}$$

or, as a planar vector field with $y = \dot{x}$:

$$\dot{x} = y,$$
$$\dot{y} = -f(x). \tag{1.8.10}$$

Such a Hamiltonian system always possesses a first integral (at least formally):

$$H(x, y) = \frac{y^2}{2} + \int f(x)\, dx \overset{\text{def}}{=} \frac{y^2}{2} + V(x), \tag{1.8.11}$$

where $V(x)$ is sometimes called a potential (energy) function, since in mechanical applications it often corresponds to a stored energy (cf. Andronov et al. [1966], Marion [1970]).

Any fixed point of (1.8.10) must lie on the $x$-axis, and correspond to a critical point of $V(x)$. Thus the real valued function $V: \mathbb{R} \to \mathbb{R}$ effectively determines the local form of the vector field and hence the flow near each fixed point. Andronov et al. [1966, Chapter 2], Nayfeh and Mook [1979] and others give exhaustive accounts of the various cases. For example, if the critical point of $V$ is nondegenerate (quadratic) then the fixed point is either a hyperbolic saddle or a center, while if the leading term in the Taylor expansion of $V$ is cubic or higher, then the fixed point is degenerate. Here we note that the special structure also enables one to obtain information on the global structure of solution curves, which are simply given by $H(x, y) = c = $ constant, or

$$y = \pm \sqrt{2(c - V(x))}, \tag{1.8.12}$$

and are thus symmetric under reflection about the $x$-axis. A major consequence of this is that, if there are two saddle points with the same energy level, corresponding to two maxima of $V(x)$, with no higher maximum between them, then they *must* be connected by heteroclinic orbits.

EXERCISE 1.8.6. Find and classify the fixed points and sketch phase portraits for the Hamiltonian systems:

(a) $\ddot{x} + x - x^2 = 0$;
(b) $\ddot{x} + x^2 + x^3 = 0$;
(c) $\ddot{x} + \sin x = 0$;
(d) $\ddot{x} + \sin x = \beta, \beta \in (0, 2)$.

Equation (1.8.5) provides an example of another special type of system. When $\lambda = 0$ the system is a *gradient vector field*

$$\dot{x} = -\frac{\partial V}{\partial x}, \quad \dot{y} = -\frac{\partial V}{\partial y}; \quad V(x, y) = \frac{\zeta}{2}(x^2 + y^2) + \frac{1}{2}\left(\frac{y^3}{3} - x^2 y\right),$$

(1.8.13)

with a sink at $(0, 0)$ (for $\zeta > 0$) and saddles at $(0, -2\zeta)$ and $(\pm\sqrt{3}\zeta, \zeta)$. For a general discussion of such fields see Hirsch and Smale [1974, pp. 199ff.]. We note that the *potential function* $V: \mathbb{R}^2 \to \mathbb{R}$ can be regarded as a Liapunov function. In this way it is possible to show that minima (resp. maxima) of $V$ correspond to sinks (resp. sources) of the $n$-dimensional system:

$$\dot{x} = -\text{grad } V(x),$$

(1.8.14)

for general potential functions $V: \mathbb{R}^n \to \mathbb{R}$. In fact any critical point of $V$, at which grad $V = 0$, must be a fixed point of (1.8.14), and the saddle points of $V$ are, of course, saddle points of (1.8.14). Now let $V^{-1}(h)$ be a level (hyper-) surface of $V$. Then, since for any point $x \in V^{-1}(h)$ at which grad $V(x) \neq 0$, the vector $-\text{grad } V(x)$ is normal to the tangent to the level surface at $x$, solution curves of (1.8.14) cross the level surfaces normally and point "downhill" in the direction of decreasing $V$.

EXERCISE 1.8.7. Show that the nonwandering set of a gradient vector field on $\mathbb{R}^2$ contains only fixed points and that no periodic or homoclinic orbits are possible. (Hint: Use $V(x)$ as a "Liapunov-like" function and modify the usual arguments to work globally.)

EXERCISE 1.8.8. Sketch the phase portraits of equation (1.8.5) for $\zeta = 0$ and $\zeta > 0$, with $\lambda < 0, = 0, > 0$. Which cases are structurally stable in the set of all two-dimensional vector fields?

The special properties of gradient vector fields enabled Palis and Smale [1970] to obtain an important general result for such systems in $n$ dimensions:

**Theorem 1.8.3.** *Gradient systems for which all fixed points are hyperbolic and all intersections of stable and unstable manifolds transversal, are structurally stable.*

For a moment we will step into $n$ dimensions ($n > 2$) to consider this result. There should be no trouble regarding hyperbolicity, but the transversal intersection condition requires some discussion. As we shall see in sketching the proof of Peixoto's theorem, saddle connections for planar systems such as $\Gamma_{12} = W^u(p_1) \cap W^s(p_2)$ of Figure 1.8.1 can be removed (=broken) by small perturbations and are thus structurally unstable. However, suppose that one has a three-dimensional system with a pair of saddle points $p_1$, $p_2$ and dim $W^u(p_1) =$ dim $W^s(p_2) = 2$. It is now possible for $W^u(p_1)$ and $W^s(p_2)$ to intersect *transversely* on an orbit $\gamma = W^u(p_1) \mathbin{\overline{\pitchfork}} W^s(p_2)$ ($\overline{\pitchfork}$ = transversal intersection), so that, at any point $q \in \gamma$, the tangent spaces $T_q W^u(p_1)$, $T_q W^s(p_2)$ span $\mathbb{R}^3$ (Figure 1.8.4). If such a transverse heteroclinic orbit exists then it is possible to show that it cannot be removed by an arbitrarily small perturbation, and is thus structurally stable. However, transverse *homoclinic* orbits cannot exist, since dim $W^u(p)$ + dim $W^s(p) \le n$ and for transversality we require dim $W^u(p_1)$ + dim $W^s(p_2) > n$ (Figure 1.8.5).

Transverse saddle connections cannot exist at all in two dimensions, since the saddle points have one-dimensional stable and unstable manifolds and a connection $\gamma = W^u(p_1) \cap W^s(p_2)$ is necessarily an open interval on which $W^u(p_1)$ and $W^s(p_2)$ are identified. The tangent space to such a curve at any point $q \in \gamma$ is thus one dimensional (Figure 1.8.3(a)).

Returning to planar flows, we recall a useful result which relates the existence of closed orbits and fixed points. This involves the (Poincaré) index of a fixed point. We start with the general idea of the index. Given a planar flow, we draw a simple closed curve $C$ not passing through any equilibrium points and consider the orientation of the vector field at a point $p = (x, y) \in C$. Letting $p$ traverse $C$ anticlockwise, the vector $(f(x, y), g(x, y))$ rotates continuously and, upon returning to the original position, must have rotated through an angle $2\pi k$ for some integer $k$. (The angle is also measured anticlockwise.) We call $k$ the *index of the closed curve $C$* and it can be shown that $k$ is independent of the form of $C$ in the sense that it is determined solely by the character of the fixed points inside $C$.

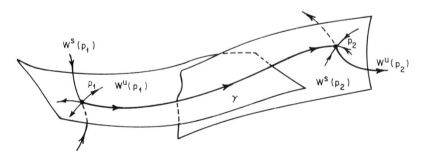

Figure 1.8.4. A transverse heteroclinic orbit in $\mathbb{R}^3$.

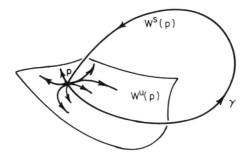

Figure 1.8.5. A nontransverse homoclinic orbit in $\mathbb{R}^3$.

If $C$ is chosen to encircle a single, isolated fixed point, $\bar{x}$, then $k$ is called *the index of $\bar{x}$*. The reader can verify the following statements either by direct examination of the vector fields (cf. Figure 1.8.6) or by evaluation of the curvilinear integral

$$k = \frac{1}{2\pi} \int_C d\left\{\arctan\left(\frac{dy}{dx}\right)\right\} = \frac{1}{2\pi} \int_C d\left\{\arctan\left(\frac{g(x,y)}{f(x,y)}\right)\right\} = \frac{1}{2\pi} \int_C \frac{f\,dg - g\,df}{f^2 + g^2},$$

(1.8.15)

as in Andronov *et al.* [1966, §V.8]:

**Proposition 1.8.4.**

 (i) *The index of a sink, a source or a center is $+1$.*
 (ii) *The index of a hyperbolic saddle point is $-1$.*
(iii) *The index of a closed orbit is $+1$.*
(iv) *The index of a closed curve not containing any fixed points is $0$.*
 (v) *The index of a closed curve is equal to the sum of the indices of the fixed points within it.*

As a direct corollary to these statements, we find

**Corollary 1.8.5.** *Inside any closed orbit $\gamma$ there must be at least one fixed point. If there is only one, then it must be a sink or a source. If all the fixed points within $\gamma$ are hyperbolic, then there must be an odd number, $2n + 1$, of which $n$ are saddles and $n + 1$ either sinks or sources.*

Degenerate fixed points having indices different from $\pm 1$ are easily constructed. The system

$$\dot{x} = x^2,$$
$$\dot{y} = -y,$$

(1.8.16)

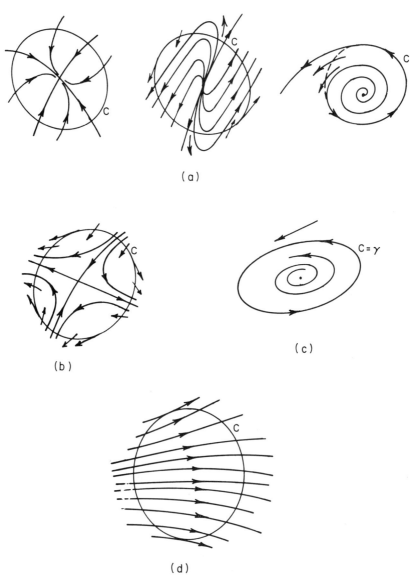

Figure 1.8.6. Indices of fixed points and closed curves. (a) Sinks and sources; (b) a hyperbolic saddle point; (c) closed orbits; (d) $C$ contains no fixed points.

for example, has a degenerate *saddle-node* of index 0 at $(0, 0)$, while that of the system

$$\dot{x} = x^2 - y^2,$$
$$\dot{y} = 2xy$$
(1.8.17)

has index 2, cf. Figure 1.8.7.

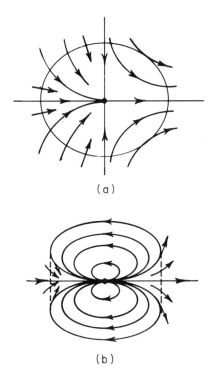

(a)

(b)

Figure 1.8.7. Indices of some nonhyperbolic equilibria. (a) The saddle-node; (b) the vector field of (1.8.17).

The following exercise shows how our second example was chosen and analyzed:

EXERCISE 1.8.9. Letting $z = x + iy$, show that the vector fields in the complex plane defined by

$$\dot{z} = z^k \quad \text{and} \quad \dot{z} = \bar{z}^k$$

have unique fixed points at $z = 0((x, y) = (0, 0))$, with indices $k$ and $-k$, respectively. (Here $\bar{z}$ denotes the complex conjugate). (Hint: Write $\dot{x} = \text{Re}(z^k)$, $\dot{y} = \text{Im}(z^k)$ and let $z = re^{i\theta}$.) Sketch the vector fields near such fixed points having indices 3 and $-3$.

A further simple but useful technique for the global approximation of solution curves is provided by the method of isoclines. Eliminating explicit time dependence from (1.8.1) we obtain the first-order system

$$\frac{dy}{dx} = \frac{g(x, y)}{f(x, y)}. \tag{1.8.18}$$

Neglecting for the moment the fact that (1.8.18) might not be well defined on $f(x, y) = 0$, we seek curves $y = h(x)$ or $x = h(y)$ on which the slope of the vector field $dy/dx = c$ is constant. Such curves are given (perhaps implicitly) by solving the equation

$$g(x, y) = cf(x, y), \tag{1.8.19}$$

and are called *isoclines*.

If a sufficiently close set of isoclines is constructed, then the solutions of (1.8.1) can be sketched fairly accurately. An example will help to illustrate the method:

$$\begin{aligned} \dot{x} &= x^2 - xy, \\ \dot{y} &= -y + x^2. \end{aligned} \tag{1.8.20}$$

We first find the two fixed points at $(x, y) = (0, 0)$ and $(1, 1)$, and ascertain that their linearized matrices are

$$Df(0, 0) = \begin{bmatrix} 0 & 0 \\ 0 & -1 \end{bmatrix} \quad \text{and} \quad Df(1, 1) = \begin{bmatrix} 1 & -1 \\ 2 & -1 \end{bmatrix}, \tag{1.8.21}$$

with eigenvalues $0$, $-1$, and $\pm i$, respectively. Thus the fixed points are both nonhyperbolic and no conclusions can be drawn from Hartman's theorem. We next note that if $x(0) = 0$ we have $\dot{x} \equiv 0$, and thus the $y$-axis is an invariant line on which the flow is governed by $\dot{y} = -y$; in fact it is the global stable manifold of $(0, 0)$. The vector field is also vertical on the line $y = x$.

We go on to seek isoclines on which $dy/dx = c \in (-\infty, \infty)$, which are obtained from

$$(-y + x^2) = c(x^2 - xy),$$

or

$$y = \frac{(1 - c)x^2}{1 - cx} \overset{\text{def}}{=} h_c(x). \tag{1.8.22}$$

Some of these curves, and the associated directions of the vector field, are sketched in Figure 1.8.8(a). In addition to plotting isoclines, it is sometimes also useful to sketch the vector field on specific lines, such as the line $y = 1$.

While the vectors sketched in Figure 1.8.8(a), together with the knowledge that the $y$-axis is invariant, give a general indication of the structure of solution curves, detailed information on the local structure near the degenerate fixed points is best obtained by the center manifold methods described in Chapter 3. Application of these techniques, which will follow as exercises and examples in that chapter, show that $(0, 0)$ is the $\omega$ limit point for all solutions starting nearby in the left-hand half plane ($x \leq 0$) and the $\alpha$ limit set for a curve of points nearby with $x > 0$, while use of the Hopf stability formula of Section 3.4, shows that $(1, 1)$ is a (weakly stable) spiral sink. We leave it to the reader to complete the global analysis:

EXERCISE 1.8.10. Prove that $(0, 0)$ is the $\omega$ limit point for all points $\{(x, y) \in \mathbb{R}^2 \mid x \leq 0\}$. Show that, if there are no closed orbits surrounding $(1, 1)$, then $(1, 1)$ is the $\omega$ limit point for all points $\{(x, y) \in \mathbb{R}^2 \mid x > 0\}$. Describe some possible structures involving closed orbits in the right-hand half plane. Can you verify that, in fact, no closed orbits exist? (cf. Figure 1.8.8(b)). (Warning: In numerical integrations, unless very small step sizes are taken, fictitious periodic orbits appear.)

As this example shows, the method of isoclines is rather tedious to apply and often gives incomplete results. Its use is now generally superceded by numerical integration of the system. However, in some cases the idea of

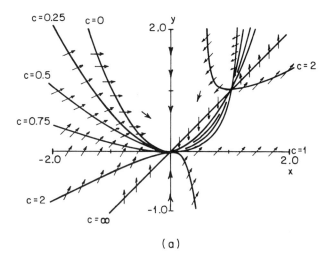

(a)

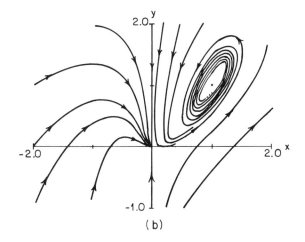

(b)

Figure 1.8.8. Isoclines and a partial phase portrait for equation (1.8.20). (a) Isoclines; (b) a numerically computed phase portrait (Runge–Kutta method, step size 0.02).

isoclines and invariant lines can be used to obtain precise information. For example, in the homogeneous cubic system

$$\dot{x} = ax^3 + bxy^2,$$
$$\dot{y} = cx^2y + dy^3,$$

(1.8.23)

which we shall meet as the normal form of a degenerate vector field in Chapter 7, we can demonstrate the existence of certain invariant lines in the phase portrait. Clearly the $x$- and $y$-axes are invariant, since $\dot{x}(t) \equiv 0$ for all $t$ if $x(0) = 0$ and similarly $\dot{y}(t) \equiv 0$ if $y(0) = 0$. On these lines, the vector fields are simply $\dot{x} = ax^3$ and $\dot{y} = dy^3$, respectively. We claim that, for suitable values of $a, b, c, d$, further invariant lines passing through the origin can exist. Let them be given by $y = \alpha x$. Then, dividing the two components of (1.8.23) we obtain

$$\frac{dy}{dx} = \frac{y}{x}\frac{(cx^2 + dy^2)}{(ax^2 + by^2)}.$$

(1.8.24)

For $y = \alpha x$ to be invariant, we also require that $dy/dx = \alpha$, so that the vector field is everywhere tangent to $y = \alpha x$. Thus, from (1.8.24) we obtain

$$\alpha = \alpha \frac{(cx^2 + d\alpha^2 x^2)}{(ax^2 + b\alpha^2 x^2)},$$

(1.8.25)

or

$$\alpha^2(d - b) = (a - c),$$

which has the two roots

$$\alpha = \pm\sqrt{(a - c)/(d - b)}$$

(1.8.26)

provided that $(a - c)$ and $(d - b)$ have the same sign. The reader should check that on such a line the flow is determined by the one-dimensional system

$$\dot{u} = \left(\frac{ad - bc}{d - b}\right)u^3.$$

(1.8.27)

EXERCISE 1.8.11. Sketch some of the phase portraits for (1.8.23) for various choices of $(a, b, c, d)$. (Refer forward to Section 7.5 if you become discouraged.)

EXERCISE 1.8.12. Use the method of isoclines to locate the limit cycle of the van der Pol oscillator

$$\ddot{x} + 2(x^2 - 1)\dot{x} + x = 0.$$

We end this section with some examples of systems with nonplanar phase spaces. The first is the inevitable pendulum, which is, in nondimensional variables,

$$\ddot{\theta} + \sin\theta = 0.$$

(1.8.28)

This provides a classical example of a system with a nonplanar phase space. The configuration variable $\theta \in [-\pi, \pi)$ is an angle and hence, defining the velocity $\dot{\theta} = v$, the phase space is seen to be the cylinder, and the system becomes

$$\begin{aligned} \dot{\theta} &= v, \\ \dot{v} &= -\sin\theta, \end{aligned} \qquad (\theta, v) \in S^1 \times \mathbb{R}. \qquad (1.8.29)$$

This conveniently avoids the embarrassment of an infinite set of distinct equilibrium points at $\theta = \pm n\pi, n = 1, 2, \ldots$, when there are only two physical rest positions at $\theta = 0$ and $\theta = \pi \equiv -\pi$. The phase portrait of (1.8.29) is easily sketched from knowledge of the first integral $H: S^1 \times \mathbb{R} \to \mathbb{R}$:

$$H(\theta, v) = \frac{v^2}{2} + (1 - \cos\theta). \qquad (1.8.30)$$

(The constant 1 need not be included, but originates in the physical problem, in which the potential energy $V(\theta)$ is of the form $mgl(1 - \cos\theta)$ for a pendulum of mass $m$ and length $l$, where $\theta$ is measured from the downward vertical, so that $V(0) = 0$.) The phase portrait is, of course, periodic in $\theta$ with period $2\pi$ (Figure 1.8.9).

The cylindrical phase space is obtained by identifying $\theta = -\pi (AA')$ and $\theta = +\pi (BB')$. A nice sketch appears in Andronov et al. [1966, p. 98]. It is important to recognize that orbits such as that marked ab in Figure 1.8.9 are in fact closed orbits which *encircle* the cylinder: such orbits correspond to rotary rather than oscillatory motions of the pendulum, and the two classes of motions are separated by the homoclinic orbits to the saddle point.

EXERCISE 1.8.13. Sketch the phase portrait for the damped pendulum $\ddot{\theta} + 2\alpha\dot{\theta} + \sin\theta = 0$, $0 \le \alpha \ll 1$, and the pendulum $\ddot{\theta} + \sin\theta = \beta$, subjected to applied torque $\beta > 0$. Consider $\beta < 1$ and $\beta > 1$. Also consider the damped pendulum with torque $\ddot{\theta} + 2\alpha\dot{\theta} + \sin\theta = \beta$. Both undamped systems possess homoclinic and periodic orbits; can the damped systems possess any such orbits? (This is quite difficult. Refer forward to Section 4.6 if you like.)

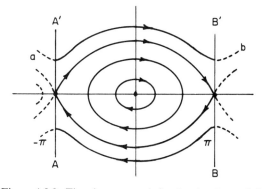

Figure 1.8.9. The phase portrait for the simple pendulum.

In addition to its application in classical mechanics, the system $\ddot{\theta} + 2\alpha\dot{\theta} + \sin\theta = \beta$ provides a discrete (single mode) approximation to the sine–Gordon equation, which is important in physics as a model of wave functions in the superconducting Josephson junction (cf. Levi *et al.* [1978]). We will return to this example in Section 4.6.

A second canonical example of a system defined on a two-dimensional manifold is the flow on a torus $T^2 = S^1 \times S^1$:

$$
\begin{aligned}
\dot{\theta} &= f(\theta, \phi); \\
\dot{\phi} &= g(\theta, \phi);
\end{aligned}
\qquad (\theta, \phi) \in T^2; f, g(2\pi)\text{-periodic in } \theta, \phi. \qquad (1.8.31)
$$

A special and important case is the linear system:

$$
\begin{aligned}
\dot{\theta} &= a, \\
\dot{\phi} &= b.
\end{aligned}
\qquad\qquad (1.8.32)
$$

As is well known, if $a$ and $b$ are rationally related then one has a continuous family of periodic orbits on $T^2$, whereas if $a/b$ is irrational, one obtains dense nonperiodic orbits. Since the irrational and rational numbers are each dense in $\mathbb{R}$, either of these two topologically distinct cases can be approximated arbitrarily closely by the other, and hence the system is structurally unstable for *all* values of $a$, $b$.

The linear flow (1.8.32) is more important than its special form might suggest, since any (nonlinear) flow on $T^2$ with no equilibrium points or closed orbits necessarily arises from a vector field $C^0$ equivalent to (1.8.32) with $a/b$ irrational. We will return to·this in Section 6.2.

Flows on two-tori also occur in linear undamped systems with two degrees of freedom, as follows. Consider the system

$$
\ddot{x} + \omega_1^2 x = 0, \qquad \ddot{y} + \omega_2^2 y = 0, \qquad (1.8.33)
$$

where we have written the equations in the canonical (normal mode) co-ordinates, so that the two modes are uncoupled. The system possesses two independent first integrals:

$$
H_1(x, \dot{x}) = \frac{\dot{x}^2}{2} + \frac{\omega_1^2 x^2}{2} = k_1, \qquad H_2(y, \dot{y}) = \frac{\dot{y}^2}{2} + \frac{\omega_2^2 y^2}{2} = k_2, \qquad (1.8.34)
$$

each of which remains constant as the four-dimensional solution vector $(x(t), \dot{x}(t), y(t), \dot{y}(t))$ evolves with time. The two integral constraints imply that solutions are confined to a two-dimensional torus which is the product of the two ellipses in $(x, \dot{x})$ and $(y, \dot{y})$ spaces given by the constraints. To see this, we perform coordinate changes to action angle variables (Goldstein [1980], Arnold [1978]):

$$
\begin{aligned}
x &= \sqrt{\frac{2I_1}{\omega_1}} \sin\theta_1, \qquad \dot{x} = \sqrt{2\omega_1 I_1} \cos\theta_1, \\
y &= \sqrt{\frac{2I_2}{\omega_2}} \sin\theta_2, \qquad \dot{y} = \sqrt{2\omega_2 I_2} \cos\theta_2,
\end{aligned}
\qquad (1.8.35)
$$

to obtain

$$\dot{I}_1 = 0, \qquad \dot{I}_2 = 0,$$
$$\dot{\theta}_1 = \omega_1, \qquad \dot{\theta}_2 = \omega_2, \tag{1.8.36}$$

which, for initial conditions $(I_1^0, I_2^0, \theta_1^0, \theta_2^0)$, has the solution

$$I_1(t) \equiv I_1^0, \, I_2(t) \equiv I_2^0,$$
$$\theta_1(t) = \omega_1 t + \theta_1^0, \qquad \theta_2(t) = \omega_2 t + \theta_2^0. \tag{1.8.37}$$

Thus, the four-dimensional phase space is filled with two-dimensional tori given by $I_1 = I_1^0, I_2 = I_2^0$ and each torus carries rational or irrational flow, depending on the ratio $\omega_1/\omega_2$. In general, $n$ degree of freedom integrable Hamiltonian systems give rise to flows on $n$-dimensional tori (cf. Arnold [1978], Goldstein [1980], for examples). The fundamental paper on toral flows by Kolmogorov [1957], which is the basis of the KAM theory, is reprinted in Abraham and Marsden [1978] (cf. Sections 4.8 and 6.2).

EXERCISE 1.8.14. Consider the Hamiltonian system on $\mathbb{R}^4$ with Hamiltonian $H(x, \dot{x}, y, \dot{y}) = \dot{x}^2/2 + \dot{y}^2/2 + x^2/2 + x^3/3 + y^2/2 - y^3/3$. Show that there are two independent integrals and describe the orbit structures of the system. (This represents a special case of the "anti-Hénon–Heiles" system (cf. Hénon and Heiles [1964], Aizawa and Saitô [1972]).)

Flows on two-tori (and, more generally, $n$-tori) also arise in studies of coupled nonconservative limit cycle oscillators. For example, consider two identical van der Pol oscillators coupled by weak linear interaction, $\beta$, with weak de-tuning, $\delta$:

$$\ddot{x} + \varepsilon(x^2 - 1)\dot{x} + x = \beta(y - x),$$
$$\ddot{y} + \varepsilon(y^2 - 1)\dot{y} + y = \beta(x - y) - \delta y, \qquad 0 \le |\delta|, |\beta| \ll \varepsilon \ll 1. \tag{1.8.38}$$

For $\delta, \beta = 0$ it is known that each van der Pol oscillator possesses an attracting limit cycle given (approximately) by

$$x(t) = 2\cos(t + \theta_1^0), \qquad \dot{x}(t) = -2\sin(t + \theta_1^0),$$
$$y(t) = 2\cos(t + \theta_2^0), \qquad \dot{y}(t) = -2\sin(t + \theta_2^0), \tag{1.8.39}$$

where $\theta_1^0$, $\theta_2^0$ are arbitrary (phase) constants determined by the initial conditions (cf. Section 2.1). The product $S^1 \times S^1$ of the two circles of radius 2 in the $(x, \dot{x})$ and $(y, \dot{y})$ planes is a two-torus $T^2 \subset \mathbb{R}^4$. However, unlike the members of the two-parameter family of two-tori in the Hamiltonian example above, this one is an *attractor*; in fact, nearby orbits approach it exponentially fast and as we shall see, it persists under perturbations. Thus a small perturbation, such as the addition of weak coupling ($\beta, \delta \ll \varepsilon$), cannot destroy the torus as a whole, which remains an attracting set.

However, since the vector field on the torus may be written

$$\dot{\theta}_1 = -1, \qquad \dot{\theta}_2 = -1, \qquad (\theta_1, \theta_2) \in T^2, \tag{1.8.40}$$

it carries linear (rational) flow, which is structurally unstable. Thus, while the torus as a whole is preserved, the structure of orbits within it changes radically when the oscillators are coupled. Rand and Holmes [1980] have studied (1.8.38) and generalizations of it, and show if $|\beta| > |\delta/2| > 0$, then there are precisely two hyperbolic periodic orbits on the torus, one an attractor and the other a repellor. Such a situation is called 1:1-phase locking or entrainment. Many other studies of phase locking have been carried out, see Nayfeh and Mook [1979] and references therein for examples.

The perturbation schemes employed typically neglect high-order terms (terms of $O(\varepsilon^2)$), but since the coupled flow with two hyperbolic orbits is structurally stable (cf. Peixoto's theorem in the next section), the result is qualitatively correct for the full system, for the addition of the neglected terms constitutes a small perturbation. However, for $|\delta/2| > |\beta|$, the approximate analysis predicts that the flow on the torus will contain no attractors or repellors, but that all orbits will either be periodic or there will be an orbit dense on $T^2$. This structurally unstable situation tells us nothing (directly) about the true flow on $T^2$. In fact one can expect a very complex sequence of bifurcations to occur directly after phase locking breaks. We shall discuss such situations in Section 6.2.

EXERCISE 1.8.15. Construct a structurally stable system on $T^2$ with two closed orbits.


# 1.9. Peixoto's Theorem for Two-Dimensional Flows

With various examples of two-dimensional flows in mind, we are now ready to state and sketch the proof of Peixoto's theorem, which represents the culmination of much previous work, in particular that of Poincaré [1899] and Andronov and Pontryagin [1937]. Letting $\mathscr{X}^r(M^2)$ denote the set of all $C^r$ vector fields on two-dimensional manifolds, we have

**Theorem 1.9.1** (Peixoto [1962]). *A $C^r$ vector field on a compact two-dimensional manifold $M^2$ is structurally stable if and only if*:

(1) *the number of fixed points and closed orbits is finite and each is hyperbolic*;
(2) *there are no orbits connecting saddle points*;
(3) *the nonwandering set consists of fixed points and periodic orbits alone*.

*Moreover, if $M^2$ is orientable, the set of structurally stable vector fields is open-dense in $\mathscr{X}^r(M^2)$.*

One can deal with planar fields provided that there is a compact set $D \subset \mathbb{R}^2$ such that the flow is directed inward (or outward) on the boundary of $D$, otherwise it is easy to construct systems with countably many fixed points or closed orbits. We also remark that, if the phase space is planar, then conditions (1) and (2) automatically imply that (3) is satisfied, since there are no limit sets possible other than fixed points, closed orbits, and homoclinic cycles, and the latter are excluded by (2).

Peixoto's theorem implies that typically a two-dimensional vector field will contain only sinks, saddles, sources, and repelling and attracting closed orbits in its invariant set, Figure 1.9.1(a). Structural stability is a generic property for two-dimensional flows on orientable manifolds. Many of the ingredients of Peixoto's theorem were proved by Andronov and his coworkers (cf. Andronov *et al.* [1966]) in the decades following 1935. Here we shall sketch the proof of the structural stability part.

The first condition (hyperbolicity of fixed points and periodic orbits) follows from a consideration of the linearized flow or of suitable Poincaré maps. It can be shown that the sets of such linear flows and maps contain open dense sets of hyperbolic flows and maps, respectively (cf. Hirsch and Smale [1974], Chapter 7). Thus, if a nonlinear flow contains, say, a non-hyperbolic fixed point, then a small perturbation suffices to render that point hyperbolic; similarly, a hyperbolic fixed point remains hyperbolic under all sufficiently small perturbations.

The second condition is demonstrated as follows. Suppose two saddles $p_1$, $p_2$ were connected, so that $W^u(p_1) \cap W^s(p_2) = \Gamma$ for the flow of the vector field

$$\dot{x}_1 = f_1(x_1, x_2),$$
$$\dot{x}_2 = f_2(x_1, x_2).$$

(cf. Figure 1.9.2). We perturb $(f_1(x_1, x_2), f_2(x_1, x_2))$ by addition of a field $(\varepsilon\phi_1(x_1, x_2), \varepsilon\phi_2(x_1, x_2))$ having compact support, vanishing outside some

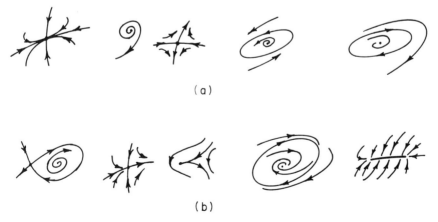

(a)

(b)

Figure 1.9.1. (a) Some structurally stable nonwandering sets on $\mathbb{R}^2$; (b) some structurally unstable nonwandering sets on $\mathbb{R}^2$.

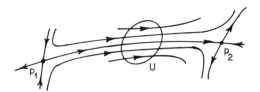

Figure 1.9.2. A saddle connection.

(small) region $U$ chosen to straddle $\Gamma$ as shown. It is easy to choose the perturbation such that orbits entering $U$ are all "pushed upward" (or downward), causing $\Gamma$ to break; Figure 1.9.3. Similarly, if the two such manifolds do not intersect, then a sufficiently small perturbation cannot cause them to intersect.

The third condition is necessary to exclude structurally unstable non-wandering sets such as the torus $T^2$ with irrational flow, as occurs in the linear flow

$$\begin{aligned} \dot{\theta} &= a, \\ \dot{\phi} &= b, \end{aligned} \qquad (\theta, \phi) \in T^2, \qquad (1.9.1)$$

with $a/b$ irrational (cf. the discussions in Section 1.8 above). Of course, if one has rational flow on $T^2$ then the torus is filled with a continuous family of nonhyperbolic closed orbits.

The proof that the set of structurally stable flows on orientable manifolds is open dense is more difficult and involves the closing lemma of Pugh [1967a, b]. We will not sketch this here, but see Palis and de Melo [1982], for example

EXERCISE 1.9.1. Sketch phase portraits for the family $\dot{x} = \mu + x^2 - xy$, $\dot{y} = y^2 - \frac{1}{2}x^2 - 1$ and show that a saddle connection exists for $\mu = 0$. What happens for $\mu > 0$; $\mu < 0$? (Cf. Guckenheimer [1973] and Section 6.1.)

Even though Peixoto's theorem guarantees that, in generic families of planar systems, the structurally stable ones occupy a set of full measure, the occurrence of infinitely many unstable (bifurcation) systems in some neighborhood cannot be excluded. The following example is due to Jacob Palis (who proposed it in connection with moduli of saddle connections, a topic

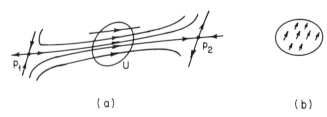

(a)                                          (b)

Figure 1.9.3. (a) A broken connection; (b) $(\varepsilon\phi_1, \varepsilon\phi_2)$.

which we do not discuss in this book). Consider the one-parameter family
$\dot{x} = f_\mu(x)$, $x \in \mathbb{R}^2$, with phase portraits as indicated in Figure 1.9.4(a)–(c)
for $\mu < 0$, $\mu = 0$, $\mu > 0$. For $\mu < 0$ there are two hyperbolic closed orbits,
and at $\mu = 0$ these coalesce in a single "semistable" orbit, which is the $\omega$ limit
set for nearby points inside it and the $\alpha$ limit set for nearby points outside.
For $\mu > 0$ no closed orbits exist. In an annular strip containing the orbits,
all we see is a local saddle-node bifurcation of closed orbits, but *globally*
the stable and unstable manifolds of the saddle points $r$ and $q$ are involved in
a crucial manner. For $\mu < 0$ the $\alpha$ limit set for points on the left-hand branch
$W_l^s(r)$ of $W^s(r)$ is the repelling closed orbit $\gamma_2$, while the $\omega$ limit set for points
on both branches of $W^u(q)$ is the attracting closed orbit $\gamma_1$. At $\mu = 0$ these
two orbits merge in $\gamma_0$, which is the $\omega$ limit set for points in $W^u(q)$ and the $\alpha$
limit set for points in $W_l^s(r)$. To see what happens to $W_l^s(r)$ and $W^u(q)$ when
$\mu > 0$, we take a local section, $\Sigma$, as indicated in Figure 1.9.4(b). For $\mu = 0$,
$\gamma_0$ pierces $\Sigma$ at $p_0$, and the sets $W_l^s(r) \cap \Sigma$ and $W_l^u(q) \cap \Sigma$, $W_r^u(q) \cap \Sigma$ are
each countable sequences of points accumulating at $p_0$, the former from
above, the latter two from below; cf. Figure 1.9.5(a), (b). Let $W_l^s(r) \cap \Sigma =$
$\{r_i\}_{i=1}^\infty$, $W_l^u(q) \cap \Sigma = \{q_i\}_{i=1}^\infty$, $W_r^u(q) \cap \Sigma = \{q_i'\}_{i=1}^\infty$.

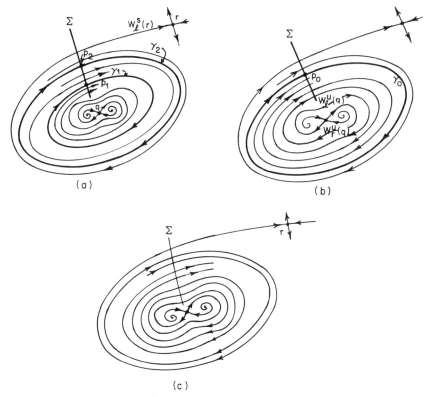

Figure 1.9.4. (a) $\mu < 0$; (b) $\mu = 0$; (c) $\mu > 0$.

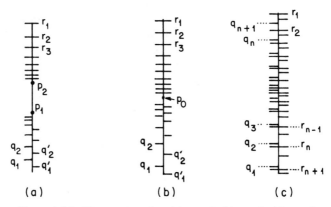

Figure 1.9.5. The cross section. (a) $\mu < 0$; (b) $\mu = 0$; (c) $\mu > 0$.

We have a Poincaré map defined on some neighborhood $U \subset \Sigma$ and, by our construction, $P^{-1}(r_i) = r_{i+1}$, $P(q_i) = q_{i+1}$, $P(q_i') = q_{i+1}'$. Clearly, for $\mu > 0$, all orbits pass through the annular region and thus leave the top of $\Sigma$ going forward, and the bottom going backward, after a finite number of iterates. (This number goes to infinity on $\mu \to 0^+$.) Thus, between $\mu = 0$ and $\mu = \varepsilon$, for any $\varepsilon > 0$, a countably infinite set of points $r_j$, $q_j$, $q_j'$, $j \geq N$, must pass each other on $\Sigma$. Hence, in the interval $\mu \in [0, \varepsilon]$, there are countably many heteroclinic saddle connection bifurcations, and the bifurcation set is a countable sequence of points $\mu_i$ accumulating on $\mu = 0$ from above. However, since the structurally unstable systems occur at isolated points $\mu_i$, we still have an open dense set of structurally stable systems in the neighborhood of $\dot{x} = f_0(x)$.

With Peixoto's theorem in mind Smale proposed that one might study systems on compact $n$-manifolds satisfying conditions (1) and (3) of Theorem 1.9.1 but with (2) suitably modified in the light of Theorem 1.8.3. Such systems are now called *Morse–Smale* systems.

**Definition.** A Morse–Smale system is one for which:

(1) the number of fixed points and periodic orbits is finite and each is hyperbolic;
(2) all stable and unstable manifolds intersect transversally;
(3) the nonwandering set consists of fixed points and periodic orbits alone.

In the definition of transversal intersection, we include the empty set, for clearly if two manifolds do not intersect (i.e., are bounded away from each other), then a small perturbation cannot cause them to intersect.

The following conjectures were then proposed:

A system is structurally stable if and only if it is Morse–Smale:
Morse–Smale systems are dense in $\mathrm{Diff}^1(M)$ or $\mathscr{X}^1(M)$;
Structurally stable systems are dense in $\mathrm{Diff}^1(M)$ or $\mathscr{X}^1(M)$.

(Here Diff$^r(M)$ (resp. $\mathscr{X}^r(M)$) denotes the set of all $C^r$ diffeomorphisms (resp. vector fields) on finite dimensional manifolds $M$.) In the following pages we shall study examples of systems which show that all three conjectures are false. All that can be salvaged is part of Conjecture 1: Morse–Smale systems *are* structurally stable (the converse is false). One of the major contributions to the fall of Conjectures 1 and 2 was Smale's construction of the horseshoe map: a two-dimensional diffeomorphism with a complicated invariant set which was suggested by certain problems in forced oscillations. Before meeting this map in Chapter 5, we shall consider some examples of three-dimensional systems, including periodically forced single degree of freedom oscillators, which have very complicated solution structures. These systems provide additional counter-examples to the conjectures above, and they are therefore of historical as well as practical interest.

# CHAPTER 2
# An Introduction to Chaos: Four Examples

In this chapter we introduce four nonlinear systems which possess fascinating properties and which are still improperly understood. We have chosen two periodically forced single degree of freedom oscillators, a three-dimensional autonomous differential equation, and a two-dimensional map. The oscillators of van der Pol [1927] and Duffing [1918] originally arose as models in electric circuit theory and solid mechanics, respectively, while the Lorenz equations (Lorenz [1963]) represent a truncation of the nonlinear partial differential equations governing convection in fluids. Finally, our map models a simple repeated impact problem (Holmes [1982]) and, in a slightly different form, resonance problems in atomic physics (Chirikov [1979], Greene [1980]). In fact the conservative, area preserving version of this map has been studied intensively as a canonical example of the transition to stochasticity and chaos in Hamiltonian systems. The range of applications for the models outlined here should suggest the pervasive importance of nonlinear systems.

Since their original proposals as models of physical processes, all four systems have been extensively studied and have taken on independent mathematical lives. Thus, in our brief surveys of these problems, we are able to give a preview of much of the material to come, including bifurcation theory for vector fields and maps, and perturbation and averaging methods. We will quote many results without proper justification or proof, but we will continually return to fill in details on these four examples as we develop our analytical tools subsequently in this book. The reader should therefore not expect to thoroughly assimilate all the material of this chapter on first reading, but should refer back to it as the examples occur in later chapters.

While each analysis proceeds in a similar spirit, the precise methods differ considerably. Almost all the tricks of applied mathematics have proved useful in the study of these problems and our account will reflect this. In particular, in each case we have illustrated typical behaviors of the systems with numerical integrations and iterations. While we can seldom prove theorems using the computer, in nonlinear dynamics it can often suggest the results which one should try to prove, and numerical simulation has been an invaluable tool in this respect. We strongly recommend that the reader perform his own numerical work on these and similar systems. Even the relatively poor precision available on a microcomputer with a video-screen output will serve to illustrate many of the interesting features of these nonlinear dynamical systems.

## 2.1. Van der Pol's Equation

Van der Pol's equation provides an example of a oscillator with nonlinear damping, energy being dissipated at large amplitudes and generated at low amplitudes. Such systems typically possess limit cycles; sustained oscillations around a state at which energy generation and dissipation balance, and they arise in many physical problems.

The original application described by van der Pol [1927] models an electrical circuit with a triode valve, the resistive properties of which change with current, the low current, negative resistance becoming positive as current increases. (Van der Pol's article is reprinted in Bellman and Kalaba [1964].) See Hayashi [1964] for more information on nonlinear oscillators as models of electrical circuits. Single degree of freedom limit cycle oscillators, similar to the unforced van der Pol system, also occur in models of wind-induced oscillations of buildings due to vortex shedding (Novak and Davenport [1970], Parkinson [1974], Hartlan and Currie [1970]), in general aeroelastic flutter problems (Dowell [1975, 1980], Holmes [1977], Holmes and Marsden [1978]), and in stability studies of both tracked and rubber tired vehicles (Cooperrider [1980], Beaman and Hedrick [1980], Taylor [1980]), as well as in certain models of chemical reactions (cf. Uppal *et al.* [1974]).

The basic system can be written in the form

$$\ddot{x} + \alpha\phi(x)\dot{x} + x = \beta p(t), \qquad (2.1.1)$$

where $\phi(x)$ is even and $\phi(x) < 0$ for $|x| < 1$, $\phi(x) > 0$ for $|x| > 1,$* $p(t)$ is $T$-periodic and $\alpha$, $\beta$ are nonnegative parameters. It will be convenient to rewrite (2.1.1) as an autonomous system

$$\dot{x} = y - \alpha\Phi(x),$$
$$\dot{y} = -x + \beta p(\theta), \qquad (x, y, \theta) \in \mathbb{R}^2 \times S^1, \qquad (2.1.2)$$
$$\dot{\theta} = 1,$$

* Additional constraints must be set on $\phi(x)$ as $|x| \to \infty$ to obtain all the results outlined below. In particular, we require $\phi(x)$ to remain strictly positive as $|x| \to \infty$ if a unique stable periodic orbit is to be obtained (cf. Stoker [1950]).

where $\Phi(x) = \int_0^x \phi(\xi)\, d\xi$ is odd and $\Phi(0) = \Phi(\pm a) = 0$ for some $a > 0$. Most of the results sketched here hold for the general system, but for simplicity we shall at first assume that $\phi$, $\Phi$, and $p$ take the forms

$$\phi(x) = x^2 - 1; \qquad \Phi(x) = \frac{x^3}{3} - x; \qquad p(t) = \cos \omega t. \qquad (2.1.3)$$

For the strongly forced problem we shall take piecewise linear functions.

## The Unforced System, $\beta = 0$

The planar system obtained by setting $\beta = 0$ in (2.1.2) is fairly simple. First suppose $\alpha \ll 1$ is a small parameter, so that (2.1.2) is a perturbation of the linear oscillator

$$\dot{x} = y, \qquad \dot{y} = -x, \qquad (2.1.4)$$

which has a phase plane filled with circular periodic orbits each of period $2\pi$. Using regular perturbation or averaging methods, we can show that precisely one of these orbits is preserved under the perturbation. Selecting the invertible transformation:

$$\begin{pmatrix} u \\ v \end{pmatrix} = \begin{pmatrix} \cos t & -\sin t \\ -\sin t & -\cos t \end{pmatrix} \begin{pmatrix} x \\ y \end{pmatrix}, \qquad (2.1.5)$$

which "freezes" the unperturbed system, we find that (2.1.2) becomes

$$\dot{u} = -\alpha \cos t[(u \cos t - v \sin t)^3/3 - (u \cos t - v \sin t)],$$
$$\dot{v} = \alpha \sin t[(u \cos t - v \sin t)^3/3 - (u \cos t - v \sin t)]. \qquad (2.1.6)$$

We note that this transformation is *orientation reversing*. There is no particular reason for this other than tradition in nonlinear oscillation studies.

We next *average* the right-hand side of (2.1.6) to approximate the functions $u$, $v$ which vary slowly because $\dot{u}$ and $\dot{v}$ are small. Integrating each function with respect to $t$ from 0 to $T = 2\pi$, holding $u$, $v$ fixed, we obtain

$$\dot{u} = \alpha u[1 - (u^2 + v^2)/4]/2$$
$$\dot{v} = \alpha v[1 - (u^2 + v^2)/4]/2. \qquad (2.1.7)$$

The averaging theory described in Section 4.1 shows that this system is correct at first order, but there is an error of $\mathcal{O}(\alpha^2)$. In polar coordinates, we therefore have

$$\dot{r} = \frac{\alpha r}{2}\left(1 - \frac{r^2}{4}\right) + \mathcal{O}(\alpha^2),$$
$$\dot{\varphi} = 0 + \mathcal{O}(\alpha^2). \qquad (2.1.8)$$

Neglecting the $\mathcal{O}(\alpha^2)$ terms, this system has an attracting circle of fixed points at $r = 2$, reflecting the existence of a one-parameter family of almost sinusoidal solutions

$$x = r(t)\cos(t + \varphi(t)), \tag{2.1.9}$$

with slowly varying amplitude $r(t) = 2 + \mathcal{O}(\alpha^2)$ and phase $\varphi(t) = \varphi^0 + \mathcal{O}(\alpha^2)$, the constant $\varphi^0$ being determined by initial conditions.

When $\alpha$ is not small the averaging procedure no longer works and other methods must be used. Hirsch and Smale [1974, Chapter 10], for example, discuss this case. However, the other limit of *large* $\alpha$ can once more be treated by perturbation methods, although this time the perturbation is singular. Letting $\hat{y} = y/\alpha$ and dropping the hats, (2.1.2) becomes

$$\begin{aligned} \dot{x} &= \alpha\left(y - \left(\frac{x^3}{3} - x\right)\right) \\ \dot{y} &= -\frac{x}{\alpha} \end{aligned} \quad \overset{\text{def}}{=} f(x, y). \tag{2.1.10}$$

Since $\alpha \gg 1 \gg 1/\alpha$, we have $|\dot{x}| \gg |\dot{y}|$ except in a neighborhood of the curve $\mathscr{C}$ given by $y = x^3/3 - x$. Thus the family $\mathscr{H}$ of horizontal lines $y = $ constant approximates the flow of (2.1.10) away from $\mathscr{C}$ increasingly well as $\alpha \to \infty$. Near $\mathscr{C}$, and in particular when $|y - (x^3/3 - x)| = (1/\alpha^2)$, both solution components are comparable and hence, after entering this *boundary layer*, solution curves turn sharply and follow $\mathscr{C}$ until they reach a critical point of $\mathscr{C}(x = \pm 1, y = \mp\frac{2}{3})$ where they must leave $\mathscr{C}$ and follow $\mathscr{H}$ to another point of $\mathscr{C}$; see Figure 2.1.1. Making these ideas precise, it is possible to find an annular region $R$ into which the vector field is directed at each point, and which must therefore contain a closed orbit, by the Poincaré–Bendixson theorem, since it contains no equilibria for $\alpha \neq 0$. To prove that this orbit is unique, we show that, since solutions spend most of their time near the stable branches of $\mathscr{C}$, where trace $Df = -\alpha(x^2 - 1) < 0$, any orbit within the annulus $R$ must be asymptotically stable; hence only one such orbit can exist. See Stoker [1950, 1980] for a review and more information.

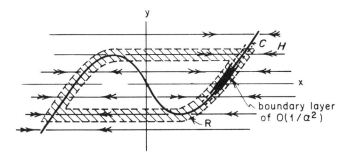

Figure 2.1.1. Relaxation oscillations.

## The Forced Oscillator

We now make our first major excursion into the world of three-dimensional dynamical systems. To start with, we suppose $\alpha, \beta \ll 1$ and use a similar transformation to that of equation (2.1.5). However, since we are interested in the (almost-) periodic forced response we use the $2\pi/\omega$ periodic transformation

$$\begin{pmatrix} u \\ v \end{pmatrix} = \begin{pmatrix} \cos \omega t & -\dfrac{1}{\omega} \sin \omega t \\ -\sin \omega t & -\dfrac{1}{\omega} \cos \omega t \end{pmatrix} \begin{pmatrix} x \\ y \end{pmatrix}, \qquad (2.1.11)$$

to obtain from (2.1.2)

$$\dot{u} = -\alpha\Phi(x) \cos \omega t - \left(\frac{\omega^2 - 1}{\omega}\right) x \sin \omega t - \frac{\beta}{\omega} \sin \omega t p(t),$$

$$\dot{v} = \alpha\Phi(x) \sin \omega t - \left(\frac{\omega^2 - 1}{\omega}\right) x \cos \omega t - \frac{\beta}{\omega} \cos \omega t p(t).$$

Thus, using $\Phi(x) = x^3/3 - x$ and $p(t) = \cos \omega t$, we have

$$\dot{u} = -\alpha\left(\frac{x^3}{3} - x\right) \cos \omega t + \left(\frac{1 - \omega^2}{\omega}\right) x \sin \omega t - \frac{\beta}{\omega} \sin \omega t \cos \omega t,$$

$$\dot{v} = \alpha\left(\frac{x^3}{3} - x\right) \sin \omega t + \left(\frac{1 - \omega^2}{\omega}\right) x \cos \omega t - \frac{\beta}{\omega} \cos^2 \omega t, \qquad (2.1.12)$$

where $x = u \cos \omega t - v \sin \omega t$. Assuming that we are near resonance, so that $|\omega^2 - 1|$, $\alpha$ and $\beta$ are *all* small, (2.1.12) can be averaged as in Sections 4.1–4.2 to yield

$$\begin{aligned} \dot{u} &= \frac{\alpha}{2}\left[ u - \sigma v - \frac{u}{4}(u^2 + v^2) \right] \\ \dot{v} &= \frac{\alpha}{2}\left[ \sigma u + v - \frac{v}{4}(u^2 + v^2) \right] - \frac{\beta}{2\omega} \end{aligned} \quad + \mathcal{O}(\alpha^2), \qquad (2.1.13)$$

where $\sigma = (1 - \omega^2)/\alpha\omega$ is $\mathcal{O}(1)$. Note that when $\omega = 1$ and $\beta = 0$, (2.1.13) reduces to (2.1.7).

Rescaling $u$ and $v$ each by a factor of 2 and letting $t \to (2/\alpha)t$, (2.1.13) can be rewritten as

$$\dot{u} = u - \sigma v - u(u^2 + v^2),$$

$$\dot{v} = \sigma u + v - v(u^2 + v^2) - \gamma, \qquad (2.1.14)$$

where $\gamma = \beta/2\alpha\omega$ and we have dropped the $\mathcal{O}(\alpha^2)$ error term.

In Chapter 4 we shall see that the time $2\pi/\omega$ flow map of the averaged equation (2.1.13) provides an approximation of the Poincaré map of (2.1.12) and hence of (2.1.1). Thus the phase portrait of (2.1.13) or, equivalently of (2.1.14), is of interest. In particular, hyperbolic fixed points of (2.1.14) correspond to periodic orbits of (2.1.1)–(2.1.2) and hyperbolic closed orbits of (2.1.14) correspond to invariant tori carrying multiply periodic solutions in (2.1.1)–(2.1.2). In the engineering literature, the former solutions are said to be (1:1) *phase locked* or *entrained,* and the latter *drifting,* although, as we shall see in subsequent chapters, such solutions are often entrained at higher order (($p:q$) phase locked).

We now describe the various phase portraits of (2.1.14) as the parameters $\sigma$, $\gamma$ vary. This description is based on the work of van der Pol [1927], Cartwright [1948], Gillies [1954], and Holmes and Rand [1978]. It is convenient first to draw the *bifurcation set* in ($\sigma$, $\gamma$) space: the set of points for which the flow of (2.1.14) is structurally unstable: Figure 2.1.2. The corresponding phase portraits are shown in Figure 2.1.3. (Detailed descriptions of bifurcations are given in Chapter 3, below.) In this two-parameter system, structural instability occurs in several distinct ways, as listed below. First we note that, in regions I and III there is a single fixed point (a sink in I, a source in III), while in region II there are two sinks and a saddle and in region IV ($=$ IVa $\cup$ IVb) a sink, a saddle, and a source.

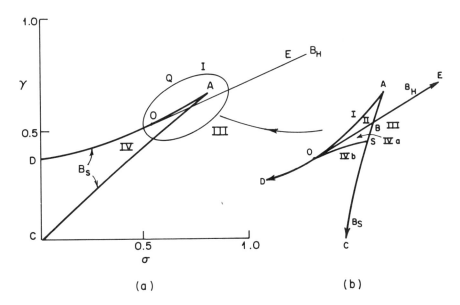

Figure 2.1.2. The bifurcation set for the averaged van der Pol equation (2.1.14), from Holmes and Rand [1978]. (a) The overall picture; (b) region $Q$ (enlarged and distorted).

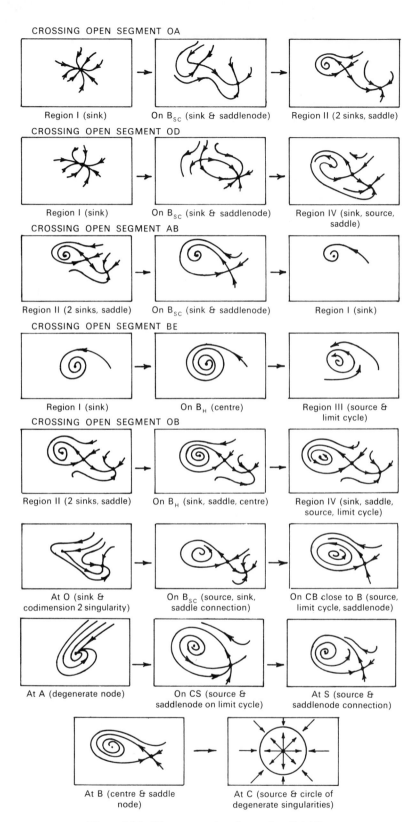

Figure 2.1.3. Phase portraits of equation (2.1.14).

We first list the bifurcations which can be obtained by conventional linear analyses:

(i) On the curves $DA$ and $AC$, marked $B_S$, the system has one hyperbolic and one nonhyperbolic fixed point: *saddle-node bifurcations* occur. These saddle-node bifurcations describe the coalescence and subsequent disappearance of two equilibria as one crosses from region II or IV into regions I or III. On $DO$ we have a sink and a saddle-node with one zero and one positive eigenvalue; on $OA$ and $AB$, a sink and a saddle-node with one zero and one negative eigenvalue; and on $BC$, a source and a saddle-node with one zero and one negative eigenvalue.

(ii) On $OE$ (one of) the sink(s) becomes nonhyperbolic, having a pair of pure imaginary eigenvalues with nonzero imaginary part: a *Hopf bifurcation* occurs. In this bifurcation a limit cycle surrounding an equilibrium point emerges from the equilibrium. Here the limit cycle is stable.

The saddle-node and Hopf bifurcations will be dealt with in Chapter 3.

In addition, more degenerate bifurcations are observed:

(iii) At the point $A: (\sigma, \gamma) = (1/\sqrt{3}, \sqrt{\frac{8}{27}})$ there is a single nonhyperbolic fixed point with one zero and one negative eigenvalue.

(iv) At the point $0: (\sigma, \gamma) = (\frac{1}{2}, \frac{1}{2})$ there is a sink and a doubly degenerate fixed point with a zero eigenvalue of multiplicity two.

(v) At the point $C: (\sigma, \gamma) = (0, 0)$ there is a source and a ring of degenerate saddle-nodes, as in the averaged unforced problem (2.1.7).

The corresponding phase portraits are sketched in the top five strips of Figure 2.1.3. The doubly degenerate bifurcations of (iii) and (iv) are considered in Chapter 7. (v) is "infinitely degenerate," since there is a continuum of fixed points in this case.

Cartwright [1948] noted that the phase portraits in region IV near $OB$ and near $OD$ are not topologically equivalent: the former has a limit cycle (born in the Hopf bifurcation) while the latter does not. She conjectured that additional bifurcation points must be present, at which the limit cycle vanishes, and that region IV must be divided into two subregions IVa and IVb. Gillies [1954] provided a partial verification of this conjecture using numerical solutions. Holmes and Rand [1978] showed that the conjecture was essentially correct and that:

(vi) There is a third bifurcation curve $OS(B_{SC})$ tangent to $B_S$ and $B_H$ at $O$, on which a homoclinic orbit to a hyperbolic saddle point exists. Approaching $B_{SC}$ from above (IVa), the period of the closed orbit grows without bound and it merges into the saddle connection, which breaks as one enters region IVb. Other, more degenerate situations occur at the points $B$ and $S$, and on $CS$ the saddle-node bifurcation already noted occurs on a closed curve. These situations are sketched in the bottom three strips of Figure 2.1.3. All these bifurcations are generic for two-parameter families and will be studied in the chapters to follow. We note that the saddle connection on $B_{SC}$ and the saddle-node on the closed curve on $CS$ represent occurrences of two of the "exotic" limit sets for two-dimensional flows introduced in Chapter 1.

Referring to Figures 2.1.2–2.1.3 we see that the *entrainment domain* in parameter space is bounded by the curves $(CB) \cup (BE)$ since above these curves (at least) one sink exists for (2.1.14) and hence the full system (2.1.2) has a stable periodic orbit of period $2\pi/\omega$. In regions IVa and III (2.1.14) has a stable hyperbolic limit cycle and hence the full system (2.1.2) has an attracting torus: a multiply periodic "drifting" solution. In regions II and IVa two attracting solutions coexist (two sinks in the former, a sink and a limit cycle in the latter), and both entrained and drifting oscillations are possible in region IVa.

In Chapter 4 we shall see that the saddle-node and Hopf bifurcations occurring for the averaged equation (2.1.14) correspond to similar bifurcations for the Poincaré map of the full system, but that the global bifurcations on $OS$ and $CS$ signal the presence of more complicated phenomena for the map. Similar phenomena, involving transverse homoclinic orbits, occur in the strongly forced van der Pol equation, to be considered now, and in the other examples of this chapter.

We now turn to the strongly forced oscillator, equations (2.1.1)–(2.1.2) with $\alpha, \beta \gg 1$. In studying this equation in connection with models of radar equipment, Cartwright and Littlewood [1945] noted that, for some parameter ranges, two distinct stable subharmonic motions, of periods $(2k \pm 1)(2\pi/\omega)$, for large $k$, were obtained.* Subsequently Levinson [1949] and Levi [1978, 1981] analyzed simplified, piecewise linear models and provided more information. Here we summarize some of the results, concentrating on Levi's geometrical picture. We note that Guckenheimer [1980a] has also devised a geometrical model which captures much of the behavior of the strongly forced system in a different parameter range.

Rescaling (2.1.2) as before, letting $\hat{y} = y/\alpha$ and dropping the hat, we have the singularly perturbed problem

$$\dot{x} = \alpha(y - \Phi(x)),$$

$$\dot{y} = \frac{1}{\alpha}(-x + \beta\rho(t)). \tag{2.1.15}$$

Henceforth, for simplicity, we assume that

$$\Phi(x) = \begin{cases} 2 + x, & x < -1, \\ -x, & |x| < 1, \\ -2 + x, & x > 1, \end{cases}$$

---

* Actually this had already been found experimentally by van der Pol and van der Mark [1927]. Also cf. Flaherty and Hoppensteadt [1978] for a more recent treatment of entrainment.

and

$$p(t) = \begin{cases} 1, & t \in [nT, (n + \tfrac{1}{2})T), \\ -1, & t \in [(n + \tfrac{1}{2})T, (n + 1)T), \end{cases} \tag{2.1.16}$$

are piecewise linear functions (cf. Levi [1981]).

Fixing $\alpha$ sufficiently large, it is again possible to find an annular "trapping region" $R \subset \mathbb{R}^2$ such that the (time periodic) vector field of (2.1.15) is always directed into $R$. Thus, taking a cross section $\Sigma = \{(x, y, t) | t = 0\}$, the Poincaré map $P_\beta$ of (2.1.15) map $R$ into itself: Figure 2.1.4. Since $P_\beta(R) \subset R$ we can define an *attracting set* $A_\beta$ as

$$A_\beta = \bigcap_{n \ge 0} P_\beta^n(R). \tag{2.1.17}$$

Now for large $\alpha$, the contraction rate in $R$ is so large that numerical work reveals only a set $A_\beta$ which appears to be a closed curve. However, since any attracting periodic orbits must lie in $A_\beta$, as Cartwright, Littlewood, and Levinson pointed out, the coexistence of two such attractors with different period implies that $A_\beta$ cannot be a simple closed curve, but must be a more complicated set. To appreciate this fully, a knowledge of rotation numbers is required. This will be provided in Section 6.2. Here we merely give an informal description of the attracting set $A_\beta$, based on Levi's work.

After a reasonable number (say $n = 50$) of iterates, the set $A_\beta^n = \bigcap_{k=0}^n P_\beta^k(R)$ is a thin annulus, the sides of which lie near the curve $y = \Phi(x)$; Figure 2.1.5. Points drift slowly down the right-hand side and up to left-hand side of $A_\beta^n$ and jump rapidly across the bottom (or top) of $A_\beta^n$; thus all points circulate clockwise, as in the unforced problem (cf. Figure 2.1.1). Motion on the sides is, in fact, a slow drift due to the small term $-x/\alpha$ in (2.1.15) coupled

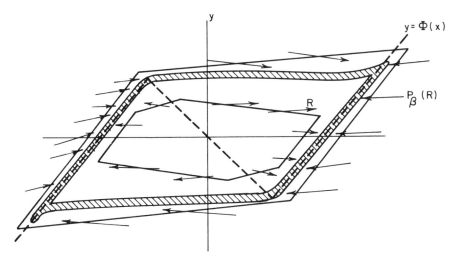

Figure 2.1.4. A trapping region for the van der Pol equations (2.1.15)–(2.1.16).

with zero mean $\mathcal{O}(1)$ oscillation due to $\beta p(t)/\alpha$. Thus, since in each cycle the points move vertically a distance

$$
\begin{aligned}
y(T) - y(0) &= \frac{1}{\alpha}\left[\int_0^T \beta p(t)\,dt - \int_0^T x(t)\,dt\right] \\
&= -\frac{1}{\alpha}\int_0^T x(t)\,dt = \mathcal{O}\!\left(\frac{1}{\alpha}\right),
\end{aligned}
\tag{2.1.18}
$$

we can select a rectangle $R^+ \subset A_\beta^n$ as shown in Figure 2.1.5 such that the upper boundary is mapped into the lower boundary by $P_\beta$. Moreover, the rapid contraction implies that every point in $R$ ultimately enters $R^+$ repeatedly under iteration of $P_\beta$. See Figure 2.1.5(a).

Points near the bottom of $R^+$ drift down and jump across to the left-hand branch of $y = \Phi(x)$ until, at $t = T/2$, points begin to move up, and the

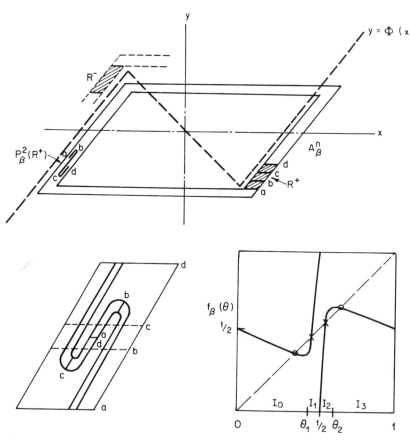

Figure 2.1.5. Levi's Poincaré maps (cf. Levi [1981]). (a) The annulus $A_\beta^n$ and the regions $R^+$, $R^-$; (b) the reduced annulus map $F_\beta$ and its one-dimensional approximation $f_\beta: S^1 \to S^1$; $\circ$ denotes a stable fixed point, $\times$ an unstable fixed point.

remaining points in $R^+$ therefore cannot jump across until $t > T$. Thus, as Levi shows, the rectangle $R^+$ is stretched, folded and bent, so that the substrips $a - b, b - c, c - d$ lie as shown in Figure 2.1.5(a) on the left-hand branch of $y = \Phi(x)$ after time $2T$. The image of each point $p \in R^+$ then drifts up without significant further distortion until it reaches a symmetrically placed rectangle, $R^-$, at time $kT \pm T/2$, depending on the position of $p \in R^+$. Here $k = \mathcal{O}(1/\alpha)$ is a large integer. The jump and stretch process then occurs again, and the image subsequently drifts down the left-hand branch to arrive in $R^+$ after $2k \pm 1$ iterates of the Poincaré map. Thus the return map $R^+$ to $R^+$ is the composition of two maps, each describing a simple jump, the folding, and the ensuing drift, which are identical due to the symmetry of the problem: $\Phi(x) = -\Phi(-x)$, $p(t + T/2) = -p(t)$. It therefore suffices to study the single time $(k \pm \frac{1}{2})T$ jump map, $F_\beta : R^+ \to R^-$, which is sketched in Figure 2.1.5(b).

To make the map continuous we identify the upper and lower edges, thus making a (small) annulus. In doing so we loose track of whether points have returned after $2k + 1$ or $2k - 1$ iterates, but we recall that points near the top take longer. Levi shows that, if the map has two stable fixed points, then one corresponds to an orbit of period $2k + 1$ and the other to an orbit of period $2k - 1$, thus explaining the observation noted earlier. However, more can be obtained: in particular, we note that the folding implies that $A_\beta$ cannot be a simple closed curve.

Since $\alpha \gg 1$, the contraction on each application of the Poincaré map is extremely rapid (points approach the attractor like $e^{-\alpha^2 t}$), so it appears reasonable to replace the "reduced" annulus map $F_\beta : R^+ \to R^-$ by a non-invertible mapping defined on a circle, $f_\beta : S^1 \to S^1$ (we will consider this process in more detail in Chapter 5). Almost all the properties of $F_\beta$, and hence of $P_\beta$, can be recovered from the one-dimensional map, as Levi shows. We sketch one such map in Figure 2.1.5(b). The segment $(\theta_1, \theta_2)$ is stretched to a length somewhat greater than 1, while the remainder of the circle is contracted and its orientation reversed. This noninvertible map has four fixed points, two stable ($|f_\beta'| < 1$) and two unstable ($|f_\beta'| > 1$). As $\beta$ varies, the whole map $f_\beta$ is effectively translated vertically and these fixed points can vanish pairwise in saddle-node bifurcations, but we shall not consider this here. Henceforth we drop the $\beta$ and simply write $f$.

Rather than attempt to analyze the map in detail we shall study it numerically, but first we show that $f$ has infinitely many unstable fixed points in the interval $[\theta_1, \theta_2]$. Letting the intervals $[0, \theta_1]$, $[\theta_1, \frac{1}{2}]$, $[\frac{1}{2}, \theta_2]$, $[\theta_2, 1]$ be denoted $I_0, I_1, I_2, I_3$, we see from Figure 2.1.5(b) that

$$f(I_0) \supset I_1,$$

$$f(I_1) \supset I_1 \cup I_2 \cup I_3,$$

$$f(I_2) \supset I_0 \cup I_1 \cup I_2, \qquad\qquad (2.1.19)$$

$$f(I_3) \supset I_2.$$

Based on this, we can write a *transition matrix*

$$A = [a_{ij}] = \begin{bmatrix} 0 & 1 & 0 & 0 \\ 0 & 1 & 1 & 1 \\ 1 & 1 & 1 & 0 \\ 0 & 0 & 1 & 0 \end{bmatrix}, \tag{2.1.20}$$

where $a_{ij} = 1$ if $f(I_i) \supset I_j$. From (2.1.19)–(2.1.20), it is possible to find orbits that move from $I_i$ to $I_j$ if $a_{ij} = 1$. (Note that this is not necessary, for example, $a_{00} = a_{33} = 0$ but (stable) fixed points exist within $I_0$ and $I_3$.) Thus, since $a_{ij} = 1$ for $i, j = 1, 2$, orbits visiting $I_1$ and $I_2$ in any prespecified order exist. We represent such an orbit $\{x_k\}_{k=0}^{\infty}$ by an infinite forward going sequence of 1's and 2's, $\{a_k\}_{k=0}^{\infty}$, chosen such that

$$\begin{aligned} a_k &= 1 \quad \text{if } f^k(x_0) \in I_1, \\ a_k &= 2 \quad \text{if } f^k(x_0) \in I_2. \end{aligned} \tag{2.1.21}$$

The representation of orbits of maps or differential equations as symbol sequences is called *symbolic dynamics* and will be developed in detail in Chapter 5, where we show that, under suitable conditions, a homeomorphism exists between the map $f$ and the shift operation

$$(\sigma\{a_k\})_j = a_{j+1},$$

defined by

$$\sigma\{a_0, a_1, a_2, \ldots\} = \{a_1, a_2, a_3, \ldots\},$$

on the space of symbol sequences. This fact permits the analysis of the dynamics of $f$ to be done by considering the various admissible sequences. In the present case, the expansion of $f$ in the interval $I_1 \cup I_2$* implies that each infinite sequence $\{a_k\}_{k=0}^{\infty}$ corresponds to a single orbit $\{x_k\}_{k=0}^{\infty}$ of $f$ and vice versa.

Since precisely one periodic orbit of $f$ corresponds to each distinct periodic symbol sequence, it is easy to enumerate these orbits. The first few are given in Table 2.1.1. To see that any such orbit, say of period $k$, is unstable, we consider the derivative $(\partial/\partial\theta)(f^k(p))$ of the function $f$ composed with itself $k$ times and evaluated at any point $p$ in the orbit. Using the chain rule repeatedly, we have

$$\frac{\partial}{\partial\theta}(f^k(p)) = \prod_{j=0}^{k-1} \frac{\partial f}{\partial\theta}(f^j(p)), \tag{2.1.22}$$

and since $\partial f/\partial\theta > 1$ for $\theta \in [\theta_1, \theta_2]$, this product must be greater than one. Thus the orbit is repelling.

Of course, many of the points in $I_1 \cup I_2 = [\theta_1, \theta_2]$, are eventually mapped out of this interval and into $I_0$ or $I_3$. Although they may subsequently

---

* The boundaries $\theta_1, \theta_2$ of $I_1, I_2$ can be chosen such that $f'(\theta) > 1$ for $\theta \in [\theta_1, \theta_2]$.

Table 2.1.1. Unstable Periodic Orbits for $f$ in $[\theta_1, \theta_2] = I_1 \cup I_2$.

| Period | Number of orbits | Symbol sequence |
|--------|------------------|-----------------|
| 1 | 2 | $111111\ldots\,;22222\ldots$ |
| 2 | 1 | $121212\ldots$ |
| 3 | 2 | $112112112\ldots;122122122\ldots$ |
| 4 | 3 | $11121112\ldots;11221122\ldots;12221222\ldots$ |
| 5 | 6 | $11112\ldots\,;11122\ldots\,;11222\ldots\,;12222\ldots\,;$ |
|   |   | $12121\ldots\,;21212\ldots$ |

reenter $I_1 \cup I_2$, most such points ultimately converge on one of the stable fixed points in $I_0$ or $I_3$. In fact the invariant set $\Lambda$ of points which remain in $I_1 \cup I_2$ is of measure zero, although it contains countably many periodic orbits and uncountably many nonperiodic orbits corresponding to non-periodic sequences $\{a_k\}_{k=0}^{\infty}$. Even though most orbits are not asymptotic to $\Lambda$, it nonetheless exerts a remarkable influence on the observable dynamics of $f$, as the numerical iterations below demonstrate.

For simplicity we select the piecewise linear map

$$f(x) = \begin{cases} (2 - x)/4, & x \in [0, \frac{18}{41}] \\ 10x - 4, & x \in [\frac{18}{41}, \frac{1}{2}] \\ 10x - 5, & x \in [\frac{1}{2}, \frac{23}{41}] \\ (3 - x)/4, & x \in [\frac{23}{41}, 1] \end{cases}, \qquad (2.1.23)$$

defined on the interval $I = [0, 1]$, taken modulo 1: Figure 2.1.6. The reader can easily check that this map is continuous, and has sinks at $x = \frac{2}{5}$, $x = \frac{3}{5}$

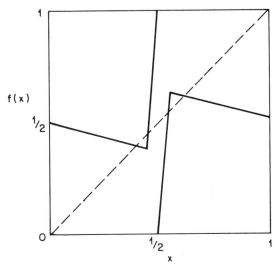

Figure 2.1.6. The piecewise linear van der Pol map $f$ of equation (2.1.23).

and sources at $x = \frac{4}{9}$, $x = \frac{5}{9}$. The fact that the map is nondifferentiable at the turning points $x = \frac{18}{41}$, $\frac{23}{41}$ is not important for our purposes in this example.

In Figure 2.1.7 we show some typical orbits $f^k(x_i)$ of points $x_i \in I$. Clearly, starting sufficiently close to either sink, the orbit simply converges on the sink, Figure 2.1.7(a, b). However, orbits which start within or enter the central region $[\frac{18}{41}, \frac{23}{41}]$ behave in a more erratic fashion, and, while almost all such orbits converge to one or the other sink, the presence of the unstable invariant set in $I_1 \cup I_2$ manifests itself in the sensitive dependence on initial conditions. In Figure 2.1.7(c)–(e) we show orbits starting close together which converge to different sinks after a transient period close to an unstable orbit of period 4.

Figure 2.1.8 illustrates the rôle of an unstable orbit of period two (the orbit $\{0, 0.5, 0, 0.5 \ldots\}$) in this respect. In Figure 2.1.8(b) we pick an initial condition in the stable manifold of this orbit and in Figure 2.1.8(a), (c)

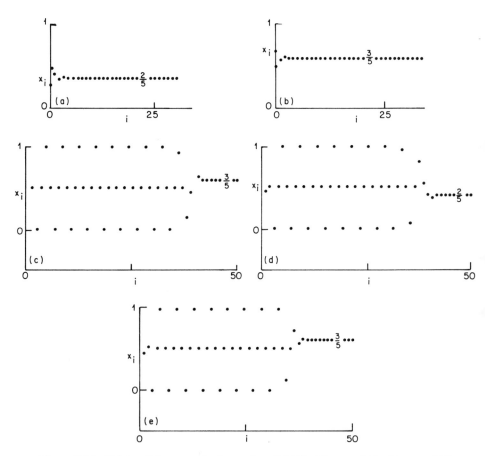

Figure 2.1.7. Orbits of the map $f$ of equation (2.1.23). (a) $x_0 = 0.29$; (b) $x_0 = 0.70$; (c) $x_0 = 0.445 + 10^{-11}$; (d) $= 0.445 + 3 \times 10^{-11}$; (e) $0.445 + 6 \times 10^{-11}$.

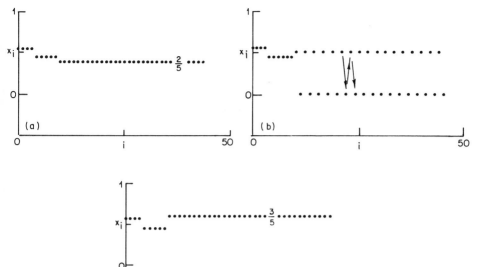

Figure 2.1.8. Orbits of the map $f$ of equation (2.1.23). (a) $x_0 = 0.5555444444$; (b) $x_0 = 0.5555444445$; (c) $x_0 = 0.5555444446$.

initial conditions very close to this are taken. Reference to (2.1.23) shows that the orbits are as follows:

(a) $0.55554444444 \rightarrow 0.555444444 \rightarrow 0.55444444 \rightarrow 0.5444444 \rightarrow$
$0.444444 \rightarrow 0.44444 \rightarrow 0.4444 \rightarrow 0.444 \rightarrow 0.44 \rightarrow 0.4 = \frac{2}{5} \ldots$.

(b) $0.5555444445 \rightarrow 0.555444445 \rightarrow 0.55444445 \rightarrow 0.5444445 \rightarrow$
$0.444445 \rightarrow 0.44445 \rightarrow 0.4445 \rightarrow 0.445 \rightarrow 0.45 \rightarrow 0.5 \rightarrow 0 \rightarrow 0.5 \rightarrow$
$0 \rightarrow 0.5 \ldots$.

(c) $0.5555444446 \rightarrow \cdots \rightarrow 0.444446 \rightarrow \cdots \rightarrow 0.46 \rightarrow 0.6 = \frac{3}{5} \ldots$.

Note that, while the computer can locate the orbit $\ldots, 0.5, 0, 0.5, 0 \ldots$, it is unstable, for

$$\left| \frac{\partial}{\partial \theta} (f^2(0)) \right| = |f'(0.5) \cdot f'(0)| = |10(-\tfrac{1}{4})| = \tfrac{5}{2} > 1.$$

In the orbits displayed above, an important feature is apparent: the orbit successively "forgets" the detail of its initial condition—the number of significant figures is effectively reduced. Any finite state machine, such as the Hewlett–Packard HP85 computer used in these computations, can only store data to finite accuracy, say $N$ significant figures. Two orbits with initial conditions differing in the $(N + 1)$st significant figure will therefore be indistinguishable in numerical work, although they may in fact have quite different behaviors: consider orbits such as those above, whose initial conditions end with 4 and 6, respectively, in the $(N + 1)$st decimal place. We shall return to this feature subsequently, when we discuss symbolic dynamics and the shift automorphism on symbol sequences in Chapter 5.

In summary, in our sketch of the van der Pol system we have seen how the relatively simple planar phase portrait of the unforced system gives way to the more complex picture of the Poincaré map associated with the periodically forced problem. Although one can recover an equivalent planar system by averaging in the case of weak forcing, the validity of such an analysis is limited. In the strongly forced problem, approximations again permit an effective reduction of dimension, but the one-dimensional map of the circle thus produced is noninvertible and displays remarkably complex dynamics. These observations are characteristic and will reappear in other examples.

## 2.2. Duffing's Equation

Duffing [1918] introduced a nonlinear oscillator with a cubic stiffness term to describe the hardening spring effect observed in many mechanical problems. Since then this equation has become, together with van der Pol's equation, one of the commonest examples in nonlinear oscillation texts and research articles. In this section we discuss a modification of the conventional Duffing equation in which the linear stiffness term is negative. Such an equation describes the dynamics of a buckled beam or plate when only one mode of vibration is considered. In particular, Moon and Holmes [1979, 1980] showed that the Duffing equation in the form

$$\ddot{x} + \delta\dot{x} - x + x^3 = \gamma \cos \omega t \qquad (2.2.1)$$

provides the simplest possible model for the forced vibrations of a cantilever beam in the nonuniform field of two permanent magnets. The experimental apparatus used by Moon and Holmes is sketched in Figure 2.2.1. A slender steel beam is clamped in a rigid framework which supports the magnets. Their attractive forces overcome the elastic forces which would otherwise keep the beam straight and, in the absence of external forcing, the beam settles with its tip close to one or the other of the magnets. There is also an unstable central equilibrium position, at which the magnetic forces cancel: its existence can easily be demonstrated as a "potential barrier" separating the two domains of attraction, if the beam is carefully displaced. The simplest model for such a potential is the symmetric one-dimensional field

$$V(x) = \frac{x^4}{4} - \frac{x^2}{2}, \qquad (2.2.2)$$

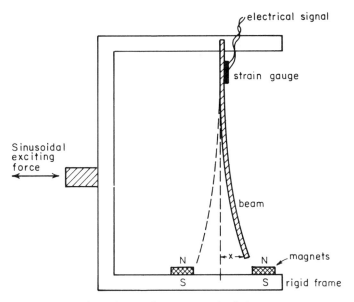

Figure 2.2.1. The magneto-elastic beam.

although the symmetry is not crucial. Here the single configuration variable $x$ represents a characteristic measure of the beam's position, say its tip displacement. The force acting on the beam is governed by the gradient of $V$, and thus, using Newton's second law, a simple model for the beam is provided by

$$\ddot{x} = -\text{grad } V, \qquad (2.2.3)$$

or

$$\ddot{x} - x + x^3 = 0. \qquad (2.2.4)$$

The dissipation due to friction, viscous damping from the surrounding air, and magnetic damping, is modelled by a linear velocity dependent term, giving the equation

$$\ddot{x} + \delta\dot{x} - x + x^3 = 0. \qquad (2.2.5)$$

Equation (2.2.5) is easy to analyze and will be found to provide a reasonable model for the motions of the beam in a stationary rigid framework. (The reader is free to add appropriate coefficients to obtain physically relevant frequencies if he wishes.) Of course, equation (2.2.5) represents only a single mode of vibration, but if the beam is sufficiently long and slender, and the magnets sufficiently powerful, vibrations *are* observed to occur primarily in the first mode (cf. Moon and Holmes [1979]).

Now we begin to shake the apparatus sinusoidally as shown, using an electromagnetic vibration generator. This is modelled by the addition of a

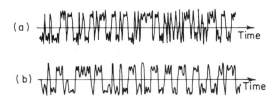

Figure 2.2.2. (a) Vibrations of the beam, and (b) a solution of equation (2.2.1).

forcing term, and thus we obtain equation (2.2.1).* In fact Moon and Holmes derived (2.2.1) as a Galërkin approximation of a set of rather general partial differential equations for an elastic beam in a nonuniform magnetic field, and a good case can be made for its relevance as the simplest possible model for the experiment. The sketch given here is only intended to make the model plausible.

An electromechanical strain gauge glued near the root of the beam measures the curvature at that point as a function of time. Since the motion is primarily in the first mode, this effectively provides a measure of the tip displacement $x(t)$. For low force amplitudes, $\gamma$, we observe periodic motions close to either one or the other magnet, but as $\gamma$ is slowly increased, a point comes at which the beam "suddenly" begins whipping back and forth in an irregular, apparently chaotic manner. This is not a transient phenomenon: with $\gamma$ and $\omega$, the forcing frequency, fixed, such chaotic motions have been observed for periods of hours, or for $\approx 10^5$ cycles of the forcing function. A short piece of such a record is shown in Figure 2.2.2, together with a typical solution of (2.2.1).

The qualitative agreement is clear, although the reader may like to ponder why the experimental record is "spikier" than the solution of (2.2.1), and appears to contain higher frequency components (this spikiness is *not* a consequence of the measurement technique). We can therefore place some confidence in our simple model, which we now discuss.

We first consider the problem without external forcing $\gamma = 0$. We introduce an additional parameter $\beta$ which measures the relative strength of the magnetic and elastic forces. An increase in $\beta$ corresponds (approximately) to an increase in strength of the magnetic forces. The system is

$$\ddot{x} + \delta\dot{x} - \beta x + x^3 = 0,$$

or, as a first-order equation

$$\dot{u} = v,$$
$$\dot{v} = \beta u - u^3 - \delta v, \tag{2.2.6}$$

and it is easy to check that, for $\beta < 0$ (weak magnets), there is a single equilibrium at $(x, \dot{x}) = 0$ while if $\beta > 0$, there are three equilibria at $x = 0$,

---

* Since the beam is subject to inertial excitation, for frame displacement $\Gamma \cos \omega t$ the force on the beam is actually of the form $\Gamma \omega^2 \cos \omega t$. We subsume the $\omega^2$ in the parameter $\gamma$.

$\pm\sqrt{\beta}$. If $\delta > 0$, the equilibria are, respectively, a sink (for $\beta < 0$) and two sinks and a saddle (for $\beta > 0$). In terms of the bifurcation analyses of Chapters 3 and 7, (2.2.6) undergoes a pitchfork bifurcation of equilibria as $\beta$ passes through zero.

To obtain global information on the phase portrait, we note that, for $\delta = 0$, the system is Hamiltonian with Hamiltonian energy

$$H(u, v) = \frac{v^2}{2} - \beta\frac{u^2}{2} + \frac{u^4}{4}. \qquad (2.2.7)$$

Since the solutions lie on level curves of $H$, we can immediately draw the phase portraits for $\delta = 0$ as in Figure 2.3.3(a). Addition of the term $-\delta v$ to the second equation directs the vector field inward in all the closed level curves (except at $v = 0$) and we therefore obtain the qualitative behaviors of Figure 2.2.3(b). In particular, a closed, simply connected region $D \subset \mathbb{R}^2$ can be found such that the vector field is directed inward everywhere on its boundary. This global stability property persists even for nonzero forcing, when the vector field is time dependent (cf. Holmes [1979a], Holmes and Whitley [1983a]). We now turn to this case.

Letting $\gamma \neq 0$ and setting $\beta = 1$, we return to equation (2.2.1), which may be rewritten as an autonomous system

$$\left.\begin{aligned} \dot{u} &= v, \\ \dot{v} &= u - u^3 - \delta v + \gamma \cos \omega\theta, \\ \dot{\theta} &= 1, \end{aligned}\right\} \qquad (u, v, \theta) \in \mathbb{R}^2 \times S^1, \qquad (2.2.8)$$

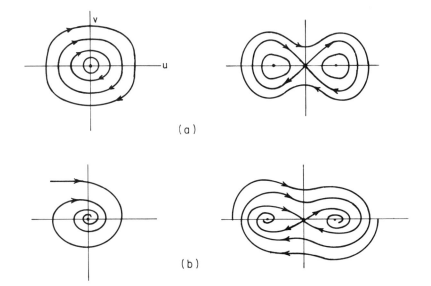

(a)

(b)

Figure 2.2.3. Duffing's equation without external forcing. (a) $\delta = 0$; (b) $\delta > 0$.

where $S^1 = \mathbb{R}/T$ is the circle of length $T = 2\pi/\omega$. We pick a cross section $\Sigma = \{(u, v, \theta) | \theta = 0\}$ and consider the Poincaré map $P: \Sigma \to \Sigma$ (the stability property referred to above, and the consequent boundedness of all solutions, ensures that $P$ is globally defined). Clearly, $P$ depends upon the parameters $\gamma$, $\delta$, $\omega$, but in what follows we shall take $\delta$ and $\omega$ as fixed, positive quantities and vary $\gamma$, and write $P = P_\gamma$.

Clearly, $P_0$ is just the time $2\pi/\omega$ flow map of the unforced problem (2.2.6) and solution curves of the latter are invariant curves for $P_0$. In particular, the separatrices of the saddle point at $(0, 0)$ for the flow are the invariant manifolds for the corresponding saddle point of $P_0$. Let us consider what happens to $P_\gamma$, and hence to the flow, as $\gamma$ is increased from zero. We first describe the results of numerical solutions and Poincaré maps plotted from them. The results presented here are due to Ueda [1981a] (cf. [1981b]); for further results see Holmes [1979a].

For small values of $\gamma$, the two sinks of (2.2.6) at $(u, v) = (\pm 1, 0)$ become small ($\mathcal{O}(\gamma)$) attracting orbits of period $2\pi/\omega$ (period one for $P_\gamma$) and the saddle point becomes a saddle-type orbit. Thus $P_\gamma$ continues to have three hyperbolic fixed points. As $\gamma$ increases, the amplitude of the orbits, particularly the stable ones, grows continuously until, depending upon the precise values of $\delta$ and $\omega$, a bifurcation occurs. Linearizing (2.2.6) at $(\pm 1, 0)$ we find that the unperturbed natural frequency, with $\delta = 0$, is $\sqrt{2}$. Moving out, the orbits surrounding $(\pm 1, 0)$ have successively longer periods, the periods tending to infinity as we approach the double homoclinic connection (Figure 2.2.3(a)), Locally, therefore, the oscillator near $(\pm 1, 0)$ behaves as a softening spring (Nayfeh and Mook [1979]), and if $\omega < \sqrt{2}$, a jump resonance occurs in which the small period 1 orbit is replaced by a relatively large period 1 orbit (Holmes [1979a]). If $\omega > \sqrt{2}$, and in particular if $\omega \approx 2\sqrt{2}$, a period 2 resonance can occur in which the fixed point of $P_\gamma$ becomes unstable and undergoes a flip bifurcation in which a stable orbit of period $4\pi/\omega$ appears (Holmes and Holmes [1981]). Such jumps and the associated fold and flip bifurcations can be studied using the averaging results of Chapter 4.

In Figure 2.2.4(a) we show a pair of relatively large period 1 orbits ($\omega = 1 < \sqrt{2}$), projected onto the $u$, $v$ plane, in addition to the associated fixed points of $P_\gamma$. As $\gamma$ continues to increase, further bifurcations can occur in which such periodic points and the corresponding periodic orbits of the flow double their periods repeatedly. These accumulate at a point at which transition from periodic to apparently chaotic nonperiodic motion like that illustrated in Figure 2.2.2 occurs. Such cascades of period doubling bifurcations have been studied extensively and have many interesting universal properties (Feigenbaum [1978, 1980], Collet and Eckmann [1980]). We discuss some of this work in Chapters 5 and 6.

In Figures 2.2.4(b) and 2.2.5(a) we show typical orbits of a single point under the Poincaré map in this chaotic parameter regime. We note that these numerically observed "strange attractor" motions exist for relatively wide

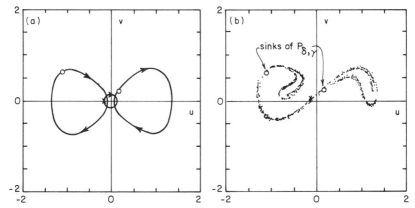

Figure 2.2.4. Orbits of the Duffing equation, $\delta = 0.25$, $\omega = 1.0$, $\gamma = 0.30$. (a) Two stable ( $\circ$ ) and one saddle-type ( $\times$ ) periodic orbits, showing fixed points of $P_\gamma$; (b) the "strange attractor." Note that the stable periodic orbits ( $\circ$ ) are close to marginal stability. Different numerical methods may yield unstable orbits.

sets of parameter values, that they can coexist with simple periodic motions, and that subharmonic motions are also observed in thin bands within the "strange attractor" region (cf. Figure 2.2.5(b)—see Holmes [1979a, Figure 7(e)] for a period 5 example). Considerably higher values of $\gamma$ give rise to simple periodic motions of large amplitude once more; Figure 2.2.5(a).

Before offering a partial interpretation and explanation of these observations, we note that the irregular time history of Figure 2.2.2 displays considerable structure when viewed in a Poincaré map (Figures 2.2.4(b) and 2.2.5(a)) and that it appears to be genuinely nonperiodic—or at least to have period longer than any observation, as in the beam experiment. In Figure 2.2.6 we show power spectra of two such motions, which display the broad

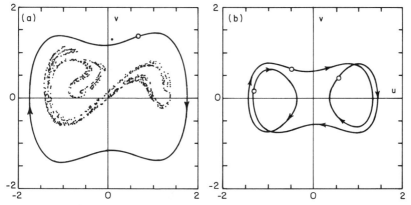

Figure 2.2.5. Orbits of the Duffing equation, $\omega = 1.0$, $\gamma = 0.30$. (a) Coexistence of a "strange attractor" and a large stable period 1 orbit, $\delta = 0.15$; (b) stable period 3 orbit, $\delta = 0.22$; the self-intersection is an artifact of the projection onto the $(u, v)$-plane.

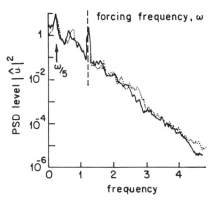

Figure 2.2.6. A power spectrum for the Duffing equation, $\omega = 1.19$, $\delta = 0.06$, $\gamma = 0.46$ ——————, $\gamma = 0.57$ ⋯⋯⋯ (from Holmes [1979a]).

frequency content of the Fourier transform $\hat{u}(\omega)$ of $u(t)$. Note the presence of peaks at the forcing frequency and some of its subharmonics.

A major key to understanding these results lies in the analysis of the invariant manifolds of the saddle point of $P_\gamma$ near $(0, 0)$. We denote this point as $p$ and its stable and unstable manifolds as $W^s(p)$, $W^u(p)$. As already noted, $W^s(p)$ and $W^u(p)$ for $\gamma = 0$ are simply the saddle separatrices of Figure 2.2.3(b) (cf. Figure 2.2.7(a)). The Poincaré map $P_0$ has three hyperbolic fixed points and the manifolds do not intersect, and therefore we can immediately conclude that $P_\gamma$ for small $\gamma$ is topologically equivalent to $P_0$, since $P_0$ is structurally stable. The numerical observations illustrate this (Figure 2.2.7(b)). As $\gamma$ increases, however, the manifolds can and do become tangent and subsequently intersect transversely; Figure 2.2.7(c), (d) (see Figure 4.5.3(b) for a quadratic contact). The partial plots of $W^s(p)$ and $W^u(p)$ of Figure 2.2.7 are produced by iterating a number of points defining a short segment of $W^s(p)$ (or $W^u(p)$) near $p$ under $P_\gamma^{-1}$ (or $P_\gamma$). This *global* bifurcation occurs in addition to, and independently of, the local period doubling or jump (fold) bifurcations of fixed points referred to above.

In Chapter 4 we shall see how to compute the location of such global homoclinic bifurcations for systems like (2.2.8), in the case that $\gamma$, $\delta$ are small and the system is close to an integrable Hamiltonian one. At the same time we shall be able to find subharmonic motions which bifurcate from the continuous families of periodic orbits of Figure 2.2.3(a).

Once the manifolds intersect, we have *transversal homoclinic orbits*, and in Chapter 5 we shall see that their presence implies the existence of a complicated nonwandering Cantor set which possesses infinitely many unstable periodic orbits of arbitrary long period as well as bounded nonperiodic motions (also see Section 2.4.4, below). Moreover, for certain parameter values near those at which homoclinic tangencies occur, the work of Newhouse [1979, 1980] shows that infinite sets of *stable* periodic orbits (called Newhouse sinks) exist. We shall discuss these in Chapter 6. All this complex

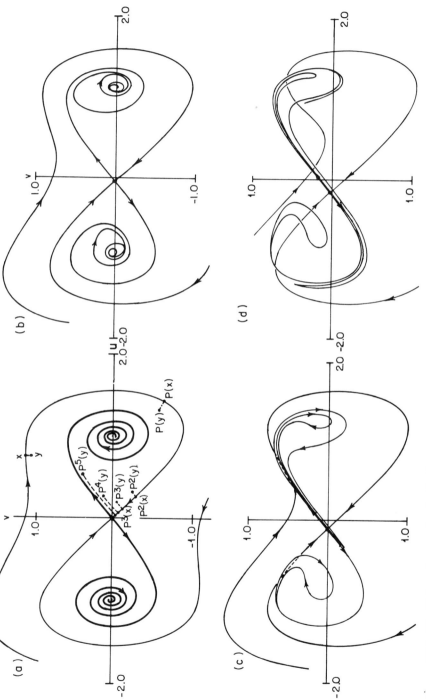

Figure 2.2.7. The Poincaré maps for the Duffing equation (2.2.8), $\delta = 0.25$, $\omega = 1.0$. (a) $\gamma = 0$; (b) $\gamma = 0.10$; (c) $\gamma = 0.20$; (d) $\gamma = 0.30$. Typical orbits shown in (a), and attracting set marked as a heavy curve.

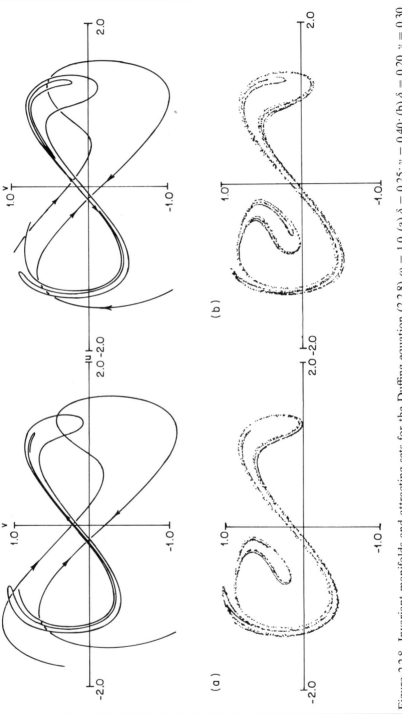

Figure 2.2.8. Invariant manifolds and attracting sets for the Duffing equation (2.2.8), $\omega = 1.0$. (a) $\delta = 0.25$; $\gamma = 0.40$; (b) $\delta = 0.20$, $\gamma = 0.30$.

behavior contributes to the attracting set, the structure of which we now turn to.

As noted above, a closed disc $D \subset \Sigma$ can be found such that $P_\gamma^n(D) \subset D$ for $n > 0$, and we can again define the closed attracting set as

$$A_\gamma = \bigcap_{n \geq 0} P_\gamma^n(D).$$

Comparing numerically obtained plots of the strange attractors with the unstable manifold of the saddle point, $W^u(p)$, we are led to the conjecture that this set $A_\gamma$ is equal to the closure of the unstable manifold; cf. Figure 2.2.8. $A_\gamma$ is certainly of zero area, since the flow of (2.2.8) contracts volume uniformly, because the divergence of the vector field is

$$\frac{\partial}{\partial u}(v) + \frac{\partial}{\partial v}(u - u^3 - \delta v + \gamma \cos \omega\theta) + \frac{\partial}{\partial \theta}(1) = -\delta < 0, \qquad (2.2.9)$$

and hence $\det |DP_\gamma| = e^{-2\pi\delta/\omega} < 1$.

For small $\gamma$ our conjecture is proved by noting that all orbits $\{P_\gamma^n(y)\}$ of $y \in D$ approach one of the sinks except those of points $x \in W^s(p)$, which approach $p$. $A_\gamma$ therefore contains the sinks and the saddle. To see that all points in $A_\gamma$ lie in the closure of $W^u(p)$, consider any curve $C$ joining a point $x \in W^s(p)$ to a point $y \in D$. (Figure 2.2.7(a)). As $n \to \infty$, the endpoints of $P^n(C)$ approach the saddle and the sink, as do all points on $C$ except those arbitrarily close to $P^n(x)$. Thus $P^n(C)$ approaches a component of $W^u(p)$, as claimed.

The situation is more complicated in the presence of homoclinic intersections. Even if our conjecture is true, the set $A_\gamma$ may contain proper subsets such as the Newhouse sinks which are attracting. What is of interest here is that, unlike the van der Pol oscillator; there are *no* low period sinks for certain parameter ranges. We conjecture that, for each integer $N < \infty$, there is an open set of parameter values $\gamma, \delta, \omega$ for which $P_\gamma$ has no attracting periodic orbits of period less than $N$. Such attractors are effectively unobservable for even moderately large values of $N$ because the characteristic width of their domains of attraction decreases dramatically (in certain cases like $e^{-N}$ (Greenspan and Holmes [1982])) and physical or numerical noise becomes dominant at these scales. Thus the asymptotic behavior of trajectories within $A_\gamma$ appears complicated. In some regimes there appear to be trajectories which are dense in $A_\gamma$. There is a substantial theoretical question as to whether this "strange attractor" of the Duffing equation is an artifact of the noise and is absent in the ideal deterministic system. We discuss in a general context the issue of the existence of strange attractors in Chapters 5 and 6.

In Section 2.4.4 we discuss a nonlinear map which displays behavior similar in many aspects to that of the Duffing Poincaré map, and in which the structure of the attracting set can be displayed more readily.

## 2.3. The Lorenz Equations

In [1963] Lorenz, a meteorologist working on the basis of earlier results due
to Salzmann [1962], presented an analysis of a coupled set of three quadratic
ordinary differential equations representing three modes (one in velocity and
two in temperature) of the Oberbeck–Boussinesq equations for fluid convection in a two-dimensional layer heated from below. The papers cited above
provide more details. The equations are

$$
\left.
\begin{aligned}
\dot{x} &= \sigma(y - x), \\
\dot{y} &= \rho x - y - xz, \\
\dot{z} &= -\beta z + xy,
\end{aligned}
\right\}
\begin{aligned}
&(x, y, z) \in \mathbb{R}^3 \\
&\sigma, \rho, \beta > 0,
\end{aligned}
\qquad (2.3.1)
$$

and contain the three parameters $\sigma$ (the Prandtl number), $\rho$ (the Rayleigh
number), and $\beta$ (an aspect ratio). This three mode truncation accurately
reflects the dominant convective properties of the fluid for Rayleigh numbers
$\rho$ near 1. In particular, when $\rho = 1$, the pure conductive solution of the partial
differential fluid equations, having zero velocity and linear temperature
gradient, becomes unstable to a solution containing steady convective rolls
or cells. With stress free boundary conditions, the Lorenz equations are a
minimal truncation of the fluid equations which embody the essential features
of this bifurcation. In Chapter 3 we present more details on this example as
an illustration of computations using the center manifold theorem.

Lorenz's [1963] analysis examines the behavior of this equation well outside the parameter domain $\rho \approx 1$, and work of Curry [1978] and
Francheschini [1982] demonstrates that, for large $\rho$, seven and fourteen mode
truncations display significantly different behavior.* Thus, as the Rayleigh
number increases, higher-order modes become important and predictions
made on the basis of the three mode truncation are of doubtful physical
relevance, in contrast to the beam problem, in which a single mode truncation does capture significant physical behavior over a wide parameter range.
Nonetheless, the equations have become of great interest to mathematicians
and physicists in recent years. In the remainder of this section we outline
some significant features of the flow of the Lorenz system. For more information, see the papers of Guckenheimer [1976], Guckenheimer and Williams
[1979], Williams [1977], Rand [1978], and the book of Sparrow [1982].

As in the Duffing equation, we fix two parameters $\sigma$ and $\beta$, and let $\rho$
vary. (The values used by Lorenz and most other investigators are $\sigma = 10$,
$\beta = \frac{8}{3}$, but similar behavior occurs for other values.) Lorenz [1963] shows
that a closed, simply connected region $D \subset \mathbb{R}^3$ can be found, containing the
origin, such that the vector field is directed everywhere inwards on the
boundary. Thus $D$ contains an attracting set $A = \bigcap_{t \geq 0} \phi_t(D)$. Moreover,

---

* Also see Marcus [1981] for more general information on modal truncations in fluid problems.

any such attractor has zero volume, since, as in the Duffing equation, the trace of the Jacobian (divergence of the vector field)

$$\frac{\partial}{\partial x}(\sigma(y - x)) + \frac{\partial}{\partial y}(\rho x - y - xz) + \frac{\partial}{\partial z}(-\beta z + xy) = -(\sigma + 1 + \beta)$$

(2.3.2)

is negative. In fact for $\rho < 1$ the origin is a hyperbolic sink and is the only attractor. At $\rho = 1$ one of the eigenvalues of the linearized system, with matrix

$$\begin{bmatrix} -\sigma & \sigma & 0 \\ \rho - z & -1 & -x \\ y & x & -\beta \end{bmatrix}_{x=y=z=0} = \begin{bmatrix} -\sigma & \sigma & 0 \\ \rho & -1 & 0 \\ 0 & 0 & -\beta \end{bmatrix}$$

(2.3.3)

is zero, the others, $\lambda = -\beta$ and $\lambda = -(1 + \sigma)$ being negative. For $\rho > 1$ two nontrivial fixed points, at

$$(x, y, z) = (\pm\sqrt{\beta(\rho - 1)}, \pm\sqrt{\beta(\rho - 1)}, \rho - 1)$$

(2.3.4)

exist, and these are sinks for $\rho \in (1, \sigma(\sigma + \beta + 3)/(\sigma - \beta - 1))$. At $\rho = 1$, a pitchfork bifurcation occurs like that of the unforced Duffing equation (2.2.6) at $\beta = 0$, and for all $\rho > 1$ the origin is a saddle point with a one-dimensional unstable manifold. At $\rho = \rho_h = \sigma(\sigma + \beta + 3)/(\sigma - \beta - 1)$ a Hopf bifurcation occurs at the nontrivial fixed points, since here the eigenvalues of the matrix are

$$\lambda = -(\sigma + \beta + 1) \quad \text{and} \quad \lambda = \pm i\sqrt{2\sigma(\sigma + 1)/(\sigma - \beta - 1)}.$$

(Following Lorenz, we assume that $\sigma > 1 + \beta$, so that imaginary roots are possible.) For $\rho > \rho_h$, the nontrivial fixed points are saddles with two-dimensional unstable manifolds. Thus for $\rho > \rho_h$ all three fixed points are unstable, but an attracting set $A = \bigcap_{n \geq 0} \phi_t(D)$ still exists, although $A$ now contains more complicated bounded solutions. We shall return to these in a moment.

It might be thought that the Hopf bifurcation occurring as $\rho$ passes through $\rho_h$ would give rise to stable periodic orbits, but in subsequent analyses (of Marsden and McCracken [1976, Chapter 4], and see Chapter 3, below), it was shown that the bifurcation is *subcritical*, so that unstable periodic orbits shrink down upon the sinks as $\rho$ increases towards $\rho_h$ and no closed orbits exist near these fixed points for $\rho > \rho_h$. In physical terms, the steady convection rolls represented by the symmetric pair of nontrivial solutions become unstable and are replaced by some other large amplitude motion. We now describe this motion, basing our description on Lorenz's original numerical work and the later geometrical analyses of Guckenheimer and Williams.

Lorenz fixed $\sigma = 10$ and $\beta = \frac{8}{3}$, so that $\rho_h \approx 24.74$. He then fixed $\rho = 28$ and integrated equation (2.3.1) numerically, choosing an initial condition close to the saddle point at $(0, 0, 0)$. A typical time history of the first 30 time units of such a solution $y(t)$, is reproduced from Lorenz's paper in Figure 2.3.1. Similar traces are observed for $x(t)$ and $z(t)$, although the latter does not undergo sign reversals. Note the growth of oscillations to an apparant threshold value after which a sign change follows and oscillations grow once more.

Lorenz found that solutions rapidly approached and thereafter apparently moved on a branched surface $S$, which we sketch in Figure 2.3.2 on the basis of three-dimensional numerical solutions due to Lanford [1977]. The boundary of $S$ is part of the unstable manifold $W^u(p)$ of the saddle point at $p = (0, 0, 0)$. In Figure 2.3.2 the first 50 loops of one "side" of $W^u(p)$ are shown; the surface is shaded and the branch is indicated. The remaining saddles, $q_\pm = (\pm\sqrt{\beta(\rho - 1)}, \pm\sqrt{\beta(\rho - 1)}, \rho - 1)$ lie in the two holes of $S$.

To get a clearer idea of the structure of the branched surface we illustrate it schematically in Figure 2.3.3, following Williams [1977]. In this scheme we replace the actual reversible three-dimensional flow by a *semiflow* on $S$; in which solutions are only defined forward in time, since, if we flow backwards, all solutions eventually reach the branch interval $[-a, a]$ and then must choose which branch to follow. The uniqueness of solutions to the initial value problem is therefore apparently violated. Lorenz used this fact to argue that the attracting set $A$, which we now call the *Lorenz attractor*, in fact has infinitely many sheets and that solutions within it do not intersect but rather move from sheet to sheet as they circulate over the apparent branch. In fact the branched surface is an artifact of rapid contraction and poor numerical resolution.

We will discuss the topological structure of the attractor $A$, and its connection with the branched manifold $S$, in Chapters 5 and 6, but here we will provide a partial analysis of the chaotic flow within $A$ in terms of a one-dimensional map, much as in the van der Pol example. We start by indicating how a connection can be made between the actual three-dimensional flow and the semiflow on $S$.

As already mentioned, numerical observations indicate that all solutions ultimately lie arbitrarily close to $S$, and thereafter pass repeatedly near the branch interval $[-a, a] \in S$. More precisely, all solutions ultimately pass transversely through a two-dimensional cross section $\Sigma$ in the neighborhood

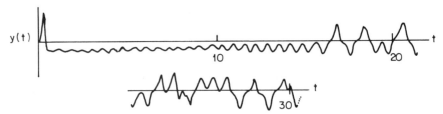

Figure 2.3.1. Numerical solution $y(t)$ of the Lorenz equation (2.3.1), $\sigma = 10$, $\rho = 28$, $b = \frac{8}{3}$,

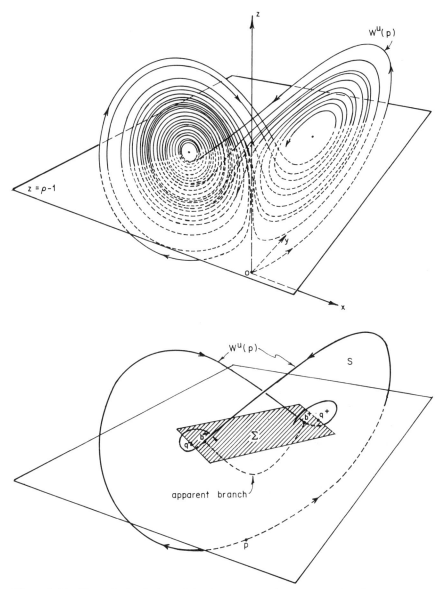

Figure 2.3.2. Numerical solution of the Lorenz equation, $\sigma = 10$, $\beta = \frac{8}{3}$, $\rho = 28$. The initial condition is chosen arbitrarily close to the saddle at $p(0, 0, 0)$, so that the solution approximates $W^u(p)$, defining the boundary of the apparent surface $S$. After Lanford [1977]. See text for description of $\Sigma$, etc.

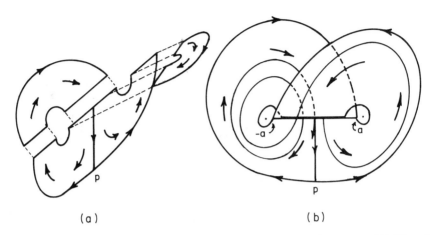

(a)                                                                (b)

Figure 2.3.3. The branched manifold of the Lorenz attractor, after Lorenz [1963] and Williams [1977]. (a) Construction; (b) the semiflow on S.

of $[-a, a]$. Thus a two-dimensional invertible Poincaré map, $P$, can be defined on $\Sigma$, as in Guckenheimer [1976]. (A suitable cross-section is provided by a strip $\Sigma$ cut from the $z = \rho - 1$ plane of Figure 2.3.2 in such a manner that $P(\Sigma) \subset \Sigma$, and hence that the intersection of $A$ with the $z = \rho - 1$ plane lies in $\Sigma$.) By projecting the orbits of $P$ in a direction transverse to $S$, we obtain an equivalent one-dimensional first return map

$$f: I \to I,$$

where $I = [-a, a]$ denotes the branch interval in the surface $S$. $f$ is a Poincaré map for the semiflow. Note that the time taken for solutions to return varies, and tends to infinity as one approaches the midpoint of $I$, the image of which is not defined, since solutions starting there flow into the saddle point $p$, and hence never return to $\Sigma$. The symmetry of the flow implies that $f$ is therefore an odd function which looks qualitatively like that of Figure 2.3.4. Note that the slope, $f'$, is everywhere greater than one, and the map has no fixed points in $I$. The expansive nature of the map ($f' > 1$) describes the growing oscillations observed numerically, and the discontinuity at $y = 0$ accounts for the sign reversals. We assume that $f(0^-) = -f(0^+) = a$, $f(a) = -f(-a) > 0$, and that $\lim_{x \to 0} f'(0) = \infty$, although we do not use the latter fact explicitly here. Also, for simplicity later, assume that $f^2(-a) > f(a)$ and $f^2(a) < f(-a)$.

As already noted, the two-dimensional Poincaré map $P$ defined on a suitable cross section $\Sigma$ is invertible since it arises from a globally defined flow. However, in the projection process, much as in the van der Pol example, we lose reversibility and the projected map $f$ is thus no longer one to one.

We stress that the geometric model of the Lorenz attractor as a one-dimensional map certainly does not capture all the details of the actual flow within $A$. Moreover, the calculations necessary to show that a foliation transverse to the sheets of $A$, carrying a uniform contraction, exists, have not

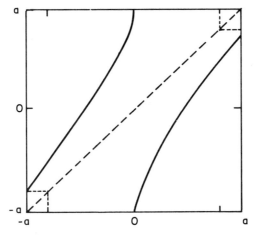

Figure 2.3.4. The Lorenz map.

been performed. Such a contracting foliation is assumed in the projection process referred to above. In subsequent chapters we will consider some of the problems involved in verifying that geometrically constructed attractors, such as this one, actually exist in specific flows, but in the remainder of this section we study the one-dimensional map $f$ in its own right. Henceforth, we refer to $f$ as the Lorenz map, although it differs from the one-dimensional map presented by Lorenz [1963].

We first note that, if $f$ has a periodic orbit of period $k$, then it is unstable, for the derivative evaluated at any point $p$ in the orbit is, by the chain rule,

$$(f^k(p))' = \prod_{j=0}^{k-1} f'(f^j(p)), \qquad (2.3.5)$$

and the product is clearly greater than one, since $f' > 1$ everywhere in $I$. Orbits of periods two and three are easy to find, for, letting $I_1 = [-a, 0]$ and $I_2 = [0, a]$ we have $f(I_1) = [f(-a), a] \supset I_2$ and $f(I_2) = [-a, f(a)] \supset I_1$ so that $f^2(I_i) = I_1 \cup I_2 = I$ for $i = 1, 2$. Similarly, $f^k(I_i) = I$ for all $k \geq 2$. Thus $f^k(I_i) \supseteq I_i$ and $f^k$ has at least two fixed points for all $k$. We immediately deduce that $f$ has orbits of all prime periods, but, while $f^k$ has a fixed point for every $k$, such a point may have least period equal to some submultiple of $k$. In fact the structure of the set of periodic orbits depends rather delicately on the precise form of $f$, and we shall not consider it further here. Instead we shall explore the sensitivity to initial conditions exhibited by this map, since here it is rather easier to understand than in the van der Pol example of Section 2.1.

As we show in Section 5.7, if $f' > \sqrt{2}$ everywhere, then any subinterval $J \subset I$ is eventually expanded under $f$ so that, for some $n$, $f^n(J)$ covers $I$ (Williams [1976, 1979]). Thus all points of $I$ are nonwandering. In fact if $f' > 1$, any two distinct points $x, y \in I$ eventually have images $f^n(x), f^n(y)$

on opposite sides of 0 (unless one lands *on* 0 and its orbit terminates). Once the points are separated in this manner, their orbits can no longer be considered close, since they behave essentially independently. The splitting apart of such orbits is evidently controlled by the positions of the preimages of the origin 0, $f^{-k}(0)$. Since $f^{-1}$ is double valued, at least over part of $I$, the number of preimages $f^{-k}(0)$ grows like $2^k$, until points leave $I$. In fact the expansion noted above implies that the set of all preimages $\bigcup_{k \geq 0} f^{-k}(0)$ lying in $I$, while having zero measure, is dense in $I$. We refer to the local expansion and consequent "independent" behavior of orbits starting arbitrarily close together as *sensitive dependence on initial conditions*, Ruelle [1979].

As we showed in Section 2.1, the van der Pol map, while possessing a complicated chaotic set containing infinitely many periodic orbits, displays relatively simple asymptotic behavior in the sense that almost all orbits converge to one or the other of the stable fixed points. In the terms of Section 1.6, the attracting set contains two simple attractors and, in spite of the presence of complex recurrent behavior almost all points in it are wandering. In contrast, in the Lorenz map, no such stable sinks or periodic orbits exist, and almost all orbits continue to move back and forth erratically, the exceptions being the (unstable) periodic orbits, orbits asymptotic to them, and orbits which land on 0 and terminate. Such orbits form a set of zero measure in $I$.

As in Section 2.1, we now illustrate the typical behavior of maps like that of Figure 2.3.4 by numerical simulation. We take the map

$$f(x) = \begin{cases} 1 - \beta|x|^\alpha; & x \in [-1, 0), \\ -1 + \beta|x|^\alpha; & x \in (0, 1], \end{cases} \tag{2.3.6}$$

defined on $I = [-1, 1]$, with the orbit terminating if $x = 0$. If we choose $\alpha < 1$, $\beta \in (1, 2)$ and $\alpha\beta > 1$, then the derivative $f' = \alpha\beta|x|^{\alpha-1} > 1$ everywhere. Note that $f'(x)$ approaches $\infty$ as $x$ approaches 0. For our computations we pick $\alpha = 1/\beta + 0.001$, so that the derivative is very close to 1 near the endpoints, reflecting the slow growth of oscillations near the spiral saddles $q^\pm$ illustrated in Figure 2.3.2. This map is shown in Figure 2.3.5.

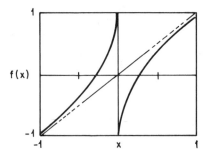

Figure 2.3.5. The map of equation (2.3.6). $\beta = 1.95$ ($\alpha = 0.514$).

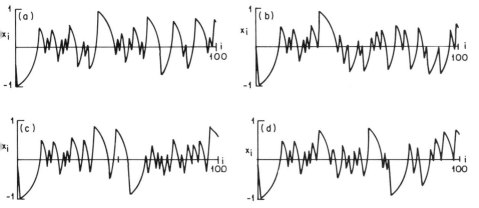

Figure 2.3.6. Orbits of the map $f$ of equation (2.3.6). (a) $\beta = 1.95$, $x_0 = 10^{-8}$; (b) $\beta = 1.95$, $x_0 = 0.9999 \times 10^{-8}$; (c) $\beta = 1.95$, $x_0 = 1.0001 \times 10^{-8}$; (d) $\beta = 1.95000001$, $x_0 = 10^{-8}$.

In Figure 2.3.6 we illustrate the sensitive dependence on initial conditions, and on variations in the parameter $\beta$. In each case the first 100 iterates are shown. For Figures 2.3.6(a)–(c) we fix $\beta = 1.95$ and take the initial condition $x_0$ as $10^{-8}$, $10^{-8} - 10^{-12}$, and $10^{-8} + 10^{-12}$, respectively. For 2.3.6(d) we take $\beta = 1.95 + 10^{-8}$ and $x_0 = 10^{-8}$. In each case the orbits diverge radically from one another after only 25 iterates.

In Figure 2.3.7 we show the first 400 iterates of the orbit of $f$ with $\beta = 1.99$, based at $x_0 = 10^{-11}$. Such an orbit corresponds to one started close to the saddle at $p = (0, 0, 0)$ in the Lorenz equations and should be compared with Lorenz's numerically obtained orbit of Figure 2.3.1. Recall that each point on Figure 2.3.7 corresponds to one oscillatory cycle in the differential equation. In these figures the discrete points of the orbit are connected by straight lines; the first few such points are indicated in Figure 2.3.7.

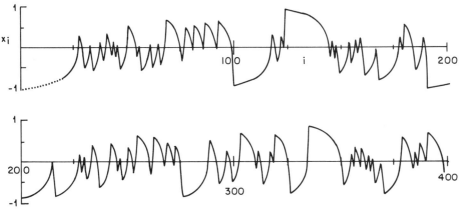

Figure 2.3.7. First 400 iterates of the orbit of $f$. $\beta = 1.99$, $x_0 = 10^{-11}$.

We end our numerical simulations by illustrating the structure of the set of preimages of 0. For simplicity, we replace $f$ by the piecewise linear map

$$f = \begin{cases} +1 + \beta x; & x \in [-1, 0), \\ -1 + \beta x; & x \in (0, 1]. \end{cases} \qquad (2.3.7)$$

The first two preimages of 0 are given by

$$\pm 1 + \beta x = 0,$$

or

$$x = \pm \frac{1}{\beta}; \qquad (2.3.8)$$

the second preimages by

$$\pm 1 + \beta x = \pm \frac{1}{\beta},$$

or

$$x = \pm \frac{1}{\beta} \pm \frac{1}{\beta^2}; \qquad (2.3.9)$$

and in general the $k$th preimages are the $2^k$ points given by

$$\sum_{j=1}^{k} \left( \pm \frac{1}{\beta^j} \right). \qquad (2.3.10)$$

In Figure 2.3.8 we show some orbits starting at such preimages. These orbits should terminate at 0 after $k$ iterates, but the finite accuracy of the computer

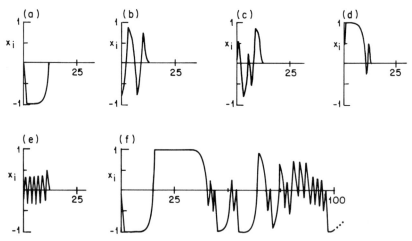

Figure 2.3.8. Orbits of $f$ of equation (2.3.7), $\beta = 2$, started in one of the preimages of 0. (a)–(e) $k = 12$, orbit terminates at $x_{12} = 0$; (f) $k = 15$ ($x_0 = 1/2^{15}$). Numerical errors prevent correct termination at $x_{15} = 0$.

(a Hewlett–Packard HP85) can only cope with preimages up to the 12th or 13th.

For the piecewise linear example, it is immediately clear that the pre-images given by (2.3.10) are dense in the interval $[-1, 1]$. This density of the stable manifold of 0 leads to an interesting conclusion, already hinted at in the sensitive dependence of $f$ on the parameter $\beta$, illustrated in Figures 2.3.6(a), (d). Not only do orbits of a given Lorenz map behave in a chaotic manner, but one-parameter families of such maps also exhibit an un-usual degree of instability themselves. In fact no map of the general type sketched in Figure 2.3.4 can correspond to a structurally stable differential equation.

To illustrate this, we consider the role of the points $\pm a$: the images $f(0\mp)$ of Figure 2.3.4. In the original equation these represent points (near $S$) at which orbits passing arbitrarily close to the saddle at $p = (0, 0, 0)$ next intersect the cross section $\Sigma$; they are in fact the first intersections of the unstable manifold of $p$ with this cross section.

We call these points $b+$ and $b-$ (Figure 2.3.2). We next consider the *preimages* of 0 under $f$, which as we have noted are dense in $I$. Since these preimages represent points lying on orbits of the flow which are asymptotic to the saddle point $p$, in the two-dimensional Poincaré map they correspond to a dense set of curves crossing the section $\Sigma$ transverse to $S$. (The projection process turns each of these curves into a point on $S$.) Whenever $b+$ and $b-$ lie in two such curves, then the unstable manifold of $p$ lies in its stable mani-fold and we consequently have a homoclinic orbit. (The symmetry of the flow implies that two such orbits exist.) Such dim $W^u(p) + $ dim $W^s(p) = 1 + 2 = 3$, this is a nontransverse, structurally unstable intersection and can therefore be destroyed by a small perturbation; cf. Figure 2.3.9. However, the density in $\Sigma$ of points on the stable manifold $W^s(p)$ implies that, if no such connection exists, one can be made by another arbitrarily small perturbation. Hence the sets of systems with and without saddle connections are each dense and neither system is structurally stable. We shall return to this in more detail in Section 5.7; also, for more detailed (and technical!) discussions of the stability of Lorenz attractors, see Guckenheimer and Williams [1979] and Robinson [1981a].

We close this section by noting that, while the qualitative features of the Lorenz attractor described here seem to persist for a wide range of $(\rho, \sigma, \beta)$, as $\rho$ increases for fixed $\sigma$ and $\beta$ the attracting motions become relatively simple once more. Robbins [1979] showed that in the limit $\rho \to \infty$ the system becomes integrable and, using the exact solutions for that case, she was able to demonstrate the existence of a pair of attracting periodic orbits for sufficiently high $\rho$. As $\rho$ decreases (in the range 100–200, for $\sigma = 10$, $\beta = \frac{8}{3}$) successive period doubling bifurcations occur in which the flow becomes progressively more complex until ultimately a strange attractor appears. See the monograph by Sparrow [1982], and the literature cited therein, for more details.

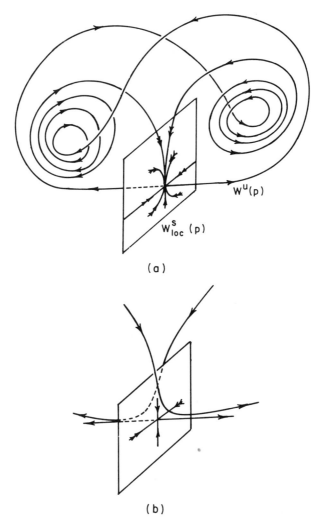

Figure 2.3.9. (a) A Lorenz system with a saddle connection; (b) the connection perturbed.

We shall consider the rôle of successive period doubling bifurcations in the creation of chaotic flows in Chapter 6, where we also discuss some of the bifurcations of the Lorenz equations which occur as $\rho$ is increased between $\rho = 1$ and $\rho_h$.

## 2.4. The Dynamics of a Bouncing Ball

As our final example, we take a two-dimensional mapping which provides a model for repeated impacts of a ball with a massive, sinusoidally vibrating table. We take the usual impact relationship (cf. Meriam [1975])

$$V(t_j) - W(t_j) = -\alpha(U(t_j) - W(t_j)), \qquad (2.4.1)$$

where $U$, $V$, and $W$ are, respectively, the absolute velocities of the approaching ball, the departing ball and the table, $0 < \alpha \leq 1$ is the coefficient of restitution, and $t = t_j$ is the time of the $j$th impact. If we further assume that the distance the ball travels between impacts under the influence of gravity, $g$, is large compared with the overall displacement of the table, then the time interval between impacts is easily approximated as

$$t_{j+1} - t_j = \frac{2V(t_j)}{g}, \tag{2.4.2}$$

and the velocity of approach at the $(j + 1)$st impact as

$$U(t_{j+1}) = -V(t_j). \tag{2.4.3}$$

Combining (2.4.1)–(2.4.3) and nondimensionalizing, we obtain the recurrence relationship relating the state of the system at the $(j + 1)$st impact to that at the $j$th in the form of a nonlinear map. Assuming that the table motion $-\beta \sin \omega t$ is sinusoidal, this can be cast in the form

$$f = f_{\alpha,\gamma}: \quad \begin{matrix} \phi_{j+1} = \phi_j + v_j, \\ v_{j+1} = \alpha v_j - \gamma \cos(\phi_j + v_j), \end{matrix} \tag{2.4.4}$$

where $\phi = \omega t$, $v = 2\omega V/g$ and $\gamma = 2\omega^2(1 + \alpha)\beta/g$.

Here $\gamma$ plays the role of force amplitude and $\alpha$ of dissipation. For more details on both the mechanical and mathematical problems, see Wood and Byrne [1981] and Holmes [1982a].

The bouncing ball problem provides an example of a physical system the analysis of which leads *directly* to a discrete dynamical system, rather than via differential equations and Poincaré maps, as in the preceding examples. However, while in this case we have a simple analytic expression for the map, the analysis of this map turns out to be just as complicated as in the previous cases, although numerical computations are much easier since the map can be iterated directly.

It is easy to verify that (2.4.4) is a smooth invertible map or diffeomorphism, its inverse being given by:

$$f^{-1}: \quad \begin{matrix} \phi_{j-1} = \phi_j - \dfrac{1}{\alpha}(\gamma \cos \phi_j + v_j), \\[2mm] v_{j-1} = \dfrac{1}{\alpha}(\gamma \cos \phi_j + v_j). \end{matrix} \tag{2.4.5}$$

Moreover, the determinant of the Jacobian matrix

$$Df = \begin{bmatrix} 1 & 1 \\ \gamma \sin(\phi_j + v_j) & \alpha + \gamma \sin(\phi_j + v_j) \end{bmatrix} \tag{2.4.6}$$

is constant $(=\alpha)$ and thus for $\alpha < 1$ the map contracts areas uniformly while in the perfectly elastic case, $\alpha = 1$, it is area preserving. The latter (Hamiltonian) case has been widely studied, mostly by physicists, on account of its relevance to certain problems in particle physics: see Chirikov [1979], Greene [1980], and Lichtenberg and Lieberman [1982], for example. In their work a slightly different coordinate system is chosen and the diffeomorphism is referred to as "the standard map." Moreover, Pustylinikov [1978] considered the mechanical problem outlined above and derived an exact mapping for the case of general periodic excitation and showed that both it and the approximate mapping (2.4.4) have open sets of initial conditions $(\phi_0, v_0)$ such that $v_n \to \infty$ as $n \to \infty$, for suitable finite values of $\gamma$ and $\alpha = 1$. Along with these unbounded motions, sets of bounded periodic motions are also found and numerical computations suggest that bounded nonperiodic motions also exist. In this section we first discuss some of the many families of periodic orbits exhibited by $f$ and then demonstrate the existence of a complicated set, the Smale horseshoe (Smale [1963, 1967]), which is to be studied in detail in Chapter 5. As in the first three sections of this chapter, we end with some numerical computations which illustrate typical behaviors of the map.

We first note that, if $\alpha < 1$, in distinction to the Hamiltonian case considered by Pustylnikov, all orbits remain bounded. From (2.4.4) we have

$$|v_{j+1}| = |\alpha v_j - \gamma \cos(\phi_j + v_j)| \le \alpha |v_j| + \gamma, \qquad (2.4.7)$$

and thus, if $|v_j| > \gamma/(1 - \alpha)$, it follows that $|v_{j+1}| < |v_j|$. Hence all orbits enter and remain within a strip bounded by $v_j = \pm \gamma/(1 - \alpha)$. This provides an example of a trapping region for a discrete dynamical system.

A second observation is important. Equation (2.4.4) is invariant under the coordinate change $\phi \to \phi + 2n\pi$, $n = \pm 1, \pm 2, \ldots$, indicating that we can take the $(\phi, v)$ phase space to be the cylinder $S^1 \times \mathbb{R}$, obtained by taking $\phi$ modulo $2\pi$. In applying our results to the mechanical problem, however, we must recall that the nondimensional time of flight, $\phi_{j+1} - \phi_j$, is only given modulo $2\pi$; cf. Holmes [1982a].

The trapping region is now a compact subset

$$D = \left\{ (\phi, v) \,\middle|\, |v| \le \varepsilon + \frac{\gamma}{1 - \alpha} \right\} \subset S^1 \times \mathbb{R},$$

and, as in the previous example, we have an attracting set

$$A = \bigcap_{n \ge 0} f^n(D).$$

(We include the $\varepsilon$ so that $D$ can be made into a closed set.)

In seeking fixed points of $f$, points $(\bar{\phi}, \bar{v})$ such that $f(\bar{\phi}, \bar{v}) = (\bar{\phi}, \bar{v})$, we use the periodicity to obtain pairs of points

$$(\bar{\phi}_n, \bar{v}_n) = \left( \arccos\left( \frac{2n\pi(\alpha - 1)}{\gamma} \right), 2n\pi \right), \qquad n = 0, \pm 1, \pm 2, \ldots, \pm N;$$

$$(2.4.8)$$

where $N$ is the greatest integer such that

$$2N\pi(1 - \alpha) < \gamma. \tag{2.4.9}$$

The stability of these fixed points is determined by the linearized map, $Df$, of equation (2.4.6). As we showed in Chapter 1, if both eigenvalues are inside the unit circle in the complex plane ($|\lambda_i| < 1$), we have a sink, if one lies outside and one inside ($|\lambda_1| < 1 < |\lambda_2|$), a saddle; and if both lie outside, a source. We note that, since $\lambda_1 \cdot \lambda_2 = \det(Df) = \alpha$, only sinks and saddles are obtained for $\alpha < 1$ (if $\alpha = 1$ we find centers and saddles). From (2.4.6) the eigenvalues are given by

$$\lambda_{1,2} = \tfrac{1}{2}\{(1 + \alpha + \gamma r) \pm \sqrt{(1 + \alpha + \gamma r)^2 - 4\alpha}\}; \qquad r = \sin(\bar{\phi}_n + \bar{v}_n). \tag{2.4.10}$$

Inserting $(\bar{\phi}_n, \bar{v}_n)$ from (2.4.8), we find that the fixed points with $\bar{\phi}_n < \pi$ ($\sin(\bar{\phi}_n + \bar{v}_n) > 0$) are all saddle points. Those with $\bar{\phi}_n > \pi$ ($\sin(\bar{\phi}_n + \bar{v}_n) < 0$) are sinks (centers) if

$$2n\pi(1 - \alpha) < \gamma < 2\sqrt{n^2\pi^2(1 - \alpha)^2 + (1 + \alpha)^2}, \tag{2.4.11}$$

and saddle points if

$$\gamma > 2\sqrt{n^2\pi^2(1 - \alpha)^2 + (1 + \alpha)^2}. \tag{2.4.12}$$

We remark that the saddles at $\bar{\phi}_n < \pi$ are *saddles of the first kind*, the associated linear maps having positive eigenvalues $0 < \lambda_1 < 1 < \lambda_2$; while the saddles at $\bar{\phi}_n > \pi$ are *saddles of the second kind*, with negative eigenvalues $\lambda_1 < -1 < \lambda_2 < 0$. Orbits approaching and leaving saddles of the latter kind do so in oscillatory manner and such fixed points are also called *reflection hyperbolic* (cf. Bernoussou [1977]). The points

$$\gamma_n = 2n\pi(1 - \alpha), \qquad \gamma_n' = 2\sqrt{n^2\pi^2(1 - \alpha)^2 + (1 + \alpha)^2} \tag{2.4.13}$$

are *bifurcation* values, at the first of which a pair of fixed points appears in a saddle-node bifurcation and at the second of which a change of stability and a period doubling or flip bifurcation occurs. Such local bifurcations of maps are studied in Chapter 3. Using the techniques of that chapter, or by direct calculation, one can show that, for $\gamma > \gamma_n'$, a periodic orbit of period 2 exists near $(\bar{\phi}_n > \pi, \bar{v}_n)$. For $\gamma$ near $\gamma_n'$ this orbit is stable, and it ultimately undergoes a second flip bifurcation, yielding an orbit of period 4. This process continues to an accumulation point $\gamma_n^{\infty}$, at which orbits of period $2^k$ for all $k$ exist. As we mentioned in Section 2.2, such countable sequences of flip bifurcations were first studied for one-dimensional maps, in which context they are now quite well understood (cf. Feigenbaum [1978], Collet and Eckmann [1980] and Section 6.8, below).

In Figure 2.4.1 we show a bifurcation diagram for the first six (including $n = 0$) of these families of fixed points. The stable period 2 motions are indicated also, and we append some sketches of physical motions of the bouncing ball corresponding to some of these orbits.

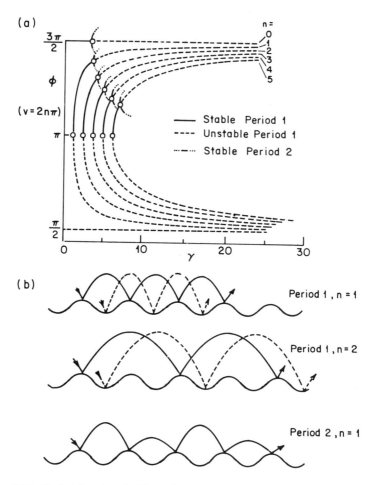

Figure 2.4.1. Period 1 and period 2 motions, $\alpha = 0.9$. The period 2 motions appear in flip bifurcations at $\gamma = \gamma'_n$. (a) Bifurcation diagram; (b) physical motions.

Rather than going on to discuss orbits of successively higher period, the derivation of which rapidly becomes impractical, we shall sketch an argument which, demonstrates, in one fell swoop, the existence of an infinite family of such orbits, as well as families of bounded nonperiodic motions. Before doing this, we note that experimental work mentioned by Wood and Byrne [1981] indicated that, while for low table velocities ($\gamma$ small) stable periodic bouncing is observed, as $\gamma$ is increased, increasingly irregular, apparently chaotic motions are seen. In the remainder of this section, and in Chapter 5, we shall show that such motions do indeed exist for the map (2.4.4), by demonstrating the existence of a Smale horseshoe (Smale [1963, 1967]): a complicated invariant set containing the infinite families of periodic and nonperiodic orbits mentioned above.

Since the set we seek is global in nature, we consider the global action of the map $f$ in that we select a closed region $Q \subset S^1 \times \mathbb{R}$ and consider its image $f(Q)$. For simplicity we take the area preserving case, $\alpha = 1$; our results will also apply to $\alpha$ sufficiently close to 1. We define $Q$ to be the parallelogram $ABCD$ bounded by the lines $\phi + v = 0 \, (AB), \phi + v = 2\pi \, (CD)$, $\phi = 0 \, (AD), \phi = 2\pi \, (BC)$. We note that $Q$ can be foliated by the family of lines $\phi + v = k$, $k \in [0, 2\pi]$ and that the images of such lines under $f$ are vertical lines $\phi = k$, $v \in [k - 2\pi - \gamma \cos k, k - \gamma \cos k]$. Finally, the images of the boundaries $\phi = 0$ and $\phi = 2\pi$ are curves $v = \phi - \gamma \cos \phi$, $v = \phi - 2\pi - \gamma \cos \phi$. Since $\alpha = 1$ and we are taking $\phi$ modulo $2\pi$, the rectangle $Q$ can be replaced by any vertical $2\pi$-translate; in Figure 2.4.2 we show the rectangle bounded by $\phi + v = 0$, $\phi + v = 2\pi$, $\phi = 0$, and $\phi = 2\pi$, together

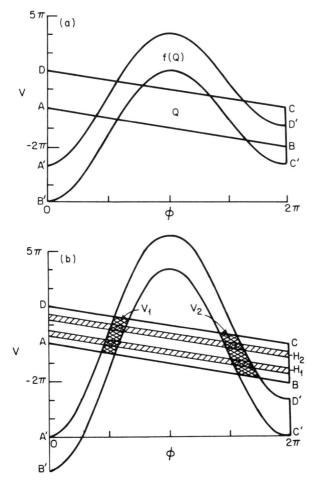

Figure 2.4.2. The creation of horseshoes as $\gamma$ increases, for the area preserving case, $\alpha = 1$. The points $A, B, C, D$ are mapped to $A', B', C', D'$. (a) $\gamma = 3\pi$; (b) $\gamma = 5\pi$.

with its images under $f_{1,\gamma}$ for $\gamma = 3\pi$ and $\gamma = 5\pi$. When $\alpha \neq 1$, images of such translated strips, will, of course, differ due to the term $\alpha v$ in equation (2.4.4); cf. Figure 2.4.3.

If we pick $\gamma$ sufficiently large, then the image $f(Q)$ intersects $Q$ in two disjoint "vertical" strips, $V_1$, $V_2$, shaded in Figures 2.4.2(b) and 2.4.3(c). The choice $\gamma > 4\pi$ is sufficient to ensure this in the area preserving case. The reader should not have too much difficulty in convincing himself that the preimages $f^{-1}(V_i)$ of these strips are a pair of disjoint "horizontal" strips $H_1$, $H_2$, connecting the vertical edges $AD$, $BC$ of $Q$ as indicated in Figure 2.4.2. Thus, schematically, we have the qualitative behavior sketched in Figure 2.4.4, in which $f$ takes the rectangle $Q$, stretches it in the vertical direction, compresses it in the horizontal direction, bends it, and replaces it intersecting $Q$ as shown. This is Smale's *horseshoe* map, the name being by now self-evident. In subsequent chapters we show that the horseshoe occurs in connection with the transversal intersections of manifolds in the Duffing and van der Pol equations. Indeed, the presence of horseshoes is essentially synonymous with what most authors appear to mean by the term "chaotic dynamics." A proper understanding of horseshoes is essential to the understanding of complex dynamics.

We shall study the horseshoe, and generalizations of it, in detail in Chapter 5. Here we merely note that, if $f$ is iterated once more, then the image $f(V_i)$ of each of the vertical strips is itself a horseshoe shaped region intersecting the original rectangle $Q$ in two thinner vertical strips. Thus the set

$$\Lambda_v^2 = Q \cap f(Q) \cap f^2(Q)$$

consists of four disjoint vertical strips. Similarly, taking inverse iterates, we find that the set

$$\Lambda_h^2 = Q \cap f^{-1}(Q) \cap f^{-2}(Q)$$

consists of four disjoint horizontal strips. In general the sets

$$\Lambda_v^n = \bigcap_{k=0}^{n} f^k(Q), \qquad \Lambda_h^n = \bigcap_{k=0}^{n} f^{-k}(Q)$$

each consist of $2^n$ vertical (or horizontal) strips. If the expansion and contraction rates are chosen uniformly, then as we show in Chapter 5, the limiting sets

$$\Lambda_v^\infty = \bigcap_{k=0}^{\infty} f^k(Q), \qquad \Lambda_h^\infty = \bigcap_{k=0}^{\infty} f^{-k}(Q)$$

each consist of an uncountable collection of lines, in fact they are each the products of a Cantor set and an interval.

Now the set $\Lambda = \Lambda_v^\infty \cap \Lambda_h^\infty = \bigcap_{k=-\infty}^{\infty} f^k(Q)$ is precisely the set of points $p \in Q$ which remain in $Q$ under all forward and backward iterates of $f$, It is thus the largest *invariant set* of $f$ contained in $Q$ (note that we have not specified clearly what happens to points which start outside $Q$ or leave $Q$, such points

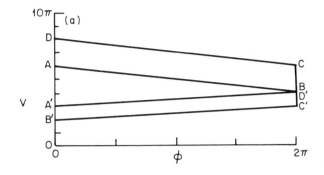

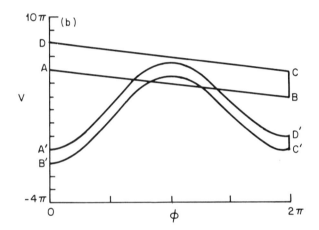

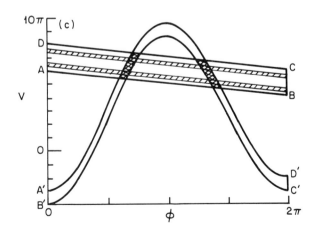

Figure 2.4.3. The creation of a horseshoe, $\alpha = 0.5$, $n = 3$. (a) $\gamma = 0$; (b) $\gamma = 3\pi = \gamma_n$; (c) $\gamma = 6\pi > \gamma_n^h$.

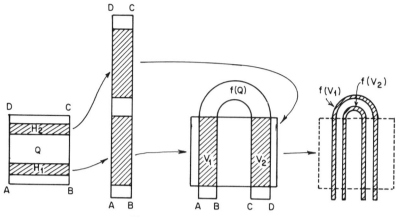

Figure 2.4.4. The horseshoe.

may reenter $Q$, but if so, we do not include them in $\Lambda$). Thus $\Lambda$ is the product of two Cantor sets (the horizontal and vertical lines of $\Lambda_v^\infty$ and $\Lambda_h^\infty$ intersect transversely) and hence is itself a *Cantor set*: a perfect, totally disconnected, closed set (cf. Hocking and Young [1961, p. 106] or Willard [1970] for a discussion of Cantor sets. Also see Chapters 5 and 6 below).

In Chapter 5 we use the method of symbolic dynamics to prove that $\Lambda$ contains the infinite families of periodic orbits referred to above, and we also derive various conditions which, if satisfied, guarantee the existence and stability of such a set in specific maps. These conditions can be checked for (2.4.4) and are found to hold for $\alpha = 1$, $\gamma > 5\pi$. If $\alpha$ is reduced, then $\gamma$ may have to be increased (cf. Figures 2.4.2–2.4.3). The relevant calculations will be sketched in Chapter 5, so we do not consider them here. In fact we are able to prove (cf. Holmes [1982a]).

**Proposition 2.4.1.** *The invariant set $\Lambda$ contains:*
(a) *a countable set of periodic orbits of all periods;*
(b) *an uncountable set of nonperiodic motions;*
(c) *a dense orbit; and*
(d) *the periodic orbits are all of saddle type and they are dense in $\Lambda$.*

*Moreover, $f|\Lambda$ is structurally stable.*

While the set $\Lambda$ is extremely complicated and contains an uncountable infinity of nonperiodic or chaotic orbits, it is not an attractor. It can, however, exert a dramatic influence on the behavior of typical orbits which pass close to it since the stable manifold $W^s(\Lambda)$, or set of orbits $\{f^n(q)\}$ asymptotic to $\Lambda$ as $n \to \infty$, behaves like an uncountable set of saddle separatrices. (In fact locally this stable manifold is just $\Lambda_h^\infty$: the product of an interval and a Cantor set.) One therefore expects orbits passing near $\Lambda$ to display an extremely sensitive dependence upon initial conditions, and exhibit a transient period of chaos before perhaps settling down to a stable periodic orbit, rather as in the Duffing equation for certain parameter ranges. Recall that such attracting orbits can coexist with the horseshoe since we have only considered the image under $f$ of a specially chosen domain $Q$ of the state space.

As in our study of the Duffing equation, we now consider the behavior of the map $f_{\alpha, \gamma}$ as the force, $\gamma$, is increased for fixed $\alpha < 1$. Specifically, we fix a region $Q$ bounded by the lines $\phi = 0, \phi = 2\pi, \phi + v = 2n\pi, \phi + v = (2n + 2)\pi$ and chosen such that $f_{\alpha, 0}(Q)$ lies entirely below $Q$; Figure 2.4.3(a). As $\gamma$ is increased, the center of the image of $Q$ rises until the image of the horizontal line $AC$ ($v = 2n\pi$), given by $v = 2n\pi\alpha - \gamma \cos \phi$, just touches $AC$ at the point $(\phi, v) = (\pi, 2n\pi)$; Figure 2.4.3(b). This occurs when $\gamma = \gamma_n = 2n\pi(1 - \alpha)$ and is, of course, the saddle-node bifurcation point for the pair of fixed points lying on $v = 2n\pi$, cf. equation (2.4.13). We already know that the sink bifurcates to a sink of period two at $\gamma'_n = 2\sqrt{n^2\pi^2(1 - \alpha)^2 + (1 + \alpha)^2}$. We can now conclude that an infinite sequence of further bifurcations must occur between $\gamma'_n$ and $\gamma^h_n$, the critical value at which the horseshoe is created (Figure 2.4.3(c)), since for $\gamma \geq \gamma^h_n$ there is a countable infinity of periodic orbits, including orbits of arbitrarily long period, in $Q$.

Work of Newhouse [1974, 1979, 1980] and Gavrilov and Silnikov [1972, 1973] indicates that, while the horseshoe is in the process of creation ($\gamma_n < \gamma < \gamma^h_n$) and the image $f_{\alpha, \gamma}(Q)$ does not yet intersect $Q$ in two disjoint strips, an infinite number of families of *stable* periodic orbits are created in saddle-node bifurcations and subsequently double their periods repeatedly in flip bifurcations as $\gamma$ increases. Thus, stable orbits with periods longer than any preassigned period can be found; such orbits are indistinguishable in practice from the bounded nonperiodic motions of the horseshoe, but their stability renders them observable. Newhouse [1979] has suggested that such orbits may constitute the hypothetical "strange attractor" observed by Hénon [1976] and many others in numerical iterations of two-dimensional maps. We will discuss these questions in more detail in Chapter 6.

We conclude once more with numerical observations, revealing some of the complex orbits exhibited by $f_{\alpha, \gamma}$. In Figure 2.4.5 we show a sequence of orbits for $\alpha = 0.8$ (fixed) and several values of $\gamma$. The stable sink of (a) has bifurcated to an orbit of period two in (b) and successively more complex orbits are seen in (c) and (d). In each case the initial conditions were chosen to be near $v = 2\pi$, and we note that, while the orbit remains close to the $n = 1$ band for low values of $\gamma$: (a), (b); for higher values it leaves the band and wanders erratically over a bounded subset of the state cylinder before settling down to a stable orbit (c), or apparently continues to wander (d). In (c) 500 iterates were required before the asymptotic behavior became clear; in (d) 5000 iterates were computed. (We note that the orbits of (c) and (d) do not correspond to physical motions of the bouncing ball, since the departure velocity $v$ becomes negative. This is a deficiency of our simple model resulting from the assumption (2.4.2).)

In Figure 2.4.6 we illustrate the condition necessary (but not sufficient) for this wandering to occur. We show partial boundaries of the domain of

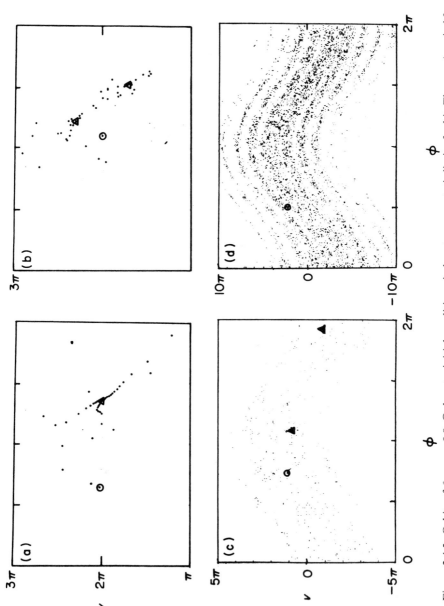

Figure 2.4.5. Orbits of $f_{\alpha,\gamma}$, $\alpha = 0.8$. $\odot$ denotes initial condition, $\triangle$ denotes asymptotic limit of orbit. The $n = 1$ period 2 motion appears for $\gamma \approx 3.813022$. (a) $\gamma = 3$; (b) $\gamma = 4$; (c) $\gamma = 5$; (d) $\gamma = 10$: no periodic limit apparent after 5000 iterates.

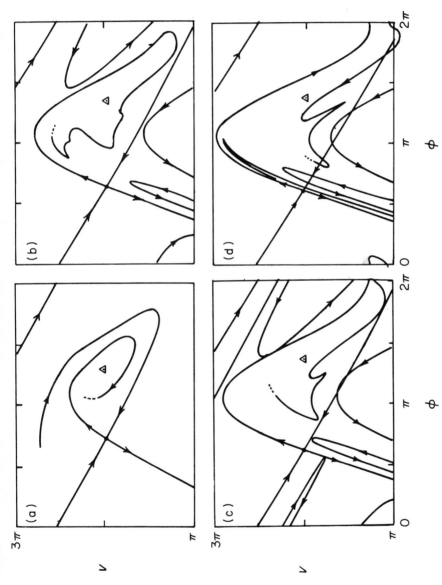

Figure 2.4.6. Stable and unstable manifold of the $n = 1$ saddle point. The $n = 1$ sink is denoted by $\triangle$, $\alpha = 0.8$. (a) $\gamma = 2$; (b) $\gamma = 3$; (c) $\gamma = 3.28$: first tangency; (d) $\gamma = 3.5$: transverse homoclinic orbits.

attraction of the period one sink in the $n = 1$ band formed by the stable manifolds of the associated saddle point. Finite segments of these manifolds are easily computed by iterating a short interval of the stable eigenvector of the linearized map (2.4.6) containing the saddle point, under the inverse map (2.4.5). The unstable manifold may similarly be found by iterating an interval of the unstable eigenvector under the map (2.4.4). When transverse intersections of stable and unstable manifolds exist, as in Figure 2.4.6(d), it is very difficult to predict the asymptotic behavior of an orbit unless the initial conditions are known extremely accurately, since any two orbits starting on different sides of the stable manifold will ultimately separate exponentially fast. The violent winding of this manifold implies that the domains of attraction have complicated boundaries with infinitely many long thin "tongues" penetrating close to other attracting orbits. Once more we encounter the sensitive dependence upon initial conditions, even when simple attractors, such as periodic orbits, ultimately capture almost all solutions.

However, in the present problem, as in the Duffing and Lorenz examples, there *appear* to be large sets of parameter values for which orbits are never asymptotic to periodic attractors. In Figure 2.4.7(a) we illustrate such an orbit, which had not displayed any recognizable asymptotic behavior after 60,000 iterates. Note, however, that the orbit does exhibit a striking global structure in that it appears to fall on a set of curves as in the Duffing example (cf. Figure 2.2.8). Magnification of regions of the phase space suggest that, as in Hénon's work [1976], this set is not finite (Figure 2.4.7(b)), but is locally the product of a smooth curve and a Cantor set. This is precisely the structure of the closure of the unstable manifolds of the horseshoes. (Locally, this is just the set $\Lambda_v^\infty \subset Q$.) In Figure 2.4.7(c) we show a portion of the unstable manifold of the saddle point at $(\phi, v) = (\pi/2, 0)$. Comparing this with Figure 2.4.7(a) it is clear that the orbit seems to approach or to lie on this manifold. In fact, if the bounded set $D = \{\phi, v \,|\, |v| < 8\pi\}$ is taken in this case ($\alpha = 0.5, \gamma = 10$) it can be shown from equation (2.4.7) that $f_{\alpha, \gamma}(D)$ is contained in $D$, and the attracting set $A$ may be defined as the intersection of all forward images of $D$,

$$A = \bigcap_{n=0}^{\infty} f_{\alpha, \gamma}^n(D), \qquad (2.4.14)$$

as in the Duffing example. Once more $A$ appears to be the closure of an unstable manifold; cf. Figure 2.4.7(c).

We note that this attracting set coexists with a stable orbit of period 2 in the $n = 3$ band near the fixed point at $(\bar{\phi}, \bar{v}) = (\arccos((6\pi(\alpha - 1))/\gamma), 6\pi)$; Figure 2.4.7(a). The domain of attraction of this motion winds back and forth in the gaps in the other attracting set. In much the same way, the Duffing equation can exhibit two distinct attractors for the same parameter values (Figure 2.2.5(a)).

These observations, and the analysis sketched here and given in more detail in subsequent chapters, implies that in the physical problem we can

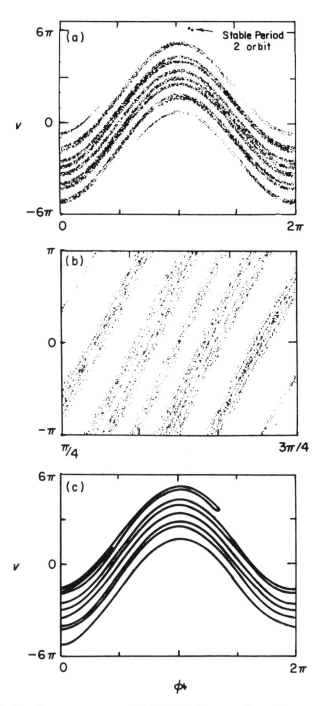

Figure 2.4.7. The "strange attractor." (a) 60,000 iterates of a point near $(\pi/2, 0)$; (b) enlargement of part of (a); (c) the unstable manifold of $(\pi/2, 0)$.

expect to see sustained, bounded, chaotic motions of the ball, which have their mathematical analogue in bounded nonperiodic orbits which remain above $v = 0$. Such motions have in fact been observed experimentally, as mentioned by Wood and Byrne [1981].

## 2.5. Conclusions: The Moral of the Tales

The sketches of the analyses and numerical simulations of the four model problems presented in this chapter suggest several things, the most striking of which is that "simple" differential equations of dimension three or greater can possess solutions of stunning complexity. Moreover, since such systems, in the form of forced oscillators or autonomous evolution equations, play an important rôle in the modelling of nonlinear processes, an understanding of typical structures of their solutions is essential. In this book we take the viewpoint that such an understanding is best achieved from a geometric or topological viewpoint. Reams of computer simulations, without some form of explanation and analysis, are not very helpful.

In the following chapters we develop a number of analytical tools which will enable us to more fully understand bifurcations, chaotic motions, and strange attractors such as those introduced in this chapter. In some cases, these "new" methods will really be conventional perturbation style analyses interpreted geometrically, but a large part of our analysis, including the concepts of hyperbolicity and the methods of symbolic dynamics, will probably be less familiar to the reader. We will continually return to these four examples to illustrate our new analytical tools, as they are introduced, and to fill in some of the many gaps in the analyses sketched here.

# CHAPTER 3

# Local Bifurcations

In this chapter, we study the local bifurcations of vector fields and maps. As we have seen, systems of physical interest typically have parameters which appear in the defining systems of equations. As these parameters are varied, changes may occur in the qualitative structure of the solutions for certain parameter values. These changes are called *bifurcations* and the parameter values are called *bifurcation values*. To the extent possible, we develop in this chapter and Chapters 6 and 7, a systematic theory which describes and permits the analysis of the typical bifurcations one encounters. We pay careful attention to the examples introduced in Chapter 2 and use these to illustrate the theory that we present.

There are evident limitations as to how far one can proceed with a systematic bifurcation theory. In parameter regions consisting of structurally unstable systems, such as those encountered in the Lorenz system, the detailed changes in the topological equivalence class of a flow can be exceedingly complicated. Many important aspects of this situation are poorly understood and lack the satisfying completeness of the structural stability theory for second-order systems. In this chapter, therefore, we shall focus upon the simplest bifurcations of individual equilibria and periodic orbits—a part of the theory which is relatively complete. Since the analysis of such bifurcations is generally performed by studying the vector field near the degenerate (bifurcating) equilibrium point or closed orbit, and bifurcating solutions are also found in a neighborhood of that limit set, these bifurcations are referred to as *local*. *Global* bifurcations, especially those which are characterized by a lack of transversality between the stable and unstable manifolds of periodic orbits and equilibria, will be discussed in Chapter 6. In Chapter 7 we consider problems in which the bifurcations are degenerate unless considered in the context of systems with two or more parameters. It turns out that even a study of local two-parameter bifurcations requires an understanding of global bifurcations, since they occur naturally in two-parameter families.

We start by considering some simple examples of bifurcations of fixed points of flows in one and two dimensions and go on to develop the general theory for dealing with bifurcations of fixed points of $n$-dimensional flows. The principal components of this theory are the center manifold and normal form theorems. At the end of the chapter we turn our attention to local bifurcations of maps and develop an analogous theory for them.

## 3.1. Bifurcation Problems

The term bifurcation was originally used by Poincaré to describe the "splitting" of equilibrium solutions in a family of differential equations. If

$$\dot{x} = f_\mu(x); \qquad x \in \mathbb{R}^n, \quad \mu \in \mathbb{R}^k \qquad (3.1.1)$$

is a system of differential equations depending on the $k$-dimensional parameter $\mu$, then the equilibrium solutions of (3.1.1) are given by the solutions of the equation $f_\mu(x) = 0$. As $\mu$ varies, the implicit function theorem implies that these equilibria are described by smooth functions of $\mu$ away from those points at which the Jacobian derivative of $f_\mu(x)$ with respect to $x$, $D_x f_\mu$ has a zero eigenvalue.* The graph of each of these functions is a *branch* of equilibria of (3.1.1). At an equilibrium $(x_0, \mu_0)$ where $D_x f_\mu$, has a zero eigenvalue, several branches of equilibria may come together, and one says that $(x_0, \mu_0)$ is a point of *bifurcation*.

As an example, consider equation (3.1.1) with $f_\mu(x) = \mu x - x^3$. Here $D_x f_\mu = \mu - 3x^2$, and the only bifurcation point is $(x, \mu) = (0, 0)$. It is easy to check that the unique fixed point $x = 0$ existing for $\mu \le 0$ is stable, that it becomes unstable for $\mu > 0$, and that the new bifurcating fixed points at $x = \pm\sqrt{\mu}$ are stable. We obtain the qualitative picture of Figure 3.1.1, in which the branches of equilibria are shown in $(x, \mu)$ space. This figure is an example of a *bifurcation diagram*.

EXERCISE 3.1.1. Investigate the bifurcations of equilibria for the system $\dot{x} = f_\mu(x)$, as $\mu$ varies near $\mu = 0$, with:

(a) $f_\mu(x) = \mu - x^2$;
(b) $f_\mu(x) = \mu x - x^2$;
(c) $f_\mu(x) = \mu^2 x - x^3$;
(d) $f_\mu(x) = \mu^2 x + x^3$;
(e) $f_\mu(x) = \mu^2 \alpha x + 2\mu x^3 - x^5$, for various $\alpha$.

---

* When the meaning is clear, we shall sometimes write $D_x f_\mu$ as $Df_\mu$ or simply $Df$, as in Chapters 1 and 2.

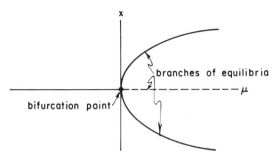

Figure 3.1.1. The bifurcation diagram for $f_\mu(x) = \mu x - x^3$. --- sources; —— sinks.

In each case find the nontrivial fixed points and investigate their stability. Also, sketch the bifurcation diagrams. Some of these bifurcations are more degenerate than others, in that small perturbations to $f_\mu$ can change the topological structure of their bifurcation diagrams. Can you identify these, and sketch some perturbed bifurcation diagrams?

Bifurcations of equilibria usually produce changes in the topological type of a flow, but there are many other kinds of changes that occur in the topological equivalence class of flows. We shall include all of these in our use of the term bifurcation:

**Definition 3.1.1.** A value $\mu_0$ of equation (3.1.1) for which the flow of (3.1.1) is not structurally stable is a *bifurcation value* of $\mu$.

This definition is not completely satisfactory because it impels one to study the finely detailed structure of flows for which complete descriptions do not exist. Consequently, attempts to construct a systematic bifurcation theory lead to very difficult technical questions, not all of which have relevance for applications of the theory. To avoid such complications, we frequently loosen the definition given above and examine only *some* of the qualitative features of a system of differential equations. We do not, however, retreat to the "static" problem of dealing only with the bifurcations of equilibria (cf. Sattinger [1973]).

Another peculiarity of this definition is that a point of bifurcation need not actually represent a change in the topological equivalence class of a flow. For example, the system $\dot{x} = -(\mu^2 x + x^3)$ has a bifurcation value $\mu = 0$, but all of the flows in this family have a globally attracting equilibrium at $x = 0$. However, arbitrary perturbations (unfoldings) do give topologically distinct flows (cf. Exercise 3.1.1).

Given a system (3.1.1), we want to draw its *bifurcation set*. This consists of the loci in $\mu$-space which correspond to systems for which structural stability breaks down in specific ways which we classify to the extent that we are able to do so. We also sometimes find it convenient to draw *bifurcation diagrams*: the loci in the $(x, \mu)$ product space of (parts of) the invariant set of (3.1.1). These invariant sets need not merely be fixed points, as in Figure 3.1.1; periodic orbits, for example, are often represented in terms of some measure ($|x|$) of their amplitude.

EXERCISE 3.1.2. Investigate the stability of the fixed point $(x, y) = (0, 0)$ for the system

$$\dot{x} = \mu_1 x - y - \mu_2 x(x^2 + y^2) - x(x^2 + y^2)^2,$$
$$\dot{y} = x + \mu_1 y - \mu_2 y(x^2 + y^2) - y(x^2 + y^2)^2,$$

and find any bifurcating families of periodic orbits that may exist (cf. Exercise 1.5.2). Draw a bifurcation diagram (a graph over $(\mu_1, \mu_2)$ space) indicating the fixed points and periodic orbits.

What is of particular interest is that there are identifiable kinds of bifurcations which appear repeatedly in many problems. Ideally, we would like to have a classification of bifurcations which produced a specific list of possibilities for each example, starting with only *general* considerations such as the number of parameters in the problem, the dimension of the phase space, and any symmetries or other special properties of the system (e.g., the forced Duffing equation and the Lorenz systems are volume contracting: this excludes a lot of types of behavior in these systems). Parts of such a classification have been developed, and we shall give a fairly complete survey of what is known in this chapter and Chapters 6 and 7.

The classification schemes are based upon concepts which have their origin in the theory of *transversality* in differential topology. The *transversality theorem* implies that when two manifolds (surfaces) of dimensions $k$ and $l$ meet in an $n$-dimensional space, then, in general, their intersection will be a manifold of dimension $(k + l - n)$. If $k + l < n$ then one does not expect intersections to occur at all. For example, two-dimensional surfaces in 3-space generally intersect along curves, while two curves in 3-space generally do not intersect. The meaning of *in general* is given in terms of function space topologies for the space of embeddings of $l$-dimensional manifolds in $n$-space. We only remark here that non-transversal intersections can be perturbed to transversal ones, but transversal intersections retain their topology under perturbation. A *general position* or *transversal intersection* of manifolds in $n$-dimensional space is one for which the tangent spaces of the intersecting manifolds span $n$-space. The dimension formula can be expressed also in terms of *codimension*. The *codimension* of an $l$-dimensional submanifold of $n$-space is $(n - l)$. Then the intersection of two submanifolds $\Sigma_1, \Sigma_2$ generally satisfies $(n - l) + (n - k) = 2n - (l + k) = n - (l + k - n)$. Therefore the codimension of $\Sigma_1 \cap \Sigma_2$ is the sum of the codimensions of $\Sigma_1$ and $\Sigma_2$ if the intersection is transversal.

As an example, consider two curves in the plane, one of which is the $x$-axis, the other being the graph of a function $f$. The two curves intersect transversally at a point $x$ if $f(x) = 0$ (the intersection condition) *and* $f'(x) \neq 0$ (transversality). We say that a transversal intersection of the curves is a *simple zero.* If $f$ has only simple zeros, then small perturbations of $f$ have the same number of zeros as $f$. In a family $f_\mu$ the simple zeros vary as smooth functions of $\mu$. (This statement is just the Implicit Function Theorem.) Nonsimple zeros do not have these properties. For example, the family $f_\mu(x) = \mu + x^2$ has a nonsimple zero at $(x, \mu) = (0, 0)$. For $\mu > 0$, the functions $f_\mu$ have no zeros at all. Note, however, that if one regards $f_\mu(x) = F(x, \mu)$ as function of *two* variables, then its graph intersects the $(x, \mu)$ coordinate plane transversally along the curve $\mu + x^2 = 0$. Thus, while the bifurcation point $(x, \mu) = (0, 0)$ corresponds to an unstable system, the bifurcation diagram corresponding to the *family of systems* is stable to small perturbations. We note that Iooss and Joseph [1981] make a strong distinction between such "turning point" or "fold" bifurcations and branching bifurcations such as that of Figure 3.1.1.

EXERCISE 3.1.3. Verify the statements in the paragraph above.

EXERCISE 3.1.4. Show that the graph of $f_\mu(x) = F(x, \mu) = x^3 - x\mu$ does not intersect the $(x, \mu)$ plane transversally everywhere. What do you think the pitchfork bifurcation diagram of Figure 3.1.1 might perturb to? (Try adding terms $v$ and $vx^2$ to $f_\mu(x)$.)

Transversality is employed in bifurcation theory in the following way. We want to study the bifurcations that occur *in general* in $k$-parameter families (3.1.1) (perhaps among a class of vector fields with symmetries or other special conditions). We do this by formulating a collection of transversality conditions (inequalities) that are met by most families at a bifurcation value $\mu_0$. At $\mu_0$, some of the conditions for structural stability will be violated and these determine the *type* of bifurcation which occurs. Let us illustrate with an example.

Consider a two-parameter system of the form (3.1.1) with a bifurcation value $\mu_0$ at which $f_\mu$ has a nonhyperbolic equilibrium $p$. We can study the linearization of $f_\mu$ at $p$ and the way the vector field $f_\mu$ changes for $\mu$ near $\mu_0$. Using transversality, we expect that the set of equilibria of (3.1.1) in $(x, \mu)$ space will form a smooth two-dimensional surface $\mathcal{M}$.

EXERCISE 3.1.5. Verify this statement.

If we examine linearizations of $f_\mu$ at the equilibria in $\mathcal{M}$, then we can formulate a transversality condition which guarantees, for example, that no linearization of $f_\mu$ has a zero eigenvalue of multiplicity greater than two and that any equilibrium which does have a zero eigenvalue of multiplicity two has a Jordan normal form with the block $\begin{pmatrix} 0 & 1 \\ 0 & 0 \end{pmatrix}$. To state this transversality condition, one defines the map of $\mathcal{M}$ into the space of square matrices which associates to $(x, \mu) \in \mathcal{M}$ the Jacobian derivative $D_x f_\mu$ at $(x, \mu)$. In the space of square matrices, there are submanifolds which correspond to various combinations of eigenvalues on the imaginary axis.

Since $\mathcal{M}$ has dimension two its image under the above map will generally meet only those submanifolds of matrices whose codimension is at most two. The set of matrices with a Jordan form having just one block $\begin{pmatrix} 0 & 1 \\ 0 & 0 \end{pmatrix}$ forms such a submanifold of codimension two.

EXERCISE 3.1.6. In the space of $2 \times 2$ matrices, find explicitly the sets of matrices with:

(1) a single zero eigenvalue;
(2) a pair of pure imaginary eigenvalues;
(3) Jordan form $\begin{pmatrix} 0 & 1 \\ 0 & 0 \end{pmatrix}$;
(4) Jordan form $\begin{pmatrix} 0 & 0 \\ 0 & 0 \end{pmatrix}$.

Show that each set is a submanifold of $\mathbb{R}^4$ and find its codimension. (Hint: Use the Implicit Function Theorem.)

If one has written down a long enough list of transversality properties, there is some hope that families of the form (3.1.1) with bifurcations of a given type will all have qualitatively similar dynamics near the bifurcation. The best definition of "qualitatively similar dynamics" is not clear. There are several alternatives and we want to pick one which is sufficiently strong but which captures the relevant examples. There is no satisfactory resolution to this dichotomy and efforts to determine which examples satisfy each of the alternative definitions involve many technical issues. Instead of working with a specific definition, we shall describe for each of the examples we consider, the dynamical features which are readily proved to persist under perturbation. The list of relevant examples which should be explored in one- and two-parameter families seems to be fairly complete (apart from situations in which there is a large group of symmetries).

Thus, using the methods outlined above, we can list the "normal forms" of the Jacobian derivatives $D_x f_\mu$ evaluated at bifurcation points $(x_0, \mu_0)$ of codimensions one and two, as follows:

*Codimension One*

(i) Simple zero eigenvalue:

$$D_x f_\mu = \begin{bmatrix} 0 & 0 \\ 0 & A \end{bmatrix};$$

(ii) Simple pure imaginary pair:

$$D_x f_\mu = \begin{bmatrix} \begin{bmatrix} 0 & -\omega \\ \omega & 0 \end{bmatrix} & 0 \\ 0 & A \end{bmatrix}.$$

*Codimension Two*

(iii) Double zero, nondiagonalizable:

$$D_x f_\mu = \begin{bmatrix} \begin{bmatrix} 0 & 1 \\ 0 & 0 \end{bmatrix} & 0 \\ 0 & A \end{bmatrix};$$

(iv) Simple zero + pure imaginary pair:

$$D_x f_\mu = \begin{bmatrix} \begin{bmatrix} 0 & -\omega & 0 \\ \omega & 0 & 0 \\ 0 & 0 & 0 \end{bmatrix} & 0 \\ 0 & A \end{bmatrix};$$

(v) Two distinct pure imaginary pairs:

$$\begin{bmatrix} \begin{bmatrix} 0 & -\omega_1 & 0 & 0 \\ \omega_1 & 0 & 0 & 0 \\ 0 & 0 & 0 & -\omega_2 \\ 0 & 0 & \omega_2 & 0 \end{bmatrix} & 0 \\ 0 & A \end{bmatrix}$$

In each case $A$ is a matrix of the appropriate dimension $((n-1) \times (n-1)$, $(n-2) \times (n-2)$ and so forth, all of whose eigenvalues have nonzero real parts.

We shall use the terms *codimension* and *unfolding* in discussing particular bifurcations. The *codimension* of a bifurcation will be the smallest dimension of a parameter space which contains the bifurcation in a persistent way. An *unfolding* of a bifurcation is a family which contains the bifurcation in a persistent way. We shall illustrate these definitions and elaborate them further in Section 3.4 when we discuss the bifurcations of a system having a simple zero eigenvalue at an equilibrium.

Before proceeding with a discussion of the local codimension one bifurcation, there are two general techniques which we discuss which have the effect of introducing coordinate systems in which computations are more easily carried out. After using these techniques one is left with a few specific systems of differential equations whose behavior determines the qualitative features of each type of bifurcation. We stress once more that these techniques are local in character, and applicable only to the bifurcations of equilibria and periodic orbits.

## 3.2. Center Manifolds

In this section we begin our development of the techniques necessary for the analysis of bifurcation problems. We discuss and state the *center manifold theorem*, which provides a means for systematically reducing the dimension of the state spaces which need to be considered when analyzing bifurcations of a given type. We use the Lorenz system and its bifurcation at $\rho = 1$ as an example which illustrates the rôle of center manifolds in bifurcation calculations. There are two analogous situations to consider: an equilibrium for a vector field and a fixed point for a diffeomorphism. The second case often arises from the Poincaré return map of a periodic orbit of a flow.

Suppose that we have a system of ordinary differential equations $\dot{x} = f(x)$ such that $f(0) = 0$. If the linearization of $f$ at the origin has no pure imaginary eigenvalues, then Hartman's theorem (Theorem 1.3.1) states that the numbers of eigenvalues with positive and negative real parts determine the topological equivalence of the flow near 0. If there are eigenvalues with zero real parts, then the flow near the origin can be quite complicated. We have already met some such examples; here are others:

EXERCISE 3.2.1. Determine the topological properties of the flows of the following systems near the origin:

(a) $\dot{x} = x^2$;

(b) $\dot{x} = y - x(x^2 + y^2)$,
    $\dot{y} = -x - y(x^2 + y^2)$,    (Hint: transform to polar coordinates);

(c) $\dot{x} = x^3 - 3xy^2$,
    $\dot{y} = -3x^2y + y^3$    (Hint: see equation (1.8.23) and Exercises 1.8.9, 1.8.11);

(d) $\dot{x} = xy + x^3$,
    $\dot{y} = -y - x^2y$.

We note that in the last example (d) one of the eigenvalues is $-1$ and hence a one-dimensional stable manifold (in this case the $y$-axis) exists. For this problem, a direct calculation shows that the $x$-axis is a second invariant set tangent to (in fact identical to) the center eigenspace $E^c$. This is an example of a *center manifold: an invariant manifold tangent to the center eigenspace*. The local dynamical behavior "transverse" to the center manifold is relatively simple, since it is controlled by the exponentially contracting (and expanding) flows in the local stable (and unstable) manifolds. Note that we cannot define the center manifold in terms of the asymptotic behavior of solutions in it (cf. equation (1.3.5)), since, as Exercise 3.2.1 shows, solutions in the center manifold can be expanding or contracting.

In general the center manifold method isolates the complicated asymptotic behavior by locating an invariant manifold tangent to the subspace spanned by the (generalized) eigenspace of eigenvalues on the imaginary axis. There are technical difficulties here that are not present in the stable manifold theorem, however. These involve the *nonuniqueness* and the *loss of smoothness* in the invariant center manifold. Before stating the main results, we illustrate these issues with a pair of examples.

Our first example is due to Kelley [1967]. Consider the system

$$\dot{x} = x^2,$$
$$\dot{y} = -y. \qquad (3.2.1)$$

The solutions to this system have the form $x(t) = x_0/(1 - tx_0)$ and $y(t) = y_0 e^{-t}$. Eliminating $t$, we obtain solution curves which are graphs of the functions $y(x) = (y_0 e^{-1/x_0})e^{1/x}$. For $x < 0$, all of these solution curves approach the origin in a way which is "flat"; that is, all of their derivatives vanish at $x = 0$. For $x \geq 0$, the only solution curve which approaches the origin is the $x$-axis. Thus the center manifold, tangent to the direction of the eigenvector belonging to 0 (the $x$-axis) is far from unique. We can obtain a $C^\infty$ center manifold by piecing together *any* solution curve in the left half plane with the positive half of the $x$-axis; cf. Figure 3.2.1. Note, however, that the only *analytic* center manifold is the $x$-axis itself.

To explain the lack of smoothness in center manifolds, we first make a simple observation about the trajectories which approach a node. Consider the linear system

$$\dot{x} = ax,$$
$$\dot{y} = by, \qquad\qquad (3.2.2)$$

with $b > a > 0$. Dividing these equations, we obtain

$$\frac{dy}{dx} = \frac{b}{a}\frac{y}{x}. \qquad\qquad (3.2.3)$$

The solutions of equation (3.2.3) are easily seen to have the form $y(x) = C|x|^{(b/a)}$. The graphs of the functions $y(x)$ are the solution curves of (3.2.2). If we extend one of these solution curves to the origin, then it fails to be infinitely differentiable if $b/a$ is not an integer and $C \neq 0$. If $r < b/a < r + 1$, then the extended curve will be $C^r$ but not $C^{r+1}$. Even if $b/a$ is an integer, the curve formed from the union of 0 and two solution curves to the right and left of 0 will only be $b/a - 1$ times differentiable in general.

We now give an example which illustrates that a center manifold may be forced to contain curves that are patched together at a node like those we

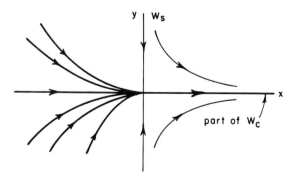

Figure 3.2.1. The phase portrait of equation (3.2.1), showing some center manifolds (heavy curves).

have just described. Consider the system

$$\dot{x} = \mu x - x^3,$$
$$\dot{y} = y, \qquad\qquad (3.2.4)$$
$$\dot{\mu} = 0,$$

in which the "parameter" $\mu$ plays the rôle of a (trivial) dependent variable.

It is easy to verify that, for the system linearized at $(x, y, \mu) = (0, 0, 0)$, the $y$-axis is an unstable subspace and the $(x, \mu)$ plane the center subspace. The equilibria of this system consist of the $\mu$-axis and the parabola $\mu = x^2$ in the $(x, \mu)$ coordinate plane, Figure 3.2.2. Since $\dot{\mu} \equiv 0$, the planes $\mu = $ constant are invariant under the flow of (3.2.4). In a plane $\mu = $ constant $\neq 0$, all of the equilibria are hyperbolic. Those on the $\mu$-axis with $\mu < 0$ and along the parabola are saddles, while those along the positive $\mu$-axis are unstable nodes. We want to find the center manifold of 0. For $\mu \leq 0$, the flow of (3.2.4) is topologically a one-parameter family of saddles, and the only choice for a center manifold comes from points in the $(x, \mu)$ coordinate plane.

When $\mu > 0$, the unstable manifolds of the saddles along the parabola (each a vertical line $x = \pm\sqrt{\mu}$) form an invariant manifold $M$ which separates $\mathbb{R}^3$ into two invariant regions. The center manifold must intersect $M$, but it can only do so by containing the parabola of equilibria. It follows that the center manifold for $\mu > 0$ must consist of the equilibria of (3.2.4) together with the stable saddle separatrices of the equilibria along the parabola. For (3.2.4), these all lie in the $(x, \mu)$ coordinate plane and the center manifold *is* the $(x, \mu)$ coordinate plane. However, if we modify the system (3.2.4) by changing the second equation to $\dot{y} = y + x^4$, then we assert (without proof) that the center manifold must still consist of the equilibria together with their saddle separatrices. However, these now no

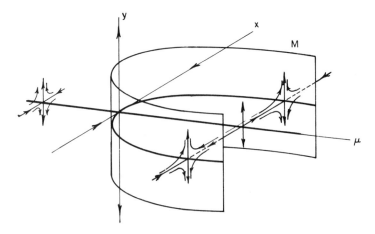

Figure 3.2.2. Invariant manifolds for equation (3.2.4).

longer fit together in a $C^\infty$ way along the curve of nodes on the positive half of the $\mu$-axis. The degree of smoothness decreases as one moves away from the origin because the linearization of (3.2.4) at the point $(0, 0, \mu_0)$ has eigenvalues in the plane $\mu = \mu_0$ which are 1 and $\mu_0$. Therefore the degree of smoothness we expect is bounded by $1/\mu_0$. If we are interested only in a $C^r$ invariant manifold with $r < \infty$, then our search for one will be successful as long as we restrict attention to a sufficiently small neighborhood of the origin (of diameter at most $1/r$ in this example).*

With these examples as motivation, we now state

**Theorem 3.2.1** (Center Manifold Theorem for Flows). *Let $f$ be a $C^r$ vector field on $\mathbb{R}^n$ vanishing at the origin ($f(0) = 0$) and let $A = Df(0)$. Divide the spectrum of $A$ into three parts, $\sigma_s, \sigma_c, \sigma_u$ with*

$$\operatorname{Re} \lambda \begin{cases} < 0 & \text{if } \lambda \in \sigma_s, \\ = 0 & \text{if } \lambda \in \sigma_c, \\ > 0 & \text{if } \lambda \in \sigma_u. \end{cases}$$

*Let the (generalized) eigenspaces of $\sigma_s, \sigma_c,$ and $\sigma_u$ be $E^s, E^c,$ and $E^u$, respectively. Then there exist $C^r$ stable and unstable invariant manifolds $W^u$ and $W^s$ tangent to $E^u$ and $E^s$ at 0 and a $C^{r-1}$ center manifold $W^c$ tangent to $E^c$ at 0. The manifolds $W^u, W^s,$ and $W^c$ are all invariant for the flow of $f$. The stable and unstable manifolds are unique, but $W^c$ need not be.†*

We illustrate the situation in Figure 3.2.3. Note that we cannot assign directions to the flow in $W^c$ without specific information on the higher-order terms of $f$ near 0.

For more information on the existence, uniqueness, and smoothness of center manifolds and for proofs of Theorem 3.2.1 and the results to follow, see Marsden and McCracken [1976], Carr [1981], and Sijbrand [1981]. Kelley's [1967] paper should also be noted, as the first publication of the full proof of Theorem 3.2.1.

One might guess that a simpler alternative to using the center manifold theorem for a system would be to project the system onto the linear subspace spanned by $E^c$. Thus, if one writes a vector field $f$ as $f = f_u + f_s + f_c$ with

---

* One can prove this by solving the first-order linear equation $dy/dx = y/(\mu x - x^3) + x^3/(\mu - x^2)$ which (with suitable boundary conditions) determines the center manifold on each "slice" $\mu = $ constant. Using the variation of constants formula, one obtains

$$y(x) = \left( \frac{x^2}{\mu - x^2} \right)^{1/2\mu} \left( C + \int \frac{x^{3-1/\mu}}{(\mu - x^2)^{1-1/2\mu}} \, dx \right)$$

and, upon obtaining a series solution for small $x$ and $\mu$, verifies the lack of smoothness referred to above. For more information and other examples, see Sijbrand [1981], van Strien [1979], and Carr [1981].

† If $f$ is $C^\infty$, then we can find a $C^r$ center manifold for any $r < \infty$.

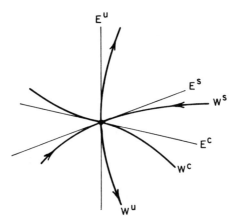

Figure 3.2.3. The stable, unstable, and center manifolds.

$f_u \in E^u$, $f_s \in E^s$, and $f_c \in E^c$, near the equilibrium one would hope that $f_c$ restricted to $E^c$ provides the correct qualitative picture of the dynamics in the center directions. The Lorenz system illustrates that this is not always the case and thus provides an instructive example of the rôle played by the center manifold calculations in a bifurcation problem.

Recall the Lorenz system (2.3.1) introduced in Chapter 2.

$$\begin{aligned}
\dot{x} &= \sigma(y - x), \\
\dot{y} &= \rho x - y - xz, \\
\dot{z} &= -\beta z + xy.
\end{aligned} \tag{3.2.5}$$

This system is a Galërkin projection of a set of partial differential equations for two-dimensional convection of the type described in Section 2.1. (We will consider the Galërkin projection and truncation procedure further in Section 7.6.) We shall study the bifurcation of (3.2.5) occurring at $(x, y, z) = 0$ and $\rho = 1$. The Jacobian derivative at 0 is the matrix.

$$\begin{pmatrix} -\sigma & \sigma & 0 \\ \rho & -1 & 0 \\ 0 & 0 & -\beta \end{pmatrix} \tag{3.2.6}$$

When $\rho = 1$, this matrix has eigenvalues $0$, $-\sigma - 1$, and $-\beta$ with eigenvectors $(1, 1, 0)$, $(\sigma, -1, 0)$, $(0, 0, 1)$. Using the eigenvectors as a basis for a new coordinate system, we set

$$\begin{pmatrix} x \\ y \\ z \end{pmatrix} = \begin{pmatrix} 1 & \sigma & 0 \\ 1 & -1 & 0 \\ 0 & 0 & 1 \end{pmatrix} \begin{pmatrix} u \\ v \\ w \end{pmatrix}, \qquad \begin{pmatrix} u \\ v \\ w \end{pmatrix} = \begin{pmatrix} \dfrac{1}{1+\sigma} & \dfrac{\sigma}{1+\sigma} & 0 \\ \dfrac{1}{1+\sigma} & \dfrac{-1}{1+\sigma} & 0 \\ 0 & 0 & 1 \end{pmatrix} \begin{pmatrix} x \\ y \\ z \end{pmatrix}.$$

$$\tag{3.2.7}$$

Under this transformation (3.2.5) becomes

$$\dot{u} = \frac{1}{1 + \sigma}\dot{x} + \frac{\sigma}{1 + \sigma}\dot{y} = \frac{\sigma}{1 + \sigma}(y - x) + \frac{\sigma}{1 + \sigma}[(x - y) - xz]$$

$$= \frac{-\sigma}{1 + \sigma}(u + \sigma v)w,$$

$$\dot{v} = \frac{1}{1 + \sigma}\dot{x} - \frac{1}{1 + \sigma}\dot{y} = \frac{\sigma}{1 + \sigma}(y - x) - \frac{1}{1 + \sigma}[(x - y) - xz]$$

$$= -(1 + \sigma)v + \frac{1}{1 + \sigma}(u + \sigma v)w,$$

$$\dot{w} = \dot{z} = -\beta z + xy = -\beta w + (u + \sigma v)(u - v), \tag{3.2.8}$$

or

$$\begin{pmatrix} \dot{u} \\ \dot{v} \\ \dot{w} \end{pmatrix} = \begin{bmatrix} 0 & 0 & 0 \\ 0 & -(1 + \sigma) & 0 \\ 0 & 0 & -\beta \end{bmatrix} \begin{pmatrix} u \\ v \\ w \end{pmatrix} + \begin{pmatrix} \dfrac{-\sigma}{1 + \sigma}(u + \sigma v)w \\ \dfrac{1}{1 + \sigma}(u + \sigma v)w \\ (u + \sigma v)(u - v) \end{pmatrix}, \tag{3.2.9}$$

so that the linear part is now in standard (diagonal) form. In the $(u, v, w)$ coordinates, the center manifold is a curve tangent to the $u$-axis. Note that the projection of the system onto the $u$-axis, obtained by setting $v = w = 0$ in the equation for $\dot{u}$, yields $\dot{u} = 0$. The $u$ axis is not invariant, however, because the equation for $\dot{w}$ includes the term $u^2$. If we make a further nonlinear coordinate change by setting $\tilde{w} = w - u^2/\beta$, however, we obtain

$$\dot{\tilde{w}} = \dot{w} - \frac{2u\dot{u}}{\beta} = -\beta\left(w - \frac{u^2}{\beta}\right) + (\sigma - 1)uv - \sigma v^2 + \frac{2\sigma}{\beta(1 + \sigma)}u(u + \sigma v)w,$$

or

$$\dot{\tilde{w}} = -\beta\tilde{w} + (\sigma - 1)uv - \sigma v^2 + \frac{2\sigma}{\beta(1 + \sigma)}u(u + \sigma v)\left(\tilde{w} + \frac{u^2}{\beta}\right). \tag{3.2.10}$$

In the $(u, v, \tilde{w})$ coordinate system, we have

$$\dot{u} = -\frac{\sigma}{1 + \sigma}(u + \sigma v)\left(\tilde{w} + \frac{u^2}{\beta}\right). \tag{3.2.11}$$

Now projection of the equation onto the $u$-axis in these coordinates gives the equation $\dot{u} = (-\sigma/\beta(1 + \sigma))u^3$. Note also that no terms of the form $u^2$ occur in the equations for $v$ and $\tilde{w}$, and thus that the $u$-axis is invariant in our transformed equations "up to second order."

Further efforts to find the center manifold can proceed by additional coordinate changes that serve to make the $u$-axis invariant for the flow.

This can be done iteratively by changes in $v$ and $\tilde{w}$ which add to these coordinates mononomials in $u$, just as $\tilde{w}$ was obtained from $w$. Additional
such coordinate changes will not change the coefficient $(-\sigma/\beta(1 + \sigma))$ of $u^3$
in the equation for $\dot{u}$, but will affect higher degree terms of the from $u^m$,
$m \geq 4$. We shall see in subsequent sections that the equation $\dot{u} =$
$(-\sigma/\beta(1 + \sigma))u^3$ along with the effect of varying $\rho$ near 1 is sufficient to
deduce the qualitative dynamics of the bifurcation in the Lorenz system
(and the fluid system we started with). It is clearly important to include the
calculation of the initial portion of the Taylor series of the center manifold
in this analysis. Failure to do so gives a misleading picture of the dynamics at
the point of bifurcation.

In studying the Lorenz example we have really been trying to approximate
the (one-dimensional) equation governing the flow in the center manifold.
We shall now develop a systematic method for performing such approximations.

The center manifold theorem implies that the bifurcating system is locally
topologically equivalent to

$$
\begin{aligned}
\dot{\tilde{x}} &= \tilde{f}(x) \\
\dot{\tilde{y}} &= -\tilde{y} ; \qquad (\tilde{x}, \tilde{y}, \tilde{z}) \in W^c \times W^s \times W^u, \qquad (3.2.12) \\
\dot{\tilde{z}} &= \tilde{z}
\end{aligned}
$$

at the bifurcation point. We now tackle the problem of computing the "reduced" vector field $\tilde{f}$. For simplicity, and because it is the most interesting
case physically, we assume that the unstable manifold is empty* and that the
linear part of the bifurcating system is in block diagonal form:

$$
\begin{aligned}
\dot{x} &= Bx + f(x, y) \\
\dot{y} &= Cy + g(x, y)
\end{aligned} ; \qquad (x, y) \in \mathbb{R}^n \times \mathbb{R}^m, \qquad (3.2.13)
$$

where $B$ and $C$ are $n \times n$ and $m \times m$ matrices whose eigenvalues have,
respectively, zero real parts and negative real parts, and $f$ and $g$ vanish,
along with their first partial derivatives, at the origin.

Since the center manifold is tangent to $E^c$ (the $y = 0$ space) we can
represent it as a (local) graph

$$
W^c = \{(x, y) | y = h(x)\}; \qquad h(0) = Dh(0) = 0, \qquad (3.2.14)
$$

where $h: U \to \mathbb{R}^m$ is defined on some neighborhood $U \subset \mathbb{R}^n$ of the origin,
Figure 3.2.4. We now consider the projection of the vector field on $y = h(x)$
onto $E^c$:

$$
\dot{x} = Bx + f(x, h(x)). \qquad (3.2.15)
$$

Since $h(x)$ is tangent to $y = 0$, the solutions of equation (3.2.15) provide a
good approximation of the flow of $\dot{\tilde{x}} = \tilde{f}(\tilde{x})$ restricted to $W^c$. In fact we have

---

* But see p. 138, below.

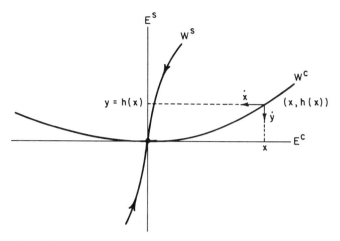

Figure 3.2.4. The center manifold and the projected vector field.

**Theorem 3.2.2** (Henry [1981], Carr [1981]). *If the origin* $x = 0$ *of* (3.2.15) *is locally asymptotically stable* (*resp. unstable*) *then the origin of* (3.2.13) *is also locally asymptotically stable* (*resp. unstable*).

This result also follows from the global linearization theory of Pugh and Shub [1970].

We now show how $h(x)$ can be calculated, or at least approximated. Substituting $y = h(x)$ in the second component of (3.2.13) and using the chain rule, we obtain

$$\dot{y} = Dh(x)\dot{x} = Dh(x)[Bx + f(x, h(x))] = Ch(x) + g(x, h(x)),$$

or

$$\mathcal{N}(h(x)) = Dh(x)[Bx + f(x, h(x))] - Ch(x) - g(x, h(x)) = 0,$$

$$(3.2.16)$$

with boundary conditions

$$h(0) = Dh(0) = 0.$$

This (partial) differential equation for $h$ cannot, of course, be solved exactly in most cases (to do so would imply that a solution of the original equation had been found), but its solution *can* be approximated arbitrarily closely as a Taylor series at $x = 0$:

**Theorem 3.2.3** (Henry [1981], Carr [1981]). *If a function* $\phi(x)$, *with* $\phi(0) = D\phi(0) = 0$, *can be found such that* $\mathcal{N}(\phi(x)) = O(|x|^p)$ *for some* $p > 1$ *as* $|x| \to 0$ *then it follows that*

$$h(x) = \phi(x) + O(|x|^p) \quad as \ |x| \to 0.$$

Thus we can approximate $h(x)$ as closely as we wish by seeking series solutions of (3.2.16). However, as we shall see in Exercise 3.2.2 below, such Taylor series expansions do not always exist, since $W^c$ may not be analytic at the origin.

To illustrate the use of Theorem 3.2.3, consider the system

$$\dot{u} = v,$$
$$\dot{v} = -v + \alpha u^2 + \beta uv,$$

(3.2.17)

where $\alpha$ and $\beta$ are to be specified subsequently. There is a unique fixed point at $(0, 0)$ and the eigenvalues of the linearized system are 0 and $-1$. Using the transformation matrix whose columns are the eigenvectors;

$$\begin{pmatrix} u \\ v \end{pmatrix} = T \begin{pmatrix} x \\ y \end{pmatrix}; \qquad T = T^{-1} = \begin{pmatrix} 1 & 1 \\ 0 & -1 \end{pmatrix},$$

(3.2.18)

we can put (3.2.17) into standard form:

$$\begin{pmatrix} \dot{x} \\ \dot{y} \end{pmatrix} = \begin{bmatrix} 0 & 0 \\ 0 & -1 \end{bmatrix} \begin{pmatrix} x \\ y \end{pmatrix} + \begin{bmatrix} 1 & 1 \\ 0 & -1 \end{bmatrix} \begin{pmatrix} 0 \\ \alpha(x + y)^2 - \beta(x + y)y \end{pmatrix},$$

or

$$\dot{x} = \alpha(x + y)^2 - \beta(xy + y^2),$$  (3.2.19a)
$$\dot{y} = -y - \alpha(x + y)^2 + \beta(xy + y^2).$$  (3.2.19b)

Since both $E^c$ and $E^s$ are one dimensional, the graph $h$ is a real valued function and (3.2.16) becomes

$$\mathcal{N}(h(x)) = h'(x)[\alpha(x + h(x))^2 - \beta(xh(x) + h^2(x))] + h(x) + \alpha(x + h(x))^2$$
$$- \beta(xh(x) + h^2(x)) = 0, \qquad h(0) = h'(0) = 0.$$  (3.2.20)

We set $h(x) = ax^2 + bx^3 + \cdots$ and substitute into (3.2.20) to find the unknown coefficients, $a, b, \ldots$. Doing so, we obtain

$$h(x) = -\alpha x^2 + \alpha(4\alpha - \beta)x^3 + \mathcal{O}(x^4),$$  (3.2.21)

and the approximation may thus be written

$$\dot{x} = \alpha(x + h(x))^2 - \beta(xh(x) + h^2(x))$$
$$= \alpha(x^2 + (\beta - 2\alpha)x^3 + (9\alpha^2 - 7\alpha\beta + \beta^2)x^4) + \mathcal{O}(x^5),$$  (3.2.22)

or, if $\alpha \neq 0$,

$$\dot{x} = \alpha x^2 + (\alpha(\beta - 2\alpha)x^3) + \mathcal{O}(x^4).$$  (3.2.23)

Note that, to second order, we obtain the same result by taking the *tangent space approximation*

$$h = 0 + \mathcal{O}(x^2)$$  (3.2.24)

in this case, since the second-order terms determine the qualitative behavior near 0 when $\alpha \neq 0$.

In the next example,

$$\dot{x} = xy, \tag{3.2.25a}$$

$$\dot{y} = -y + \alpha x^2, \tag{3.2.25b}$$

the tangent space approximation does not determine the stability near 0, since if $y = h(x) = 0$, then $\dot{x} = 0$, as in the Lorenz example. In this problem $h$ is determined by

$$h'(x)[xh(x)] + h(x) - \alpha x^2 = 0, \tag{3.2.26}$$

and, letting $h = ax^2 + bx^3 + \cdots$, we obtain

$$h = \alpha x^2 + \mathcal{O}(x^4). \tag{3.2.27}$$

Thus the reduced system is

$$\dot{x} = \alpha x^3 + \mathcal{O}(x^5). \tag{3.2.28}$$

The two examples therefore have local phase portraits (near $(0,0)$) as shown in Figure 3.2.5.

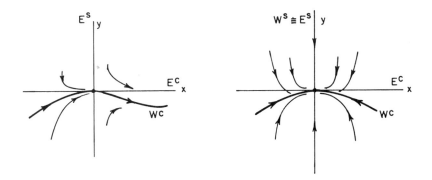

Figure 3.2.5. Center manifolds for the two examples. (a) Equation (3.2.19), $\alpha > 0$; (b) equation (3.2.25), $\alpha < 0$.

EXERCISE 3.2.2. Find a family of center manifolds for the system

$$\dot{x} = -x^3,$$
$$\dot{y} = -y + x^2,$$

by solving the equation $dy/dx - y/x^3 = -1/x$ exactly by variation of constants to obtain

$$y(x) = e^{-1/2x^2}\left\{ C + \tfrac{1}{2}\ln\left(\frac{1}{2x^2}\right) + \sum_{n=1}^{\infty} \frac{1}{2^n \cdot n! \, x^{2n}} \right\}.$$

Verify that attempts to approximate this $C^\infty$, nonanalytic manifold by the power series method fail. What happens?

EXERCISE 3.2.3. Verify that the origin for the system of equation (1.8.20)

$$\dot{x} = x^2 - xy,$$
$$\dot{y} = -y + x^2,$$

has the local stability type indicated in Figure 1.8.8(b).

EXERCISE 3.2.4. Find the approximation $\dot{x} = Bx + f(x, h(x))$ for the reduced system on $W^c$ near the origin, in the following cases:

(a) $\dot{x} = \alpha x^2 - y^2,$
$\quad \dot{y} = -y + x^2 + xy;$

(b) $\dot{x} = -y + xz,$
$\quad \dot{y} = x + yz,$
$\quad \dot{z} = -z - (x^2 + y^2) + z^2.$

In case (a) first suppose $\alpha \neq 0$, and then let $\alpha = 0$. In both cases carry the series approximations for $h$ sufficiently far to determine the stability of the reduced system.

EXERCISE 3.2.5. Starting with equation (3.2.9), and writing the center manifold as $(v, w) = (h_1(u), h_2(u))$, use the power series method to approximate a center manifold for the Lorenz equations (with $\rho = 1$) up to third order. Hence verify the "ad hoc" calculations leading to equation (3.2.11) and show that the reduced system is $\dot{u} = (-\sigma/\beta(1 + \sigma))u^3 + \mathcal{O}(u^4)$.

We next note a simple extension to the center manifold method which is useful when dealing with parametrized families of systems. In equation (3.2.13) suppose that the matrices $B$, $C$ and functions of $f$, $g$ depend upon a $k$-vector of parameters, $\mu$, and write the extended system as

$$\dot{x} = B_\mu x + f_\mu(x, y),$$
$$\dot{y} = C_\mu y + g_\mu(x, y), \qquad (x, y) \in \mathbb{R}^n \times \mathbb{R}^m, \qquad (3.2.29)$$
$$\dot{\mu} = 0, \qquad\qquad\qquad \mu \in \mathbb{R}^k,$$

(cf. the example of equation (3.2.4)). At $(x, y, \mu) = (0, 0, 0)$ (3.2.29) has an $n + k$-dimensional center manifold tangent to $(x, \mu)$ space, which may be approximated as the power series (in $x$ and $\mu$) of a graph $h: \mathbb{R}^n \times \mathbb{R}^k \to \mathbb{R}^m$ precisely as above. The invariance properties of center manifolds guarantee that any small solutions bifurcating from $(0, 0, 0)$ must lie in any center manifold and thus we may follow the local evolution of bifurcating families of solutions in this suspended family of center manifolds.

As an example we consider a quadratic Duffing's equation

$$\dot{u} = v,$$
$$\dot{v} = \beta u - u^2 - \delta v, \qquad (3.2.30)$$

for $\delta > 0$ and $\beta$ a variable parameter near 0. At $\beta = 0$ the linearized system has eigenvalues 0 and $-\delta$ at $(u, v) = (0, 0)$, and, using the transformation

$$\begin{pmatrix} u \\ v \end{pmatrix} = \begin{bmatrix} 1 & 1 \\ 0 & -\delta \end{bmatrix}\begin{pmatrix} x \\ y \end{pmatrix}, \qquad \begin{pmatrix} x \\ y \end{pmatrix} = \begin{bmatrix} 1 & 1/\delta \\ 0 & -1/\delta \end{bmatrix}\begin{pmatrix} u \\ v \end{pmatrix}, \qquad (3.2.31)$$

(3.2.30) can be rewritten as the suspended system

$$\begin{pmatrix} \dot{x} \\ \dot{y} \end{pmatrix} = \left[ \begin{bmatrix} 0 & 0 \\ 0 & -\delta \end{bmatrix} + \frac{\beta}{\delta} \begin{bmatrix} 1 & 1 \\ -1 & -1 \end{bmatrix} \right] \begin{pmatrix} x \\ y \end{pmatrix} + \frac{1}{\delta} \begin{pmatrix} -(x+y)^2 \\ (x+y)^2 \end{pmatrix} \qquad (\dot{\beta} = 0),$$

or

$$\dot{x} = \frac{\beta}{\delta}(x+y) - \frac{1}{\delta}(x+y)^2,$$

$$\dot{\beta} = 0, \qquad\qquad\qquad\qquad\qquad\qquad (3.2.32)$$

$$\dot{y} = -\delta y - \frac{\beta}{\delta}(x+y) + \frac{1}{\delta}(x+y)^2.$$

We seek a center manifold

$$y = h(x, \beta) = ax^2 + bx\beta + c\beta^2 + \mathcal{O}(3), \qquad (3.2.33)$$

where $\mathcal{O}(3)$ means terms of orders $x^3$, $x^2\beta$, $x\beta^2$, and $\beta^3$. Equation (3.2.16) is in this case

$$\left(\frac{\partial h}{\partial x}, \frac{\partial h}{\partial \beta}\right)\left(\begin{matrix}(\beta/\delta)(x+h) - (1/\delta)(x+h)^2 \\ 0\end{matrix}\right) + \delta h + \frac{\beta}{\delta}(x+h) - \frac{1}{\delta}(x+h)^2 = 0,$$

$$(3.2.34)$$

and substituting (3.2.33) in we obtain

$$(2ax + b\beta, \ldots)\left(\begin{matrix}(\beta/\delta)(x + \cdots) + \cdots \\ 0\end{matrix}\right) + \delta(ax^2 + bx\beta + c\beta^2)$$

$$+ \frac{\beta}{\delta}(x + ax^2 + bx\beta + c\beta^2)$$

$$- \frac{1}{\delta}(x + \cdots)^2 = \mathcal{O}(3).$$

Equating powers of $x^2$, $x\beta$, and $\beta^2$, we find that

$$a = \frac{1}{\delta^2}, \qquad b = -\frac{1}{\delta^2}, \qquad c = 0;$$

and thus

$$y = \frac{1}{\delta^2}(x^2 - \beta x) + \mathcal{O}(3). \qquad (3.2.35)$$

The reduced system, which determines stability, is therefore given by

$$\dot{x} = \frac{\beta}{\delta}\left(x + \frac{1}{\delta^2}(x^2 - \beta x)\right) - \frac{1}{\delta}(x + \cdots)^2 + \mathcal{O}(3), \qquad (\dot{\beta} = 0),$$

or

$$\dot{x} = \frac{\beta}{\delta}\left(1 - \frac{\beta}{\delta^2}\right)x - \frac{1}{\delta}\left(1 - \frac{\beta}{\delta^2}\right)x^2 + \mathcal{O}(3) \qquad (\dot{\beta} = 0), \qquad (3.2.36)$$

and, *for sufficiently small* $\beta$ ($|\beta| < \delta^2$), we obtain the bifurcation diagram of Figure 3.2.6, the correctness of which may be checked by direct calculation. The suspended family of center manifolds for this example is indicated in Figure 3.2.7.

EXERCISE 3.2.6. Approximate a center manifold $(v, w) = (h_1(u, \rho), h_2(u, \rho))$ for the Lorenz equations (3.2.5) when $\rho$ is near 1. (Use the same transformations (3.2.7) but keep $\rho$ as a variable parameter in the matrix (3.2.6).)

EXERCISE 3.2.7. Verify that for the Duffing equation (2.2.6) with $|\beta|$ small and $\delta > 0$ fixed, the tangent space approximation to the center manifold as a graph over $E^c \times$ ($\beta$-space) yields a qualitatively correct bifurcation diagram.

We remark finally in this section that there is a center manifold theorem for diffeomorphisms at a fixed point corresponding to the one we have stated for flows at an equilibrium. At a fixed point $p$ of a diffeomorphism $G$, there are invariant manifolds corresponding to the generalized eigenspaces of $DG(p)$ for eigenvalues which lie inside, on, and outside the unit circle. For a statement of the theorem, see Marsden and McCracken [1976].

Center manifolds can be approximated in much the same way as for flows. Assuming that our discrete system is in the form

$$\begin{aligned} x_{n+1} &= Bx_n + F(x_n, y_n), \\ y_{n+1} &= Cy_n + G(x_n, y_n), \end{aligned} \qquad (3.2.37)$$

where all the eigenvalues of $B$ are on the unit circle and all those of $C$ within the unit circle, we again seek the center manifold as a graph $y = h(x)$. Substitution into (3.2.37) yields

$$\begin{aligned} y_{n+1} = h(x_{n+1}) &= h(Bx_n + F(x_n, h(x_n))) \\ &= Ch(x_n) + G(x_n, h(x_n)), \end{aligned}$$

or

$$\mathcal{N}(h(x)) = h(Bx + F(x, h(x))) - Ch(x) - G(x, h(x)) = 0; \qquad (3.2.38)$$

and we can again approximate by power series methods.

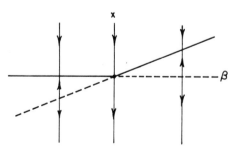

Figure 3.2.6. The bifurcation diagram for equation (3.2.30).

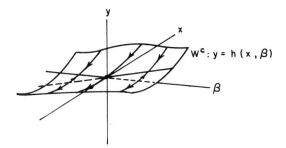

Figure 3.2.7. The family of center manifolds for equation (3.2.30).

As an example, consider the map

$$x_{n+1} = x_n + x_n y_n,$$
$$y_{n+1} = \lambda y_n - x_n^2, \qquad 0 < \lambda < 1. \qquad (3.2.39)$$

Letting $y = h(x) = ax^2 + bx^3 + O(x^4)$ and substituting in (3.2.38), we have

$$a(x + x(ax^2 + \cdots))^2 + b(x + x(ax^2 + \cdots))^3 - \lambda(ax^2 + bx^3) + x^2 = \mathcal{O}(x^4),$$

or

$$ax^2 + bx^3 - \lambda ax^2 - \lambda bx^3 + x^2 = \mathcal{O}(x^4),$$

so that

$$a = \frac{1}{\lambda - 1}, \qquad (3.2.40)$$

and

$$b = 0.$$

Thus we obtain

$$y = \frac{x^2}{\lambda - 1} + \mathcal{O}(x^4) \qquad (3.2.41)$$

for the center manifold, and

$$x_{n+1} = x_n + \frac{x_n^3}{\lambda - 1} \qquad (3.2.42)$$

for the reduced system. Since $\lambda - 1 < 0$, the zero solution of (3.2.42), and hence of (3.2.39), is locally asymptotically stable.

EXERCISE 3.2.8. Compute the reduced systems on the center manifold to third order for the following maps and describe their bifurcations.

(a) $(x, y) \mapsto (x + xy - y^2, \frac{1}{2}y + x^2)$;
(b) $(x, y) \mapsto (x + x^3 + \alpha xy, \frac{1}{2}y - x^2)$, for various $\alpha$;
(c) $(x, y) \mapsto (y, -\frac{1}{2}x + \frac{3}{2}y - y^3)$.

In the discussions above we have assumed the unstable manifold is empty at the bifurcation point. If this is not the case, then we must deal with a system of the form

$$\dot{x} = Bx + f(x, y_s, y_u),$$
$$\dot{y}_s = C_s y + g_s(x, y_s, y_u), \qquad (x, y_s, y_u) \in \mathbb{R}^n \times \mathbb{R}^{n_s} \times \mathbb{R}^{n_u}, \quad (3.2.43)$$
$$\dot{y}_u = C_u z + g_u(x, y_s, y_u),$$

where the eigenvalues of $B$ all have unit modulus, as before, while those of $C_s$ have negative real parts and those of $C_u$ positive real parts. Again one seeks a center manifold as a graph over $U \subset E^c \sim \mathbb{R}^n$: $(y_s, y_u) = (h_s(x), h_u(x))$.

Then, writing the vectors $\begin{pmatrix} y_s \\ y_u \end{pmatrix}$ as $y$ and $\begin{pmatrix} g_s \\ g_u \end{pmatrix}$ as $g$ and the matrix $\begin{bmatrix} C_s & 0 \\ 0 & C_u \end{bmatrix}$

as $C$, we can proceed just as before. Maps can be treated similarly: parameters can be added and extended systems formed just as for flows.

EXERCISE 3.2.9. Find center manifolds to second order and reduced systems to third order for the following problems:

(a) $\dot{x} = x^2 y + \alpha z^2, \dot{y} = -y + x^2 + zy, \dot{z} = z - y^2 + xy$;
(b) $(x, y, z) \mapsto (-x + yz, y \to -\frac{1}{2}y + x^2, z \to 2z - xy)$.

## 3.3. Normal Forms

In this section we continue the development of technical tools which provide the basis for our study of the qualitative properties of flows near a bifurcation. We assume that the center manifold theorem has been applied to a system and henceforth we restrict our attention to the flow within the center manifold; that is, to the approximating equation (3.2.15). We shall try to find additional coordinate transformations which simplify the analytic expression of the vector field on the center manifold. The resulting "simplified" vector fields are called *normal forms*. Analysis of the dynamics of the normal forms yields a qualitative picture of the flows of each bifurcation type.

The idea of introducing successive coordinate transformations to simplify the analytic expression of a general problem is a powerful one. It forms the basis of the Kolmogorov–Arnold–Moser (KAM) theory for studying quasiperiodic phenomena (cf. Sections 4.8 and 6.2) and is also used in the averaging method, to be described in Chapter 4 (in fact, normal forms may be computed by averaging, cf. Chow and Mallet-Paret [1977]). We shall also occasionally assume elsewhere in the analysis of flows near a hyperbolic equilibrium that the system has been linearized there by a coordinate transformation. For the codimension two bifurcations discussed in Chapter 7 and the Hopf bifurcation discussed in Section 4 of this chapter, the normal forms make some symmetry properties of the bifurcation apparent. Exploitation of this symmetry is of considerable help in the analysis of the dynamics.

We now describe in more detail the problem of calculating the normal forms. We start with a system of differential equations

$$\dot{x} = f(x), \tag{3.3.1}$$

which has an equilibrium at 0. (In (3.3.1) we omit explicit reference to the parameter $\mu$.) We would like to find a coordinate change $x = h(y)$ with $h(0) = 0$ such that the system (3.3.1) becomes "as simple as possible." In the $y$-coordinates, we have

$$Dh(y)\dot{y} = f(h(y))$$

or

$$\dot{y} = (Dh(y))^{-1}f(h(y)). \tag{3.3.2}$$

The best that one can hope for is that (3.3.2) will be linear. Formally (this means in terms of power series), one can try to iteratively find a sequence of coordinate transformations $h_1$, $h_2$, ... which remove terms of increasing degree from the Taylor series of (3.3.2) at the origin. The normal form procedure systematizes these calculations without, however, giving the strongest results in all cases. In general, "as simple as possible" means that all inessential terms have been removed (up to some degree) from the Taylor series. When the procedure is applied to a hyperbolic equilibrium, then one obtains the formal part of Hartman's linearization theorem, as we now explain. After this digression, we shall return to nonhyperbolic bifurcating equilibria.

Assume for the moment that $Df(0)$ has distinct (but possibly complex) eigenvalues $\lambda_1, \ldots, \lambda_n$ and that an initial linear change of coordinates has diagonalized $Df(0)$. Then (3.3.1) written in coordinates becomes

$$\begin{aligned}
\dot{x}_1 &= \lambda_1 x_1 + g_1(x_1, \ldots, x_n) \\
\dot{x}_2 &= \lambda_2 x_2 + g_2(x_1, \ldots, x_n) \quad \text{or} \quad \dot{x} = \Lambda x + g(x), \\
&\;\;\vdots \\
\dot{x}_n &= \lambda_n x_n + g_n(x_1, \ldots, x_n)
\end{aligned} \tag{3.3.3}$$

where the functions $g_i$ vanish to second order at the origin. We would like to find a coordinate change $h$ of the form identity plus higher order terms, which has the property that (3.3.2) has non-linear terms which vanish to higher order than those of $g$. If $k$ is the smallest degree of a nonvanishing derivative of some $g_i$, we try to find a transformation $h$ of the form

$$x = h(y) = y + P(y), \tag{3.3.4}$$

with $P$ a polynomial of degree $k$, so that the lowest degree of the nonlinear terms in the transformed equation (3.3.2) is $(k + 1)$). Now (3.3.2) takes the form

$$\dot{y} = (I + DP(y))^{-1}f(y + P(y)). \tag{3.3.5}$$

We want to expand this expression, retaining *only terms of degree k and lower.* Denoting the terms of $g_i$ of degree $k$ by $g_i^k$ and $P(y)$ by $(P_1(y), \ldots, P_n(y))$, we have

$$\dot{y}_i = \lambda_i y_i + \lambda_i P_i(y) + g_i^k(y) - \sum_{j=1}^{n} \frac{\partial P_i}{\partial y_j} \lambda_j y_j. \qquad (3.3.6)$$

We have used in this formula the fact that $(I + DP)^{-1} = I - DP$, modulo terms of degree $k$ and higher. (To compute (3.3.5) modulo terms of degree $(k + 1)$ we only need $(I + DP)^{-1}$ modulo terms of degree $k$ because $f$ has degree 1.) Therefore, we want to find a $P$ which satisfies the equation

$$\lambda_i P_i(y) - \sum_j \frac{\partial P_i}{\partial y_j} \lambda_j y_j = -g_i^k(y). \qquad (3.3.7)$$

We observe that the operator which associates to $P$ the left-hand side of (3.3.7) is linear in the coefficients of $P$. In addition, if $P_i$ is the monomial $y_1^{a_1} \cdots y_n^{a_n}$, then $(\partial P_i / \partial y_j) \lambda_j y_j = a_j \lambda_j P_i$ and the left-hand side of (3.3.7) becomes $(\lambda_i - \sum_j a_j \lambda_j) P_i$ and hence the monomials are eigenvectors for the operator with eigenvalues $\lambda_i - \sum_j a_j \lambda_j$. We conclude that $P$ can be found satisfying (3.3.7) provided that *none of the sums* $\lambda_i - \sum_j a_j \lambda_j$ *is zero when* $a_1, \ldots, a_n$ *are nonnegative integers with* $\sum_j a_j = k$. If there is no equation $\lambda_i - \sum_j a_j \lambda_j = 0$ which is satisfied for nonnegative integers $a_j$ with $\sum_j a_j \geq 2$, then the equation can be linearized to any desired algebraic order.

The question of $C^\infty$ or analytic linearization is more difficult than the formal problem considered above. When all the eigenvalues have real part with the same sign (0 is a source or a sink), Poincaré was able to solve the analytic problem. When 0 is a saddle and there are eigenvalues of opposite sign, the problem has *small divisors*. The analytic linearization problem depends on arithmetic *diophantine* conditions satisfied by the eigenvalues. It was solved by Siegel [1952], while the $C^\infty$ linearization problem was solved later by Sternberg [1958]. Nonetheless, the observations above should help to explain why Hartman's theorem is only able to guarantee linearization via a homeomorphism.

EXERCISE 3.3.1. Show that the system

$$\dot{x} = x + o(|x|, |y|),$$

$$\dot{y} = -y + o(|x|, |y|)$$

cannot be linearized by a smooth coordinate change. In particular, show that cubic terms of the forms $\begin{pmatrix} x^2 y \\ 0 \end{pmatrix}$ and $\begin{pmatrix} 0 \\ xy^2 \end{pmatrix}$ cannot be removed.

For bifurcation theory we are specifically interested in equilibria at which there are eigenvalues with zero real parts. At such equilibria, the linearization problem cannot be solved and there are (nonlinear) *resonance terms* in $f$ which cannot be removed by coordinate changes. The normal form theorem formulates systematically how well one can do, using the procedure analogous to that outlined above to solve the linearization problem for hyperbolic equilibria. The key observations which form the basis of the computations are: (1) that the solvability depends only on the linear part of the vector field; and (2) that the problem can be reduced to a sequence of linear equations to be solved. The final result is a Taylor series for the vector field which contains *only* the essential resonant terms.

If $L = Df(0)x$ denotes the linear part of (3.3.1) at $x = 0$, then $L$ induces a map ad $L$ on the linear space $H_k$ of vector fields whose coefficients are homogeneous polynomials of degree $k$. The map ad $L$ is defined by

$$\text{ad } L(Y) = [Y, L] = DLY - DYL, \tag{3.3.8}$$

where $[\cdot, \cdot]$ denotes the *Lie bracket* operation (Abraham and Marsden [1978], Choquet-Bruhat *et al.* [1977]). In component form, we have

$$[Y, L]^i = \sum_{j=1}^{n} \left( \frac{\partial L^i}{\partial y_j} Y^j - \frac{\partial Y^i}{\partial y_j} L^j \right). \tag{3.3.9}$$

The main result is

**Theorem 3.3.1.** *Let $\dot{x} = f(x)$ be a $C^r$ system of differential equations with $f(0) = 0$ and $Df(0)x = L$. Choose a complement $G_k$ for ad $L(H_k)$ in $H_k$, so that $H_k = \text{ad } L(H_k) + G_k$. Then there is an analytic change of coordinates in a neighborhood of the origin which transforms the system $\dot{x} = f(x)$ to $\dot{y} = g(y) = g^{(1)}(y) + g^{(2)}(y) + \cdots + g^{(r)}(y) + R$, with $L = g^{(1)}(y)$ and $g^{(k)} \in G_k$ for $2 \le k \le r$ and $R_r = o(|y|^r)$.*

PROOF. We give a constructive proof which can be used to implement the calculations of normal forms in examples. The procedure follows the pattern in our discussion of the linearization problem. We use induction and assume that $\dot{x} = f(x)$ has been transformed so that the terms of degree smaller than $s$ lie in the complementary subspace $G_i$, $2 \le i < s$. We then introduce a coordinate transformation of the form $x = h(y) = y + P(y)$, where $P$ is a homogeneous polynomial of degree $s$ whose coefficients are to be determined. Substitution then gives the equation

$$(I + DP(y))\dot{y} = f^{(1)}(y) + f^{(2)}(y) + \cdots + f^{(s)}(y) + Df(0)P(y) + o(|y|^s). \tag{3.3.10}$$

The terms of degree smaller than $s$ are unchanged by this transformation, while the new terms of degree $s$ are

$$f^{(s)}(y) + DLP(y) - DP(y)L = f^{(s)}(y) + \text{ad } L(P)(y), \tag{3.3.11}$$

where $L(y) = f^{(1)}(y)$. Clearly a suitable choice of $P$ will make

$$f^s(y) + \text{ad } L(P)(y)$$

lie in $G_s$ as desired.

In the computations above we have neglected higher-order terms $(O(|y|^s))$ at each stage. This is fine if, as at present, we merely wish to derive the normal forms of vector fields with specific linear parts $L$, but *general* nonlinear parts. However, the reader should realize that the successive transformations introduce additional higher-order terms at each stage, all of which must be retained (up to some order) if *specific* coefficients of the terms in a given normal form are to be computed. For example, in our study of the Hopf bifurcation in Section 3.4 we shall see that, while quadratic terms can be completely removed from the normal form, they nonetheless make an important contribution to the coefficients of the two cubic terms which cannot be removed.

We now illustrate the normal form procedure for a planar system which has an equilibrium with eigenvalues $\pm i$. In the appropriate linear coordinate system, $DL$ is given by $\begin{pmatrix} 0 & -1 \\ +1 & 0 \end{pmatrix}$, and we have $L = \begin{pmatrix} -y \\ x \end{pmatrix}$. We compute the action of ad $L$ on $H_2$ for the monomial vector fields in the following manner. For each basis vector $Y_i$ of $H_2$ we have

$$\text{ad } L(Y_i) = DLY_i - DY_iL$$

$$= \begin{bmatrix} 0 & -1 \\ 1 & 0 \end{bmatrix} \begin{pmatrix} Y_i^1 \\ Y_i^2 \end{pmatrix} - \begin{bmatrix} \dfrac{\partial Y_i^1}{\partial x} & \dfrac{\partial Y_i^1}{\partial y} \\ \dfrac{\partial Y_i^2}{\partial x} & \dfrac{\partial Y_i^2}{\partial y} \end{bmatrix} \begin{pmatrix} -y \\ x \end{pmatrix}; \qquad (3.3.12)$$

thus, for example,

$$\text{ad } L\begin{pmatrix} x^2 \\ 0 \end{pmatrix} = \begin{pmatrix} 2xy \\ x^2 \end{pmatrix}; \qquad \text{ad } L\begin{pmatrix} 0 \\ xy \end{pmatrix} = \begin{pmatrix} -xy \\ y^2 - x^2 \end{pmatrix}, \text{ etc.} \qquad (3.3.13)$$

In what follows we denote the components $Y^1$, $Y^2$ of the vector field $Y$ by the first-order partial operators $\partial/\partial x$, $\partial/\partial y$, so that, for example, $2xy(\partial/\partial x) + x^2(\partial/\partial y)$ represents the vector field $\begin{pmatrix} 2xy \\ x^2 \end{pmatrix}$.

Taking the basis $x^2$, $xy$, $y^2$ in each component $\partial/\partial x$, $\partial/\partial y$, we therefore obtain the table for $H_2$:

| | $x^2$ | $xy$ | $y^2$ |
|---|---|---|---|
| $\dfrac{\partial}{\partial x}$ | $2xy\dfrac{\partial}{\partial x} + x^2\dfrac{\partial}{\partial y}$ | $(y^2 - x^2)\dfrac{\partial}{\partial x} + xy\dfrac{\partial}{\partial y}$ | $-2xy\dfrac{\partial}{\partial x} + y^2\dfrac{\partial}{\partial y}$ |
| $\dfrac{\partial}{\partial y}$ | $-x^2\dfrac{\partial}{\partial x} + 2xy\dfrac{\partial}{\partial y}$ | $-xy\dfrac{\partial}{\partial x} + (y^2 - x^2)\dfrac{\partial}{\partial y}$ | $-y^2\dfrac{\partial}{\partial x} - 2xy\dfrac{\partial}{\partial y}$ |

Thus, in terms of the basis $\{x^2(\partial/\partial x), xy(\partial/\partial x), y^2(\partial/\partial x), x^2(\partial/\partial y), xy(\partial/\partial y), y^2(\partial/\partial y)\}$, ad $L$ has the following matrix, in which the components of ad $L(Y_i)$

appear in the column corresponding to the $Y_i$ in the row above the matrix:

$$
\begin{array}{c|cccccc}
 & x^2\frac{\partial}{\partial x} & xy\frac{\partial}{\partial x} & y^2\frac{\partial}{\partial x} & x^2\frac{\partial}{\partial y} & xy\frac{\partial}{\partial y} & y^2\frac{\partial}{\partial y} \\
\hline
x^2(\partial/\partial x) & 0 & -1 & 0 & -1 & 0 & 0 \\
xy(\partial/\partial x) & +2 & 0 & -2 & 0 & -1 & 0 \\
y^2(\partial/\partial x) & 0 & +1 & 0 & 0 & 0 & -1 \\
x^2(\partial/\partial y) & +1 & 0 & 0 & 0 & -1 & 0 \\
xy(\partial/\partial y) & 0 & +1 & 0 & +2 & 0 & -2 \\
y^2(\partial/\partial y) & 0 & 0 & +1 & 0 & +1 & 0
\end{array}
$$

This matrix is nonsingular (its determinant is $-9$), so *all* quadratic terms of a vector field with linear part $-y(\partial/\partial x) + x(\partial/\partial y)$ can be removed by coordinate transformation.

For cubic terms, similar calculations lead to the table:

| | $x^3$ | $x^2y$ |
|---|---|---|
| $\dfrac{\partial}{\partial x}$ | $+3x^2y\dfrac{\partial}{\partial x} + x^3\dfrac{\partial}{\partial y}$ | $(2xy^2 - x^3)\dfrac{\partial}{\partial x} + x^2y\dfrac{\partial}{\partial y}$ |
| $\dfrac{\partial}{\partial y}$ | $-x^3\dfrac{\partial}{\partial x} + 3x^2y\dfrac{\partial}{\partial y}$ | $-x^2y\dfrac{\partial}{\partial x} + (2xy^2 - x^3)\dfrac{\partial}{\partial y}$ |

| | $xy^2$ | $y^3$ |
|---|---|---|
| $\dfrac{\partial}{\partial x}$ | $(y^3 - 2x^2y)\dfrac{\partial}{\partial x} + xy^2\dfrac{\partial}{\partial y}$ | $-3xy^2\dfrac{\partial}{\partial x} + y^3\dfrac{\partial}{\partial y}$ |
| $\dfrac{\partial}{\partial y}$ | $-xy^2\dfrac{\partial}{\partial x} + (y^3 - 2x^2y)\dfrac{\partial}{\partial y}$ | $-y^3\dfrac{\partial}{\partial x} - 3xy^2\dfrac{\partial}{\partial y}$ |

The matrix of ad $L$ on $H_3$ is therefore

$$
\begin{array}{c|cccccccc}
 & x^3\frac{\partial}{\partial x} & x^2y\frac{\partial}{\partial x} & xy^2\frac{\partial}{\partial x} & y^3\frac{\partial}{\partial x} & x^3\frac{\partial}{\partial y} & x^2y\frac{\partial}{\partial y} & xy^2\frac{\partial}{\partial y} & y^3\frac{\partial}{\partial y} \\
\hline
x^3(\partial/\partial x) & 0 & -1 & 0 & 0 & -1 & 0 & 0 & 0 \\
x^2y(\partial/\partial x) & +3 & 0 & -2 & 0 & 0 & -1 & 0 & 0 \\
xy^2(\partial/\partial x) & 0 & +2 & 0 & -3 & 0 & 0 & -1 & 0 \\
y^3(\partial/\partial x) & 0 & 0 & +1 & 0 & 0 & 0 & 0 & -1 \\
x^3(\partial/\partial y) & +1 & 0 & 0 & 0 & 0 & -1 & 0 & 0 \\
x^2y(\partial/\partial y) & 0 & +1 & 0 & 0 & +3 & 0 & -2 & 0 \\
xy^2(\partial/\partial y) & 0 & 0 & +1 & 0 & 0 & +2 & 0 & -3 \\
y^3(\partial/\partial y) & 0 & 0 & 0 & +1 & 0 & 0 & +1 & 0
\end{array}
$$

We observe that the vectors $(3, 0, 1, 0, 0, 1, 0, 3)$ and $(0, 1, 0, 3, -3, 0, -1, 0)$ are left eigenvectors for zero for this matrix. Therefore, ad $L(H_3)$ has a complement which is at least two dimensional. Further calculation easily shows that columns 1, 2, 3, 4, 5, and 8 are linearly independent, so that ad $L(H_3)$ does have dimension 6. A basis for the complement $G'_3$ can be chosen to be the vector fields $(x^2 + y^2)(x(\partial/\partial x) + y(\partial/\partial y))$ and $(x^2 + y^2)(-y(\partial/\partial x) + x(\partial/\partial y))$. Expressed in terms of systems of differential equations, we have shown that the normal form theorem gives a coordinate transformation which transforms the system

$$\dot{x} = -y + o(|x|, |y|),$$
$$\dot{y} = x + o(|x|, |y|),$$

(3.3.14)

into the system

$$\dot{u} = -v + (au - bv)(u^2 + v^2) + \text{higher-order terms},$$
$$\dot{v} = u + (av + bu)(u^2 + v^2) + \text{higher-order terms},$$

(3.3.15)

for suitable constants $a$ and $b$. This result plays an important rôle in the analysis of the Hopf bifurcation in Section 4 of this chapter.

In general, denoting the matrix of ad $L$ on $H_k$ by $M_k$, we have that ad $L(H_k)$ is the *column space* of $M_k$ (with the appropriate assignment of vector fields). Thus, if $M_k$ has zero eigenvalues, elementary facts from linear algebra imply that a complementary space $G_k$ to ad $L(H_k)$ can be chosen to be $G_k = \text{span}\{e_j\}$, where the $e_j$ are left eigenvectors of 0 for $M_k$. (Each $e_j$ is automatically orthogonal to every column of $M_k$.) However, this choice of $G_k$ is far from unique, and may not be the most convenient. In the example above, for instance, rather than taking the vector fields $(3x^2 + y^2)x(\partial/\partial x) + (x^2 + 3y^2)y(\partial/\partial y)$ and $(x^2 + 3y^2)y(\partial/\partial x) - (3x^2 + y^2)x(\partial/\partial y)$, corresponding to the left eigenvectors given, it is better to take $G'_3 = \text{span}\{(1, 0, 1, 0, 0, 1, 0, 1), (0, -1, 0, -1, 1, 0, 1, 0)\}$, giving the normal form of (3.3.15). The reader can check that these latter vectors, together with the eight columns of $M_3$, do span $H_3$. Moreover, as we shall see in Section 3.4, they have a particularly simple representation in polar coordinates.

We also remark that, if the linear part $L$ of the original vector field is symmetric for some compact group action (i.e., $L$ is equivariant), then $G_k$ can also be chosen to be invariant under the same action. This fact will play an important role in our discussion of multiple bifurcations in Chapter 7. We already see that in the pure imaginary eigenvalue example above, the normal form (3.3.15) truncated at cubic terms is invariant under arbitrary rotations, just as is the linear field $L = \begin{pmatrix} -y \\ x \end{pmatrix}$. Successive applications of the normal form procedure show that the invariance persists to all algebraic orders.

EXERCISE 3.3.2. Show that the system

$$\dot{x} = y + o(|x|, |y|),$$
$$\dot{y} = o(|x|, |y|)$$

can be transformed to

$$\dot{u} = v + au^2$$
$$\dot{v} = bu^2 \qquad + o(|u|^2, |v|^2),$$

or to

$$\dot{u} = v$$
$$\dot{v} = au^2 + buv \qquad + o(|u|^2, |v|^2).$$

Find a basis for $G_3$ and give the normal form up to the third order (cf. Chapter 7).

To end this section we discuss the rôle of parameters in normal form calculations. As in the computation of center manifolds for parametrised systems, we again employ the trick of extending the system $\dot{x} = f(x, \mu)$ to the larger system

$$\dot{x} = f(x, \mu),$$
$$\dot{\mu} = 0. \qquad (3.3.16)$$

One can perform the normal form calculations in this larger system while insisting that the coordinate transformations $H(x, \mu)$ all be of the form $H(x, \mu) = (h(x, \mu), \mu)$. These transformations necessarily leave the equation $\dot{\mu} = 0$ invariant and will transform the system $\dot{x} = f(x, \mu)$ in a $\mu$-dependent way. In practice, these calculations proceed as before, but the coefficients are regarded as power series in the parameters $\mu$.

The normal form theorem described in this section is far from being the final word concerning the question of when two vector fields can be transformed into one another by a smooth change of coordinates. Apart from the Siegel and Sternberg linearization theorem referred to above, Takens [1973a] gives results concerning this question which apply to vector fields having a single zero eigenvalue or a single pair of pure imaginary eigenvalues. For equilibria with more degenerate linear parts little appears to be known.

## 3.4. Codimension One Bifurcations of Equilibria

In this section we describe the simplest bifurcations of equilibria. These are represented by the following four differential equations which depend on a single parameter $\mu$:

$$\dot{x} = \mu - x^2 \quad (saddle\text{-}node), \qquad (3.4.1)$$
$$\dot{x} = \mu x - x^2 \quad (transcritical), \qquad (3.4.2)$$
$$\dot{x} = \mu x - x^3 \quad (pitchfork), \qquad (3.4.3)$$

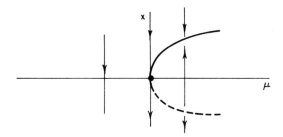

Figure 3.4.1. Saddle-node bifurcation.

and

$$\begin{cases} \dot{x} = -y + x(\mu - (x^2 + y^2)) \\ \dot{y} = x + y(\mu - (x^2 + y^2)) \end{cases} \quad (Hopf). \tag{3.4.4}$$

The bifurcation diagrams for these four equations are depicted in Figures 3.4.1–3.4.4. Each of the equations (3.4.1)–(3.4.4) arises naturally in a suitable context as determining the local qualitative behavior of the *generic* bifurcation of an equilibrium. Our purpose here is to describe in detail how, and under what conditions, one can reduce the study of the general equation (3.1.1) to one of these four specific examples.

## The Saddle-Node

Consider a system of equations

$$\dot{x} = f_\mu(x), \tag{3.4.5}$$

with $x \in \mathbb{R}^n$, $\mu \in \mathbb{R}$, and $f_\mu$ smooth. Assume that at $\mu = \mu_0$, $x = x_0$, (3.4.5) has an equilibrium at which there is a zero eigenvalue (for the linearization). Usually, this zero eigenvalue will be simple, and the center manifold theorem allows us to reduce the study of this kind of bifurcation problem to one in which $x$ is one dimensional. More precisely, using the ideas of Section 3.2,

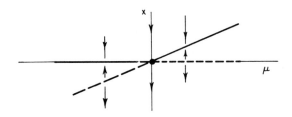

Figure 3.4.2. Transcritical bifurcation.

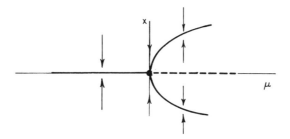

Figure 3.4.3. Pitchfork bifurcation (supercritical).

we can find a two-dimensional center manifold $\Sigma \subset \mathbb{R}^n \times \mathbb{R}$ passing through $(x_0, \mu_0)$ such that

(1) The tangent space of $\Sigma$ at $(x_0, \mu_0)$ is spanned by an eigenvector of $0$ for $Df_{\mu_0}(x_0)$ and a vector parallel to the $\mu$-axis.
(2) For any finite $r$, $\Sigma$ is $C^r$ if restricted to a small enough neighborhood of $(x_0, \mu_0)$.
(3) The vector field of (3.4.5) is tangent to $\Sigma$.
(4) There is a neighborhood $U$ of $(x_0, \mu_0)$ in $\mathbb{R}^n \times \mathbb{R}$ such that all trajectories contained entirely in $U$ for all time lie in $\Sigma$.

(Note: The center manifold theorem allows one to formulate stronger properties than (4) which describe the qualitative structure of trajectories which remain close to $(x_0, \mu_0)$ in forward time or in backwards time, cf. Carr [1981].)

Restricting (3.4.5) to $\Sigma$, we obtain a one-parameter family of equations on the one-dimensional curves $\Sigma_\mu$ in $\Sigma$ obtained by fixing $\mu$ (cf. Figure 3.2.7). This one-parameter family is our reduction of the bifurcation problem.

Let us now formulate transversality conditions for a system (3.4.5) with $n = 1$, which yield the saddle-node bifurcation. We have $(df_{\mu_0}/dx)(x_0) = 0$, but we take $(\partial f_{\mu_0}/\partial \mu)(x_0) \neq 0$ as a transversality condition. The implicit function theorem then implies that the equilibria of (3.4.5) form a curve which will be tangent to the line $\mu = \mu_0$. An additional transversality condition $(d^2 f_{\mu_0}/dx^2)(x_0) \neq 0$ implies that the curve of equilibria has a *quadratic* tangency with $\mu = \mu_0$ and locally lies to one side of this line.

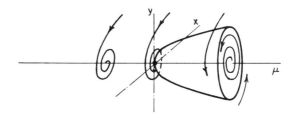

Figure 3.4.4. Hopf bifurcation (supercritical).

This information is already sufficient to imply that the local phase portraits of this system are topologically equivalent to those of a family $\dot{x} = \pm(\mu - \mu_0) \pm (x - x_0)^2$. However, we can also formulate these transversality conditions for an $n$-dimensional system without recourse to the center manifold reduction. The following theorem states the necessary conditions (cf. Sotomayor [1973]).

**Theorem 3.4.1.** *Let* $\dot{x} = f_\mu(x)$ *be a system of differential equations in* $\mathbb{R}^n$ *depending on the single parameter* $\mu$. *When* $\mu = \mu_0$, *assume that there is an equilibrium for which the following hypotheses are statisfied:*

(SN1)   $D_x f_{\mu_0}(p)$ *has a simple eigenvalue* 0 *with right eigenvector* $v$ *and left eigenvector* $w$. $D_x f_{\mu_0}(p)$ *has* $k$ *eigenvalues with negative real parts and* $(n - k - 1)$ *eigenvalues with positive real parts* (*counting multiplicity*).

(SN2)   $w((\partial f_\mu/\partial \mu)(p, \mu_0)) \neq 0$.

(SN3)   $w(D_x^2 f_{\mu_0}(p)(v, v)) \neq 0$.

*Then there is a smooth curve of equilibria in* $\mathbb{R}^n \times \mathbb{R}$ *passing through* $(p, \mu_0)$, *tangent to the hyperplane* $\mathbb{R}^n \times \{\mu_0\}$. *Depending on the signs of the expressions in* (SN2) *and* (SN3), *there are no equilibria near* $(p, \mu_0)$ *when* $\mu < \mu_0$ ($\mu > \mu_0$) *and two equilibria near* $(p, \mu_0)$ *for each parameter value* $\mu > \mu_0$ ($\mu < \mu_0$). *The two equilibria for* $\dot{x} = f_\mu(x)$ *near* $(p, \mu_0)$ *are hyperbolic and have stable manifolds of dimensions* $k$ *and* $k + 1$, *respectively. The set of equations* $\dot{x} = f_\mu(x)$ *which satisfy* (SN1)–(SN3) *is open and dense in the space of* $C^\infty$ *one-parameter families of vector fields with an equilibrium at* $(p, \mu_0)$ *with a zero eigenvalue.*

This formal (and formidable) theorem merely expresses the fact that the "generic" saddle node bifurcation is qualitatively like the family of equations $\dot{x} = \mu - x^2$ in the direction of the zero eigenvector, with hyperbolic behavior in the complementary directions. Hypotheses (SN2) and (SN3) are the transversality conditions which control the nondegeneracy of the behavior with respect to the parameter and the dominant effect of the quadratic nonlinear term.

EXERCISE 3.4.1. Consider the variational van der Pol equations (equation (2.1.14))

$$\dot{u} = u - \sigma v - u(u^2 + v^2),$$
$$\dot{v} = \sigma u + v - v(u^2 + v^2) - \gamma.$$

Find the locus of saddle-node bifurcations in $(\sigma, \gamma)$ space. At which point(s) do conditions (SN1) and (SN3) fail? (This involves tedious calculations, since a cubic equation must be solved for the fixed points.)

The results obtained from Theorem 3.4.1 are limited in two different ways. On the one hand, it is possible that more quantitative information about the flows near bifurcation can be extracted. For example, one can use the system $\dot{x} = \mu - x^2$ to give estimates of how rapid the convergence to the various equilibria are. Higher-order terms in the Taylor expansion of an equation can be used to refine these estimates. This is an aspect of the theory of differential equations which we do not pursue further in this book because our attention is to focus on geometric issues rather than analytic ones. In this regard, the reader should be reminded that we often do not strive to state the strongest or most general theorem for a given situation but rather aim to illustrate the phenomena and methods of analysis in the simplest ways.

The second limitation of Theorem 3.4.1 is that there may be global changes in a phase portrait associated with a saddle-node bifurcation. Consider, for example, the flows depicted in Figure 3.4.5, which we have already met in the van der Pol example of Section 2.1 (cf. Figure 2.1.3). Here a saddle-node in a two-dimensional system occurs, with the coalescense of a sink and a saddle. After the bifurcation, there is a new periodic orbit which has appeared because the unstable separatrix at the saddle-node lies in the two-dimensional stable manifold of the bifurcating equilibrium. This is an example of a *global bifurcation* phenomenon that cannot be reduced to the study of a neighborhood of an equilibrium or a fixed point in a return map. We return to global bifurcation problems in Chapter 6.

## Transcritical and Pitchfork Bifurcations

The importance of the saddle-node bifurcation is that all bifurcations of one-parameter families at an equilibrium with a zero eigenvalue can be perturbed to saddle-node bifurcations. Thus one expects that the zero eigenvalue bifurcations encountered in applications will be saddle-nodes. If they are not, then there is probably something special about the formulation of the problem which restricts the context so as to prevent the saddle-node from occurring. The *transcritical* bifurcation is one example which illustrates how the setting of the problem can rule out the saddle-node bifurcation.

In classical bifurcation theory, it is often assumed that there is a trivial solution from which bifurcation is to occur. Thus, the systems (3.4.5) are assumed to satisfy $f_\mu(0) = 0$ for all $\mu$, so that $x = 0$ is an equilibrium for all

Figure 3.4.5. A saddle node occurring on a closed curve leads to global bifurcation.

parameter values. Since the saddle-node families contain parameter values for which there are *no* equilibria near the point of bifurcation, this situation is qualitatively different. To formulate the appropriate transversality conditions we look at the one-parameter families which satisfy the constraint that $f_\mu(0) = 0$ for all $\mu$. This prevents hypothesis (SN2) of Theorem 3.4.1 from being satisfied. If we replace this condition by the requirement that $w((\partial^2 f / \partial \mu\, \partial x)(v)) \neq 0$ at $(0, \mu_0)$, then the phase portraits of the family near the bifurcation will be topologically equivalent to those of Figure 3.4.2, and we have a *transcritical bifurcation* or *exchange of stability*.

A second setting in which the saddle-node does not occur involves systems which have a symmetry. Many physical problems are formulated so that the equations defining the system do have symmetries of some kind. For example, the Duffing equation is invariant under the transformation $(x, y) \rightarrow (-x, -y)$ and the Lorenz equation is symmetric under the transformation $(x, y, z) \rightarrow (-x, -y, z)$. In one dimension, a differential equation (3.4.5) is symmetric or *equivariant* with respect to the symmetry $x \rightarrow -x$ if $f_\mu(-x) = -f_\mu(x)$. Thus the equivariant vector fields are ones for which $f_\mu$ is an odd function of $x$. In particular, all such equations have an equilibrium at 0. The transcritical bifurcation cannot occur in these systems, however, because an odd function $f_\mu$ cannot satisfy the condition $\partial^2 f_\mu / \partial x^2 \neq 0$ required by the transcritical bifurcation (cf. SN3). If this condition is replaced by the transversality hypothesis $\partial^3 f_\mu / \partial x^3 \neq 0$, then one obtains the *pitchfork bifurcation*. At the point of bifurcation, the stability of the trivial equilibrium changes, and a new *pair* of equilibria (related by the symmetry) appear to one side of the point of bifurcation in parameter space, as in Figure 3.4.3. We leave to the reader the formulation of results analogous to Theorem 3.4.1 for the transcritical and pitchfork bifurcations (cf. Sotomayor [1973]).

We note that the direction of the bifurcation and the stability of the branches in these examples is determined by the sign of $\partial^2 f_\mu / \partial x^2$ or $\partial^3 f_\mu / \partial x^3$. In the last case, if $\partial^3 f_\mu / \partial x^3$ is negative, then the branches occur "above" the bifurcation value and we have a *supercritical* pitchfork bifurcation, whereas we have a *subcritical* bifurcation if it is positive.

EXERCISE 3.4.2. Compute the normal form for the Lorenz equations at the bifurcation of the origin when $\rho = 1$, through terms of third degree. Note that the bifurcation is a pitchfork even though the analytic expression of the Lorenz equations involves only quadratic terms (cf. Exercise 3.2.5).

EXERCISE 3.4.3. Analyze the pitchfork bifurcation which takes place for the variational van der Pol equation at $(\sigma, \gamma) = (1/\sqrt{3}, \sqrt{8/27})$.

## Hopf Bifurcations

Consider now a system (3.4.5) with a parameter value $\mu_0$ and equilibrium $p(\mu_0)$ at which $Df_{\mu_0}$ has a simple pair of pure imaginary eigenvalues, $\pm i\omega$, $\omega > 0$, and no other eigenvalues with zero real part. The implicit function theorem guarantees (since $Df_{\mu_0}$ is invertible) that for each $\mu$ near $\mu_0$ there will be an equilibrium $p(\mu)$ near $p(\mu_0)$ which varies smoothly with $\mu$. Nonetheless, the dimensions of stable and unstable manifolds of $p(\mu)$ do change if the eigenvalues of $Df(p(u))$ cross the imaginary axis at $\mu_0$. This qualitative change in the local flow near $p(\mu)$ must be marked by some other local changes in the phase portraits not involving fixed points.

A clue to what happens in the generic bifurcation problem involving an equilibrium with pure imaginary eigenvalues can be gained from examining linear systems in which there is a change of this type. For example, consider the system

$$
\begin{aligned}
\dot{x} &= \mu x - \omega y, \\
\dot{y} &= \omega x + \mu y,
\end{aligned}
\tag{3.4.6}
$$

whose solutions have the form

$$
\begin{pmatrix} x(t) \\ y(t) \end{pmatrix} = e^{\mu t} \begin{pmatrix} \cos \omega t & -\sin \omega t \\ \sin \omega t & \cos \omega t \end{pmatrix} \begin{pmatrix} x_0 \\ y_0 \end{pmatrix}.
\tag{3.4.7}
$$

When $\mu < 0$, solutions spiral into the origin, and when $\mu > 0$, solutions spiral away from the origin. When $\mu = 0$, all solutions are periodic. Even in a one-parameter family of equations, it is highly special to find a parameter value at which there is a whole family of periodic orbits, but there is still a surface of periodic orbits which appears in the general problem.

The normal form theorem gives us the required information about how the generic problem differs from the system (3.4.6). By smooth changes of coordinates, the Taylor series of degree 3 for the general problem can be brought to the following form (cf. Equation (3.3.15))

$$
\begin{aligned}
\dot{x} &= (d\mu + a(x^2 + y^2))x - (\omega + c\mu + b(x^2 + y^2))y, \\
\dot{y} &= (\omega + c\mu + b(x^2 + y^2))x + (d\mu + a(x^2 + y^2))y,
\end{aligned}
\tag{3.4.8}
$$

which is expressed in polar coordinates as

$$
\begin{aligned}
\dot{r} &= (d\mu + ar^2)r, \\
\dot{\theta} &= (\omega + c\mu + br^2).
\end{aligned}
\tag{3.4.9}
$$

Since the $\dot{r}$ equation in (3.4.9) separates from $\theta$, we see that there are periodic orbits of (3.4.8) which are circles $r = \text{const.}$, obtained from the nonzero solutions of $\dot{r} = 0$ in (3.4.9). If $a \neq 0$ and $d \neq 0$ these solutions lie along the parabola $\mu = -ar^2/d$. This implies that the surface of periodic orbits has a quadratic tangency with its tangent plane $\mu = 0$ in $\mathbb{R}^2 \times \mathbb{R}$. The content of the Hopf bifurcation theorem is that the qualitative properties of (3.4.8) near the origin remain unchanged if higher-order terms are added to the system:

**Theorem 3.4.2** [Hopf [1942]]. *Suppose that the system* $\dot{x} = f_\mu(x)$, $x \in \mathbb{R}^n$, $\mu \in \mathbb{R}$ *has an equilibrium* $(x_0, \mu_0)$ *at which the following properties are satisfied*:

(H1) $D_x f_{\mu_0}(x_0)$ *has a simple pair of pure imaginary eigenvalues and no other eigenvalues with zero real parts.*

*Then* (H1) *implies that there is a smooth curve of equilibria* $(x(\mu), \mu)$ *with* $x(\mu_0) = x_0$. *The eigenvalues* $\lambda(\mu)$, $\bar{\lambda}(\mu)$ *of* $D_x f_{\mu_0}(x(\mu))$ *which are imaginary at* $\mu = \mu_0$ *vary smoothly with* $\mu$. *If, moreover,*

(H2)                                      $$\frac{d}{d\mu}(\operatorname{Re} \lambda(\mu))|_{\mu = \mu_0} = d \neq 0,$$

*then there is a unique three-dimensional center manifold passing through* $(x_0, \mu_0)$ *in* $\mathbb{R}^n \times \mathbb{R}$ *and a smooth system of coordinates (preserving the planes* $\mu = $ const.*) for which the Taylor expansion of degree 3 on the center manifold is given by* (3.4.8). *If* $a \neq 0$, *there is a surface of periodic solutions in the center manifold which has quadratic tangency with the eigenspace of* $\lambda(\mu_0)$, $\bar{\lambda}(\mu_0)$ *agreeing to second order with the paraboloid* $\mu = -(a/d)(x^2 + y^2)$. *If* $a < 0$, *then these periodic solutions are stable limit cycles, while if* $a > 0$, *the periodic solutions are repelling.*

This theorem can be proved by a direct application of the center manifold and normal form theorems given above (cf. Marsden and McCracken [1976]).

EXERCISE 3.4.4. Find the Hopf bifurcations which occur in the variational van der Pol equations (2.1.14).

EXERCISE 3.4.5. In the Duffing equation,

$$\ddot{x} + \mu\dot{x} + (x - x^3) = 0,$$

a bifurcation with pure imaginary eigenvalues occurs at $\mu = 0$ when the system changes from having negative to positive dissipation, but it is degenerate. Why? Compute the normal form through terms of third degree. What modifications might be made to the system to make the bifurcation "generic"; i.e., satisfy all of the hypotheses of the Hopf theorem?

For large systems of equations, computation of the normal form (3.4.8) and the cubic coefficient $a$, which determines the stability, can be a substantial undertaking.

In a two-dimensional system of the form

$$\begin{pmatrix} \dot{x} \\ \dot{y} \end{pmatrix} = \begin{pmatrix} 0 & -\omega \\ \omega & 0 \end{pmatrix} \begin{pmatrix} x \\ y \end{pmatrix} + \begin{pmatrix} f(x, y) \\ g(x, y) \end{pmatrix}, \tag{3.4.10}$$

with $f(0) = g(0) = 0$ and $Df(0) = Dg(0) = 0$, the normal form calculation which we sketch in the appendix to this section, yields

$$a = \frac{1}{16}[f_{xxx} + f_{xyy} + g_{xxy} + g_{yyy}] + \frac{1}{16\omega}[f_{xy}(f_{xx} + f_{yy})$$

$$- g_{xy}(g_{xx} + g_{yy}) - f_{xx}g_{xx} + f_{yy}g_{yy}], \tag{3.4.11}$$

where $f_{xy}$ denotes $(\partial^2 f/\partial x\,\partial y)(0, 0)$, etc.* In applying this formula to systems of dimension greater than two, however, the reader should recall that the quadratic terms which play a rôle in the center manifold calculations can affect the value of $a$. One cannot find $a$ by simply projecting the system of equations onto the eigenspace of $\pm i\omega$, but must approximate the center manifold at least to quadratic terms (cf. Exercise 3.2.4(b) and Exercise 3.4.8 below). As an example of the use of this algorithm, and of the importance of quadratic terms in the determination of the leading cubic nonlinear term of the normal form, we return to the problem of Section 1.8, Equation (1.8.20). Under the change of coordinates $(x, y) = (\xi + 1, \eta + 1)$, this system becomes

$$\begin{pmatrix} \dot{\xi} \\ \dot{\eta} \end{pmatrix} = \begin{bmatrix} 1 & -1 \\ 2 & -1 \end{bmatrix} \begin{pmatrix} \xi \\ \eta \end{pmatrix} + \begin{pmatrix} \xi^2 - \xi\eta \\ \xi^2 \end{pmatrix}. \tag{3.4.12}$$

The eigenvalues are $\lambda = \pm i$, and, under a further transformation

$$\begin{pmatrix} \xi \\ \eta \end{pmatrix} = \begin{bmatrix} 0 & 1 \\ -1 & 1 \end{bmatrix} \begin{pmatrix} u \\ v \end{pmatrix}, \qquad \begin{pmatrix} u \\ v \end{pmatrix} = \begin{bmatrix} 1 & -1 \\ 1 & 0 \end{bmatrix} \begin{pmatrix} \xi \\ \eta \end{pmatrix}, \tag{3.4.13}$$

we obtain the system in "standard form":

$$\dot{u} = -v + uv - v^2,$$
$$\dot{v} = u + uv. \tag{3.4.14}$$

Clearly, all third derivatives are identically zero and we have

$$f_{uu} = 0, \qquad f_{uv} = 1, \qquad f_{vv} = -2,$$
$$g_{uu} = 0, \qquad g_{uv} = 1, \qquad g_{vv} = 0, \tag{3.4.15}$$

leading to the coefficient

$$a = \tfrac{1}{16}[1(-2)] = -\tfrac{1}{8} < 0, \tag{3.4.16}$$

so that the fixed point $(1, 1)$ is a (weakly) stable sink as claimed in Section 1.8.

EXERCISE 3.4.6. Calculate the stability type of the degenerate fixed point at $(x, y) = (0, 0)$ for the system

$$\dot{x} = -y + \alpha y^2 + \beta x^2 y,$$
$$\dot{y} = x - \gamma y^2 + \delta xy - y^3.$$

How does stability depend upon the values of the coefficients $\alpha$, $\beta$, $\gamma$, $\delta$?

EXERCISE 3.4.7. Show that the system $\ddot{x} + \mu\dot{x} + vx + x^2\dot{x} + x^3 = 0$ undergoes Hopf bifurcations on the lines $B_1\{\mu = 0 | v > 0\}$ and $B_2\{\mu = v | \mu, v < 0\}$. Show that the former is supercritical and occurs at the fixed point $(0, 0)$ while the latter is subcritical and occurs simultaneously at the fixed points $(x, \dot{x}) = (\pm\sqrt{-v}, 0)$. Attempt to sketch

* Chow and Mallet-Paret [1977] present an alternative, but equivalent formula derived by the method of averaging.

the phase portraits of this system for various values of $(\mu, \nu) \in \mathbb{R}^2$. What is the form of the degenerate singularity occurring at $(x, \dot{x}) = (0, 0)$ when $(\mu, \nu) = (0, 0)$? (The global aspects of this problem are difficult and require the use of techniques discussed in Chapters 4 and 7. See Section 7.3 in particular.)

EXERCISE 3.4.8 (For the computationally minded). For $\sigma = 10$, $\beta = \frac{8}{3}$, the nonzero equilibria of the Lorenz equation undergo a Hopf bifurcation with $24 < \rho < 25$. Compute the cubic coefficient which determines stability in this example. (Cf. Marsden and McCracken [1976] for a discussion of this problem, but beware, there are some mistakes in their derivation.)

We end this section by noting that Allwright [1977] and Mees [1981] have obtained Hopf bifurcation criteria by means of harmonic balance and the use of a Liapunov function approach.

## Appendix to Section 3.4: Derivation of Stability Formula (3.4.11)

If the reduced (approximate) system has a purely imaginary pair of eigenvalues $\lambda$, $\bar{\lambda} = \pm i\omega$, then it can be conveniently represented as a single complex equation:

$$\dot{z} = \lambda z + h(z, \bar{z}), \tag{3.4.17}$$

where

$$z = x + iy \quad \text{and} \quad \lambda = i\omega.$$

The normal form (3.4.8) becomes, at $\mu = 0$,

$$\dot{w} = \lambda w + c_1 w^2 \bar{w} + c_2 w^3 \bar{w}^2 + \cdots + c_k w^{k+1} \bar{w}^k + O(|w|^{2k+3})$$
$$\stackrel{\text{def}}{=} \lambda w + \hat{h}(w, \bar{w}), \tag{3.4.18}$$

where the complex coefficients are of the form

$$c_j = a_j + ib_j, \tag{3.4.19}$$

and an overbar denotes complex conjugation.

EXERCISE 3.4.9. Check the assertions above.

Since in polar coordinates we have

$$\dot{r} = a_1 r^3 + a_2 r^5 + \cdots,$$
$$\dot{\theta} = \omega + b_1 r^2 + b_2 r^4 + \cdots, \tag{3.4.20}$$

the first nonvanishing coefficients $a_j, b_j$ determine the stability (and local amplitude growth) of the periodic orbit and the amplitude dependent modification to its period.

Thus far we have merely recast our system into complex form. Now, following Hassard and Wan [1978], we will show how this form enables us to calculate the leading coefficient, $a_1 = \text{Re}(c_1)$, relatively simply. The computations are considerably easier than those of Marsden and McCracken [1976]. To transform (3.4.17) to (3.4.18) we use the near identity transformation

$$z = w + \psi(w, \overline{w}), \qquad \psi = O(|w|^2). \tag{3.4.21}$$

Substituting (3.4.21) in (3.4.17) and using (3.4.18) we obtain

$$\lambda(w\psi_w - \psi) + \overline{\lambda w \psi_{\overline{w}}} = h(w + \psi, \overline{w} + \overline{\psi}) - \hat{h}(w, \overline{w})(1 + \psi_w) - \overline{\hat{h}(w, \overline{w})}\psi_{\overline{w}}, \tag{3.4.22}$$

where subscripts denote partial differentiation. We now express $\psi$ as a Taylor series (with $\psi_{jk} = \partial \psi^{j+k}/\partial w^j \partial \overline{w}^k$):

$$\psi(w, \overline{w}) = \sum_{2 \le j+k \le 3} \psi_{jk} \frac{w^j \overline{w}^k}{j! k!} + O(|w|^4). \tag{3.4.23}$$

Next, using the fact that the normal form $\hat{h}(w, \overline{w}) = c_1 w^2 \overline{w} + O(|w|^5)$ and substituting (3.4.23) in (3.4.22), we obtain

$$\lambda \psi_{ww} \frac{w^2}{2} + \overline{\lambda} \psi_{w\overline{w}} w \overline{w} + (2\overline{\lambda} - \lambda)\psi_{\overline{w}\overline{w}} \frac{\overline{w}^2}{2} = h_{ww} \frac{w^2}{2} + h_{w\overline{w}} w \overline{w} + h_{\overline{w}\overline{w}} \frac{\overline{w}^2}{2}$$

$$+ O(|w|^3). \tag{3.4.24}$$

Equating coefficients yields the leading terms in the transformation

$$\psi_{ww} = \frac{h_{ww}}{\lambda} = -\frac{ih_{ww}}{\omega}, \qquad \psi_{w\overline{w}} = \frac{h_{w\overline{w}}}{\overline{\lambda}} = \frac{ih_{w\overline{w}}}{\omega},$$

$$\psi_{\overline{w}\overline{w}} = \frac{h_{\overline{w}\overline{w}}}{(2\overline{\lambda} - \lambda)} = \frac{ih_{\overline{w}\overline{w}}}{3\omega}. \tag{3.4.25}$$

We now carry out the expansion to one higher order and equate the coefficients of the normal form term $w^2 \overline{w}$. The reader can check that, for this term, the coefficient on the left-hand side of (3.4.22) vanishes identically and the right-hand side therefore becomes

$$h_{ww}\psi_{w\overline{w}} + h_{w\overline{w}}\left(\frac{\psi_{ww}}{2} + \overline{\psi}_{w\overline{w}}\right) + h_{\overline{w}\overline{w}} \frac{\overline{\psi}_{\overline{w}\overline{w}}}{2} + \frac{h_{ww\overline{w}}}{2} - c_1 = 0,$$

or, using (3.4.25)

$$c_1 = \frac{i}{2\omega}(h_{ww}h_{w\overline{w}} - 2|h_{w\overline{w}}|^2 - \tfrac{1}{3}|h_{\overline{w}\overline{w}}|^2) + \frac{h_{ww\overline{w}}}{2}. \tag{3.4.26}$$

Hence we have

$$2a_1 = 2\,\text{Re}(c_1) = h_{ww\overline{w}}^R - \frac{1}{\omega}(h_{ww}^R h_{w\overline{w}}^I + h_{w\overline{w}}^R h_{ww}^I), \tag{3.4.27}$$

where the superscripts $R$, $I$ denote real and imaginary parts, respectively. A similar expression can be found for $b_1$. In this way we see precisely how the third-order terms are modified by our transformation $\phi = \text{id} + \psi$, with which we have removed the second-order terms. In Hassard and Wan's [1978] paper the second (fifth-order) coefficient $c_2$ is calculated, and the system is embedded in a higher-dimensional problem, so that one also has additional terms arising from the center manifold approximation.

We stress that these calculations can be carried out for the original system in real variables, but that they are considerably more cumbersome in that form. However, since we are typically working with systems in real form, it is convenient to express $a_1$ in terms of the real functions $f$, $g$ of Equation (3.4.8). Expanding $f$ and $g$ in Taylor series and taking real and imaginary parts of the complex valued function $h$ (and its series) we find that the relevant terms in (3.4.27) may be expressed as

$$
\left.
\begin{aligned}
h^R_{ww\bar{w}} &= \tfrac{1}{8}(f_{xxx} + f_{xyy} + g_{xxy} + g_{yyy}), \\
h^R_{ww} &= \tfrac{1}{4}(f_{xx} - f_{yy} + 2g_{xy}), \\
h^I_{ww} &= \tfrac{1}{4}(g_{xx} - g_{yy} - 2f_{xy}), \\
h^R_{w\bar{w}} &= \tfrac{1}{4}(f_{xx} + f_{yy}), \\
h^I_{w\bar{w}} &= \tfrac{1}{4}(g_{xx} + g_{yy}).
\end{aligned}
\right\} \tag{3.4.28}
$$

EXERCISE 3.4.10. Verify (3.4.28) and find expressions for $|h_{w\bar{w}}|^2$, $|h_{\bar{w}\bar{w}}|^2$ and $h^I_{ww\bar{w}}$ in terms of $f$ and $g$, so that you can calculate $b_1 = \text{Im}(c_1)$.

The stability formula (3.4.11) can now be derived by substitution of the expressions of (3.4.28) into (3.4.27):

$$
16a_1 = (f_{xxx} + f_{xyy} + g_{xxy} + g_{yyy})
$$

$$
+ \frac{1}{\omega}[f_{xy}(f_{xx} + f_{yy}) - g_{xy}(g_{xx} + g_{yy}) - f_{xx}g_{xx} + f_{yy}g_{yy}]. \tag{3.4.29}
$$

We end by noting that the normal form for the parametrized Hopf bifurcation is neatly expressed in complex variables as

$$
\dot{w} = \lambda w + c_1 w^2 \bar{w} + O(|w|^5), \tag{3.4.30}
$$

when $\lambda = \mu + i\omega$; cf. Arnold [1972].

## 3.5. Codimension One Bifurcations of Maps and Periodic Orbits

In this section we consider the simplest bifurcations for periodic orbits. The strategy that we adopt involves computing Poincaré return maps and then trying to repeat the results of Section 3.4 for these discrete dynamical systems. There are some additional complications that introduce new subtleties to some of these problems. In practice, computations of the bifurcations of periodic orbits from a defining system of equations are substantially more difficult than those for equilibria because one must first integrate the equations near the periodic orbit to find the Poincaré return map before further analysis can proceed. Thus, the results obtained here have been most frequently applied:

(1) in comparison with numerical calculations;
(2) directly to discrete dynamical systems defined by a mapping; or
(3) in perturbation situations close to ones in which a system can be explicitly integrated.

The third category will form the subject of Chapter 4. In view of these computational difficulties, in this section we shall focus upon the geometric aspects of these bifurcations.

There are three ways in which a fixed point $p$ of a discrete mapping $f: \mathbb{R}^n \to \mathbb{R}^n$ may fail to be hyperbolic: $Df(p)$ may have an eigenvalue $+1$, an eigenvalue $-1$, or a pair of complex eigenvalues $\lambda, \bar{\lambda}$ with $|\lambda| = 1$. (If $Df(p)$ has an eigenvalue $\mu$ at the fixed point $p$, we say $p$ has eigenvalue $\mu$.) The bifurcation theory for fixed points with eigenvalue 1 is completely analogous to the bifurcation theory for equilibria with eigenvalue 0. The generic one-parameter family has a two-dimensional center manifold (including the parameter direction) on which it is topologically equivalent to the *saddle-node* family defined by the map

$$f_\mu(x) = x + \mu - x^2. \tag{3.5.1}$$

The same considerations of constraint and symmetry as discussed in Section 4 alter the generic picture, giving either *transcritical* or *pitchfork* bifurcations. Rather than working out examples in detail, we leave the computation of the following exercises to the reader:

EXERCISE 3.5.1. Show that the map $x \to \mu - x^2$ undergoes a saddle-node bifurcation at $(x, \mu) = (-\frac{1}{2}, -\frac{1}{4})$. On which side of the bifurcation value $\mu = -\frac{1}{4}$ do the fixed points lie?

EXERCISE 3.5.2. Show that the map $x \to \mu x(1 - x)$ undergoes a transcritical bifurcation at $(x, \mu) = (0, 1)$.

EXERCISE 3.5.3. Show that the map $(x, y) \to (y, -\frac{1}{2}x + \mu y - y^3)$ undergoes a pitchfork bifurcation at $(x, y, \mu) = (0, 0, \frac{3}{2})$. Is it sub- or supercritical? Approximate a suspended center manifold near $(0, 0, \frac{3}{2})$ and sketch the bifurcation diagram (cf. Exercise 3.2.8(c)).

Bifurcations with eigenvalue $-1$ do not have an analogue for equilibria, while the theory for complex eigenvalues is more subtle than that of the Hopf bifurcation for flows.

Eigenvalues with $-1$ are associated with *flip* bifurcations, also referred to as *period doubling* or *subharmonic* bifurcations. Using a center manifold reduction, we restrict our attention to one-dimensional mappings $f_\mu$ and assume that $\mu$ is a one-dimensional parameter. If 0 is a fixed point of $f_{\mu_0}: \mathbb{R} \to \mathbb{R}$ with eigenvalue $-1$, then the Taylor expansion of $f_{\mu_0}$ to degree 3 is

$$f_{\mu_0}(x) = -x + a_2 x^2 + a_3 x^3 + R_3(x), \quad \text{with } R_3(x) = o(|x^3|). \quad (3.5.2)$$

The implicit function theorem guarantees that there will be a smooth curve $(x(\mu), \mu)$ of fixed points in the plane passing through $(0, \mu_0)$, so, apart from a change of stability, we must look for changes in the dynamical behavior elsewhere. Composing $f_{\mu_0}$ with itself, we find

$$f_{\mu_0}^2(x) = -(-x + a_2 x^2 + a_3 x^3) + a_2(-x + a_2 x^2)^2 + a_3(-x)^3 + \tilde{R}_3$$
$$= x - (2a_2^2 + 2a_3)x^3 + \tilde{\tilde{R}}_3. \quad (3.5.3)$$

Since $f_{\mu_0}^2$ has eigenvalue $+1$, its fixed points need not vary smoothly and we expect that there may be fixed points of $f_\mu^2$ near $(0, \mu_0)$ which are *not* fixed points of $f_\mu$. Such points are evidently periodic orbits of period 2. Examining the Taylor series of $f_{\mu_0}^2(x)$, we see that the coefficient of the quadratic term is zero, and thus the bifurcation behavior has similarities with the pitchfork, the primary difference being that the new orbits which appear are not fixed points but have period 2. The ideas outlined above lead to the following result, the completion of the proof of which we leave to the reader:

**Theorem 3.5.1.** *Let $f_\mu: \mathbb{R} \to \mathbb{R}$ be a one-parameter family of mappings such that $f_{\mu_0}$ has a fixed point $x_0$ with eigenvalue $-1$. Assume*

(F1)    $\left( \dfrac{\partial f}{\partial \mu} \dfrac{\partial^2 f}{\partial x^2} + 2 \dfrac{\partial^2 f}{\partial x \, \partial \mu} \right) = \dfrac{\partial f}{\partial \mu} \dfrac{\partial^2 f}{\partial x^2} - \left( \dfrac{\partial f}{\partial x} - 1 \right) \dfrac{\partial^2 f}{\partial x \, \partial \mu} \neq 0 \quad$ at $(x_0, \mu_0)$;

(F2)    $a = \left( \dfrac{1}{2} \left( \dfrac{\partial^2 f}{\partial x^2} \right)^2 + \dfrac{1}{3} \left( \dfrac{\partial^3 f}{\partial x^3} \right) \right) \neq 0 \quad$ at $(x_0, \mu_0)$.

*Then there is a smooth curve of fixed points of $f_\mu$ passing through $(x_0, \mu_0)$, the stability of which changes at $(x_0, \mu_0)$. There is also a smooth curve $\gamma$ passing through $(x_0, \mu_0)$ so that $\gamma - \{(x_0, \mu_0)\}$ is a union of hyperbolic period 2 orbits. The curve $\gamma$ has quadratic tangency with the line $\mathbb{R} \times \{\mu_0\}$ at $(x_0, \mu_0)$.*

Here the quantity (F1) is the $\mu$-derivative of $f'$ along the curve of the fixed points. It plays the rôle of the nondegeneracy conditions SN2 and H2 in Theorems 3.4.1 and 3.4.2. In (F2) the sign of $a$ determines the stability and direction of bifurcation of the orbits of period 2. If $a$ is positive, the orbits are stable; if $a$ is negative they are unstable. We note that cubic terms $(\partial^3 f / \partial x^3)$ are necessary for the determination of $a$.

Figure 3.5.1 shows the bifurcation diagram for the family

$$f_\mu(x) = -(1 + \mu)x + x^3. \quad (3.5.4)$$

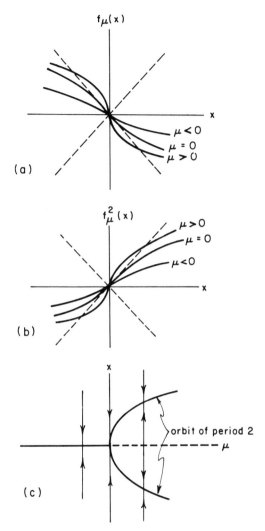

Figure 3.5.1. The flip bifurcation for equation (3.5.4). (a) Graphs of $f_\mu(x)$; (b) graphs of $f_\mu^2(x)$; (c) the bifurcation diagram.

As an example we consider the one-dimensional quadratic map

$$f_\mu : x \to \mu - x^2 \qquad (3.5.5)$$

of Exercise 3.5.1. The upper branch of equilibria is given (for $\mu > -\frac{1}{4}$) by

$$x = -\tfrac{1}{2} + \sqrt{\tfrac{1}{4} + \mu}. \qquad (3.5.6)$$

Linearizing along that branch, we find

$$\frac{\partial f}{\partial x} = -2x = 1 - \sqrt{1 + 4\mu}, \qquad (3.5.7)$$

and evidently $\partial f /\partial x = -1$ at $\mu = \frac{3}{4}$, and hence $(x_0, \mu_0) = (\frac{1}{2}, \frac{3}{4})$ is a candidate for a flip bifurcation point. In this example it is easy to check that conditions F1 and F2 of Theorem 3.5.1 are met and hence that a flip occurs.

The stability of the period two orbits in this example is determined by noting that the second and third derivatives of $f$ at $(x_0, \mu_0)$ are, respectively

$$\frac{\partial^2 f}{\partial x^2}(x_0, \mu_0) = -2 \quad \text{and} \quad \frac{\partial^3 f}{\partial x^3} \equiv 0, \qquad (3.5.8)$$

and hence that the quantity $a$ of (F2) is positive: the flip is supercritical.

EXERCISE 3.5.4. Show that the examples of Exercises 3.5.2 and 3.5.3 both undergo flip bifurcations at $\mu = 3$. Are they subcritical or supercritical?

EXERCISE 3.5.5. Verify that the map $x \to -(1 + \mu)x - x^2 + \beta x^3$ can undergo both sub- and supercritical flip bifurcations, depending upon the value of $\beta$.

We make one final remark about the relationship of a return map $P$ with eigenvalue $-1$ at a fixed point $p$, to the continuous flow around the corresponding periodic orbit. The trajectories of $P$ alternate from one side of $p$ to the other along the direction of the eigenvector to $-1$ (cf. Section 1.4, Table 1). This means that the two-dimensional center manifold for the periodic orbit is twisted around the periodic orbit like Mobius band around its center line. The map $P$ which glues the two ends of a strip together reverses orientation around $p$. One cannot embed a Mobius strip in an orientable two-dimensional manifold (such as the plane), so that the flip bifurcation cannot occur in such systems (cf. Exercise 1.5.3(b)). As we shall see, however, flip bifurcations can and do occur in flows of dimension $\geq 3$, regardless of whether the associated Poincaré map is orientation preserving or not.

EXERCISE 3.5.6. Build a geometric model of a flow in $\mathbb{R}^3$ which has a flip bifurcation. Visualize the invariant manifolds of the periodic orbits near the bifurcation. At the bifurcation point, must the (un)stable manifold of the periodic orbit be twisted like a Mobius band? What can you say about the invariant manifolds of the periodic orbit with the longer period?

We now turn to bifurcations of a periodic orbit at which there are complex eigenvalues $\lambda, \bar{\lambda}$ with $|\lambda| = 1$. Analogy with the theory of the Hopf bifurcation suggests that orbits near the bifurcation will be present which encircle the fixed point. An individual orbit of a discrete mapping cannot fill an entire circle and the bifurcation structure is more complicated than that which can be deduced from a search for new periodic orbits. Indeed, there are flows near the bifurcation which have *no* new periodic orbits near the bifurcating one but have *quasiperiodic* orbits instead. A more subtle analysis is required to capture these. Before one reaches this portion of the analysis, however, there is another difficulty to contend with.

Let us assume that we have a transformation $f: \mathbb{R}^2 \to \mathbb{R}^2$ so that the origin is a fixed point and $Df(0)$ is the matrix which is rotation by the angle $2\pi\theta$:

$$\begin{pmatrix} \cos 2\pi\theta & -\sin 2\pi\theta \\ \sin 2\pi\theta & \cos 2\pi\theta \end{pmatrix}. \tag{3.5.9}$$

We want to perform normal form calculations which simplify the higher-order terms in the Taylor series of $f$ by using nonlinear coordinate transformations. As in the case of flows, the calculations are simplified if they are complexified (cf. the Appendix to Section 3.4). If we regard $(x, y)$ as each being complex, then the eigenvectors of $Df(0)$ are $\begin{pmatrix} 1 \\ i \end{pmatrix}$ and $\begin{pmatrix} 1 \\ -i \end{pmatrix}$ with eigenvalues $e^{2\pi i\theta}$ and $e^{-2\pi i\theta}$ and coordinates $z$ and $\bar{z}$, respectively. Suppose we now want to alter the Taylor expansion at degree $k$ by a real coordinate transformation of the form

$$h(z, \bar{z}) = \text{id} + \text{terms of degree } k.$$

Since the $\bar{z}$ coordinate of the image of $h$ is the complex conjugate of the $z$ coordinate of the image of $h$, it suffices to compute the $z$ coordinate of the image of the conjugated mapping $hfh^{-1}$. If the Taylor expansion of the $z$ coordinate of the image of $f$ of degree $k$ is

$$f(z, \bar{z}) = e^{2\pi i\theta}z + f_2 + f_3 + \cdots + R_k \tag{3.5.10}$$

and the $z$ coordinate of $h(z, \bar{z})$ is

$$h(z, \bar{z}) = z + P_k(z, \bar{z}), \tag{3.5.11}$$

then the $z$ coordinate of the Taylor expansion of the conjugated mapping $hfh^{-1}$ is

$$e^{2\pi i\theta}z + f_2 + \cdots + f_k + P_k(e^{2\pi i\theta}z, e^{-2\pi i\theta}\bar{z}) - e^{2\pi i\theta}P_k(z, \bar{z}). \tag{3.5.12}$$

Thus we can remove from $f_k$ terms which can be expressed in the form

$$(P_k(e^{2\pi i\theta}z, e^{-2\pi i\theta}\bar{z}) - e^{2\pi i\theta}P_k(z, \bar{z})). \tag{3.5.13}$$

If we denote by the expression (3.5.13), ad $Df(P_k(z, \bar{z}))$, then ad $Df$ defines a mapping on the space of vector valued homogeneous polynomials which is linear and diagonalizable with the eigenfunctions $\begin{pmatrix} z^l\bar{z}^{k-l} \\ 0 \end{pmatrix} \begin{pmatrix} 0 \\ z^l\bar{z}^{k-l} \end{pmatrix}$ in $(z, \bar{z})$ coordinates. We note that

$$\text{ad } Df\begin{pmatrix} z^l\bar{z}^{k-l} \\ 0 \end{pmatrix} = (e^{2\pi i(2l-k)\theta} - e^{2\pi i\theta})\begin{pmatrix} z^l\bar{z}^{k-l} \\ 0 \end{pmatrix},$$

$$\tag{3.5.14}$$

$$\text{ad } Df\begin{pmatrix} 0 \\ z^l\bar{z}^{k-l} \end{pmatrix} = (e^{2\pi i(2l-k)\theta} + e^{2\pi i\theta})\begin{pmatrix} 0 \\ z^l\bar{z}^{k-l} \end{pmatrix}.$$

The zero eigenvalues of Ad $Df$ occur when $(2l - k)\theta \equiv \pm\theta \pmod 1$. If $\theta$ is irrational, then this can happen only when $k$ is odd and $l = (k \pm 1)/2$. The zero eigenvectors for these values have the form $(z\bar{z})^l \begin{pmatrix} z \\ 0 \end{pmatrix}$ and $(z\bar{z})^l \begin{pmatrix} 0 \\ \bar{z} \end{pmatrix}$. Expressed in real terms these represent mappings of the form $(x^2 + y^2)^l \cdot g(x, y)$ where $g$ is a linear mapping with matrix $\begin{pmatrix} a & -b \\ b & a \end{pmatrix}$. When $\theta$ is irrational, therefore, the normal forms of $f$ are analogues of the normal forms computed for the Hopf bifurcation for flows. However, *if $\theta$ is rational, then there are additional resonant terms which come from other solutions of the equation $(2l - k)\theta \equiv \pm\theta \pmod 1$*. The denominator of $\theta$ determines the lowest degree at which these terms may appear.

We have already met the cases $\theta = 0$ (saddle-node) and $\theta = \frac{1}{2}$ (flip). In addition, when $\theta = \pm\frac{1}{3}$ or $\theta = \pm\frac{1}{4}$, then there are resonant terms of degree 2 or 3, respectively, in the normal forms. When $\theta = \pm\frac{1}{3}$, these have the complex form $\bar{z}^2 \begin{pmatrix} 1 \\ 0 \end{pmatrix}$ and $\bar{z}^2 \begin{pmatrix} 0 \\ 1 \end{pmatrix}$, while if $\theta = \pm\frac{1}{4}$ these have the form $\bar{z}^3 \begin{pmatrix} 1 \\ 0 \end{pmatrix}$ and $\bar{z}^3 \begin{pmatrix} 0 \\ 1 \end{pmatrix}$. This means that the bifurcation structures associated with fixed points that are third and fourth roots of unity are special. Arnold [1977] and Takens [1974b] present analyses of these cases. If one assumes that $\lambda$ is not a third or fourth root of unity, then it is possible to proceed with a general analysis of Hopf bifurcation for periodic orbits (secondary Hopf bifurcation), and we have:

**Theorem 3.5.2.** *Let $f_\mu \colon \mathbb{R}^2 \to \mathbb{R}^2$ be a one-parameter family of mappings which has a smooth family of fixed points $x(\mu)$ at which the eigenvalues are complex conjugates $\lambda(\mu), \bar{\lambda}(\mu)$. Assume*

(SH1)          $|\lambda(\mu_0)| = 1$   but $\lambda^j(\mu_0) \neq 1$   for $j = 1, 2, 3, 4$.

(SH2)                              $\dfrac{d}{d\mu}(|\lambda(\mu_0)|) = d \neq 0.$

*Then there is a smooth change of coordinates $h$ so that the expression of $hf_\mu h^{-1}$ in polar coordinates has the form*

$$hf_\mu h^{-1}(r, \theta) = (r(1 + d(\mu - \mu_0) + ar^2), \theta + c + br^2) + \text{higher-order terms}.$$

$$(3.5.15)$$

*(Note: $\lambda$ complex and (SH2) imply $|\arg(\lambda)| = c$ and $d$ are nonzero.) If, in addition*

(SH3)                                        $a \neq 0.$

*Then there is a two-dimensional surface $\Sigma$ (not necessarily infinitely differentiable) in $\mathbb{R}^2 \times \mathbb{R}$ having quadratic tangency with the plane $\mathbb{R}^2 \times \{\mu_0\}$*

which is invariant for $f$. If $\Sigma \cap (\mathbb{R}^2 \times \{\mu\})$ is larger than a point, then it is a simple closed curve.

As in the case of flows, the signs of the coefficients $a$ and $d$ determine the direction and stability of the bifurcating periodic orbits; $c$ and $b$ give asymptotic information on rotation numbers, as outlined below.

Marsden and McCracken [1976] contains Lanford's exposition of Ruelle's proof of this theorem using the technique of graph transforms. The theorem states that (outside the strong resonance cases $\lambda^3 = 1$ and $\lambda^4 = 1$), something like the limit cycles of the Hopf theorem appear in the phase portrait of $f_\mu$. These are simple closed curves which bound the basin of attraction or repulsion of a fixed point. Theorem 3.5.2 does not however address the question of describing the dynamics within $\Sigma$. In all of its details, this is a difficult problem which involves the introduction of *rotation numbers* and consideration of subtle *small divisor problems*. These are discussed in Section 6.2. Here we comment only that if $b \neq 0$ in (3.5.15) then it can be proved that there will be a complicated pattern of periodic and quasiperiodic dynamical behavior on $\Sigma$. To study this one must examine the global bifurcations of diffeomorphisms whose state space is the circle.

A stability formula, giving an expression for the coefficient $a$ in the normal form (3.5.15), can be derived in much the same way as for flows (cf. Section 3.4). For details, see Iooss and Joseph [1981] or Wan [1978]. Assuming that the bifurcating system (restricted to the center manifold) is in the form

$$\begin{pmatrix} x \\ y \end{pmatrix} \mapsto \begin{bmatrix} \cos(c) & -\sin(c) \\ \sin(c) & \cos(c) \end{bmatrix} \begin{pmatrix} x \\ y \end{pmatrix} + \begin{pmatrix} f(x, y) \\ g(x, y) \end{pmatrix}, \qquad (3.5.16)$$

with eigenvalues $\lambda, \bar{\lambda} = e^{\pm ic}$, one obtains

$$a = -\mathrm{Re}\left[\frac{(1 - 2\lambda)\bar{\lambda}^2}{1 - \lambda} \xi_{11}\xi_{20}\right] - \tfrac{1}{2}|\xi_{11}|^2 - |\xi_{02}|^2 + \mathrm{Re}\,(\bar{\lambda}\xi_{21}),$$

where

$$\xi_{20} = \tfrac{1}{8}[(f_{xx} - f_{yy} + 2g_{xy}) + i(g_{xx} - g_{yy} - 2f_{xy})],$$
$$\xi_{11} = \tfrac{1}{4}[(f_{xx} + f_{yy}) + i(g_{xx} + g_{yy})],$$
$$\xi_{02} = \tfrac{1}{8}[(f_{xx} - f_{yy} - 2g_{xy}) + i(g_{xx} - g_{yy} + 2f_{xy})],$$

and

$$\xi_{21} = \tfrac{1}{16}[(f_{xxx} + f_{xyy} + g_{xxy} + g_{yyy}) + i(g_{xxx} + g_{xyy} - f_{xxy} - f_{yyy})].$$

$$(3.5.17)$$

We end, as usual with an example. Consider the delayed logistic equation (Maynard-Smith [1971], Pounder and Rogers [1980], Aronson *et al.* [1980, 1982]):

$$F_\mu : (x, y) \to (y, \mu y(1 - x)). \qquad (3.5.18)$$

This map has fixed points at $(x, y) = (0, 0)$ and $(x, y) = ((\mu - 1)/\mu, (\mu - 1)/\mu)$. The reader can check that, for $\mu > 1$, $(0, 0)$ is a saddle point.* The matrix of the map linearized at the other non zero-fixed point is

$$DF\left(\frac{\mu - 1}{\mu}, \frac{\mu - 1}{\mu}\right) = \begin{bmatrix} 0 & 1 \\ 1 - \mu & 1 \end{bmatrix}, \tag{3.5.19}$$

which has eigenvalues

$$\lambda_{1,2} = \tfrac{1}{2}(1 \pm \sqrt{5 - 4\mu}). \tag{3.5.20}$$

For $\mu > \tfrac{5}{4}$, these eigenvalues are complex conjugate and may be written

$$\lambda, \bar{\lambda} = (\mu - 1)e^{\pm ic}, \quad \text{where } \tan c = \sqrt{4\mu - 5}. \tag{3.5.21}$$

It is now easy to check that hypotheses (SH1) and (SH2) of Theorem 3.5.2 hold, since, at $\mu = 2$, $\lambda, \bar{\lambda} = e^{\pm i\pi/3}$ are sixth roots of unity, while

$$\frac{d}{d\mu} |\lambda(\mu)|_{\mu=2} = 1.$$

To compute $a$ from Equation (3.5.17), and hence check (SH3), we set $\mu = 2$ in (3.5.18) and apply the changes of coordinates

$$(\bar{x}, \bar{y}) = (x - \tfrac{1}{2}, y - \tfrac{1}{2}),$$

and

$$\begin{pmatrix} u \\ v \end{pmatrix} = \begin{bmatrix} -1/\sqrt{3} & 2/\sqrt{3} \\ 1 & 0 \end{bmatrix} \begin{pmatrix} \bar{x} \\ \bar{y} \end{pmatrix}; \quad \begin{pmatrix} \bar{x} \\ \bar{y} \end{pmatrix} = \begin{bmatrix} 0 & 1 \\ \sqrt{3}/2 & \tfrac{1}{2} \end{bmatrix} \begin{pmatrix} u \\ v \end{pmatrix}, \tag{3.5.22}$$

which translate the bifurcating equilibrium to the origin, and bring the linear part into normal form. Under these transformations (3.5.18) becomes

$$\begin{pmatrix} u \\ v \end{pmatrix} \rightarrow \begin{bmatrix} \tfrac{1}{2} & -\sqrt{3}/2 \\ \sqrt{3}/2 & \tfrac{1}{2} \end{bmatrix} \begin{pmatrix} u \\ v \end{pmatrix} - \begin{bmatrix} 2uv + 2v^2 \\ 0 \end{bmatrix} \tag{3.5.23}$$

with eigenvalues $\lambda, \bar{\lambda} = \tfrac{1}{2} \pm i(\sqrt{3}/2)$. The nonlinear terms are quadratic and we have the following

$$\begin{aligned} f_{uu} &= 0, & f_{uv} &= -2, & f_{vv} &= -4, \\ g_{uu} &= 0, & g_{uv} &= 0, & g_{vv} &= 0. \end{aligned} \tag{3.5.24}$$

We therefore obtain

$$\xi_{20} = \tfrac{1}{8}[4 + 4i] = \frac{1}{2} + \frac{i}{2},$$

$$\xi_{11} = \tfrac{1}{4}[-4 + 0i] = -1,$$

$$\xi_{02} = \tfrac{1}{8}[4 - 4i] = \frac{1}{2} - \frac{i}{2}, \tag{3.5.25}$$

$$\xi_{21} = 0,$$

---

* The map is noninvertible on the line $y = 0$, and one eigenvalue of $(0, 0)$ is $0$. This noninvertibility does not concern us in studying the Hopf bifurcation, however, since the latter occurs away from $y = 0$.

and substitution into the formula for $a$ yields

$$a = \frac{\sqrt{3} - 7}{4} < 0. \qquad (3.5.26)$$

Since $(d/d\mu)(|\lambda(\mu)|)_{\mu=2} = d = 1 > 0$, we deduce from (3.5.15) that the bifurcation is supercritical and hence that an attracting invariant closed curve exists, surrounding $(x, y) = (\frac{1}{2}, \frac{1}{2})$ for $\mu > 2$ and $|\mu - 2|$ small.

EXERCISE 3.5.7 (Grand Finale). Show that the two-parameter family of maps

$$(x, y) \to (y, \mu_1 y + \mu_2 - x^2)$$

can undergo saddle-node, flip, and Hopf bifurcations. Find the bifurcation sets in $(\mu_1, \mu_2)$ space on which these occur and discuss their stability types. Can double eigenvalues occur? What do you think might happen near those points, or near the points at which the linearized map has complex eigenvalues of modulus 1 with $\lambda^j = 1$ for $j = 3, 4$? (This example is taken from Whitley [1982].)

# CHAPTER 4

# Averaging and Perturbation from a Geometric Viewpoint

In this chapter we describe some classical methods of analysis which are particularly applicable to problems in nonlinear oscillations. While these methods might be familiar to the reader who has studied nonlinear mechanics and perturbation theory, the present geometrical approach and the stress on obtaining approximations to Poincaré maps will probably be less familiar.

We start with the averaging method, originally due to Krylov and Bogoliubov [1934], which is particularly useful for weakly nonlinear problems or small perturbations of the linear oscillator. We show that, under suitable conditions, global information, valid on semi-infinite time intervals, can be obtained by this approach. Generally, in perturbation methods one starts with an (integrable) system whose solutions are known completely, and studies small perturbations of it. Since the unperturbed and perturbed vector fields are close, one might expect that solutions will also be close, but as we shall see, this is not generally the case, in that the unperturbed systems are often structurally unstable. As we have seen, arbitrarily small perturbations of such systems can cause radical qualitative changes in the structure of solutions. However, these changes are generally associated with limiting, asymptotic behavior and one does usually find that unperturbed and perturbed solutions remain close for *finite* times. Moreover, in this chapter we shall show that such finite time results, together with ideas from dynamical systems theory, do enable us to make deductions about the asymptotic behavior of solutions and the structure of the nonwandering set for the perturbed system.

Averaging is applicable to systems of the form

$$\dot{x} = \varepsilon f(x, t); \qquad x \in \mathbb{R}^n, \qquad \varepsilon \ll 1, \qquad (4.0.1)$$

where $f$ is $T$-periodic in $t$. In such a system the $T$-periodic forcing contrasts with the "slow" evolution of solutions on the average due to the $\mathcal{O}(\varepsilon)$ vector field. In the first four sections we will show how weakly nonlinear oscillators of the form

$$\ddot{x} + \omega^2 x = \varepsilon f(x, \dot{x}, t) \qquad (4.0.2)$$

can be recast in the standard form (4.0.1) and averaging applied. In this analysis we essentially deal with small perturbations of the linear oscillator $\ddot{x} + \omega^2 x = 0$, which is an example of an integrable Hamiltonian system.

We continue with a description of Melnikov's [1963] method for dealing with perturbations of general integrable Hamiltonian systems. Here one typically starts with a strongly nonlinear system,

$$\dot{x} = f(x), \qquad x \in \mathbb{R}^{2n}, \qquad (4.0.3)$$

and adds weak dissipation and forcing:

$$\dot{x} = f(x) + \varepsilon g(x, t). \qquad (4.0.4)$$

While we are primarily concerned with periodically forced two-dimensional systems, we state the averaging results in the more general $n$-dimensional context, since they are no more difficult in that form. For Melnikov's method we restrict ourselves to two-dimensional problems, although some $n$-dimensional, and even infinite-dimensional, generalizations are available. We comment on extensions of this nature towards the end of the chapter. We also outline some of the theory of area preserving maps of the plane arising as Poincaré maps in time-periodic single degree of freedom Hamiltonian systems and in time-independent two degree of freedom systems.

## 4.1. Averaging and Poincaré Maps

There are many versions of the averaging theorem. Our account is based on the versions due to Hale [1969; Chapter V, Theorem 3.2], and Sanders and Verhulst [1982], who give a very full discussion from the viewpoint of asymptotics. We consider systems of the form

$$\dot{x} = \varepsilon f(x, t, \varepsilon); \qquad x \in U \subseteq \mathbb{R}^n, \quad 0 \leq \varepsilon \ll 1, \qquad (4.1.1)$$

where $f: \mathbb{R}^n \times \mathbb{R} \times \mathbb{R}^+ \to \mathbb{R}^n$ is $C^r, r \geq 2$, bounded on bounded sets, and of period $T > 0$ in $t$. We normally restrict ourselves to a bounded set $U \subset \mathbb{R}^n$. The associated *autonomous averaged system* is defined as

$$\dot{y} = \varepsilon \frac{1}{T} \int_0^T f(y, t, 0) \, dt \overset{\text{def}}{=} \varepsilon \bar{f}(y). \qquad (4.1.2)$$

In this situation we have

**Theorem 4.1.1** (The Averaging Theorem). *There exists a $C^r$ change of co-ordinates $x = y + \varepsilon w(y, t, \varepsilon)$ under which (4.1.1) becomes*

$$\dot{y} = \varepsilon \bar{f}(y) + \varepsilon^2 f_1(y, t, \varepsilon), \tag{4.1.3}$$

*where $f_1$ is of period $T$ in $t$. Moreover*

(i) *If $x(t)$ and $y(t)$ are solutions of (4.1.1) and (4.1.2) based at $x_0, y_0$, respectively, at $t = 0$, and $|x_0 - y_0| = \mathcal{O}(\varepsilon)$, then $|x(t) - y(t)| = \mathcal{O}(\varepsilon)$ on a time scale $t \sim 1/\varepsilon$.*

(ii) *If $p_0$ is a hyperbolic fixed point of (4.1.2) then there exists $\varepsilon_0 > 0$ such that, for all $0 < \varepsilon \le \varepsilon_0$, (4.1.1) possesses a unique hyperbolic periodic orbit $\gamma_\varepsilon(t) = p_0 + \mathcal{O}(\varepsilon)$ of the same stability type as $p_0$.*

(iii) *If $x^s(t) \in W^s(\gamma_\varepsilon)$ is a solution of (4.1.1) lying in the stable manifold of the hyperbolic periodic orbit $\gamma_\varepsilon = p_0 + \mathcal{O}(\varepsilon)$, $y^s(t) \in W^s(p_0)$ is a solution of (4.1.2) lying in the stable manifold of the hyperbolic fixed point $p_0$ and $|x^s(0) - y^s(0)| = \mathcal{O}(\varepsilon)$, then $|x^s(t) - y^s(t)| = \mathcal{O}(\varepsilon)$ for $t \in [0, \infty)$. Similar results apply to solutions lying in the unstable manifolds on the time interval $t \in (-\infty, 0]$.*

**Remarks.** Conclusions (ii) and (iii) generalize to more complicated hyperbolic sets. In particular, Hale [1969] shows that if (4.1.2) has a hyperbolic closed orbit $\Gamma$, then (4.1.1) has a hyperbolic invariant torus $T_\Gamma$. Generalizations to almost periodic functions $f$ are also available (Hale [1969]). Conclusion (iii) implies that the averaging theorem can be used to approximate stable and unstable manifolds in bounded sets and generally to study the global structure of the Poincaré map of (4.1.1), as our examples will demonstrate.

PROOF. We will sketch the first two parts of the proof using standard results from differential equations; for the last parts it is more convenient to use the ideas of Poincaré maps and invariant manifolds. We start by explicitly computing the change of coordinates. Let

$$f(x, t, \varepsilon) = \bar{f}(x) + \tilde{f}(x, t, \varepsilon) \tag{4.1.4}$$

be split into its mean, $\bar{f}$, and oscillating part $\tilde{f}$. Let

$$x = y + \varepsilon w(y, t, \varepsilon), \tag{4.1.5}$$

without yet choosing $w$. Differentiating (4.1.5) and using (4.1.1) and (4.1.4) we have

$$[I + \varepsilon D_y w]\dot{y} = \dot{x} - \varepsilon \frac{\partial w}{\partial t}$$

$$= \varepsilon \bar{f}(y + \varepsilon w) + \varepsilon \tilde{f}(y + \varepsilon w, t, \varepsilon) - \varepsilon \frac{\partial w}{\partial t},$$

* $\gamma_\varepsilon$ may be a trivial periodic orbit, $\gamma_\varepsilon(t) \equiv p_0$, cf. Example 1 on p. 171 below.

or

$$\dot{y} = \varepsilon[I + \varepsilon D_y w]^{-1}\left[\tilde{f}(y + \varepsilon w) + \hat{f}(y + \varepsilon w, t, \varepsilon) - \frac{\partial w}{\partial t}\right]. \quad (4.1.6)$$

Expanding (4.1.6) in powers of $\varepsilon$ and choosing $w$ to be the anti-derivative of $\hat{f}$:

$$\frac{\partial w}{\partial t} = \hat{f}(y, t, 0), \quad (4.1.7)$$

we obtain

$$\dot{y} = \varepsilon\tilde{f}(y) + \varepsilon^2\left[D_y f(y, t, 0)w(y, t, 0) - D_y w(y, t, 0)\tilde{f}(y) + \frac{\partial\hat{f}}{\partial\varepsilon}(y, t, 0)\right] + \mathcal{O}(\varepsilon^3)$$

$$\overset{\text{def}}{=} \varepsilon\tilde{f}(y) + \varepsilon^2 f_1(y, t, \varepsilon), \quad (4.1.8)$$

as required.

To obtain conclusion (i) we use a version of Gronwall's lemma:

**Lemma 4.1.2** (cf. Coddington and Levinson [1955], p. 37). *If $u$, $v$, and $c \geq 0$ on $[0, t]$, $c$ is differentiable, and*

$$v(t) \leq c(t) + \int_0^t u(s)v(s)\,ds,$$

*then*

$$v(t) \leq c(0)\exp\int_0^t u(s)\,ds + \int_0^t c'(s)\left[\exp\int_s^t u(\tau)\,d\tau\right]ds.$$

To prove the lemma, let $R(t) = \int_0^t u(s)v(s)\,ds$ and first show that $R' - uR \leq uc$. After integrating this differential inequality and some manipulation, including integration by parts, one obtains the result.

Now consider equations (4.1.2) and (4.1.3). Integrating and subtracting, we have

$$y_\varepsilon(t) - y(t) = y_{\varepsilon0} - y_0 + \varepsilon\int_0^t [\tilde{f}(y_\varepsilon(s)) - \tilde{f}(y(s))]\,ds$$

$$+ \varepsilon^2\int_0^t f_1(y_\varepsilon(s), s, \varepsilon)\,ds,$$

where $y_\varepsilon(t)$ is a solution of (4.1.3) based at $y_{\varepsilon0}$. Letting $y_\varepsilon - y = \zeta$, $L$ be the Lipschitz constant of $\tilde{f}$ and $C$ the maximum value of $f_1$, this becomes

$$|\zeta(t)| \leq |\zeta(0)| + \varepsilon L\int_0^t |\zeta(s)|\,ds + \varepsilon^2 Ct. \quad (4.1.9)$$

Applying Gronwall's lemma, with $c(t) = |\zeta(0)| + \varepsilon^2 Ct$ and $u(s) = \varepsilon L$, we have

$$|\zeta(t)| \leq |\zeta(0)|e^{\varepsilon Lt} + \varepsilon^2 C\int_0^t e^{\varepsilon L(t-s)}\,ds$$

$$\leq \left[|\zeta(0)| + \frac{\varepsilon C}{L}\right]e^{\varepsilon Lt}. \quad (4.1.10)$$

Thus, if $|y_{\varepsilon 0} - y_0| = \mathcal{O}(\varepsilon)$, we conclude that $|y_\varepsilon(t) - y(t)| = \mathcal{O}(\varepsilon)$ for $t \in [0, 1/\varepsilon L]$. Finally, via the transformation (4.1.5) we have

$$|x(t) - y_\varepsilon(t)| = \varepsilon w(y_\varepsilon, t, \varepsilon) = \mathcal{O}(\varepsilon)$$

and, using the triangle inequality

$$|x(t) - y(t)| \leq |x(t) - y_\varepsilon(t)| + |y_\varepsilon(t) - y(t)|,$$

we obtain the desired result.

To prove (ii) we consider the Poincaré maps $P_0$, $P_\varepsilon$ associated with (4.1.2) and (4.1.3). Rewriting these latter systems as

$$\dot{y} = \varepsilon \bar{f}(y); \qquad\qquad\qquad \dot{\theta} = 1, \qquad\qquad (4.1.11)$$

$$\dot{y} = \varepsilon \bar{f}(y) + \varepsilon^2 f_1(y, \theta, \varepsilon); \qquad \dot{\theta} = 1, \qquad\qquad (4.1.12)$$

where $(y, \theta) \in \mathbb{R}^n \times S^1$, and $S^1 = R/T$ is the circle of length $T$, we define a global cross section $\Sigma = \{(y, \theta) | \theta = 0\}$, and the first return or time $T$ Poincaré maps* $P_0: U \to \Sigma$, $P_\varepsilon: U \to \Sigma$ are then defined for (4.1.11), (4.1.12) in the usual way, where $U \subseteq \Sigma$ is some open set. Note that $P_\varepsilon$ is $\varepsilon^2$-close to $P_0$ since $T$ is fixed independent of $\varepsilon$. If $p_0$ is a hyperbolic fixed point for (4.1.2), then it is also a hyperbolic fixed point of $DP_0(p_0)$ since $DP_0(p_0) = e^{\varepsilon TDf(p)}$. Therefore, $\lim_{\varepsilon \to 0}(1/\varepsilon)[e^{\varepsilon TDf(p_0)} - Id] = TDf(p_0)$ is invertible. Since $P_\varepsilon$ is $\varepsilon$-close to $P_0$, we also have $\lim_{\varepsilon \to 0}(1/\varepsilon)[DP_\varepsilon(p_0) - Id] = TDf(p_0)$. The implicit function theorem implies that the zeros of $(1/\varepsilon)[DP_\varepsilon(p_0) - Id]$ form a smooth curve $(p_\varepsilon, \varepsilon)$ in $\mathbb{R}^n \times \mathbb{R}$. The $p_\varepsilon$ are fixed points of $P_\varepsilon$, and the eigenvalues of $DP_\varepsilon(p_\varepsilon)$ are $\varepsilon^2$ close to those of $DP_0(p_0)$ since $p_\varepsilon = p_0 + \mathcal{O}(\varepsilon)$ and $DP_\varepsilon(p_\varepsilon) = \exp[\varepsilon T(Df(p_\varepsilon) + \varepsilon^2 Df_1(p_\varepsilon))] = \exp[\varepsilon TDf(p_0)] + \mathcal{O}(\varepsilon^2)$. Thus (4.1.12) has a periodic orbit $\gamma_\varepsilon$ $\varepsilon$-close to $p_0$, and via the change of coordinates (4.1.5), equation (4.1.1) has a similar orbit.

We remark that all that is required for the existence of a periodic orbit in (4.1.1) is the absence of any eigenvalues equal to one in the spectrum of $DP_0(p_0)$. However, the stability types of $p_0$ and $\gamma_\varepsilon$ may not correspond if any eigenvalues of $DP_0(p_0)$ lie on the unit circle.

To prove (iii), suppose that (4.1.2) has a hyperbolic saddle point $p_0$ and consider solutions $y(t) \in W^s(p_0)$ and the corresponding solutions $y_\varepsilon(t) \in W^s(\gamma_\varepsilon)$ of the full system (4.1.3). The cases in which $p_0$ is a sink or source and $W^s$ is replaced by $W^u$ can be dealt with similarly. The proof is divided into two parts; an outer region in which the averaged vector field $\varepsilon \bar{f}(y)$ is large in comparison with the remainder term $\varepsilon^2 f_1(y, t)$, and an inner region in which the "perturbations" $\varepsilon^2 f_1$ and $\varepsilon \bar{f}$ are of comparable order. For more details see Sanders and Verhulst [1982]. We fix a $\delta$ neighborhood, $U_\delta$, of $p_0$, so that, outside $U_\delta$, we have $|\bar{f}(y))| \gg \varepsilon |f_1(y, t, \varepsilon)|$. As above, standard Gronwall estimates show that $|y_\varepsilon - y_0| = \mathcal{O}(\varepsilon)$ for times of order $1/\varepsilon$

---

* In $P_0$, the subscript "0" indicates that the $\mathcal{O}(\varepsilon^2)$ term is removed, *not that* $\varepsilon = 0$ in (4.1.11). The notation, $P_0$ for the $\mathcal{O}(\varepsilon)$ map and $P_\varepsilon$ for the full map, will be used throughout this and the following three sections.

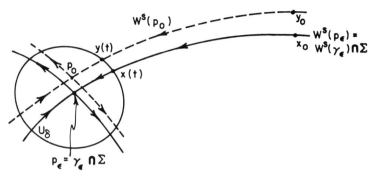

Figure 4.1.1. Validity of averaging on semi-infinite time intervals.

outside $U_\delta$. On the other hand, in $U_\delta$ the (local) stable manifold theorem guarantees that the stable manifold $W^s_{loc}(\gamma_\varepsilon)$ is $\varepsilon$, $C^r$ close to $W^s_{loc}(p_0) \times [0, T]$. Moreover, within $W^s_{loc}(\gamma_\varepsilon)$ and $W^s_{loc}(p_0)$ solutions are contracting towards $\gamma_\varepsilon$ and $p_0$, respectively, this contraction being dominated by an exponential term of the form $e^{-\lambda t}$. Using this fact, we can prove that, if $y_\varepsilon$ and $y_0$ enter $U_\delta$ within $\mathcal{O}(\varepsilon)$, they remain within $\mathcal{O}(\varepsilon)$ for all forward time, Figure 4.1.1. Piecing the two estimates together and using the transformation (4.1.5) as above, we obtain the desired result.                                      □

We note that Sanders [1980] and Murdock and Robinson [1980] (cf. Robinson [1981b]) give proofs of part (iii) of this theorem. In the proof of the last part of the theorem we are using the smooth dependence of (local) invariant manifolds on parameters. Thus statement (iii) also follows directly from the "big" invariant manifold theorem of Hirsch et al. [1977, Theorem 4.1].

## 4.2. Examples of Averaging

EXAMPLE 1. Consider the scalar system

$$\dot{x} = \varepsilon x \sin^2 t. \tag{4.2.1}$$

Here $f(x, t, \varepsilon) = \bar{f}(x) + \tilde{f}(x, t, \varepsilon) = x/2 - (x/2) \cos 2t$ and we have

$$\frac{\partial w}{\partial t} = -\frac{y}{2} \cos 2t,$$

or

$$w = -\frac{y}{4} \sin 2t. \tag{4.2.2}$$

Note that the $t$-independent term which could appear in the anti-derivative is generally taken to be zero. From (4.1.8), the transformed system is

$$\dot{y} = \varepsilon \frac{y}{2} + \varepsilon^2 \left[ (\tfrac{1}{2} - \tfrac{1}{2}\cos 2t)\left(-\frac{y}{4}\sin 2t\right) - (-\tfrac{1}{4}\sin 2t)\left(\frac{y}{2}\right)\right] + \mathcal{O}(\varepsilon^3),$$

or

$$\dot{y} = \varepsilon \frac{y}{2} + \varepsilon^2 \frac{y}{16}\sin 4t + \mathcal{O}(\varepsilon^3). \tag{4.2.3}$$

Here the autonomous averaged equation is simply

$$\dot{y} = \varepsilon \frac{y}{2}. \tag{4.2.4}$$

The exact solution of (4.2.1) with initial value $x(0) = x_0$ is easily found to be

$$x(t) = x_0\, e^{\varepsilon((t/2) - \sin(2t)/4)}. \tag{4.2.5}$$

Comparing this with the solution of the averaged equation

$$y(t) = y_0\, e^{\varepsilon t/2}, \tag{4.2.6}$$

we see that

$$x(t) - y(t) = e^{\varepsilon t/2}[\,|x_0 - y_0| - \varepsilon x_0 \sin(2t)/4 + \mathcal{O}(\varepsilon^2)], \tag{4.2.7}$$

in agreement with conclusion (i) of the theorem. Here the hyperbolic source $y = 0$ of (4.2.4) corresponds to a trivial hyperbolic periodic orbit $x \equiv 0$ of (4.2.1), and, letting $t \to -\infty$ in (4.2.5)–(4.2.7), we see that $x(t), y(t) \to 0$ and hence $|x(t) - y(t)| \to 0$, in agreement with conclusions (ii) and (iii).

EXERCISE 4.2.1. Study the system $\dot{x} = -\varepsilon x \cos t$ by the method of averaging. Does it have a hyperbolic limit set? Compare the averaged and exact solutions.

EXERCISE 4.2.2. Repeat the averaging analysis for $\dot{x} = \varepsilon(-x + \cos^2 t)$. In particular check the validity of conclusions (ii) and (iii) of the theorem.

EXERCISE 4.2.3. Study the nonlinear systems

$$\dot{x} = \varepsilon(x - x^2)\sin^2 t$$

and

$$\dot{x} = \varepsilon(x \sin^2 t - x^2/2)$$

by the method of averaging. What do you notice about their solutions?

EXAMPLE 2 (Weakly nonlinear forced oscillations). In many weakly nonlinear oscillator problems, the second-order equation to be studied takes the form

$$\ddot{x} + \omega_0^2 x = \varepsilon f(x, \dot{x}, t), \tag{4.2.8}$$

where $f$ is $T$ periodic in $t$. In particular, if $f$ is sinusoidal with frequency $\omega \approx k\omega_0$, we have a system close to a *resonance of order k*. In such a situation, our expectation of finding an almost sinusoidal response of frequency $\omega/k$ prompts the use of the invertible van der Pol transformation, which recasts (4.2.8) into the form (4.1.1) which can then be averaged. We set

$$\begin{pmatrix} u \\ v \end{pmatrix} = A \begin{pmatrix} x \\ \dot{x} \end{pmatrix}, \qquad A = \begin{bmatrix} \cos\left(\dfrac{\omega t}{k}\right) & -\dfrac{k}{\omega}\sin\left(\dfrac{\omega t}{k}\right) \\[3mm] -\sin\left(\dfrac{\omega t}{k}\right) & -\dfrac{k}{\omega}\cos\left(\dfrac{\omega t}{k}\right) \end{bmatrix},$$

$$\tag{4.2.9}$$

$$A^{-1} = \begin{bmatrix} \cos\left(\dfrac{\omega t}{k}\right) & -\sin\left(\dfrac{\omega t}{k}\right) \\[3mm] -\dfrac{\omega}{k}\sin\left(\dfrac{\omega t}{k}\right) & -\dfrac{\omega}{k}\cos\left(\dfrac{\omega t}{k}\right) \end{bmatrix},$$

under which (4.2.8) becomes

$$\dot{u} = -\frac{k}{\omega}\left[\left(\frac{\omega^2 - k^2\omega_0^2}{k^2}\right)x + \varepsilon f(x, \dot{x}, t)\right]\sin\left(\frac{\omega t}{k}\right),$$

$$\dot{v} = -\frac{k}{\omega}\left[\left(\frac{\omega^2 - k^2\omega_0^2}{k^2}\right)x + \varepsilon f(x, \dot{x}, t)\right]\cos\left(\frac{\omega t}{k}\right), \tag{4.2.10}$$

in which $x, \dot{x}$ can be written as functions of $u, v,$ and $t$ via (4.2.9). If $\omega^2 - k^2\omega_0^2 = \mathcal{O}(\varepsilon)$, then (4.2.10) is in the correct form for averaging.

As a specific example, we take the standard Duffing equation which is considered in almost every text on nonlinear oscillations:

$$\ddot{x} + \omega_0^2 x = \varepsilon[\gamma \cos \omega t - \delta\dot{x} - \alpha x^3], \tag{4.2.11}$$

where $\omega_0^2 - \omega^2 = \varepsilon\Omega$, i.e., we are close to order one resonance. Setting $k = 1$ in (4.2.9) we obtain the transformed system

$$\dot{u} = \frac{\varepsilon}{\omega}[\Omega(u \cos \omega t - v \sin \omega t) - \omega\delta(u \sin \omega t + v \cos \omega t)$$

$$+ \alpha(u \cos \omega t - v \sin \omega t)^3 - \gamma \cos \omega t]\sin \omega t,$$

$$\dot{v} = \frac{\varepsilon}{\omega}[\Omega(u \cos \omega t - v \sin \omega t) - \omega\delta(u \sin \omega t + v \cos \omega t)$$

$$+ \alpha(u \cos \omega t - v \sin \omega t)^3 - \gamma \cos \omega t]\cos \omega t. \tag{4.2.12}$$

Averaging (4.2.12) over one period $T = 2\pi/\omega$, we obtain

$$\dot{u} = \frac{\varepsilon}{2\omega} \left[ -\omega\delta u - \Omega v - \frac{3\alpha}{4}(u^2 + v^2)v \right] \stackrel{\text{def}}{=} \varepsilon f_1(u, v),$$

$$\dot{v} = \frac{\varepsilon}{2\omega} \left[ \Omega u - \omega\delta v + \frac{3\alpha}{4}(u^2 + v^2)u - \gamma \right] \stackrel{\text{def}}{=} \varepsilon f_2(u, v)$$

(4.2.13)

or, in polar coordinates; $r = \sqrt{u^2 + v^2}$, $\phi = \arctan(v/u)$:

$$\dot{r} = \frac{\varepsilon}{2\omega} \left[ -\omega\delta r - \gamma \sin \phi \right],$$

$$r\dot{\phi} = \frac{\varepsilon}{2\omega} \left[ \Omega r + \frac{3\alpha}{4} r^3 - \gamma \cos \phi \right].$$

(4.2.14)

Perturbation methods carried to $\mathcal{O}(\varepsilon)$ give precisely the same result (cf. Nayfeh and Mook [1979], Section 4.1.1).

Recalling the transformation

$$x = u(t) \cos \omega t - v(t) \sin \omega t \equiv r(t) \cos(\omega t + \phi(t)),$$

we see that the slowly varying amplitude, $r$, and phase, $\phi$, of the solution of (4.2.11) are given, to first order, by solutions of the averaged system (4.2.14). It is therefore important to find the equilibrium solutions of or fixed points of (4.2.14), which, by the averaging theorem and the transformation (4.2.9), correspond to steady, almost sinusoidal solutions of the original equation. Fixing $\alpha$, $\delta$, and $\gamma$ and plotting the fixed points $\bar{r}$, $\bar{\phi}$ of (4.2.14) against $\Omega$ or $\omega/\omega_0$, we obtain the frequency response curve familiar to engineers: see Figure 4.2.1. We shall consider the "jump" bifurcation phenomenon below. For more details on the Duffing equation see Nayfeh and Mook [1979], or, for bifurcation details, Holmes and Rand [1976]. The stability types of the branches of steady solutions shown in Figure 4.2.1 are obtained by consideration of the eigenvalues of the linearized averaged equation, and we invite the reader to check our assertions.

In Figure 4.2.2 we show typical phase portraits for (4.2.13)–(4.2.14), obtained by numerical integration for parameter values for which three hyperbolic fixed points coexist. In Figure 4.2.3(a) we reproduce the stable and unstable manifolds of the saddle point, under the orientation reversing transformation (4.2.9) (with $k = 1$) applied at $t = 0$: $x = u$, $\dot{x} = -\omega v$. According to Theorem 4.1.1, these manifolds should approximate the stable and unstable manifolds of the Poincaré map of the full system (4.2.11), which we show in Figure 4.2.3(b). These latter manifolds were also computed numerically. We note that the agreement is good, but remark that it deteriorates as $\omega$ moves away from $\omega_0$ ($\Omega$ increases). For more examples, see Fiala [1976].

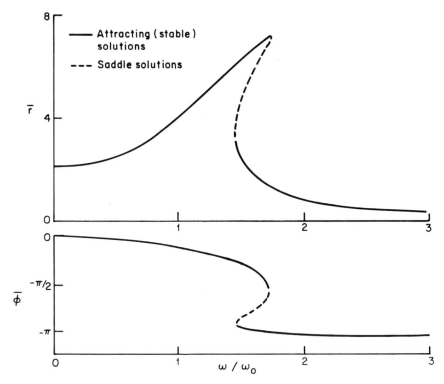

Figure 4.2.1. Frequency response function for the Duffing equation: $\varepsilon\alpha = 0.05$, $\varepsilon\delta = 0.2$, $\varepsilon\gamma = 2.5$.

EXERCISE 4.2.4. Carry out averaging for the "original" van der Pol equation

$$\ddot{x} + \frac{\alpha}{\omega}(x^2 - 1)\dot{x} + x = \alpha\gamma \cos \omega t,$$

with $1 - \omega^2 = \alpha\sigma = \mathcal{O}(\alpha) \ll 1$. Show that the averaged equation takes the form given in (2.1.13) with $\beta = \alpha\gamma$. Check as many of the assertions made in Section 2.1, concerning the averaged system, as you can.

EXERCISE 4.2.5. Consider equation (4.2.11) with $\varepsilon\gamma = \bar{\gamma} = O(1)$ and $\omega \approx 3\omega_0$ and apply the method of averaging to study the subharmonics of order three. In this case the transformation (4.2.9) should be replaced by

$$\begin{pmatrix} u \\ v \end{pmatrix} = A\begin{pmatrix} x + B \cos(\omega t - \phi) \\ \dot{x} - \omega B \sin(\omega t - \phi) \end{pmatrix},$$

where

$$A = \begin{pmatrix} \cos\left(\dfrac{\omega t}{3}\right) & -\dfrac{3}{\omega}\sin\left(\dfrac{\omega t}{3}\right) \\[3mm] -\sin\left(\dfrac{\omega t}{3}\right) & -\dfrac{3}{\omega}\cos\left(\dfrac{\omega t}{3}\right) \end{pmatrix}. \tag{4.2.15}$$

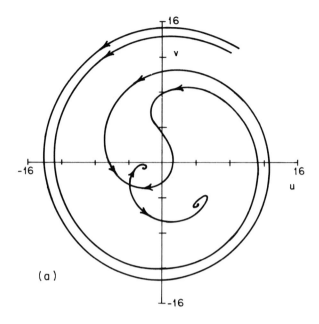

(a)

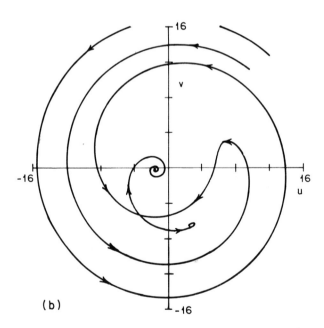

(b)

Figure 4.2.2. Phase portraits of equation (4.2.13): $\varepsilon\alpha = 0.05$, $\varepsilon\delta = 0.2$, $\omega_0 = 1$, $\varepsilon\gamma = 2.5$. (a) $\omega = 1.5$; (b) $\omega = 1.65$.

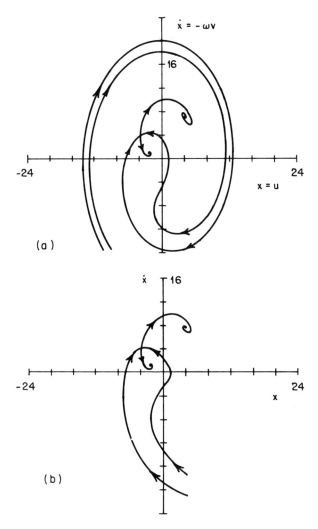

$\dot{x} = -\omega v$

16

-24

24

$x = u$

(a)

$\dot{x}$  16

-24

24

$x$

(b)

Figure 4.2.3. A Poincaré map for the Duffing equation compared with the time $T$ flow map of the averaged equation: $\varepsilon\alpha = 0.05$, $\varepsilon\delta = 0.2$, $\omega_0 = 1$, $\varepsilon\gamma = 2.5$, $\omega = 1.5$. (a) Invariant manifolds of the saddle point for the averaged equation (4.2.13); (b) invariant manifolds of the saddle point for the Poincaré map.

EXERCISE 4.2.5 (continued). Note that we must include a component at the forcing frequency $\omega$. The amplitude $B$ and phase $\phi$ of this are determined (to first order) by solving (4.2.11) for the first harmonic. Letting $x = B\cos(\omega t - \phi) + \mathcal{O}(\varepsilon)$ and substituting into (4.2.11) we obtain $B \approx \bar{\gamma}/(\omega_0^2 - \omega^2)$, $\phi \approx 0$. (This is like the linear oscillator solution, why?) The rest is up to you, but see Hale [1969, pp. 202–204] if you need help. For another example see Holmes and Holmes [1981].

EXERCISE 4.2.6. Repeat Exercise 4.2.5 with $\varepsilon\gamma = \mathcal{O}(\varepsilon)$ as before. Do you find any subharmonics at first order in the averaged equation?

Note that the transformation of equation (4.2.15) takes a solution of the form

$$x(t) = \bar{u} \cos\left(\frac{\omega t}{3}\right) - \bar{v} \sin\left(\frac{\omega t}{3}\right) - B \cos(\omega t - \phi) \qquad (4.2.16)$$

into $(\bar{u}, \bar{v})$. Thus, a solution with an almost sinusoidal fundamental component of frequency $\omega$ and a subharmonic component of frequency $\omega/3$ becomes almost constant under this transformation, much as an almost sinusoidal (single frequency) solution behaves under the transformation (4.2.9). In each case the transformation is chosen not only to render the correct form for averaging, but also in the light of our *expectation* about the form of the solution. This severe disadvantage of having to know what to look for before starting is shared by most perturbation methods.

The reader should note that a subharmonic of period $kT$, corresponding to a cycle of period $k$ in the Poincaré map, corresponds to a set of $k$ fixed points for the averaged equation. Averaging is carried out over the shortest common period $kT$ and thus the time $kT$ flow map for the averaged system approximates the true time $kT$ map for the real system; the Poincaré map iterated $k$ times.

In closing, we remark that in some cases second or even higher-order averaging is required if the results of first order averaging are inconclusive. In such a case the average,

$$\bar{f}_1(z) = \frac{1}{T} \int_0^T f_1(z, t, 0) \, dt,$$

of the second-order term is computed and the averaged equation is obtained after a second transformation $y = z + \varepsilon^2 w(z, t, \varepsilon)$:

$$\dot{z} = \varepsilon \bar{f}(z) + \varepsilon^2 \bar{f}_1(z) + \mathcal{O}(\varepsilon^3).$$

For more information and examples, see Sanders and Verhulst [1982] and Holmes and Holmes [1981]; an example is also given in Section 4.7, below. Also, Chow and Hale [1982] show how higher-order averaging can be carried out conveniently using a Lie series approach, thus underlining the basic relationship between averaging and normal forms. For additional information and examples, see Cushman and co-workers [1980, 1982, 1983], Churchill *et al.* [1982], and Deprit [1982].

## 4.3. Averaging and Local Bifurcations

Suppose that we have a one-parameter family of systems similar to (4.1.1):

$$\dot{x} = \varepsilon f_\mu(x, t, \varepsilon); \qquad \mu \in \mathbb{R}, \qquad (4.3.1)$$

with the associated family of averaged systems

$$\dot{y} = \varepsilon \bar{f}_\mu(y), \qquad (4.3.2)$$

and suppose that (4.3.2) undergoes a bifurcation as $\mu$ varies. We wish to know if (4.3.1) undergoes a similar bifurcation. For the simple codimension one bifurcations, the answer is yes, with some qualifications.

**Theorem 4.3.1.** *If at $\mu = \mu_0$ (4.3.2) undergoes a saddle-node or a Hopf bifurcation, then, for $\mu$ near $\mu_0$ and $\varepsilon$ sufficiently small, the Poincaré map of (4.3.1) also undergoes a saddle-node or a Hopf bifurcation.*

PROOF. Fix $\varepsilon > 0$, small. Again we use the $\varepsilon$-closeness of the Poincaré maps $P_0^\mu$ and $P_\varepsilon^\mu$ of (4.3.2) and (4.3.1), respectively. For the saddle-node we consider the bifurcation equations (cf. the proof of Theorem 4.1.1):

$$\frac{1}{\varepsilon}(Id - P_0^\mu)(y) = 0 \quad \text{and} \quad \frac{1}{\varepsilon}(Id - P_\varepsilon^\mu)(x) = 0. \tag{4.3.3a,b}$$

By hypothesis, the averaged system has a pair of fixed points $y_+$, $y_-$ which coalesce at $\mu_0$ in a smooth (locally parabolic) arc in $(y, \mu)$ space (Figure 4.3.1). There is a local change of coordinates near the point $(y_\pm(\mu_0), \mu_0) \in \mathbb{R}^n \times \mathbb{R}$ under which points in these branches can be put into the form $y_\pm(\mu) = (\pm c\sqrt{\mu_0 - \mu}, 0) \in \mathbb{R} \times \mathbb{R}^{n-1} = \mathbb{R}^n$. Since $P_\varepsilon^\mu$ is $\varepsilon^2$-close to $P_0^\mu$, it follows that, in the same coordinates, $P_\varepsilon^\mu$ has a pair of branches of equilibria $x_\pm(\mu) = (\pm c\sqrt{\mu_0 - \mu} + \mathcal{O}(\varepsilon), \mathcal{O}(\varepsilon))$, and since the matrices $DP_\varepsilon^\mu(x_\pm(\mu))$ and $DP_0^\mu(y_\pm(\mu))$ are also close, the linearized stability of the branches is the same for both systems.

For the Hopf bifurcation, we note that, since $P_0^\mu$ has a (locally unique) curve of equilibria $\bar{y}(\mu)$ and the spectrum of $DP_0^\mu(\bar{y}(\mu))$ does not contain 1, a nearby curve of equilibria $\bar{x}(\mu) = \bar{y}(\mu) + \mathcal{O}(\varepsilon)$ persists for $P_\varepsilon^\mu$. Since the spectra of $DP_0^\mu(\bar{y}(\mu))$ and $DP_\varepsilon^\mu(\bar{x}(\mu))$ are likewise close, a (simple) pair of complex eigenvalues must pass through the unit circle for the latter problem near $\mu = \mu_0$. It remains only to check that the nonresonance condition is met (SH1 of Theorem 3.5.2). (The stability type of the bifurcation (SH2 and SH3) will clearly be preserved under small perturbations.) Suppose that the relevant eigenvalues of $D\bar{f}_{\mu_0}(\bar{y}(\mu_0))$ are $\pm i\omega$. The eigenvalues of $DP_0^{\mu_0}(\bar{y}(\mu_0))$ are therefore $e^{\pm i\omega T}$ and those of the perturbed linearized map $DP_\varepsilon^{\mu_\varepsilon}(\bar{x}(\mu_\varepsilon))$ at its nearby bifurcation point are of the form $e^{\pm i(\varepsilon\omega T + \mathcal{O}(\varepsilon^2))}$. We therefore require $e^{im\varepsilon\omega T} \neq 1$ for $m = 1, 2, 3, 4$, a condition which is easily satisfied if $\varepsilon$ is sufficiently small, since $T$ is fixed. $\square$

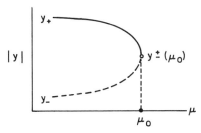

Figure 4.3.1. A saddle-node bifurcation.

This result can be applied to both the Duffing and van der Pol examples of the preceding section. In particular, it follows that the saddle-node bifurcations occurring at the fold or "jump resonance" points of Figure 4.2.1 in the averaged Duffing equation correspond to saddle-node bifurcations of periodic orbits for the full system, and that a Hopf bifurcation in which an attracting invariant circle is created occurs for the Poincaré map of the van der Pol equation (cf. Section 2.1).

Extension to multiparameter systems and more complicated bifurcations can be made. For example, there is a point in $\Omega, \gamma$ space for which the Duffing example of Section 4.2 has a degenerate fixed point which can split into one, two, or three points—as in the perturbed pitchfork bifurcation. At this point the two folds of the frequency response function of Figure 4.2.1 coalesce in a point of vertical tangency. It follows that, at a point near this, the full system has a bifurcation in which three periodic orbits coalesce. However, the averaging method cannot correctly capture all such codimension two bifurcations, primarily due to subtle, global effects involving homoclinic orbits. These effects are introduced in the next section.

# 4.4. Averaging, Hamiltonian Systems, and Global Behavior: Cautionary Notes

We have seen that the averaging theorem provides a method of approximating the true Poincaré map $P_\varepsilon$, and does so essentially by replacing $P_\varepsilon$, an $n$-dimensional map, with an $n$-dimensional vector field $\bar{f}$. Thus, one must be careful in interpretation. By Theorem 4.1.1, local behavior carries over; for example, fixed points of $\bar{f}$ correspond to periodic orbits of (4.1.1) and hence to fixed (or periodic) points of $P_\varepsilon$. However, global behavior does not carry over directly, since the generic properties of $n$-dimensional vector fields and $n$-dimensional maps are quite different. Thus, if one detects a certain global feature in $\bar{f}$, the corresponding global feature in $P_\varepsilon$ may not be similar. However, under certain conditions global behavior does carry over:

**Theorem 4.4.1.** *If the time $T$ flow map $P_0$ of (4.1.2), restricted to a bounded domain $D \subset \mathbb{R}^n$, has a limit set consisting solely of hyperbolic fixed points and all intersections of stable and unstable manifolds are transverse, then the corresponding Poincare map $P_\varepsilon|_D$ of (4.1.1) is topologically equivalent to $P_0|_D$ for $\varepsilon > 0$ sufficiently small.*

PROOF. This result is essentially a corollary to the main averaging Theorem 4.1.1, following directly from conclusions (ii) and (iii) of that theorem. In the terminology of dynamical systems, the result follows from the structural stability of the flow of (4.1.2) for $\varepsilon > 0$. However, naive perturbation arguments do not work directly, since if one lets $\varepsilon \to 0$ in (4.1.3) then the unperturbed system, $\dot{x} = 0$, is totally degenerate. On the other hand, rescaling time $\tau = \varepsilon t$ and considering the equation

$$y' = \bar{f}(y) + \varepsilon f_1\left(y, \frac{\tau}{\varepsilon}, \varepsilon\right), \qquad (4.4.1)$$

where $y' = dy/d\tau$, we have a perturbation of variable period $\varepsilon T$. However, this does not affect the Gronwall estimates used in the proof of Theorem 4.1.1, and one can obtain the required results as in the proof of that theorem. $\square$

Returning to the Duffing equation example of Section 4.2, we can use Bendixson's criterion to check that the phase portrait of (4.2.13) contains no closed loops, i.e., no closed orbits or homoclinic orbits. We verify that

$$\text{trace } Df = \frac{\partial f_1}{\partial u} + \frac{\partial f_2}{\partial v} = -\delta < 0 \qquad (4.4.2)$$

for all $(u, v) \in \mathbb{R}^2$. Hence the phase portrait can only contain fixed points and, since it is planar and no homoclinic loops occur, and there is only one saddle point, the stable and unstable manifolds of this saddle cannot intersect. Thus the hypotheses of Theorem 4.4.1 are satisfied and $P_\varepsilon$ is topologically equivalent to $P_0$ for $\varepsilon$ sufficiently small, provided we are not at a saddle-node bifurcation point. Here the reader should refer back to Figure 4.2.3.

EXERCISE 4.4.1. Verify that the systems of Exercises 4.2.2–4.2.3 also satisfy the conditions of Theorem 4.4.1.

In many averaged equations, periodic orbits appear in addition to fixed points. For example, we have already seen that the averaged van der Pol equations (2.1.14) undergo a Hopf bifurcation and that the resulting periodic orbit subsequently vanishes in a saddle connection. This situation is more delicate, but under certain conditions the periodic orbit of the averaged equations persists as an invariant torus for the full system.

**Theorem 4.4.2.** *If (4.1.2) has a hyperbolic periodic orbit $\gamma_0$, then the flow of the suspended system (4.1.12) has a hyperbolic invariant torus $T_\varepsilon$ near $\gamma_0 \times S^1$. Equivalently, the Poincaré map $P_\varepsilon$ of (4.1.1) has an invariant closed curve $\gamma_\varepsilon$ near $\gamma_0$.*

For a proof, see Hale [1969]. The normal hyperbolicity of the invariant set $\gamma_0 \times S^1$ guarantees its persistence for small nonzero $\varepsilon$ (cf. Hirsch *et al.* [1977]). However, we note that, while smooth invariant closed curves exist for both $P_0$ and $P_\varepsilon$, the dynamics of $P_\varepsilon|_{\gamma_\varepsilon}$ are generally quite complex due to resonance effects. To appreciate this, consider (4.4.1) with rescaled time $\tau = \varepsilon t$, the perturbation now having variable period $\varepsilon T$. If, for $\varepsilon = 0$, (4.4.1) has an orbit of period $\bar{\tau}$, then, as $\varepsilon \to 0$, the resonance relationship

$$\bar{\tau} = \frac{m\varepsilon T}{n}, \qquad m, n \in \mathbb{Z}, \qquad (4.4.3)$$

is satisfied countably many times. Thus, from the general theory of maps on the circle, one expects the smooth closed curve $\gamma_\varepsilon$ to contain sets of periodic points whose periods $m \sim 1/\varepsilon$ depend on $\varepsilon$, and which appear and vanish in countable sequences of bifurcations as $\varepsilon \to 0$. The analysis of such resonant motions depends delicately on small denominators and further discussion of it will be postponed until Chapter 6.

A second, and potentially more serious, problem arises in the interpretation of averaged equations. If the original system (4.1.1) is Hamiltonian with energy $\varepsilon H(u, v, t, \varepsilon)$, then the transformation (4.1.5) can be chosen to be canonical (Goldstein [1980]), so that the transformed system (4.1.3) is also Hamiltonian. In particular, it should be clear that the averaged system (4.1.2) is Hamiltonian. If $x \in \mathbb{R}^2$, so that (4.1.2) is a two-dimensional autonomous system, then its solution curves are level curves of the averaged Hamiltonian function

$$\bar{H}(u, v) = \frac{1}{T} \int_0^T H(u, v, t, 0) \, dt. \tag{4.4.4}$$

If the averaged system has a compact level curve containing a saddle point then necessarily this curve is the nontransversal intersection (in fact the identification) of stable and unstable manifolds for the unperturbed Poincaré map and we cannot expect it to be preserved upon restoration of the time-dependent term $\varepsilon^2 f_1$.

Once more consider the Duffing equation of Section 4.2 as an example and let $\delta = 0$. The original system is now Hamiltonian, and, directly from (4.2.13) we can see that the averaged Hamiltonian is

$$\varepsilon \bar{H}(u, v) = -\frac{\varepsilon}{4\omega} \left[ \Omega(u^2 + v^2) + \frac{3\alpha}{8} (u^2 + v^2)^2 - 2\gamma u \right]. \tag{4.4.5}$$

The corresponding phase portrait is shown in Figure 4.4.1. Note, in addition to the double homoclinic loop, the three continuous families of periodic orbits. Since these level curves are invariant curves for the (unperturbed) Poincaré map $P_0$ of the averaged system, in the generic case we expect the homoclinic loops to break, yielding transverse intersections of the perturbed stable and unstable manifolds. We also expect the resonant closed curves to break up in some complicated way. None of the closed orbits is hyperbolic and we have to resort to the Kolmogorov–Arnold–Moser theory to prove that "sufficiently irrational" ones close enough to the elliptic centers are preserved for $P_\varepsilon$, with $\varepsilon$ small. We shall return to the general topic of perturbation of integrable Hamiltonian systems in Section 4.8 after discussing Melnikov's method.

Even when the original problem is non-Hamiltonian, homoclinic bifurcations can arise in an averaged equation. For example, as noted in Section 2.1, such a bifurcation occurs for the averaged van der Pol system with weak damping and forcing, the sequence of phase portraits of Figure

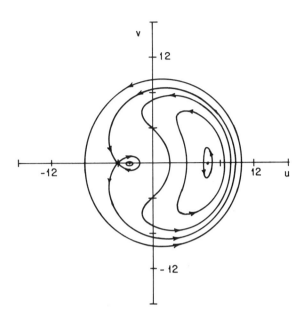

Figure 4.4.1. The phase portrait for the averaged Duffing equation: $\varepsilon\alpha = 0.05$, $\varepsilon\gamma = 2.5$, $\delta = 0$, $\omega = 1.5$, $\omega_0 = 1$.

4.4.2 being obtained. The general theory sketched above shows that, for $\varepsilon$ sufficiently small and depending upon the other parameters, the phase portraits (Figure 4.4.2(a), (c)) are preserved for the full system, since the time $T$ flow maps for the averaged system in these cases is structurally stable. However, as we approach the parameter value for which the homoclinic bifurcation (Figure 4.4.2(b)) occurs, the $\varepsilon$-range for which the conclusions of Theorem 4.4.2 hold, shrinks. Clearly, at the bifurcation point, we cannot expect the stable and unstable manifolds of the saddle for the perturbed map $P_\varepsilon$ to be identified, as in Figure 4.4.2(b), and in the generic case one expects to find transversal intersections occurring near the bifurcation point for the unperturbed system, with (generic) quadratic tangencies of the manifolds at discrete points occurring at the boundaries of the region, see Figure 4.4.3.

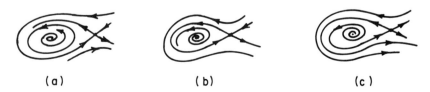

(a)                    ( b)                    (c )

Figure 4.4.2. A homoclinic bifurcation for the averaged autonomous van der Pol equation (cf. Figure 2.1.3).

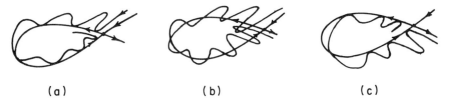

$$(a) \qquad\qquad\qquad (b) \qquad\qquad\qquad (c)$$

Figure 4.4.3. A generic perturbation of the degenerate homoclinic bifurcation for the map $P_\varepsilon^\mu$.

Much of this book, including the rest of this chapter and Chapters 5 and 6, will be concerned with the detection and analysis of transverse homoclinic points of two-dimensional maps, the bifurcations leading to them, and the complicated dynamics resulting from them.

## 4.5. Melnikov's Method: Perturbations of Planar Homoclinic Orbits

In this and the next section, we develop a method which enables us to study the Poincaré map for time-periodic systems of the form

$$\dot{x} = f(x) + \varepsilon g(x, t); \qquad x = \binom{u}{v} \in \mathbb{R}^2, \qquad\qquad (4.5.1)$$

where $g$ is of (fixed) period $T$ in $t$. Equivalently, we have the suspended system:

$$\left.\begin{array}{l} \dot{x} = f(x) + \varepsilon g(x, \theta) \\ \dot{\theta} = 1 \end{array}\right\}; \qquad (x, \theta) \in \mathbb{R}^2 \times S^1.$$

Here $f(x)$ is a Hamiltonian vector field defined on $\mathbb{R}^2$ or some subset thereof, and $\varepsilon g(x, t)$ is a small perturbation which need not be Hamiltonian itself. Many physical problems, such as the buckled beam of Section 2.2, can be expressed in the form (4.5.1), but we must admit that our main reason for studying systems of this type is that they present one of the few cases in which *global* information on specific systems can be obtained analytically. We also remark that, after a time scale change $t \to \varepsilon t$, the averaged system of equation (4.1.3) also falls into this class, but as we point out in Section 4.7, there are difficulties in applying the Melnikov analysis directly to averaged systems, since the period of $g$ is then $\mathcal{O}(1/\varepsilon)$ (cf. equation (4.4.1)). Various extensions to the method are outlined in Section 4.8, which enable it to be applied to a wider class of systems.

The basic ideas to be introduced here are due to Melnikov [1963]. More recently Chow *et al.* [1980] have obtained similar results using alternative methods, and Holmes and Marsden [1981, 1982a, b, 1983a] have applied the method to certain infinite-dimensional flows arising from partial differential equations and to multidegree of freedom autonomous Hamiltonian systems. The basic idea is to make use of the globally computable solutions of the unperturbed *integrable* system in the computation of perturbed solutions. To do this we must first ensure that the perturbation calculations are uniformly valid on arbitrarily long or semi-infinite time intervals.

First we make our assumptions precise. We consider systems of the form (4.5.1) where

$$f = \begin{pmatrix} f_1(x) \\ f_2(x) \end{pmatrix}, \qquad g = \begin{pmatrix} g_1(x, t) \\ g_2(x, t) \end{pmatrix}$$

are sufficiently smooth ($C^r, r \geq 2$) and bounded on bounded sets, and $g$ is $T$-periodic in $t$. For simplicity, we assume that the unperturbed system is Hamiltonian with $f_1 = \partial H/\partial v$, $f_2 = -\partial H/\partial u$. (The non-Hamiltonian case is considered by Melnikov [1963] and Holmes [1980b], cf. Exercise 4.5.1.) In general we shall restrict ourselves to a bounded region $D \subset \mathbb{R}^2$ of the phase space. Our specific assumptions on the unperturbed flow are:

A1 For $\varepsilon = 0$ (4.5.1) possesses a homoclinic orbit $q^0(t)$, to a hyperbolic saddle point $p_0$.
A2 Let $\Gamma^0 = \{q^0(t) | t \in \mathbb{R}\} \cup \{p_0\}$. The interior of $\Gamma^0$ is filled with a continuous family of periodic orbits $q^\alpha(t)$, $\alpha \in (-1, 0)$. Letting $d(x, \Gamma^0) = \inf_{q \in \Gamma^0} |x - q|$ we have $\lim_{\alpha \to 0} \sup_{t \in \mathbb{R}} d(q^\alpha(t), \Gamma^0) = 0$.
A3 Let $h_\alpha = H(q^\alpha(t))$ and $T_\alpha$ be the period of $q^\alpha(t)$. Then $T_\alpha$ is a differentiable function of $h_\alpha$ and $dT_\alpha/dh_\alpha > 0$ inside $\Gamma^0$.

We note that A2 and A3 imply that $T_\alpha \to \infty$ monotonically as $\alpha \to 0$. We illustrate the situation in Figure 4.5.1.

Many of the results to follow can be proved under less restrictive assumptions. In particular, for this section we only require A1. In what follows we outline the main ideas of the proof but omit some details. See Greenspan [1981] for more information.

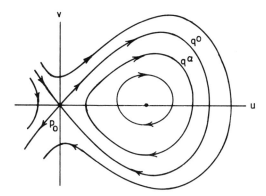

Figure 4.5.1. The unperturbed system.

As in Section 4.1 we define a Poincaré map $P_\varepsilon^{t_0}: \Sigma^{t_0} \to \Sigma^{t_0}$, where $\Sigma^{t_0} = \{(x, t)\,|\,t = t_0 \in [0, T]\} \subset \mathbb{R}^2 \times S^1$ is the global cross section at time $t_0$ for the suspended autonomous flow of (4.5.1). Note that we will need to vary the "section time" $t_0$ in what follows.

Assumption A1 immediately implies that the unperturbed Poincaré map $P_0^{t_0}$ has a hyperbolic saddle point $p_0$ and that the closed curve $\Gamma^0 = W^u(p_0) \cap W^s(p_0)$ is filled with nontransverse homoclinic points for $P_0^{t_0}$.* We expect this highly degenerate structure to break under the perturbation $\varepsilon g(x, t)$, and perhaps to yield transverse homoclinic orbits or no homoclinic points at all. The goal of this section is the development of a method to determine what happens in specific cases. In particular, this method will enable us to prove the existence of transverse homoclinic points and homoclinic bifurcations in important physical examples, and hence, using the results of Chapter 5, to prove that horseshoes and chaotic motions occur. This is one of the few analytical methods available for the detection and study of chaotic motions.

We start with two basic perturbation results:

**Lemma 4.5.1.** *Under the above assumptions, for $\varepsilon$ sufficiently small, (4.5.1) has a unique hyperbolic periodic orbit $\gamma_\varepsilon^0(t) = p_0 + \mathcal{O}(\varepsilon)$. Correspondingly, the Poincaré map $P_\varepsilon^{t_0}$ has a unique hyperbolic saddle point $p_\varepsilon^{t_0} = p_0 + \mathcal{O}(\varepsilon)$.*

PROOF. This is a straightforward application of the implicit function theorem, our assumptions implying that $DP_0^{t_0}(p_0)$ does not contain 1 in its spectrum and hence that $\mathrm{Id} - DP_0^{t_0}(p_0)$ is invertible and there is a smooth curve of fixed points $(p_\varepsilon^{t_0}, \varepsilon)$ in $(x, \varepsilon)$ space passing through $(p_0, 0)$.  □

**Lemma 4.5.2.** *The local stable and unstable manifolds $W_{\mathrm{loc}}^s(\gamma_\varepsilon)$, $W_{\mathrm{loc}}^u(\gamma_\varepsilon)$ of the perturbed periodic orbit are $C^r$-close to those of the unperturbed periodic orbit $p_0 \times S^1$. Moreover, orbits $q_\varepsilon^s(t, t_0)$, $q_\varepsilon^u(t, t_0)$ lying in $W^s(\gamma_\varepsilon)$, $W^u(\gamma_\varepsilon)$ and based on $\Sigma^{t_0}$ can be expressed as follows, with uniform validity in the indicated time intervals.*

$$
\begin{aligned}
q_\varepsilon^s(t, t_0) &= q^0(t - t_0) + \varepsilon q_1^s(t, t_0) + \mathcal{O}(\varepsilon^2), & t \in [t_0, \infty); \\
q_\varepsilon^u(t, t_0) &= q^0(t - t_0) + \varepsilon q_1^u(t, t_0) + \mathcal{O}(\varepsilon^2), & t \in (-\infty, t_0].
\end{aligned}
\tag{4.5.2}
$$

PROOF. The existence of the perturbed manifolds follows from invariant manifold theory (cf. Nitecki [1971], Hartman [1973], or Hirsch *et al.* [1977]). As in the proof of the averaging theorem, we fix a $\nu$-neighborhood ($0 \le \varepsilon \ll \nu \ll 1$) $U_\nu$ of $p_0$ inside which the local perturbed manifolds and their tangent spaces are $\varepsilon$-close to those of the unperturbed flow (or Poincaré map). A standard Gronwall estimate shows that perturbed orbits starting within $\mathcal{O}(\varepsilon)$ of $q^0(0)$ remain within $\mathcal{O}(\varepsilon)$ of $q^0(t - t_0)$ for finite times and hence that

---

* Here the subscript 0 implies that we set $\varepsilon = 0$ in (4.5.1).

one can follow any such orbit from an arbitrary point near $q^0(0)$ on $\Gamma^0$ *outside* $U_v$ to the boundary of $U_v$, at, say $t = t_1$. Once in $U_v$, if the perturbed orbit $q_\varepsilon^s$ is selected to lie in $W^s(\gamma_\varepsilon)$, then its behavior is governed by the exponential contraction associated with the linearized system. Moreover, the perturbation theory of invariant manifolds implies that

$$|q_\varepsilon^s(t_1, t_0) - q^0(t_1 - t_0)| = \mathcal{O}(\varepsilon),$$

since the perturbed manifold is $C^r$-close to the unperturbed manifold. Straightforward estimates then show that

$$|q_\varepsilon^s(t, t_0) - q^0(t - t_0)| = \mathcal{O}(\varepsilon)$$

for $t \in (t_1, \infty)$. Reversing time, one obtains a similar result for $q_\varepsilon^u(t, t_0)$ (cf. Figure 4.1.1). □

Sanders [1980, 1982] was the first to work out the details of the asymptotics of solutions in the perturbed manifolds.

This lemma implies that solutions lying in the stable manifold are *uniformly* approximated, for $t \geq 0$, by the solution $q_1^s$ of the first variational equation:

$$\dot{q}_1^s(t, t_0) = Df(q^0(t - t_0))q_1^s(t, t_0) + g(q^0(t - t_0), t). \quad (4.5.3)$$

A similar expression holds for $q_1^u(t, t_0)$ with $t \leq t_0$. We can thus use regular perturbation theory to approximate solutions in the stable and unstable manifolds of the perturbed system. Note that the initial time, $t_0$, appears explicitly, since solutions of the perturbed systems are not invariant under arbitrary translations in time ((4.5.1) is nonautonomous for $\varepsilon \neq 0$).

We next define the *separation of the manifolds* $W^u(p_\varepsilon^{t_0})$, $W^s(p_\varepsilon^{t_0})$ *on the section* $\Sigma^{t_0}$ *at the point* $q^0(0)$ as

$$d(t_0) = q_\varepsilon^u(t_0) - q_\varepsilon^s(t_0), \quad (4.5.4)$$

where $q_\varepsilon^u(t_0) \stackrel{\text{def}}{=} q_\varepsilon^u(t_0, t_0)$, $q_\varepsilon^s(t_0) \stackrel{\text{def}}{=} q_\varepsilon^s(t_0, t_0)$ are the unique points on $W^u(p_\varepsilon^{t_0})$, $W^s(p_\varepsilon^{t_0})$ "closest" to $p_\varepsilon^{t_0}$ and lying on the normal

$$f^\perp(q^0(0)) = (-f_2(q^0(0)), f_1(q^0(0)))^T$$

to $\Gamma^0$ at $q^0(0)$. The $C^r$ closeness of the manifolds to $\Gamma^0$, and Lemma 4.5.2, then imply that

$$d(t_0) = \varepsilon \frac{f(q^0(0)) \wedge (q_1^u(t_0) - q_1^s(t_0))}{|f(q^0(0))|} + \mathcal{O}(\varepsilon^2). \quad (4.5.5)$$

Here the wedge product is defined by $a \wedge b = a_1 b_2 - a_2 b_1$ and $f \wedge (q_1^u - q_1^s)$ is the projection of $q_1^u - q_1^s$ onto $f^\perp$; cf. Figure 4.5.2. Finally, we define the *Melnikov function*

$$M(t_0) = \int_{-\infty}^{\infty} f(q^0(t - t_0)) \wedge g(q^0(t - t_0), t) \, dt. \quad (4.5.6)$$

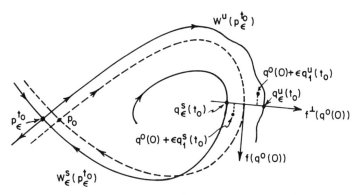

Figure 4.5.2. The perturbed manifolds and the distance function.

**Theorem 4.5.3.** *If $M(t_0)$ has simple zeros and is independent of $\varepsilon$, then, for $\varepsilon > 0$ sufficiently small, $W^u(p_\varepsilon^{t_0})$ and $W^s(p_\varepsilon^{t_0})$ intersect transversely. If $M(t_0)$ remains away from zero then $W^u(p_\varepsilon^{t_0}) \cap W^s(p_\varepsilon^{t_0}) = \varnothing$.*

**Remark.** This result is important, since it permits us to test for the existence of transverse homoclinic orbits in specific differential equations. As we show in Chapter 5, the presence of such orbits implies, via the Smale–Birkhoff theorem, that some iterate $(P_\varepsilon^{t_0})^N$ of the Poincaré map has an invariant hyperbolic set: a Smale horseshoe. As noted in Section 2.4, a horseshoe contains a countable infinity of (unstable) periodic orbits, an uncountable set of bounded, nonperiodic orbits, and a dense orbit. The sensitive dependence on initial conditions which it engenders in the flow of the differential equation is of great practical interest.

PROOF. Consider the time-dependent distance function

$$\Delta(t, t_0) = f(q^0(t - t_0)) \wedge (q_1^u(t, t_0) - q_1^s(t, t_0))$$
$$\overset{\text{def}}{=} \Delta^u(t, t_0) - \Delta^s(t, t_0), \tag{4.5.7}$$

and note that $d(t_0) = \varepsilon \Delta(t_0, t_0)/|f(q^0(0))| + \mathcal{O}(\varepsilon^2)$, from (4.5.5). We compute the derivative

$$\frac{d}{dt} \Delta^s(t, t_0) = Df(q^0(t - t_0))\dot{q}^0(t - t_0) \wedge q_1^s(t, t_0)$$
$$+ f(q^0(t - t_0)) \wedge \dot{q}_1^s(t, t_0).$$

Using (4.5.3) and the fact that $\dot{q}^0 = f(q^0)$, this yields

$$\dot{\Delta}^s = Df(q^0)f(q^0) \wedge q_1^s + f(q^0) \wedge (Df(q^0)q_1^s + g(q^0, t))$$
$$= \text{trace } Df(q^0)\Delta^s + f(q^0) \wedge g(q^0, t). \tag{4.5.8}$$

But, since $f$ is Hamiltonian, trace $Df \equiv 0$, and integrating (4.5.8) from $t_0$ to $\infty$ we have

$$\Delta^s(\infty, t_0) - \Delta^s(t_0, t_0) = \int_{t_0}^{\infty} f(q^0(t - t_0)) \wedge g(q^0(t - t_0), t) \, dt. \tag{4.5.9}$$

However, $\Delta^s(\infty, t_0) = \lim_{t \to \infty} f(q^0(t - t_0)) \wedge q_1^s(t, t_0)$ and $\lim_{t \to \infty} q^0(t - t_0)$ $= p_0$, so that $\lim_{t \to \infty} f(q^0(t - t_0)) = 0$ while $q_1^s(t, t_0)$ is bounded, from Lemma 4.5.2. Thus $\Delta^s(\infty, t_0) = 0$ and (4.5.9) gives us a formula for $\Delta^s(t_0, t_0)$. A similar calculation gives

$$\Delta^u(t_0, t_0) = \int_{-\infty}^{t_0} f(q_0(t - t_0)) \wedge g(q^0(t - t_0), t) \, dt, \qquad (4.5.10)$$

and addition of (4.5.9) and (4.5.10) and use of (4.5.5) yields

$$d(t_0) = \frac{\varepsilon M(t_0)}{|f(q^0(0))|} + \mathcal{O}(\varepsilon^2). \qquad (4.5.11)$$

Since $|f(q^0(0))| = 0(1)$, $M(t_0)$ provides a good measure of the separation of the manifolds at $q^0(0)$ on $\Sigma^{t_0}$. We recall that the vector $f^{\perp}(q^0(0))$ and its base point $q^0(0)$ are fixed on the section $\Sigma^{t_0}$ and that, as $t^0$ varies, $\Sigma^{t_0}$ sweeps around $\mathbb{R}^2 \times S^1$. Thus, if $M(t_0)$ oscillates about zero with maxima and minima independent of $\varepsilon$, then, from (4.5.4)–(4.5.5), $q^u(t_0)$ and $q^s(t_0)$ must change their orientation with respect to $f^{\perp}(q^0(0))$ as $t^0$ varies. We require that $M$ be independent of $\varepsilon$ to ensure that $\varepsilon$ can be chosen sufficiently small so that the $\mathcal{O}(\varepsilon^2)$ error in (4.5.11) is dominated by the term $\varepsilon M/|f(q^0)|$. (In Section 4.7 we shall see that $M$ can depend upon $\varepsilon$ in certain cases, and this leads to considerable difficulties.)

This implies that there must be a time $t_0 = \tau$ such that $q_\varepsilon^s(\tau) = q_\varepsilon^u(\tau)$ and we have a homoclinic point $q \in W^s(p_\varepsilon^\tau) \cap W^u(p_\varepsilon^\tau)$. But since all the Poincaré maps $P^{t_0}$ are equivalent, $W^s(p_\varepsilon^{t_0})$ and $W^u(p_\varepsilon^{t_0})$ must intersect for all $t_0 \in [0, T]$. Moreover, if the zeros are simple $(dM/dt_0 \neq 0)$, then it follows that the intersections are transversal. Conversely, if no zeros exist, then $q_\varepsilon^u(t_0)$ and $q_\varepsilon^s(t_0)$ retain the same orientation and hence the manifolds do not intersect.

$\square$

**Remarks.** 1. We note that $M(t_0)$ is $T$-periodic in $t_0$, as it should be, since the maps $P_\varepsilon^{t_0}$ and $P_\varepsilon^{t_0 + T}$ are identical, and thus $d(t_0) = d(t_0 + T)$. In computing $M(t_0)$ we are effectively standing at a fixed point $q^0(0)$ on a moving cross section $\Sigma^{t_0}$ and watching the perturbed manifolds oscillate as $t_0$ varies. In his analysis, Greenspan [1981] fixes the section and moves the base point $q^0(0)$ along the unperturbed loop $\Gamma^0$. However, the two approaches are equivalent, since the perturbed solutions are dominated by the one-parameter family of unperturbed orbits $q^0(t - t_0)$ lying in $\Gamma^0$ for which translations in time $(t_0)$ and along $\Gamma^0$ are indistinguishable.

2. If the perturbation $g$ is derived from a (time-dependent) Hamiltonian function $G(u, v)$: $g_1 = \partial G/\partial v$, $g_2 = -\partial G/\partial u$; then we have

$$M(t_0) = \int_{-\infty}^{\infty} \{H(q^0(t - t_0)), G(q^0(t - t_0), t)\} \, dt, \qquad (4.5.12)$$

where $\{H, G\}$ denotes the Poisson bracket (Goldstein [1980]):

$$\{H, G\} = \frac{\partial H}{\partial u} \frac{\partial G}{\partial v} - \frac{\partial H}{\partial v} \frac{\partial G}{\partial u}. \tag{4.5.13}$$

This is a natural formula if one recalls that the first variation of the unperturbed Hamiltonian energy $H$ will be obtained by integrating its evolution equation,

$$\dot{H} = \{H, G\}, \tag{4.5.14}$$

along the unperturbed orbit $q^0(t - t_0)$; cf. Arnold [1964].

3. If $g = g(x)$ is not explicitly time dependent, then we have, using Green's theorem,

$$\int_{-\infty}^{\infty} f(q^0(t - t_0)) \wedge g(q^0(t - t_0))\, dt = \int_{-\infty}^{\infty} (f_1 g_2 - f_2 g_1)\, dt$$

$$= \int_{-\infty}^{\infty} (g_2(u^0, v^0)\dot{u}^0 - g_1(u^0, v^0)\dot{v}^0)\, dt$$

$$= \int_{\Gamma^0} g_2(u, v)\, du - g_1(u, v)\, dv$$

$$= \int_{\text{int } \Gamma^0} \text{trace } Dg(x)\, dx. \tag{4.5.15}$$

Thus the formula obtained in Andronov *et al.* [1971] is a special (planar) case of the more general Melnikov function which describes the "splitting" of the perturbed saddle separatrices. (Also see Section 6.1.)

4. We end by noting that the change of variables $t \to t + t_0$ puts the Melnikov integral (4.5.6) into the form

$$M(t_0) = \int_{-\infty}^{\infty} f(q^0(t)) \wedge g(q^0(t), t + t_0)\, dt, \tag{4.5.16}$$

which is often more convenient in calculations.

EXERCISE 4.5.1. Suppose that assumption A1 made above holds, but that $\dot{x} = f(x)$ is not Hamiltonian, so that trace $Df \neq 0$. Derive a Melnikov function for this case.

We now turn to the case in which the perturbation $g = g(x, t; \mu)$ depends upon parameters $\mu \in \mathbb{R}^k$. For simplicity we take $k = 1$.

**Theorem 4.5.4.** *Consider the parametrized family* $\dot{x} = f(x) + \varepsilon g(x, t; \mu)$, $\mu \in \mathbb{R}$ *and let hypotheses* A1–A3 *hold. Suppose that the Melnikov function* $M(t_0, \mu)$ *has a (quadratic) zero* $M(\tau, \mu_b) = (\partial M/\partial t_0)(\tau, \mu_b) = 0$ *but* $(\partial^2 M/\partial t_0^2)(\tau, \mu_b) \neq 0$ *and* $(\partial M/\partial \mu)(\tau, \mu_b) \neq 0$. *Then* $\mu_B = \mu_b + \mathcal{O}(\varepsilon)$ *is a bifurcation value for which quadratic homoclinic tangencies occur in the family of systems.*

PROOF. By hypothesis, using (4.5.5), we have

$$d(t_0, \mu) = \varepsilon\{\alpha(\mu - \mu_b) + \beta(t_0 - \tau)^2\} + \mathcal{O}(\varepsilon|\mu - \mu_b|^2) + \mathcal{O}(\varepsilon^2), \quad (4.5.17)$$

where we have expanded in a Taylor series about $(t_0, \mu) = (\tau, \mu_b)$, and $\alpha$, $\beta$ are finite constants. Taking $\varepsilon$ sufficiently small we find that $d(t_0, \mu)$ has a quadratic zero with respect to $t_0$ for some $\mu_B$ near $\mu_b$, and hence that $W^u(p^\tau_\varepsilon)$, $W^s(p^\tau_\varepsilon)$ have a quadratic tangency near $q^0(0)$ on $\Sigma^\tau$.          □

This result is important, since it permits us to verify in specific examples one of the hypotheses of Newhouse's [1979] theorem on wild hyperbolic sets (see Chapter 6).

As an example, we apply the results of this section to the Duffing equation with negative linear stiffness and weak sinusoidal forcing and damping, introduced in Chapter 2. Written in the form (4.5.1), we have

$$\dot{u} = v,$$
$$\dot{v} = u - u^3 + \varepsilon(\gamma \cos \omega t - \delta v), \quad (4.5.18)$$

where the force amplitude $\gamma$, frequency $\omega$, and the damping $\delta$ are variable parameters and $\varepsilon$ is a small (scaling) parameter. For $\varepsilon = 0$ the system has centers at $(u, v) = (\pm 1, 0)$ and a hyperbolic saddle at $(0, 0)$. The level set

$$H(u, v) = \frac{v^2}{2} - \frac{u^2}{2} + \frac{u^4}{4} = 0 \quad (4.5.19)$$

is composed of two homoclinic orbits, $\Gamma^0_+$, $\Gamma^0_-$ and the point $p_0 = (0, 0)$. The unperturbed homoclinic orbits based at $q^0_\pm(0) = (\pm\sqrt{2}, 0)$ are given by

$$q^0_+(t) = (\sqrt{2}\ \text{sech}\ t, -\sqrt{2}\ \text{sech}\ t\ \tanh t),$$
$$q^0_-(t) = -q^0_+(t). \quad (4.5.20)$$

We will compute the Melnikov function for $q^0_+$ (the computation for $q^0_-$ is identical). Using the form (4.5.16), we have

$$M(t_0) = \int_{-\infty}^{\infty} v^0(t)[\gamma \cos \omega(t + t_0) - \delta v^0(t)]\ dt$$

$$= -\sqrt{2}\gamma \int_{-\infty}^{\infty} \text{sech}\ t\ \tanh t\ \cos \omega(t + t_0)\ dt$$

$$- 2\delta \int_{-\infty}^{\infty} \text{sech}^2 t\ \tanh^2 t\ dt. \quad (4.5.21)$$

The integrals can be evaluated (the first by the method of residues) to yield

$$M(t_0; \gamma, \delta, \omega) = -\frac{4\delta}{3} + \sqrt{2}\gamma\pi\omega\ \text{sech}\left(\frac{\pi\omega}{2}\right) \sin \omega t_0. \quad (4.5.22)$$

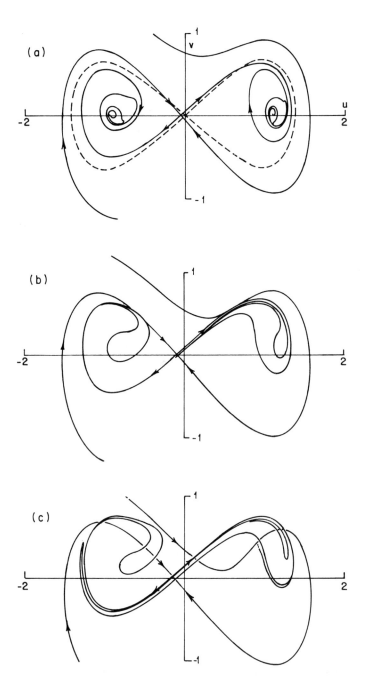

Figure 4.5.3. Poincaré maps for the Duffing equation, showing stable and unstable manifolds of the saddle point near $(0, 0)$, $\omega = 1.0$, $\varepsilon\delta = 0.25$. (a) $\varepsilon\gamma = 0.11$; (b) $\varepsilon\gamma = 0.19$; (c) $\varepsilon\gamma = 0.30$.

If we define

$$R^0(\omega) = \frac{4 \cosh(\pi\omega/2)}{3\sqrt{2}\,\pi\omega},\qquad\qquad (4.5.23)$$

then it follows from Theorem 4.5.3 that if $\gamma/\delta > R^0(\omega)$, $W^s(p_\varepsilon)$ intersects $W^u(p_\varepsilon)$ for $\varepsilon$ sufficiently small, and if $\gamma/\delta < R^0(\omega)$, $W^s(p_\varepsilon) \cap W^u(p_\varepsilon) = \varnothing$. Moreover, since $M(t_0; \gamma, \delta, \omega)$ has quadratic zeros when $\gamma/\delta = R^0(\omega)$, Theorem 4.5.4 implies that there is a bifurcation curve in the $\gamma$, $\delta$ plane for each fixed $\omega$, tangent to $\gamma = R^0(\omega)\delta$ at $\gamma = \delta = 0$, on which quadratic homoclinic tangencies occur. (To use Theorem 4.5.4 directly, we fix $\omega$ and $\delta$, for example, and vary $\gamma$.) We show some Poincaré maps of equation (4.5.17) in Figure 4.5.3. These were computed numerically by Ueda [1981a]. The unperturbed double homoclinic loop $\Gamma^0_+ \cup \{0, 0\} \cup \Gamma^0_-$ is also shown for reference on Figure 4.5.3(a). Note that the (first) tangency appears to occur about $\varepsilon\gamma = 0.19$, in comparison with a theoretical value of $0.188$ from (4.5.23)! We will continue our study of this example in the next section.

EXERCISE 4.5.2. Use the Melnikov method to compute bifurcation curves near which quadratic homoclinic tangencies occur for the plane pendulum with mixed constant and oscillating torque excitation (the undamped sine–Gordon equation):

$$\dot\theta = v,$$

$$\dot v = -\sin\theta + \varepsilon(\alpha + \gamma\cos t).$$

EXERCISE 4.5.3. Show that the Hamiltonian system with time-dependent Hamiltonian perturbation

$$H(p, q; t) = \frac{p^2 + q^2}{2} - \frac{q^3}{3} + \frac{\varepsilon q^2 \cos t}{2}$$

has transverse homoclinic orbits for all $\varepsilon \neq 0$, small.

# 4.6. Melnikov's Method: Perturbations of Hamiltonian Systems and Subharmonic Orbits

We continue to assume that the system (4.5.1) satisfies hypotheses A1–A3 of Section 4.5 and now turn to the family of periodic orbits $q^\alpha(t)$ lying within $\Gamma^0$. We wish to know if any of these will be preserved under the perturbation $\varepsilon g(x, t)$. Once more we start with a perturbation lemma:

**Lemma 4.6.1.** Let $q^\alpha(t - t_0)$ be a periodic orbit of the unperturbed system based on $\Sigma^{t_0}$, with period $T_\alpha$. Then there exists a perturbed orbit $q_\varepsilon^\alpha(t, t_0)$, not necessarily periodic, which can be expressed as

$$q_\varepsilon^\alpha(t, t_0) = q^\alpha(t - t_0) + \varepsilon q_1^\alpha(t, t_0) + \mathcal{O}(\varepsilon^2), \qquad (4.6.1)$$

uniformly in $t \in [t_0, t_0 + T_\alpha]$, for $\varepsilon$ sufficiently small and all $\alpha \in (-1, 0)$.

PROOF. The proof relies heavily upon the geometrical structure of the perturbed stable and unstable manifolds established in Lemma 4.5.2. We again fix a neighborhood $U_v$ of the fixed point $p_0$ and take a curve of initial conditions $q^\alpha(0) \subset \Sigma^0$ not lying in $U_v$ and with $\lim_{\alpha \to 0} q^\alpha(0) = q^0(0)$. Any orbit $q^\alpha(t - t_0)$ starting on such a curve takes a finite time to reach the boundary of $U_v$ as $t$ increases or decreases and hence we have

$$|q_\varepsilon^\alpha(t, t_0) - q^\alpha(t - t_0)| = \mathcal{O}(\varepsilon)$$

for $t \in (t_0 - t_1, t_0 + t_2)$, say. Once in $U_v$, the unperturbed and perturbed orbits may take arbitrary long to pass through and exit, since, as $\alpha \to 0$, they pass arbitrarily close to the saddle point $p_0$ (or to $\gamma_\varepsilon$). However, for fixed $\varepsilon = \varepsilon_0$ we can find a set of orbits *lying sufficiently close to the stable and unstable manifolds* which remain within $\mathcal{O}(\varepsilon)$ of those manifolds until they enter a $c\varepsilon$ neighborhood $U_{c\varepsilon}$ of $p_0$, $c\varepsilon \ll v$, which is chosen to contain $\gamma_\varepsilon$; Figure 4.6.1. This implies that we can extend our estimate uniformly to $t \in (t_0 - t_3, t_0 + t_4)$, where $t_3 + t_4$ is the time required for the unperturbed orbit $q^\alpha$ to pass from the boundary of $U_{c\varepsilon}$ and return to it. It remains to check that $q_\varepsilon^\alpha(t, t_0)$ and $q^\alpha(t - t_0)$ remain within $\mathcal{O}(\varepsilon)$ during the arbitrarily long passage through $U_{c\varepsilon}$. This follows from the shear on the flow near $p_0$, assured by assumptions A2 and A3 of Section 4.5; one has to check that $q_\varepsilon^\alpha$ exits from $U_{c\varepsilon}$ "at the correct time," and since orbits passing near $\gamma_\varepsilon$ take arbitrarily long to exit, and those near the boundary of $U_{c\varepsilon}$ arbitrarily short times, at least one orbit can be found for any given (unperturbed) time of passage. It follows that an initial condition $q_\varepsilon^\alpha(t_0) \in \Sigma^{t_0}$ can be picked within $\mathcal{O}(\varepsilon)$ of $q^\alpha(0)$ and $q^0(0)$ such that the orbit $q_\varepsilon^\alpha(t_0)$ will remain within $\mathcal{O}(\varepsilon)$ of $q^\alpha(t - t_0)$ (and $q_\varepsilon^s(t - t_0)$) until it reaches $U_{c\varepsilon}$. It will then "transfer its allegiance" to

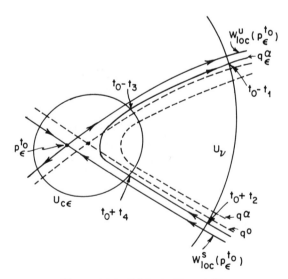

Figure 4.6.1. Orbits in $U_v$ and $U_{c\varepsilon}$.

$q_\varepsilon^u(t - t_0)$ until it once more reaches an $\varepsilon$-neighborhood of $q^\alpha(0)$. Throughout it remains within $\mathcal{O}(\varepsilon)$ of $q^\alpha(t - t_0)$. This takes care of orbits with unperturbed periods $T_\alpha$ *larger* than some $T'_\alpha = T'_\alpha(\varepsilon_0)$, depending on $\varepsilon_0$. For orbits with periods shorter than $T'_\alpha(\varepsilon_0)$, a standard Gronwall estimate ensures $\varepsilon$-closeness. Then we have our result for all $T_\alpha$ and $\varepsilon = \varepsilon_0$. But since $f$ and $g$ are $C^r$, the solutions will vary smoothly in $\varepsilon$ and thus the result holds for all $0 < \varepsilon \le \varepsilon_0$. □

We next define the *subharmonic Melnikov function*. Letting $q^\alpha(t - t_0)$ be a periodic orbit of period $T_\alpha = mT/n$, with $m$ and $n$ relatively prime, and where $T$ is the period of the perturbation, we set

$$M^{m/n}(t_0) = \int_0^{mT} f(q^\alpha(t)) \wedge g(q^\alpha(t), t + t_0)\, dt. \tag{4.6.2}$$

**Theorem 4.6.2.** *If $M^{m/n}(t_0)$ has simple zeros and is independent of $\varepsilon$ and $dT_\alpha/dh_\alpha \ne 0$, then for $0 < \varepsilon \le \varepsilon(n)$ (4.5.1) has a subharmonic orbit of period $mT$. If $n = 1$ then the result is uniformly valid in $0 < \varepsilon \le \varepsilon(1)$.*

PROOF. A calculation similar to that of Theorem 4.5.3 shows that

$$f(q^\alpha(0)) \wedge (q_\varepsilon^\alpha(t_0 + mT, t_0) - q_\varepsilon^\alpha(t_0, t_0)) = \varepsilon \int_{t_0}^{t_0 + mT} f(q^\alpha(t - t_0))$$

$$\wedge\, g(q^\alpha(t - t_0), t)dt + \mathcal{O}(\varepsilon^2)$$

$$= \varepsilon \int_0^{mT} f(q^\alpha(t)) \wedge \frac{g(q^\alpha(t), t + t_0)\, dt}{|f(q^\alpha(0))|} + \mathcal{O}(\varepsilon^2).$$

Thus, if $M^{m/n}(t_0)$ has a zero, then there is a perturbed orbit $q_\varepsilon^\alpha(t, t_0)$ which leaves $q_\varepsilon^\alpha(t_0)$ and returns to $\Sigma^{t_0}$ at $q_\varepsilon^\alpha(t_0 + mT)$ with the vector $q_\varepsilon^\alpha(t_0 + mT)$ $- q_\varepsilon^\alpha(t_0) \subset \Sigma^{t_0}$ parallel to $f(q^\alpha(0))$ to $\mathcal{O}(\varepsilon)$. Letting $M^\beta(t_0) = \int_0^{mT} f(q^\beta) \wedge g(q^\beta, t + t_0)\, dt$ for $\beta$ near $\alpha$, it is clear that $M^\beta$ depends smoothly on $\beta$. From assumption A3, for $\beta < \alpha$, $T_\beta < T_\alpha$ and for $\beta > \alpha$, $T_\beta > T_\alpha$; it therefore follows that we can find perturbed orbits $q_\varepsilon^{\beta_1}, q_\varepsilon^{\beta_2}, \beta_1 < \alpha < \beta_2$ such that the vectors $q_\varepsilon^{\beta_i}(t_0 + mT) - q_\varepsilon^{\beta_i}(t_0) \subset \Sigma^{t_0}$ are parallel to $f(q^{\beta_i}(0))$ to $\mathcal{O}(\varepsilon)$, but that they have opposite orientations. Thus there exists a curve of initial conditions connecting $q_\varepsilon^{\beta_1}(t_0)$ to $q_\varepsilon^{\beta_2}(t_0)$ which is mapped back to the section $\Sigma^{t_0}$ under $m$ iterates of $P_\varepsilon^{t_0}$ as indicated in Figure 4.6.2. We have established that the map $(P_\varepsilon^{t_0})^m$, truncated at $\mathcal{O}(\varepsilon)$, has a fixed point near $q^\alpha(0)$.

Next, anticipating the material of Sections 4.7–4.8, we introduce local coordinates $(h, \phi)$ near $q^\alpha(0)$, with $\phi$ constant along curves normal to unperturbed orbits and $h$ constant on each unperturbed orbit $q^\beta(t)$. In these coordinates, the unperturbed map can be written

$$(P_0^{t_0})^m \begin{pmatrix} h \\ \phi \end{pmatrix} = \begin{pmatrix} h \\ \omega(h) + \phi \end{pmatrix};$$

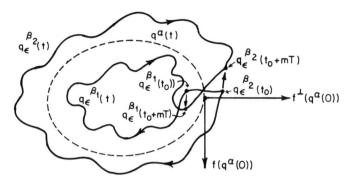

Figure 4.6.2. The existence of a fixed point "to $\mathcal{O}(\varepsilon)$."

where $\omega(0) = 0$ and $\omega'(h) < 0$, from assumption A3. The perturbed map then takes the form

$$(P_\varepsilon^{t_0})^m \begin{pmatrix} h \\ \phi \end{pmatrix} = \begin{pmatrix} h + \varepsilon F(h, \phi) \\ \omega(h) + \phi + \varepsilon G(h, \phi) \end{pmatrix} + \mathcal{O}(\varepsilon^2). \qquad (4.6.3)$$

Now we assert that the Melnikov function determines $\partial F/\partial \phi$. To see this, we note that the variable $t_0$ in $M^{m/n}(t_0)$ plays precisely the same rôle as $\phi$. Since $M^{m/n}$ measures the $\mathcal{O}(\varepsilon)$ variation in the component of the orbit parallel to $f^\perp(q^\alpha(0))$, we have

$$\frac{\partial F}{\partial \phi} = \frac{\partial M^{m/n}}{\partial t_0} \bigg/ |f(q^\alpha(0))|.$$

Consequently, if $M^{m/n}$ has a *simple* zero, we have $\partial F/\partial \phi \neq 0$ near $q^\alpha(0)$.

We have already established that $(P_\varepsilon^{t_0})^m$ has a fixed point at $\mathcal{O}(\varepsilon)$. To verify that this persists for the map when the higher-order terms ($\mathcal{O}(\varepsilon^2)$) are added, we need merely show that

$$\det|Id - (DP_\varepsilon^{t_0})^m| \neq 0.$$

Now from (4.6.3), we have

$$\det|Id - (DP_\varepsilon^{t_0})^m| = \varepsilon\omega'(h)\frac{\partial F}{\partial \phi} + \mathcal{O}(\varepsilon^2).$$

Since $\omega'(h) < 0$ by assumption A3, and the simple zero of $M^{m/n}$ guarantees $\partial F/\partial \phi \neq 0$, we conclude that $[Id - (DP_\varepsilon^{t_0})^m]$ is invertible. The implicit function theorem then guarantees that $(P_\varepsilon^{t_0})^m$ has a fixed point near $q^\alpha(0)$, and hence that there is a subharmonic of order $m/n$. The nonuniformity for $n > 1$ arises because Lemma 4.6.1 applies only to orbits of duration $T_\alpha = mT$ making one pass through $U_\nu(p)$, and "ultrasubharmonics" of period $mT/n$ make $n$ passes through $U_\nu(p)$. ☐

**Remark.** We note that, if $M^{m/n}$ has no zeros, then all solutions move either inward or outward across the unperturbed orbit and the perturbed map has *no* fixed points.

We will study the stability of the subharmonics found by Theorem 4.6.2 in the next section.

We also have a bifurcation result analogous to Theorem 4.5.4:

**Theorem 4.6.3.** *Consider the parametrized family* $\dot{x} = f(x) + \varepsilon g(x, t; \mu)$, $\mu \in \mathbb{R}$, *and let hypotheses* A1–A3 *hold. Suppose that* $M^{m/n}(t_0, \mu)$ *has a quadratic zero* $M^{m/n} = \partial M^{m/n}/\partial t_0 = 0$; $\partial^2 M^{m/n}/\partial t_0^2$, $\partial M^{m/n}/\partial \mu \neq 0$ *at* $\mu = \mu_b$. *Then* $\mu_{m/n} = \mu_b + \mathcal{O}(\varepsilon)$ *is a bifurcation value at which saddle-nodes of periodic orbits occur.*

PROOF. The proof is similar to that of Theorem 4.5.4 using ideas in the proof of Theorem 4.6.2. ☐

The next result is related to one obtained by Chow *et al.* [1980]. It implies that the homoclinic bifurcation is the limit of a countable sequence of subharmonic saddle-node bifurcations.

**Theorem 4.6.4.** *Let* $M^{m/1}(t_0) = M^m(t_0)$, *then*

$$\lim_{m \to \infty} M^m(t_0) = M(t_0). \tag{4.6.4}$$

PROOF. We must show that the integral

$$M^m(t_0) = \int_{-mT/2}^{mT/2} f(q^\alpha(t - t_0)) \wedge g(q^\alpha(t - t_0), t)\, dt \tag{4.6.5}$$

converges to

$$M(t_0) = \int_{-\infty}^{\infty} f(q^0(t - t_0)) \wedge g(q^0(t - t_0), t)\, dt, \tag{4.6.6}$$

as $m \to \infty$ and $\alpha(m) \to 0$. (Note that the periodicity of $M^m(t_0)$ implies that we can change the limits from $0 \to mT$ to $-mT/2 \to mT/2$.) Letting $\Gamma^\alpha = \{q^\alpha(t)\,|\,t \in [0, T_\alpha)\}$ and $\Gamma^0 = \{q^0(t)\,|\,t \in \mathbb{R}\} \cup \{p_0\}$, we select a neighborhood $U_\nu(p)$ such that the arc lengths of $\Gamma^0 \cap U_\nu(p)$, $\Gamma^\alpha \cap U_\nu(p)$ are each less than $\nu$. Choose $\tau$ such that $q^0(-\tau)$ and $q^0(\tau)$ both lie within $U_\nu$. Then, for $\alpha$ close enough to zero, $q^\alpha(\pm\tau)$ also lie in $U_\nu$. We have

$$
M(t_0) - M^m(t_0) = \left[ \int_{-\tau}^{\tau} f(q^0) \wedge g(q^0, t)\, dt - \int_{-\tau}^{\tau} f(q^\alpha) \wedge g(q^\alpha, t)\, dt \right]
$$
$$
+ \left[ \int_{-\infty}^{-\tau} f(q^0) \wedge g(q^0, t)\, dt + \int_{\tau}^{\infty} f(q^0) \wedge g(q^0, t)\, dt \right.
$$
$$
- \int_{-mT/2}^{-\tau} f(q^\alpha) \wedge g(q^\alpha, t)\, dt
$$
$$
\left. - \int_{\tau}^{mT/2} f(q^\alpha) \wedge g(q^\alpha, t)\, dt \right]. \tag{4.6.7}
$$

The smoothness of $f$ and $g$ and continuity of solutions with respect to initial conditions implies that, given $v > 0$, there is an $\alpha_0 < 0$ such that, for $\alpha \in [\alpha_0, 0)$ the first [bracketed] term of (4.6.7) is less than $v$. Clearly, as $\alpha \to 0^-$, $m \to \infty$. The second term may be expressed as the difference between two integrals over arcs of $\Gamma^0$ and $\Gamma^\alpha$. Using the arc-length increments $ds = \sqrt{\dot{u}^{02} + \dot{v}^{02}}\, dt = |f(q^0)|\, dt$ on $\Gamma^0$ and $ds = |f(q^\alpha)|\, dt$ on $\Gamma^\alpha$, the second term becomes

$$\int_{q^0(-\tau)}^{q^0(\tau)} f(q^0) \wedge g(q^0, t) \frac{ds}{|f(q^0)|} - \int_{q^\alpha(-\tau)}^{q^\alpha(\tau)} f(q^\alpha) \wedge g(q^\alpha, t) \frac{ds}{|f(q^\alpha)|}. \quad (4.6.8)$$

Our assumptions on $g$ imply that $\sup_{q \in U_v(p); t \in \mathbb{R}} |g(q^\alpha, t)| = K < \infty$, and thus that the second term is bounded by $2Kv$. Hence, for $\alpha \in [\alpha_0, 0)$,

$$|M(t_0) - M^m(t_0)| < (2K + 1)v,$$

and thus $|M(t_0) - M^m(t_0)| \to 0$ as $\alpha \to 0$. $\qquad\qquad\qquad \square$

To illustrate the use of the subharmonic analysis, we return to the Duffing equation (4.5.18). When $\varepsilon = 0$, within each of the homoclinic loops $\Gamma^0_\pm$ there is a one-parameter family of periodic orbits which may be written

$$q^k_+(t) = \left(\frac{\sqrt{2}}{\sqrt{2 - k^2}} dn\left(\frac{t}{\sqrt{2 - k^2}}, k\right), \frac{-\sqrt{2}k^2}{2 - k^2} sn\left(\frac{t}{\sqrt{2 - k^2}}, k\right)\right.$$

$$\left. \times\ cn\left(\frac{t}{\sqrt{2 - k^2}}, k\right)\right),$$

$$q^k_-(t) = -q^k_+(t), \quad (4.6.9)$$

where $sn$, $cn$, and $dn$ are the Jacobi elliptic functions and $k$ is the elliptic modulus. As $k \to 1$, $q^k_\pm \to q^0_\pm \cup \{0, 0\}$ and as $k \to 0$ $q^k_\pm \to (\pm 1, 0)$. We have selected our initial conditions at $t = t_0$ to be

$$q^k_\pm(0) = \left(\pm\sqrt{\frac{2}{2 - k^2}}, 0\right). \quad (4.6.10)$$

We note that the Hamiltonian (4.5.19) can be rewritten within $\Gamma^0_+$ (or $\Gamma^0_-$) in terms of the elliptic modulus $k$:

$$H(q^k) = \frac{k^2 - 1}{(2 - k^2)^2} \overset{\text{def}}{=} h_k. \quad (4.6.11)$$

Moreover, the period of these orbits is given by

$$T_k = 2K(k)\sqrt{2 - k^2}, \quad (4.6.12)$$

where $K(k)$ is the complete elliptic integral of the first kind. $T_k$ increases monotonically in $k$ with $\lim_{k \to 0} T_k = \sqrt{2}\pi$, $\lim_{k \to 1} T_k = \infty$, and

$$\frac{dT_k}{dh_k} = \frac{dT_k/dk}{dH/dk} > 0 \quad (4.6.13)$$

and

$$\lim_{k \to 1} \frac{dT_k}{dh_k} = \infty. \qquad (4.6.14)$$

Thus assumptions A2 and A3 of Section 4.5 are satisfied along with A1.

We now compute the subharmonic Melnikov function for the resonant periodic orbits. We will only consider those within $\Gamma_+^0 : q_+^k(t - t_0)$. The resonance condition is, from (4.6.12)

$$2K(k)\sqrt{2 - k^2} = \frac{2\pi m}{\omega n}, \qquad (4.6.15)$$

and for each choice of $m$, $n$ with $2\pi m/\omega n > \sqrt{2}\pi$, (4.6.15) can be solved uniquely to give $k = k(m, n)$ and hence a unique resonant orbit $q_+^{k(m, n)}$. Computing

$$M^{m/n}(t_0; \gamma, \delta, \omega) = \int_0^{mT} v^{k(m, n)}(t)[\gamma \cos \omega(t + t_0) - \delta v^{k(m, n)}(t)] \, dt, \qquad (4.6.16)$$

using the Fourier series for (4.6.9) and Remark 3 following Theorem 4.5.3, we obtain

$$M^{m/n}(t_0; \gamma, \delta, \omega) = -\delta J_1(m, n) + \gamma J_2(m, n, \omega) \sin \omega t_0, \qquad (4.6.17)$$

where

$$J_1(m, n) = \tfrac{2}{3}[(2 - k^2(m, n))2E(k(m, n)) - 4k'^2(m, n)K(k(m, n))]/$$
$$(2 - k^2(m, n))^{3/2},$$

and

$$J_2(m, n, \omega) = \begin{cases} 0; & n \neq 1 \\ \sqrt{2}\pi\omega \operatorname{sech} \dfrac{\pi m K'(k(m, 1))}{K(k(m, 1))}; & n = 1. \end{cases}$$

Here $E(k)$ is the complete elliptic integral of the second kind and $k'$ is the complementary elliptic modulus $k'^2 = 1 - k^2$. Defining

$$R^m(\omega) = \frac{J_1(m, 1)}{J_2(m, 1, \omega)}, \qquad (4.6.18)$$

we conclude from Theorems 4.6.2 and 4.6.3 that if $\gamma/\delta > R^m(\omega)$, then there is a pair of subharmonics of order $m$ (period $2\pi m/\omega$) which appear on a bifurcation curve tangent to $\gamma = R^m(\omega)\delta$ at $\gamma = \delta = 0$. In this example no ultrasubharmonics occur, at least in these $\mathcal{O}(\varepsilon)$ calculations.

Routine computations verify that

$$\lim_{m \to \infty} M^{m/1}(t_0; \gamma, \delta, \omega) = M(t_0, \gamma, \delta, \omega), \qquad (4.6.19)$$

that the limit is approached from below and that the rate of convergence is extremely rapid. In Figure 4.6.3 we show some of the bifurcation curves $R^m(\omega)$.

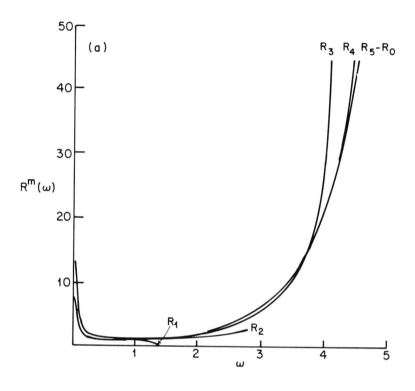

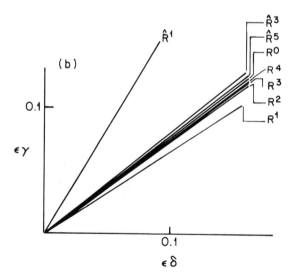

Figure 4.6.3. Bifurcation curves for the homoclinic orbits and the subharmonics of the Duffing equation. (a) $R^m(\omega)$ (subharmonic orbits within $\Gamma_0^{\pm}$); (b) the slopes $R^m(\omega)$, $\omega = 1$. Note rapid convergence of $R^m(\omega)$ and $\hat{R}^m(\omega)$ to $R^0(\omega)$.

Similar computations can be carried out for the orbits lying outside $\Gamma_+^0 \cup \{(0, 0)\} \cup \Gamma_-^0$, and one obtains a sequence of bifurcation curves whose tangents $\gamma = \hat{J}_1(m, 1)\delta/\hat{J}_2(m, 1, \omega) = \hat{R}^m(\omega)\delta$ accumulate on $\gamma = R^0(\omega)\delta$ from above. For more information, see Greenspan [1981] or Greenspan and Holmes [1982].

EXERCISE 4.6.1. Repeat the subharmonic analysis carried out above for the damped, forced plane pendulum

$$\dot{\theta} = v,$$

$$\dot{v} = -\sin \theta + \varepsilon(\gamma \cos t - \delta v).$$

Note that, for $\varepsilon = 0$, there are *two* families of periodic orbits, those with energy

$$H(\theta, v) = \frac{v^2}{2} + (1 - \cos \theta) < 2$$

(oscillations), and those with energy $H(\theta, v) > 2$ (rotations). Also, see the example below.

As our second and final example of this section we consider the plane pendulum with weak constant torque, $\alpha$, and damping, $\delta$:

$$\dot{\theta} = v,$$
$$\dot{v} = -\sin \theta + \varepsilon(\alpha - \delta v); \qquad \alpha, \delta \geq 0. \tag{4.6.20}$$

This system arizes in models of synchronous electric motors, single point Josephson junctions in superconductivity (Levi *et al.* [1978]) and in many other applications. The unperturbed system has the familiar phase portrait of Figure 4.6.4, with a pair of homoclinic orbits given by

$$(\theta_\pm^0(t), v_\pm^0(t)) = (\pm 2 \arctan(\sinh t), \pm 2 \operatorname{sech} t). \tag{4.6.21}$$

We wish to compute values of $\alpha$, $\delta$ for which one (or both) of these orbits is (are) preserved. On the upper orbit, we obtain a time-independent Melnikov function which depends on the two parameters $\alpha$, $\delta$:

$$M^+(\alpha, \delta) = \int_{-\infty}^{\infty} v_+^0(t)[\alpha - \delta v_+^0(t)]\, dt.$$

Using the fact that $v = d\theta/dt$, this becomes

$$M^+(\alpha, \delta) = \alpha \int_{-\pi}^{\pi} d\theta - 4\delta \int_{-\infty}^{\infty} \operatorname{sech}^2 t\, dt$$

$$= 2\pi\alpha - 8\delta. \tag{4.6.22}$$

On the lower orbit the first integral runs from $\pi$ to $-\pi$ and we have

$$M^-(\alpha, \delta) = -2\pi\alpha - 8\delta. \tag{4.6.23}$$

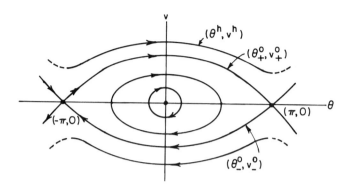

Figure 4.6.4. The unperturbed plane pendulum.

Here the Melnikov function is time independent (as it must be, since (4.6.20) is an autonomous system). Theorem 4.5.4 must therefore be modified appropriately. In the present case, fixing $\delta > 0$, for example, we have a simple zero of $M^+(\alpha, \delta)$ at $\alpha = 4\delta/\pi$. An argument like that used in the proof of Theorem 4.5.4 then implies that the upper homoclinic orbit survives in a homoclinic bifurcation on a unique curve tangent to the line

$$\alpha = \frac{4}{\pi}\delta \qquad\qquad (4.6.24)$$

at $\delta = 0$, while for $\alpha$, $\delta > 0$ the lower orbit is always broken. Moreover, since $M^+ > 0$ for $\alpha > (4/\pi)\delta$ and $M^+ < 0$ for $\alpha < (4/\pi)\delta$, from (4.5.5) and (4.5.11) we conclude that the upper branch of $W^s(\pi, 0)$ lies below (resp. above) the upper branch of $W^u(-\pi, 0)$ in these two cases, cf. Figure 4.6.5, below. Note that, for all $\alpha$, $\delta > 0$, $M^- < 0$, and the lower branch of the stable manifold $W^s(-\pi, 0)$ always lies below the lower branch of $W^u(\pi, 0)$, cf. Figure 4.6.5, below.

We next ask if any other nonwandering or invariant curves are maintained under perturbation. We first remark that no such sets persist for $v < 0$, since in the lower half of the phase space the perturbation $\begin{pmatrix} 0 \\ -2\alpha v + \beta \end{pmatrix}$ is directed entirely upward and all solution curves therefore have a net upward drift. Also, none of the closed orbits encircling the center are preserved for $\alpha \neq 0$ (see Exercise 4.6.2, below). The closed orbits lying above the upper connection and encircling the phase cylinder can, however, be maintained. It is easy to show that, for $\alpha$ sufficiently large, at least one such closed orbit exists and that it is an attractor. One simply finds two curves encircling the cylinder and bounding a band $B$ above the upper saddle connection such that the vector field is directed into $B$. Then, since $B$ contains no fixed points it must contain at least one attracting closed orbit, such as the one marked $\Gamma$ in Figure 4.6.5(c), the points $a$ and $a'$ being identified (cf. Levi et al. [1978]).

To prove that this closed orbit is unique, at least for small $\alpha$, $\beta$ we perturb the exact elliptic integral solution given by

$$\frac{(v^h(t))^2}{2} + (1 - \cos \theta^h(t)) = h, \tag{4.6.25}$$

for $h > 2$, choosing the positive (upper) branch. Note that this reduces to the homoclinic orbit when $h = 2$. From equation (4.6.25) we obtain

$$\frac{d\theta^h}{dt} = \sqrt{2h - 2(1 - \cos \theta^h)} = \sqrt{2h} \left(1 - \frac{2}{h} \sin^2\left(\frac{\theta^h}{2}\right)\right)^{1/2}. \tag{4.6.26}$$

The period of such an unperturbed closed curve is twice the time taken for a solution leaving $\theta = 0$ to reach $\theta = \pi$:

$$\sqrt{2h}\, T(h) = 2 \int_0^\pi \frac{d\theta^h}{(1 - (2/h) \sin^2(\theta^h/2))^{1/2}};$$

or letting $\theta^h = 2\phi$,

$$T(h) = \frac{4}{\sqrt{2h}} \int_0^{\pi/2} \frac{d\phi}{(1 - (2/h) \sin^2 \phi)^{1/2}} \frac{4}{\sqrt{2h}} K\left(\sqrt{\frac{2}{h}}\right), \tag{4.6.27}$$

where $K(m)$ is the complete elliptic integral of the first kind. Note that $T(h) \to \infty$ as $h \to 2$, and $T(h) \to 0$ as $h \to \infty$, since $K(0) = \pi/2$.

Although we can calculate the unperturbed solution in terms of elliptic functions as

$$(\theta^h(t), v^h(t)) = \left(2 \arcsin\left(sn\left(\frac{\sqrt{2h}}{2} t, \sqrt{\frac{2}{h}}\right)\right), \sqrt{2h}\, dn\left(\frac{\sqrt{2h}}{2} t, \frac{2}{h}\right)\right), \tag{4.6.28}$$

we do not explicitly need it in this form. The subharmonic Melnikov function is given by

$$M(\alpha, \delta, h) = \int_0^{T(h)} v^h(t)[\alpha - \delta v^h(t)]\, dt, \tag{4.6.29}$$

and using $v = d\theta/dt$ and (4.6.26), the integrals can be rewritten as

$$M(\alpha, \delta, h) = \alpha \int_{-\pi}^\pi d\theta - \delta \int_{-\pi}^\pi v^h(t)\, d\theta$$

$$= 2\pi\alpha - 2\delta \int_0^\pi \sqrt{2h} \left(1 - \left(\frac{2}{h}\right) \sin^2\left(\frac{\theta}{2}\right)\right)^{1/2} d\theta$$

$$= 2\pi\alpha - 4\sqrt{2h}\,\delta \int_0^{\pi/2} \left(1 - \left(\frac{2}{h}\right) \sin^2 \phi\right)^{1/2} d\phi$$

$$= 2\pi\alpha - 4\sqrt{2h}\,\delta E\left(\sqrt{\frac{2}{h}}\right), \tag{4.6.30}$$

where $E(m)$ is the complete elliptic integral of the second kind. Note that, at $h = 2$, $E(1) = 1$ and equation (4.6.30) reduces to equation (4.6.22); moreover, $E(2/h)$ increases monotonically as $h$ increases and $E(2/h) \to \pi/2$ as $h \to \infty$. We conclude that, for $\varepsilon$ sufficiently small and each choice of $\alpha, \delta$ with $\alpha > 4\delta/\pi$ there is precisely one value of $h$ (say $h'$) for which $M(\alpha, \delta, h') = 0$, and that $M(\alpha, \delta, h) < 0$ (resp. $> 0$) for $h > h'$ (resp. $h < h'$). It follows that there is a unique attracting closed orbit encircling the phase cylinder for each $\delta \in (0, \pi\alpha/4)$. Note that $h'$ increases as $\alpha/\delta$ increases. See Figure 4.6.5.

EXERCISE 4.6.2. Show that none of the closed orbits encircling the center in Figure 4.6.4 is preserved under the given perturbation when $\alpha \neq 0$. (Hint: Use Bendixson's criterion, with care.)

EXERCISE 4.6.3. Use Melnikov's method to study the bifurcations to subharmonics and homoclinic orbits for the damped, parametrically excited pendulum:

$$\ddot{\theta} + \varepsilon\delta\dot{\theta} + (1 + \varepsilon\gamma \cos \omega t) \sin \theta = 0.$$

EXERCISE 4.6.4. Consider the system

$$\dot{x} = y,$$
$$\dot{y} = x - x^3 + \alpha y + \beta x^2 y, \qquad \alpha, \beta \geq 0, \text{ small.}$$

Show that there are pairs $(\alpha, \beta)$ such that the homoclinic orbits occurring for $\alpha = \beta = 0$ are preserved, and that the bifurcation set is given approximately by $\alpha = 4\beta/5$. Show that, for $\alpha < 4\beta/5$ and near that value there are two closed orbits surrounding the three fixed points, while for $\alpha \in (4\beta/5, \beta)$ there are three closed orbits, one surrounding each node and one surrounding all three fixed points. (Hint: See Carr [1981], Takens [1974b], Holmes and Rand [1980], Greenspan and Holmes [1983], or Knobloch and Proctor [1981].)

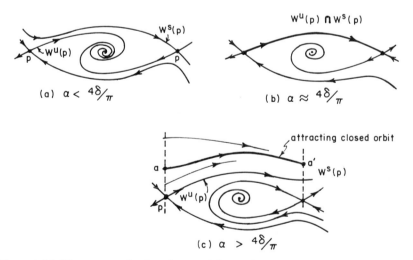

Figure 4.6.5. Phase portraits for the pendulum with torque $\alpha$ and dissipation $\delta$. (a) $\alpha < 4\delta/\pi$; (b) $\alpha \approx 4\delta/\pi$; (c) $\alpha > 4\delta/\pi$.

## 4.7. Stability of Subharmonic Orbits

In this section we outline another approach to the study of subharmonic motions which yields stability results as well as information of a more global nature. The assumptions of Section 4.5 are again made and we start with a symplectic transformation to action angle coordinates (Goldstein [1980]) within the interior of the homoclinic loop $\Gamma^0$ (Melnikov [1963]):

$$I = I(u, v), \qquad \theta = \theta(u, v), \tag{4.7.1a}$$

with inverse

$$u = U(I, \theta), \qquad v = V(I, \theta). \tag{4.7.1b}$$

The new coordinates $I$ and $\theta$ are "nonlinear polar coordinates," chosen in such a way that, for the unperturbed planar Hamiltonian system, $I(t)$ remains constant on solutions while $\theta(t)$ (mod $2\pi$) increases linearly. Such a transformation can be made in any region containing an elliptic center and filled with a continuous family of periodic orbits. In the present case the transformation becomes singular at the homoclinic orbit $\Gamma^0$. For general information on action angle coordinates and for formulae for their derivation, see Goldstein [1980] or Arnold [1978]. Here we note the important fact that, under this transformation, the unperturbed Hamiltonian $H(u, v)$ becomes

$$H(U(I, \theta), V(I, \theta)) \equiv H(I), \tag{4.7.2}$$

i.e., $H$ is independent of $\theta$. We illustrate the effect of this transformation for an example in Figure 4.7.1, below.

Under the transformation, equation (4.5.1) becomes

$$\dot{I} = \varepsilon\left(\frac{\partial I}{\partial u} g_1 + \frac{\partial I}{\partial v} g_2\right) \stackrel{\text{def}}{=} \varepsilon F(I, \theta, t),$$

$$\dot{\theta} = \Omega(I) + \varepsilon\left(\frac{\partial \theta}{\partial u} g_1 + \frac{\partial \theta}{\partial v} g_2\right) \stackrel{\text{def}}{=} \Omega(I) + \varepsilon G(I, \theta, t), \tag{4.7.3}$$

where we have used the fact that $H(I)$ is independent of $\theta$, and we define $\Omega(I) \equiv \partial H/\partial I$: the angular frequency of the closed orbit with action $I$ and energy $H(I)$.

As in Section 4.6 we choose a resonant orbit with period $mT/n$ and action $I^{m,n}$, and perturb from it, using the rotation transformation

$$I = I^{m,n} + \sqrt{\varepsilon} h,$$

$$\theta = \Omega(I^{m,n})t + \phi = \frac{2\pi n}{mT} t + \phi, \tag{4.7.4}$$

$$\stackrel{\text{def}}{=} \Omega^{m,n} t + \phi.$$

Here $h$ and $\phi$ represent perturbations from the solutions $q^{\alpha(m,n)}(t)$ of period $mT/n$ which fill the resonant torus of the suspended unperturbed problem.

Substituting (4.7.4) in (4.7.3) and expanding in a Taylor series in $\varepsilon$, we obtain the leading terms

$$\begin{aligned} \dot{h} &= \sqrt{\varepsilon} F(I^{m,n}, \Omega^{m,n} t + \phi, t) \\ \dot{\phi} &= \sqrt{\varepsilon} \Omega'(I^{m,n}) h \end{aligned} + \mathcal{O}(\varepsilon), \qquad (4.7.5)$$

where $\Omega' = \partial\Omega/\partial I$. Our assumption that $\partial T/\partial h \neq 0$ guarantees that $\Omega' \neq 0$.

Now by the chain rule we have

$$\frac{\partial I}{\partial u} = \frac{\partial I}{\partial H}\frac{\partial H}{\partial u} = \frac{1}{\Omega}\frac{\partial H}{\partial u} = -\frac{1}{\Omega} f_2,$$

and                                                                                  (4.7.6)

$$\frac{\partial I}{\partial v} = \frac{1}{\Omega} f_1,$$

and thus the first component of (4.7.5) becomes, using the explicit form of $F$ from (4.7.3),

$$F = \frac{\sqrt{\varepsilon}}{\Omega^{m,n}} (f_1 g_2 - f_2 g_1). \qquad (4.7.7)$$

If $F$ and $\Omega'$ are bounded, the averaging theorem can be applied to (4.7.5) to remove the explicit $t$ dependence at $\mathcal{O}(\sqrt{\varepsilon})$, and yield to $\mathcal{O}(\sqrt{\varepsilon})$:

$$\begin{aligned} \dot{h} &= \frac{\sqrt{\varepsilon}}{\Omega^{m,n}} \frac{1}{mT} \int_0^{mT} f(q^\alpha(t)) \wedge g\left(q^\alpha(t), t + \frac{\phi}{\Omega^{m,n}}\right) dt, \\ \dot{\phi} &= \sqrt{\varepsilon} \Omega'(I^{m,n}) h, \end{aligned}$$

or

$$\begin{aligned} \dot{h} &= \frac{\sqrt{\varepsilon}}{2\pi n} M^{m/n}\left(\frac{\phi}{\Omega^{m,n}}\right), \\ \dot{\phi} &= \sqrt{\varepsilon} \Omega'(I^{m,n}) h, \end{aligned} \qquad (4.7.8)$$

where $q^\alpha$ is the resonant orbit of period $mT/n$. We note that (4.7.8) is a (time-independent) Hamiltonian system with Hamiltonian

$$\bar{H}(h, \phi) = \sqrt{\varepsilon} \left\{ \frac{1}{2\pi n} \int M^{m/n}\left(\frac{\phi}{\Omega^{m,n}}\right) d\phi - \Omega'(I^{m,n}) \frac{h^2}{2} \right\}. \qquad (4.7.9)$$

Straightforward analysis reveals that (4.7.8) has fixed points at $h = 0$ and values of $\phi$ for which $M^{m/n} = 0$, and that these fixed points are saddles if $(\partial/\partial\phi)(M^{m/n}) < 0$ and centers if $(\partial/\partial\phi)(M^{m/n}) > 0$, since by hypothesis $\partial T/\partial h, \partial T/\partial I > 0$ and hence $\Omega' < 0$. The averaging theorem then implies that the full system has saddle-type orbits near the saddle points of (4.7.8) and periodic orbits near the centers, whose stability types are not determined by this $\mathcal{O}(\sqrt{\varepsilon})$ truncation. We therefore obtain a result similar to Theorem 4.6.2, but note that, if (as in the Duffing example) $|\Omega'| \to \infty$ as $m \to \infty$ for fixed $n$, then the region in $\varepsilon$ for which averaging is valid is not uniform and shrinks as $m$ increases.

We shall not consider the details of the general computation of the $\mathcal{O}(\varepsilon)$ terms here. Morosov [1973], Greenspan [1981], and Greenspan and Holmes [1982, 1983] provide examples. The computations typically involve special functions and are even more awkward than those of the Duffing example in Section 4.6. A simple example will follow below. However, using an observation due to Chow et al. [1980] we can determine the stability of the second ("center-like") set of orbits, at least for $\varepsilon$ sufficiently small.

**Proposition 4.7.1.** *If* trace $Dg < 0$ *(resp.* $> 0$) *then the periodic orbits of* (4.7.5), *if any exist, are saddles and sinks (resp. saddles and sources), and the Poincaré map* $P_\varepsilon$ *cannot possess any invariant closed curves.*

PROOF. Since the unperturbed system is Hamiltonian, the Poincaré map $P_0$ is area preserving and $\det|DP| = 1$. Suppose that the perturbation is "Hamiltonian plus damping," so that tr $Dg < 0$ (as in the Duffing example (4.5.18)). Then, since the perturbed three-dimensional flow contracts volumes like $e^{\text{tr } Dg \cdot t}$, the perturbed map must contract areas $(\det|DP_\varepsilon| < 1$ everywhere). Thus all periodic points of $P$ are either sinks or saddles, since the products of their eigenvalues $\lambda_1 \cdot \lambda_2 = \det[DP_\varepsilon] < 1$. Moreover, no simple invariant closed curves can exist, since their interiors would be reduced in area by application of $P_\varepsilon$. Analogous results clearly hold for $\det[DP] > 1$, in which case we have saddles or sources as stated. $\qquad\qquad\square$

Note that, if saddles and sinks, say, are created as one passes a bifurcation curve such as one of those of Figure 4.6.3, then the sink may subsequently undergo a period doubling bifurcation in which it becomes a saddle, throwing off a sink of period 2. This occurs, at least for sufficiently large $m$, in the Duffing example (cf. Greenspan and Holmes [1982]), but must be proved by methods other than the averaging analysis outlined above. We will return to this question in Chapter 6.

Now we give an example from Greenspan and Holmes [1983], in which the action angle transformation is trigonometric and the computations are straightforward. Our example also demonstrates the use of averaging at both the first and second order, cf. Sections 4.1–4.2.

We consider the system

$$\dot{u} = v(1 - (u^2 + v^2)) + \varepsilon[\delta u - u(u^2 + v^2) + \gamma u \cos t],$$
$$\dot{v} = -u(1 - (u^2 + v^2)) + \varepsilon[\delta v - v(u^2 + v^2)]. \tag{4.7.10}$$

Using the transformation to action angle coordinates for the linear oscillator

$$u = \sqrt{2I} \sin \theta, \qquad v = \sqrt{2I} \cos \theta,$$
$$I = \frac{u^2 + v^2}{2}, \qquad \theta = \arctan(v/u), \tag{4.7.11}$$

(4.7.10) becomes

$$\dot{I} = \varepsilon[2\delta I - 4I^2 + 2\gamma I \sin^2 \theta \cos t],$$
$$\dot{\theta} = (1 - 2I) + \varepsilon[\gamma \sin \theta \cos \theta \cos t], \tag{4.7.12}$$

the associated unperturbed Hamiltonian now being $H = I - I^2$.
   The period $T(I)$ of the unperturbed orbits is

$$T(I) = \frac{2\pi}{1 - 2I}, \tag{4.7.13}$$

and the unperturbed phase plane is as in Figure 4.7.1. We will study perturbations of the resonant orbit of period $4\pi$, with action $I = I^{2,1} = \frac{1}{4}$. Following (4.7.4), we make the transformation

$$I = \tfrac{1}{4} + \sqrt{\varepsilon}h,$$
$$\theta = \frac{t}{2} + \phi, \tag{4.7.14}$$

to obtain, after some trigonometrical expansion:

$$\dot{h} = \sqrt{\varepsilon}\left[\frac{\delta}{2} - \frac{1}{4} + \frac{\gamma}{4}\left(\cos t + \frac{\sin 2\phi}{2}\sin 2t - \frac{\cos 2\phi}{2}(1 + \cos 2t)\right)\right]$$
$$+ \varepsilon\left[2\delta - 2 + \gamma\left(\cos t + \frac{\sin 2\phi}{2}\sin 2t - \frac{\cos 2\phi}{2}(1 + \cos 2t)\right)\right]h$$
$$+ 4\varepsilon^{3/2}h^2$$

$$\dot{\phi} = -\sqrt{\varepsilon}2h + \varepsilon\left[\frac{\gamma}{4}(\cos 2\phi \sin 2t + \sin 2\phi(1 + \cos 2t)\right]. \tag{4.7.15}$$

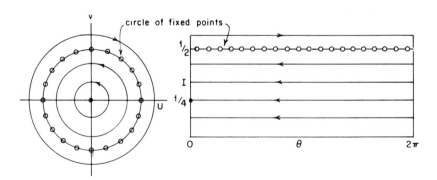

Figure 4.7.1. The unperturbed phase plane in $(u, v)$ and $(I, \theta)$ coordinates.

To average the $\mathcal{O}(\sqrt{\varepsilon})$ terms of (4.7.15), we pick the transformation

$$h = \bar{h} + \sqrt{\varepsilon}\frac{\gamma}{4} \int \left( \cos t + \frac{\sin 2\bar{\phi}}{2} \sin 2t - \frac{\cos 2\bar{\phi}}{2} \cos 2t \right) dt$$

$$= \bar{h} + \frac{\sqrt{\varepsilon}\gamma}{4} \left( \sin t - \frac{\sin 2\bar{\phi}}{4} \cos 2t - \frac{\cos 2\bar{\phi}}{4} \sin 2t \right); \qquad \phi = \bar{\phi}$$

$$(4.7.16)$$

(cf. equations (4.1.5)–(4.1.7). Applying (4.7.16) and using (4.1.8), the reader can check that (4.7.15) becomes

$$\dot{\bar{h}} = \sqrt{\varepsilon} \left[ \frac{\delta}{2} - \frac{1}{4} - \frac{\gamma}{4} \cos 2\bar{\phi} \right]$$

$$+ \varepsilon \left[ 2\delta - 2 + \gamma \left( \cos t + \frac{\sin 2\bar{\phi}}{2} \sin 2t - \frac{\cos 2\bar{\phi}}{2} (1 + \cos t) \right) \right.$$

$$\left. - \frac{\gamma}{2} \left( \frac{\cos 2\bar{\phi}}{2} \cos 2t - \frac{\sin 2\bar{\phi}}{2} \sin 2t \right) \right] \bar{h} + \mathcal{O}(\varepsilon^{3/2})$$

$$\dot{\bar{\phi}} = -\sqrt{\varepsilon}2\bar{h} + \varepsilon \left[ \frac{\gamma}{4} (\cos 2\bar{\phi} \sin t + \sin 2\bar{\phi}(1 + \cos 2t)) \right.$$

$$\left. - \gamma \left( \sin t - \frac{\sin 2\bar{\phi}}{4} \cos 2t - \frac{\cos 2\bar{\phi}}{4} \sin 2t \right) \right] + \mathcal{O}(\varepsilon^{3/2}). \quad (4.7.17)$$

Now if we truncate at $\mathcal{O}(\sqrt{\varepsilon})$, the autonomous averaged system is Hamiltonian, and thus, in view of the discussion above and in Section 4.4, only limited conclusions can be drawn. We therefore proceed to average the $\mathcal{O}(\varepsilon)$ terms, implicitly using a second transformation

$$(\bar{h}, \bar{\phi}) \rightarrow (\bar{\bar{h}}, \bar{\bar{\phi}}) + \mathcal{O}(\varepsilon).$$

Dropping the double bars, we obtain, to $\mathcal{O}(\varepsilon)$:

$$\dot{h} = \sqrt{\varepsilon} \left( \frac{\delta}{2} - \frac{1}{4} - \frac{\gamma}{4} \cos 2\phi \right) + \varepsilon \left[ 2\delta - 2 - \frac{\gamma}{2} \cos 2\phi \right] h,$$

$$\dot{\phi} = -\sqrt{\varepsilon}2h + \varepsilon \frac{\gamma}{4} \sin 2\phi. \qquad (4.7.18)$$

We leave the analysis of (4.7.18) to the reader; but see Figure 4.7.2:

EXERCISE 4.7.1. Show that, if $|\gamma| > 2(\delta - \frac{1}{2})$, equation (4.7.18) has four fixed points, two of which are saddles. Find the stability type of the other two; does it depend upon the value of $\delta$? What can you say about the global structure of solutions of (4.7.18) and about the bifurcations occurring as $\delta, \gamma$ vary? Verify that Figure 4.7.2 is correct. What do these results imply for the Poincaré map of the original system (4.7.10)?

(Hint: Some judicious rescaling *almost* puts (4.7.17) into the form of a perturbed plane pendulum, as in the second example of Section 4.6, but this is still a substantial exercise.)

EXERCISE 4.7.2. Compute the Melnikov function from equation (4.7.10) directly and compare it with the leading term of equation (4.7.18).

Finally we note a characteristic problem which obstructs the direct application of Melnikov's method in connection with averaging or normal form calculations. This difficulty was first pointed out by Sanders [1980] after being overlooked by Holmes [1980b]. Suppose that one has an averaged

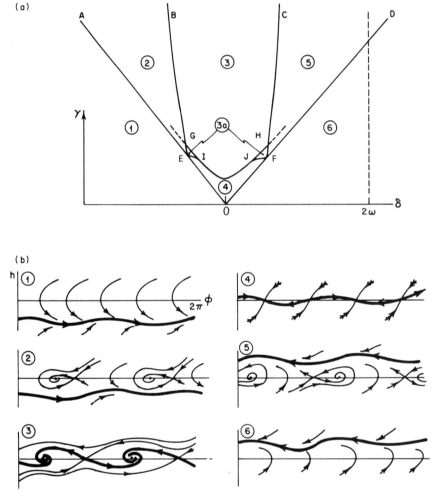

Figure 4.7.2. The behavior of equation (4.7.18). (a) The bifurcation set; (b) structurally stable phase portraits for $\gamma$, $\delta$ in regions $1 - 6$

system (4.1.3) which is "Hamiltonian to first order," i.e., $\dot{y} = \varepsilon \bar{f}(y)$ is Hamiltonian but the "perturbation" $\varepsilon^2 f_1(y, t, \varepsilon)$ may or may not be Hamiltonian. For simplicity, let $y \in \mathbb{R}^2$. Rescaling time, $t \rightarrow t/\varepsilon$, we have

$$\dot{y} = \bar{f}(y) + \varepsilon f_1\left(y, \frac{t}{\varepsilon}, \varepsilon\right), \qquad y \in \mathbb{R}^2. \tag{4.7.19}$$

Suppose further that, for $\varepsilon = 0$, (4.7.19) has a homoclinic orbit $q^0(t)$, as in the undamped Duffing example of Section 4.4. We compute a Melnikov integral

$$M\left(\frac{t_0}{\varepsilon}, \varepsilon\right) = \int_{-\infty}^{\infty} \bar{f}(q^0(t)) \wedge f_1\left(q^0(t), \frac{t + t_0}{\varepsilon}, \varepsilon\right) dt \tag{4.7.20}$$

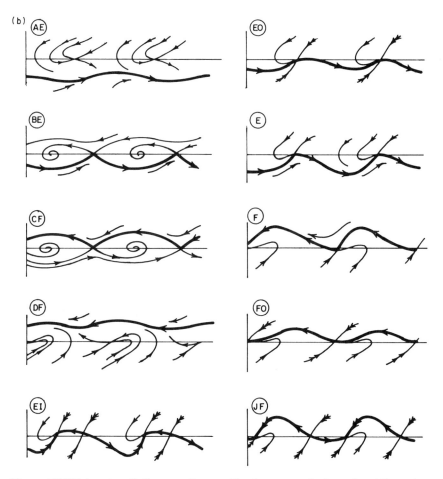

Figure 4.7.2(b) (*continued*). Structurally unstable phase portraits for $\gamma$, $\delta$ on bifurcation curves.

as before, but note now that $M$ depends upon $\varepsilon$. Moreover, the relatively rapid oscillation (period $\varepsilon T$) of the explicitly time-dependent terms of $f_1$ leads in general to an exponentially small quantity after integration:

$$\max_{t_0 \in [0, T]} M\left(\frac{t_0}{\varepsilon}, \varepsilon\right) \sim e^{-c/\varepsilon} \qquad (4.7.21)$$

(cf. Sanders [1980, 1982]). Thus Theorem 4.5.3 does not apply, at least without a careful study of the errors involved in approximation of the true distance function $d(t_0)$ by $M(t_0)$ in (4.5.5). One's immediate desire to apply Melnikov theory to the averaged Hamiltonian system of the Duffing or van der Pol examples is, unhappily, thwarted. However, the time-independent terms of $f_1$, containing no rapid oscillation, provide constant $\mathcal{O}(\varepsilon^k)$ contributions to the Melnikov function, unless tr $Df_1 \equiv 0$ (cf. equation (4.5.15)). Thus we can deduce that (time-independent) damping of arbitrarily small algebraic order is sufficient to break the homoclinic connections between periodic orbits in the limit as $\varepsilon \to 0$ (cf. Sanders [1980]). For more information see Neishtadt [1984] and Holmes *et al.* [1986].

# 4.8. Two Degree of Freedom Hamiltonians and Area Preserving Maps of the Plane

Most of this book is concerned with dissipative dynamical systems and with the structure of the nonwandering and attracting sets occurring in such systems. However, as we have seen in this chapter, it is often useful to regard such systems as perturbations of Hamiltonian systems, since the existence of energy integrals or other constants of the motion enables us to obtain global information on the structure of solutions. So far we have, in the main, restricted discussion to single degree of freedom Hamiltonian systems, but since systems of two and more degrees of freedom played such an important rôle in the development of dynamical systems theory in classical mechanics (cf. Poincaré [1899], Birkhoff [1927], Moser [1973]), and since many of the ideas are directly relevant to higher-dimensional dissipative problems, we sketch some of the theory in this final section. The general results are interesting in their own right and, recent developments of the Melnikov method having made the study of specific systems easier, this seems an appropriate place for a Hamiltonian interlude. Moreover, much of the recent physics literature on "chaos" has addressed the problem of Hamiltonian systems, and we wish to place the present book in relation to that tradition. See Lichtenberg and Lieberman [1982] or Chirikov [1979] for general background and additional references. For more information on Hamiltonian mechanics see Goldstein [1980] or Arnold [1978].

Although the following ideas can be generalized to multidegree of freedom systems, here we shall consider a two degree of freedom system with Hamiltonian $H(q_1, p_1, q_2, p_2)$; the associated equations of motion (Hamilton's equations), being

$$\dot{q}_1 = \frac{\partial H}{\partial p_1}, \qquad \dot{q}_2 = \frac{\partial H}{\partial p_2},$$
$$\dot{p}_1 = -\frac{\partial H}{\partial q_1}, \qquad \dot{p}_2 = -\frac{\partial H}{\partial q_2}. \tag{4.8.1}$$

Here the $q_i$ are generalized coordinates and the $p_i$ their conjugate momenta. As in the single degree of freedom case, it is easy to check that the total derivative of $H$ vanishes,

$$\frac{dH}{dt} = \sum_{i=1}^{2} \left( \frac{\partial H}{\partial q_i} \dot{q}_i + \frac{\partial H}{\partial p_i} \dot{p}_i \right) \equiv 0, \tag{4.8.2}$$

and thus that $H$ is constant on solution curves. More importantly (4.8.2) implies that each three-dimensional manifold

$$H(p_1, p_2, q_1, q_2) = h = \text{const.} \tag{4.8.3}$$

is invariant for the flow of (4.8.1) and thus that, for a given total energy $h$, the flow is essentially *three* dimensional. This enables us to construct a two-dimensional (local) cross section and an associated Poincaré map in specific cases, and to study the dynamics of (4.8.1) in terms of this two-dimensional map.

To clarify these ideas, we shall take a specific class of systems and suppose that two of the variables, say $q_2$ and $p_2$, can be expressed as action-angle coordinates, i.e., that, as in Section 4.7, there is an invertible, canonical (symplectic) change of coordinates (Goldstein [1980]), perhaps only locally defined:

$$I = I(q_2, p_2), \qquad q_2 = Q_2(\theta, I),$$
$$\theta = \theta(q_2, p_2), \qquad p_2 = P_2(\theta, I), \tag{4.8.4}$$

where $Q_2$ and $P_2$ are $2\pi$ periodic in $\theta$, under which the Hamiltonian $H(q_1, p_1, q_2, p_2)$ becomes

$$H(q_1, p_1, Q_2(I, \theta), P_2(I, \theta)) \overset{\text{def}}{=} H(q, p, \theta, I) = h. \tag{4.8.5}$$

(Henceforth we drop the subscripts on $q_1, p_1$.) Hamilton's equations for (4.8.5) are

$$\dot{q} = \frac{\partial H}{\partial p}, \qquad \dot{\theta} = \frac{\partial H}{\partial I},$$
$$\dot{p} = -\frac{\partial H}{\partial q}, \qquad \dot{I} = -\frac{\partial H}{\partial \theta}. \tag{4.8.6}$$

We now assume that $\partial H/\partial I \neq 0$ in some region of phase space, and hence that (4.8.5) can be inverted in that region to solve for $I$ in terms of $q$, $p$, $\theta$, and $h$:

$$I = L(q, p, \theta; h). \tag{4.8.7}$$

Having eliminated explicit dependence upon $I$ in the manifold $H = h$, we next eliminate the variable conjugate to $H$, namely, $t$. Since $\dot\theta = \partial H/\partial I$ is strictly positive (or negative), $\theta(t)$ increases (decreases) continually with $t$. Thus $\theta(t)$ can be inverted to eliminate explicit $t$-dependence. We write $q' = dq/d\theta$, $p' = dp/d\theta$, so that

$$q' = \dot q/\dot\theta = \frac{\partial H}{\partial p} \Big/ \frac{\partial H}{\partial I}, \qquad p' = \dot p/\dot\theta = -\frac{\partial H}{\partial q} \Big/ \frac{\partial H}{\partial I}. \tag{4.8.8}$$

Differentiating (4.8.5) implicitly, using (4.8.7), gives

$$\frac{\partial H}{\partial q} + \frac{\partial H}{\partial I}\frac{\partial L}{\partial q} = 0, \qquad \frac{\partial H}{\partial p} + \frac{\partial H}{\partial I}\frac{\partial L}{\partial p} = 0, \tag{4.8.9}$$

and (4.8.8)–(4.8.9) together yield

$$\left.\begin{array}{l} q' = -\dfrac{\partial L}{\partial p}(q, p, \theta; h) \\[2mm] p' = \dfrac{\partial L}{\partial q}(q, p, \theta; h) \end{array}\right\} ; \qquad (q, p, \theta) \in D \times S^1. \tag{4.8.10}$$

We call the $2\pi$-periodic two-dimensional family of systems (4.8.10) the *reduced Hamiltonian systems*. Such a system exists on each energy manifold $H = h$, and in each region $D \times S^1$ of phase space, for which our assumption $\partial H/\partial I \neq 0$ is valid. The idea of *reduction*, used here, is discussed by Birkhoff [1927] and was certainly known to Poincaré [1890–1899]; also see Whittaker [1959, Chapter 12].

We next take a cross section

$$\Sigma_{h^0}^{\theta_0} = \{(q, p, \theta) \in D \times S^1 | \theta = \theta_0 \in [0, 2\pi]; h = h^0\} \tag{4.8.11}$$

and consider the Poincaré map $P_{h^0}^{\theta_0}: U \to \Sigma_{h^0}^{\theta_0}$ induced by the solutions of (4.8.10) on some subset $U \subseteq \Sigma_{h^0}^{\theta_0}$. We have now reduced our four-dimensional flow to a (family of) two-dimensional maps. We remark that in much numerical work on two degree of freedom Hamiltonian systems a slightly different Poincaré map is constructed (cf. Hénon and Heiles [1964], Lichtenberg and Lieberman [1982]). Remaining in the "old" coordinate system $(q_1, p_1, q_2, p_2)$ one fixes (say) $q_2 = 0$, and constructs a map from $(q_1, p_1)$ space to itself by recording the (numerical) values of $q_1(t)$, $p_1(t)$ whenever $q_2(t) = 0$ and $p_2(t) > 0$. As above, globally defined maps are usually not obtained, for the $(q_1, p_1)$ plane is not necessarily transverse to the vector field everywhere, but the maps thus obtained are equivalent to our map.

As an example, we take the linear two degree of freedom system considered in Section 1.8:

$$H(q_1, p_1, q_2, p_2) = \frac{p_1^2 + \omega_1^2 q_1^2}{2} + \frac{p_2^2 + \omega_2^2 q_2^2}{2}. \tag{4.8.12}$$

Using action angle coordinates, $q_2 = \sqrt{(2I/\omega_2)} \sin \theta$, $p_2 = \sqrt{2\omega_2 I} \cos \theta$, and dropping the subscripts for $q_1, p_1$, this becomes

$$H(p, q, I, \theta) = \frac{p^2 + \omega_1^2 q^2}{2} + \omega_2 I = h. \tag{4.8.13}$$

Clearly $\partial H/\partial I = \omega_2 > 0$ everywhere and thus (4.8.13) can be inverted to yield

$$I = \left( h - \frac{p^2 + \omega_1^2 q^2}{2} \right) \bigg/ \omega_2, \tag{4.8.14}$$

provided that $h > 0$. Thus the reduced systems (4.8.10) are

$$q' = \frac{1}{\omega_2} p,$$
$$\qquad\qquad (q, p) \in D_h \subset \mathbb{R}^2, \tag{4.8.15}$$
$$p' = -\frac{\omega_1^2}{\omega_2} q,$$

where

$$D_h = \{(q, p) \in \mathbb{R}^2 \mid 0 \le p^2 + \omega_1^2 q^2 < 2h\}.$$

Since $H$ does not depend explicitly on $\theta$, the reduced system is autonomous in this case and the Poincaré map can be easily constructed. Taking the section $\Sigma_h^0$, the solution $q(\theta)$, $p(\theta)$ of (4.8.15) based at $q^0$, $p^0$, is given by

$$q(\theta) = q^0 \cos\left(\frac{\omega_1}{\omega_2}\theta\right) + \frac{\omega_2}{\omega_1} p^0 \sin\left(\frac{\omega_1}{\omega_2}\theta\right),$$
$$\tag{4.8.16}$$
$$p(\theta) = -\frac{\omega_1}{\omega_2} q^0 \sin\left(\frac{\omega_1}{\omega_2}\theta\right) + p^0 \cos\left(\frac{\omega_1}{\omega_2}\theta\right),$$

and thus we obtain the linear Poincaré map

$$P_h^0\binom{q}{p} = \begin{bmatrix} \cos(2\pi\Omega) & \frac{1}{\Omega}\sin(2\pi\Omega) \\ -\Omega\sin(2\pi\Omega) & \cos(2\pi\Omega) \end{bmatrix} \binom{q}{p}, \tag{4.8.17}$$

where $\Omega = \omega_1/\omega_2$. Thus the unique fixed point $(q, p) = (0, 0)$ is an elliptic center surrounded by a family of closed curves filled with periodic points if $\Omega$ is rational, and dense orbits if $\Omega$ is irrational. These curves are, of course, intersections with $\Sigma^0$ of the family of two-tori described in the outline of this example in Section 1.8, and the fixed point $(q, p) = (0, 0)$ corresponds to one of the "normal modes" of the two oscillator system.

EXERCISE 4.8.1. Consider the "nonlinear harmonic" oscillator with Hamiltonian $H = \frac{1}{3}(p_1^2 + q_1^2)^{3/2} + \frac{1}{2}(p_2^2 + q_2^2)$. Show that the reduced system can be written as

$$q_1' = p_1\sqrt{p_1^2 + q_1^2},$$

$$p_1' = -q_1\sqrt{p_1^2 + q_1^2},$$

and hence deduce that the Poincaré map has two dense sets of closed curves, one set filled with periodic orbits and the other with dense orbits.

EXERCISE 4.8.2. Find a reduced system for the "pendulum-oscillator" model with Hamiltonian

$$H(q_1, p_1, q_2, p_2) = \frac{p_1^2}{2} + (1 - \cos q_1) + \frac{p_2^2 + \omega_2^2 q_2^2}{2}$$

and discuss the associated Poincaré map. (We return to this example later in this section.)

The examples we have discussed so far are *completely integrable*, in the sense that there are two *independent* functions ($H$ is one of them) which remain invariant under the flow of Hamilton's equations. For the first example the second function can be taken to be the action

$$I = (p_2^2 + \omega_2^2 q_2^2)/2\omega_2$$

of the second oscillator, and solutions lie on the two-dimensional tori which are the intersections of

$$H = h^0 \quad \text{and} \quad I = I^0. \tag{4.8.18}$$

Similar observations apply to the examples in Exercises 4.8.1–4.8.2. However, as Poincaré and Birkhoff realized, very few two degree of freedom systems possess two independent integrals and the classical Hamilton–Jacobi theory, which seeks such integrals, fails in most cases. Moreover, attempts to approximate second integrals by averaging, perturbation or normal form calculations are not uniformly valid in such cases. We refer the reader to Lichtenberg and Lieberman [1982] for an account of such perturbation methods and shall content ourselves here with showing how Melnikov's method, together with reduction, can be used to prove the nonexistence of second (analytic) integrals of motion in specific examples. In our discussion of this method, we also make some general observations about the two-dimensional Poincaré map $P_{h^0}^{\theta_0}$ of the reduced system. We shall drop the sub- and superscripts $h^0$ and $\theta_0$ henceforth and write $P_{h^0}^{\theta_0} = P_\varepsilon$ or $P_0$, the subscripts indicating the absence or presence of the perturbation $\varepsilon H^1$ in equation (4.8.20) below.

First we note that $P$ is an area preserving diffeomorphism, since

$$DP = e^{2\pi Df}, \quad \text{where } Df = \begin{bmatrix} -\dfrac{\partial^2 L}{\partial q\,\partial p} & -\dfrac{\partial^2 L}{\partial p^2} \\[2mm] \dfrac{\partial^2 L}{\partial q^2} & \dfrac{\partial^2 L}{\partial p\,\partial q} \end{bmatrix} \tag{4.8.19}$$

and trace $Df \equiv 0$. (The volume preservation of Liouville's theorem appears as area preservation of $P$.)

To continue, we assume that our Hamiltonian is a (small) perturbation $H^\varepsilon$ of an integrable Hamiltonian $H^0$. For simplicity we take a system of the form

$$H^\varepsilon(q, p, \theta, I) = F(q, p) + G(I) + \varepsilon H^1(q, p, \theta, I), \qquad (4.8.20)$$

where $H^1$ is $2\pi$ periodic in $\theta$ and the unperturbed system $H^0(p, q, \theta) = F(q, p) + G(I)$ decouples directly into two independent systems with integrals $F$ and $G$ (or equivalently, $H^0$ and $I$). As before, we make the nondegeneracy assumption that

$$\Omega(I) \stackrel{\text{def}}{=} \frac{\partial G}{\partial I} \left( = \frac{\partial H^0}{\partial I} \right) \qquad (4.8.21)$$

is nonzero, and, specifically, that $\Omega > 0$ for $I > 0$. This implies that, for $\varepsilon$ small, the equation $H^\varepsilon = F + G + \varepsilon H^1 = h$ is invertible and can be solved for $I$ much as above. However, here the presence of the small parameter $\varepsilon$ permits us to compute the inverse function $L$ of equation (4.8.7) explicitly as a power series in $\varepsilon$. We find

$$I = L^\varepsilon(q, p, \theta; h) = L^0(q, p; h) + \varepsilon L^1(q, p, \theta; h) + \mathcal{O}(\varepsilon^2),$$

where

$$\left. \begin{array}{l} L^0 = G^{-1}(h - F(q, p)), \\[2ex] L^1 = -\dfrac{H^1(q, p, \theta, L^0(q, p; h))}{\Omega(L^0(q, p; h))}. \end{array} \right\} \qquad (4.8.22)$$

EXERCISE 4.8.3. Verify (4.8.22).

Using (4.8.22), the reduced Hamiltonian system becomes

$$q' = -\frac{\partial L^0}{\partial p}(q, p; h) - \varepsilon \frac{\partial L^1}{\partial p}(q, p, \theta; h) + \mathcal{O}(\varepsilon^2),$$

$$p' = \frac{\partial L^0}{\partial p}(q, p; h) + \varepsilon \frac{\partial L^1}{\partial q}(q, p, \theta; h) + \mathcal{O}(\varepsilon^2). \qquad (4.8.23)$$

Since $H^1$ is $2\pi$ periodic in $\theta$, so is $L^1$, and thus (4.8.23) is a system of precisely the form studied by Melnikov's method in the preceding sections.

Specifically, let us assume (as above) that some (compact) region of the phase plane of the (uncoupled) $F(q, p)$ system is filled with periodic orbits whose periods vary continuously with respect to the $F$-energy. Each such orbit is a level set of $F$: $F(q, p) = h^\alpha$, and consequently, provided that the total energy $h > h^\alpha$, the unperturbed ($\varepsilon = 0$) autonomous system (4.8.23) has a corresponding closed orbit given by

$$L^0(p, q, h) = G^{-1}(h - h^\alpha) \stackrel{\text{def}}{=} I^\alpha; \qquad (4.8.24)$$

moreover, the period, $T^\alpha$, of these orbits varies continuously with $I^\alpha$.

Now the Poincaré map $P_0$ associated with (4.8.23) for $\varepsilon = 0$ is simply its time $2\pi$ flow map, and hence has a continuous family of invariant closed curves filled with points of period $m$ if $T^\alpha = 2\pi m/n$ for $m$, $n \in Z$ and filled with dense orbits if $T^\alpha/2\pi$ is irrational; Figure 4.8.1 (cf. Exercise 4.8.1). For simplicity, let us assume that $T^\alpha$ increases as $l^\alpha$ increases, so that the average angle turned by points under $P_0$ decreases as one moves out across the invariant curves, as we indicate in Figure 4.8.1.

Transforming $(p, q)$ to a second set of action angle variables $J$, $\phi$, the unperturbed system becomes

$$\dot{J} = 0,$$
$$\dot{\phi} = \Lambda(J),$$

(4.8.25)

where $\Lambda(J) = 2\pi/T^\alpha$, and the Poincaré map can be written

$$(J, \phi) \overset{P_0}{\longmapsto} (J, \phi + 2\pi\Lambda(J)).$$

(4.8.26)

By assumption, $\Lambda$ decreases as $J$ increases and thus $P_0$ is referred to as a *twist mapping*. The perturbation $\varepsilon L^1$ will modify this map to

$$(J, \phi) \overset{P_\varepsilon}{\longmapsto} (J + \varepsilon f(J, \phi), \phi + 2\pi\Lambda(J) + \varepsilon g(J, \phi)),$$

(4.8.27)

where $f$ and $g$ are bounded functions, each $2\pi$ periodic in $\phi$, and $P_\varepsilon$ still preserves area.

Before we develop our method of computing certain features of $P_\varepsilon$, we outline some general (and powerful) results on perturbations of area preserving maps like $P_0$. For more information and background, see Arnold and Avez [1968, especially §19–21 and §34, Moser [1973], and Arnold [1978]. The most important result is the celebrated Kolmogorov–Arnold–Moser (KAM) theorem, which asserts that, for sufficiently small $\varepsilon$, "most" of the closed curves $J = $ constant of $P_0$ are preserved for $P_\varepsilon$. Various versions of this theorem were obtained by Kolmogorov [1954], Moser [1962], Arnold [1963a, b], and Rüssman [1970], and it was later extended to $n$ degree of

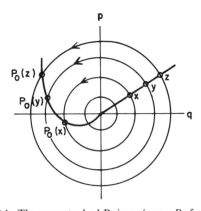

Figure 4.8.1. The unperturbed Poincaré map $P_0$ for (4.8.23).

freedom systems with $(2n - 2)$-dimensional Poincaré maps (cf. Arnold and Avez [1968], Arnold [1978]). Rather than giving it in its full generality we shall state it in the context of the present two-dimensional example:

**Theorem 4.8.1** (KAM). *If $\Lambda'(J) \neq 0$ and $\varepsilon$ is sufficiently small, then the perturbed map $P_\varepsilon$ has a set of invariant closed curves of positive Lebesgue measure $\mu(\varepsilon)$ close to the original set $J = J^\alpha$; moreover, $\mu(\varepsilon)/\mu(J) \to 1$ as $\varepsilon \to 0$. The surviving invariant closed curves are filled with dense irrational orbits.*

This result implies that, for the original Hamiltonian system (4.8.20) a measurable set of tori carrying irrational flow are preserved for small $\varepsilon \neq 0$.

In terms of the original unperturbed Hamiltonian

$$H^0(q, p, I) = F(q, p) + G(I) = F(J) + G(I),$$

the nondegeneracy condition $\Lambda'(J) \neq 0$ becomes,

$$G''(I)(F'(J))^2 + F''(J)(G'(I))^2 \neq 0 \quad \text{for each } I + J = h \quad (4.8.28)$$

(cf. Arnold [1978]).

EXERCISE 4.8.4. Verify that satisfaction of equation (4.8.28) implies that $\Lambda'(J) \neq 0$. If the unperturbed system $H^0(I, J)$ is not separable, show that the nondegeneracy condition is

$$\det \begin{vmatrix} H^0_{II} & H^0_{IJ} & H^0_I \\ H^0_{IJ} & H^0_{JJ} & H^0_J \\ H^0_I & H^0_J & 0 \end{vmatrix} \neq 0. \quad (4.8.29)$$

We note that the conditions (4.8.28)–(4.8.29) are for *isoenergetic nondegeneracy*, and ensure that the frequency ratio on the unperturbed invariant tori varies as one crosses the tori in each fixed energy surface (Arnold [1978]).

EXERCISE 4.8.5. Show that the Poincaré map of the reduced system of Exercise 4.8.1 can be written in the form (4.8.26), with $\Lambda(J) = \sqrt{2J}$. Also show that (4.8.28) is satisfied for the original system. (Hint: Let $q_2 = \sqrt{2I} \sin \theta$, $p_2 = \sqrt{2I} \cos \theta$, $q_1 = \sqrt{2J} \sin \phi$, $p_1 = \sqrt{2J} \cos \phi$.)

The general picture (mainly built up from numerical experiments) is that the "sufficiently irrational" closed curves survive for small $\varepsilon$ for *all* perturbations, but that as $\varepsilon$ increases they typically disappear one by one until no closed curves exist close to the original ones (although others may appear). To give a more precise idea of what we mean by "sufficiently irrational," we give a more specific result due to Moser:

**Theorem 4.8.2** (The Twist Theorem (Moser [1973], §2)). *Consider a small area preserving perturbation $P_\varepsilon$ of an area preserving two-dimensional map $P_0$. Assume $\Lambda(J)$ is $C^r$, $r \geq 5$, and $|\Lambda'(J)| \geq v > 0$ on an annulus*

$R = \{(J, \phi) | a \leq J \leq b\}$. *Then there exists a $\delta$ depending on $\varepsilon$ and $\Lambda(J)$, such that, if the perturbed map $P_\varepsilon$ satisfies*

$$\sup_{(J, \phi) \in R} \{\|\varepsilon F\|_r + \|\varepsilon G\|_r\} < \nu\delta,^* \qquad (4.8.30)$$

*then P possesses an invariant curve $\Gamma_\varepsilon$ of the form*

$$J = J^0 + U(\psi), \qquad \phi = \phi + V(\psi) \qquad (4.8.31)$$

*in R, where U and V are $C^1$, of period $2\pi$ and*

$$\|U\|_1 + \|V\|_1 < \varepsilon, \qquad (4.8.32)$$

*and $a < J^0 < b$. Moreover, the induced mapping $P_\varepsilon|_{\Gamma_\varepsilon}$ on this curve is given by*

$$\phi \to \phi + 2\pi\lambda \qquad (4.8.33)$$

*where $\lambda$ is irrational and satisfies the infinite set of relations*

$$\left| \lambda - \frac{n}{m} \right| \geq \gamma m^{-\alpha}, \qquad (4.8.34)$$

*for some $\gamma$, $\alpha > 0$ and all integers $m, n > 0$. Each choice of $\lambda$ satisfying (4.8.34) in the range of $\Lambda(J)$ gives rise to such an invariant curve.*

In both these results the fact that $P_\varepsilon$ is area preserving is crucial, since if it contracted or expanded areas, invariant closed curves might not exist. We shall return to the question of rotation numbers in our discussion of maps on the circle in Section 6.2.

The persistence of sets of invariant closed curves has an important implication for the stability of motion of the original system within each energy manifold $H = h$, for their existence implies that some of the original nested family of invariant two-dimensional tori $J$ = constant are preserved and hence provide boundaries in the three-dimensional energy manifold across which solutions cannot pass. The perturbed solutions are either confined on such tori or trapped between pairs of such tori and hence cannot wander arbitrarily far in phase space. We remark that, while in systems of $n \geq 3$ degrees of freedom, sets of the analogous $n$-tori are likewise preserved, they no longer act as boundaries in the $(2n - 1)$-dimensional energy manifolds, and solutions can consequently "leak out" and wander across phase space. This process is now known as *Arnold diffusion*; for more information see Arnold [1964], Lieberman [1980], Lichtenberg and Lieberman [1982], and Holmes and Marsden [1982b].

We now consider what happens to the "resonant" tori—those for which the condition (4.8.34) fails. We choose an invariant closed curve $\Gamma_0$ of $P_0$ with action $J^\alpha$ such that $\Lambda(J^\alpha) = n/m$ $(T^\alpha = 2\pi m/n)$. This curve is filled

---

* $\|F\|_r$ denotes the $C^r$ norm $\sum_{k=0}^r |D^k F(J, \phi)|$.

with degenerate periodic points of period $m$ at each of which the linearized map has the Jordan form

$$DP_0^m = \begin{bmatrix} 1 & 0 \\ 2\pi\Lambda'(J) & 1 \end{bmatrix}, \tag{4.8.35}$$

resulting in a shear-like motion (cf. Figure 4.8.1). To see what happens to this ring of fixed points, we consider the behavior of two invariant curves $J^{\beta_1}$, $J^{\beta_2}$ on either side of $J^\alpha$ ($J^{\beta_1} < J^\alpha < J^{\beta_2}$). Under $P_0^m$, points in $J^{\beta_1}$ rotate through an angle greater than $2\pi$, while those in $J^{\beta_2}$ rotate less than $2\pi$. Fixing $\beta_1$ and $\beta_2$, and taking $\varepsilon$ sufficiently small (depending on $m$) we see that this behavior persists for $P_\varepsilon^m$. It follows that there is a point $J(\phi, \varepsilon) \in (J^{\beta_1}, J^{\beta_2})$ on each radius $\phi = $ constant which rotates through exactly $2\pi$ under $P_\varepsilon^m$, and hence moves only in a radial direction. Since the perturbation $P_\varepsilon$ is smooth, these points $J(\phi, \varepsilon)$ form a smooth closed curve $\Gamma_\varepsilon$ which converges on $\Gamma_0$ {$J = J^\alpha$} as $\varepsilon \to 0$ (cf. the proof of Theorem 4.6.2).

Now the map is area preserving, and hence $\Gamma_\varepsilon$ and $P_\varepsilon(\Gamma_\varepsilon)$ enclose the same area and evidently must intersect. Each intersection point is a fixed point of the perturbed map $P_\varepsilon^m$. Poincare [1899] proved the existence of $2km$ such points for some (unknown) integer $k$; see Figure 4.8.2. (It should be clear that, in the generic case, the intersections of $\Gamma_\varepsilon$ and $P_\varepsilon^m(\Gamma_\varepsilon)$ are transversal and that there must therefore be an even number (at least $2m$) of them.)

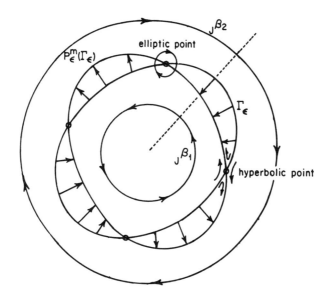

Figure 4.8.2. The map $P_\varepsilon^m$, showing the curve $\Gamma_\varepsilon$, its image $P_\varepsilon^m(\Gamma_\varepsilon)$ and the fixed points (after Arnold and Avez [1968]). In this picture we have modded out by $2\pi m$, so that points appear to rotate clockwise outside the resonant circle but anticlockwise inside it.

The stability types of these fixed points follows from a consideration of the behavior of points nearby. We first note that the eigenvalues of the perturbed map linearized at the fixed points must have product $\lambda_1 \lambda_2 = 1$, since $\det |DP_\varepsilon^m| = 1$. Thus the degenerate (parabolic) points with matrix (4.8.35) split into either hyperbolic fixed points ($0 < \lambda_1 < 1 < \lambda_2$) or elliptic fixed points ($\lambda_2 = \bar{\lambda}_1, |\lambda_i| = 1$). Assuming that the intersections of $\Gamma_\varepsilon$ and $P_\varepsilon^m(\Gamma_\varepsilon)$ are transversal, it is easy to see from Figure 4.8.2 that precisely half are elliptic centers and half hyperbolic saddles, since points either circulate around the fixed point or leave its neighborhood "monotonically" (cf. Arnold and Avez [1968], §20).

EXERCISE 4.8.6. Prove this last statement by letting $P_\varepsilon^m$ be the time 1 flow map of a two-dimensional vector field and using index theory (cf. Section 1.8).

The invariant manifolds of the saddle points are evidently confined between neighboring "irrational" invariant curves which are preserved for $P_\varepsilon$, and these invariant manifolds must intersect (if they did not the map could not preserve area). In general, some of these intersections will be transverse. Thus, in each "resonant band" near the unperturbed curve $J^\alpha$, we expect to find a complicated set of invariant curves, such as those which we attempt to sketch in Figure 4.8.3. Zehnder [1973] established this result in the generic case. Such regions are called *stochastic layers* in the physics literature or are sometimes referred to as *homoclinic tangles*. As we show in Chapter 5, associated with each transversal homoclinic intersection point, there is a further countable family of hyperbolic periodic points, and thus the dynamics in the stochastic layer is very complicated.

EXERCISE 4.8.7. Within the framework sketched above, discuss the area preserving map $f: (\phi, v) \to (\phi + v, v - \gamma \cos(\phi + v))$ of Section 2.4 for small values of $\gamma$. Find some of

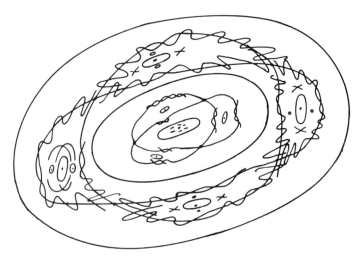

Figure 4.8.3. Resonant curves generically break into stochastic layers.

the orbits of period 2 (and possibly 3) by direct calculation and compare them with the hyperbolic and elliptic orbits which are expected to bifurcate from the appropriate resonant closed curves. (See Greene [1980] for a mega-attack on this problem.)

The results we have summarized above rely on non-constructive arguments to conclude that there is transversal intersection behavior. In contrast, we end this section, and the chapter, by showing how Melnikov's method can be used to compute what actually happens in a resonant band in specific examples. Churchill [1982] was among the first to suggest such computations. Returning to equation (4.8.23), we fix the total energy $h$ and select $I^\alpha$ such that the periodic orbit $q^\alpha(\theta)$, $p^\alpha(\theta)$ of the unperturbed system

$$ q' = -\frac{\partial L^0}{\partial p}, \qquad p' = \frac{\partial L^0}{\partial q}, \tag{4.8.36} $$

has period $T^\alpha = 2\pi m/n$ in $\theta$ and is thus in resonance with the $2\pi$ periodic perturbation $L^1$. A direct application of Melnikov's method (Theorem 4.6.2 and Sections 4.6–4.7) together with the fact that the Poincaré map is area preserving, now gives the following result (we leave the proof as an exercise):

**Theorem 4.8.3.** *If the subharmonic Melnikov function*

$$ M^{m/n}(\theta^0) = \int_0^{2\pi m} \{L^0(q^\alpha(\theta), p^\alpha(\theta); h), L^1(q^\alpha(\theta), p^\alpha(\theta), \theta + \theta^0; h\} \, d\theta \tag{4.8.37} $$

*is independent of $\varepsilon$ and has $j$ simple zeros in $\theta^0 \in [0, 2\pi m/n)$, then the resonant closed curve $L^0 = I^\alpha$ for the unperturbed Poincaré map $P_0$ breaks into a set of $2k = j/m$ periodic orbits each of period $m$. Moreover, $j$ is necessarily an even multiple of $m$ and precisely $k$ of the periodic orbits are hyperbolic and $k$ elliptic.*

As in (4.5.12)–(4.5.13), $\{L^0, L^1\} = (\partial L^0/\partial q)(\partial L^1/\partial p) - (\partial L^0/\partial p)(\partial L^1/\partial q)$ is the Poisson bracket of $L^0$ and $L^1$. In examples, the calculation of $L^0$ and $L^1$ using (4.8.22) is often awkward and it is more convenient to state the result in terms of the original Hamiltonian $H^\varepsilon = F + G + \varepsilon H^1$. A direct calculation (using (4.8.22)) yields

$$ \{L^0, L^1\} = \frac{1}{\Omega^2(L^0)} \{F, H^1\}, \tag{4.8.38} $$

where $\{F, H^1\} = (\partial F/\partial q)(\partial H^1/\partial p) - (\partial F/\partial p)(\partial H^1/\partial q)$. Thus, using the fact that, on the unperturbed solution $(q^\alpha, p^\alpha)$, $\theta = \Omega(L^0(q^\alpha, p^\alpha; h))t + \theta^0 = \Omega(I^\alpha)t + \theta^0$, we can replace (4.8.37) by

$$ M^{m/n}(\theta^0) = \frac{1}{\Omega(I^\alpha)} \int_0^{2\pi m/\Omega(I^\alpha)} \{F(q^\alpha(t), p^\alpha(t)), H^1(q^\alpha(t), p^\alpha(t), \Omega(I^\alpha)t + \theta^0; I^\alpha\} \, dt. $$

$$ \tag{4.8.39} $$

EXERCISE 4.8.8. Verify that (4.8.38) is correct, and prove Theorem 4.8.3.

If the $F$ system has a homoclinic orbit $(q^0(t), p^0(t))$ corresponding to a level curve $F = h^0$, then an analogue of Theorem 4.5.3 yields

**Theorem 4.8.4.** *Consider a two degree of freedom Hamiltonian system of the form (4.8.20), and assume that $F$ contains a homoclinic orbit $(q^0(t), p^0(t))$ connecting a hyperbolic saddle to itself (or $F$ has a homoclinic cycle). Suppose $\Omega(I) = G'(I) > 0$ for $I > 0$. Let $h^0 = F(q^0, p^0)$ be the energy of the homoclinic orbit and let $h > h^0$ and $I^0 = G^{-1}(h - h^0)$ be constants. Let $\{F, H^1\}(t + \theta^0)$ denote the Poisson bracket of $F(q^0, p^0)$ and $H^1(q^0, p^0, \Omega(I^0)t + \theta^0; I^0)$ evaluated at $q^0(t)$ and $p^0(t)$. Define*

$$M(\theta^0) = \int_{-\infty}^{\infty} \{F, H^1\}(t + \theta^0)\, dt, \qquad (4.8.40)$$

*and assume that $M(\theta^0)$ has a simple zero and is independent of $\varepsilon$. Then for $\varepsilon > 0$ sufficiently small the Hamiltonian system corresponding to (4.8.20) has transverse homoclinic orbits on the energy surface $H^\varepsilon = h$.*

The analysis of transverse homoclinic orbits and the discussion of the horseshoe to follow in Chapter 5 then implies:

**Corollary 4.8.5** (cf. Theorem 5.3.5). *The system has a hyperbolic invariant set\* $\Lambda$ in its dynamics on the energy surface $H = h$; $\Lambda$ possesses a dense orbit and thus the system possesses no global analytic second integral.*

(Ziglin [1982] has recently published a similar result.)

If the hypotheses of Theorems 4.8.3 and 4.8.4 are met, then the unperturbed and perturbed maps $P_0$ and $P_\varepsilon$ (cf. (4.8.26)–(4.8.27)) have the structures of invariant curves sketched in Figure 4.8.4.

As an example, we consider the weakly coupled pendulum-linear oscillator system (cf. Holmes and Marsden [1982a]), with Hamiltonian

$$H^\varepsilon(p, q, x, y) = \frac{p^2}{2} + (1 - \cos q) + \frac{y^2 + \omega^2 x^2}{2} + \varepsilon \frac{(x - q)^2}{2}, \qquad (4.8.41)$$

(cf. Exercise 4.8.2). Transforming the $(x, y)$-system to action angle coordinates, with

$$x = \sqrt{\frac{2I}{\omega}}\, \sin\theta, \qquad y = \sqrt{2\omega I}\, \cos\theta,$$

(4.8.41) becomes

$$H^\varepsilon = \frac{p^2}{2} + (1 - \cos q) + \omega I + \frac{\varepsilon}{2}\left(\sqrt{\frac{2I}{\omega}}\, \sin\theta - q\right)^2$$

$$= F(q, p) + G(I) + \varepsilon H^1(q, p, I, \theta). \qquad (4.8.42)$$

---

\* See Section 5.2 below.

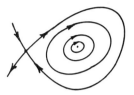 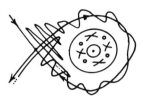

Figure 4.8.4. Invariant curves of the unperturbed and perturbed Poincaré maps. (a) $P_0$; (b) $P_\varepsilon$.

The $F$ system has a (pair of) homoclinic orbit(s)

$$(q^0(t), p^0(t)) = (\pm 2 \arctan(\sinh(t)), \pm 2 \operatorname{sech} t), \qquad (4.8.43)$$

with energy

$$F(q, p) = h^0 = 2; \qquad (4.8.44)$$

we therefore pick $h > 2$ and let

$$I^0 = G^{-1}(h - h^0) = \frac{1}{\omega}(h - 2). \qquad (4.8.45)$$

The Melnikov function may now be evaluated, using

$$\{F, H^1\} = \frac{\partial F}{\partial q}\frac{\partial H^1}{\partial p} - \frac{\partial F}{\partial p}\frac{\partial H^1}{\partial q}$$
$$= p\left(\sqrt{\frac{2I}{\omega}}\sin\theta - q\right), \qquad (4.8.46)$$

and evaluating the bracket on the orbit $(q^0(t), p^0(t), \theta(t) = \omega t + \theta^0; I(t) = I^0)$:

$$M(\theta^0) = \int_{-\infty}^{\infty} 2\operatorname{sech}(t)(-2\arctan(\sinh t) \pm \frac{\sqrt{2(h-2)}}{\omega}\sin(\omega t + \theta^0))\, dt. \qquad (4.8.47)$$

Here " $+$ " refers to the upper branch of the homoclinic orbit ($p > 0$) and " $-$ " to the lower ($p < 0$). The first integrand of (4.8.47) is odd and vanishes, the second may be evaluated by the method of residues to yield

$$M(\theta^0) = \frac{\pm 2\pi\sqrt{2(h-2)}}{\omega}\operatorname{sech}\left(\frac{\pi\omega}{2}\right)\sin\theta^0. \qquad (4.8.48)$$

Since $M(\theta^0)$ has simple zeros, the hypotheses of Theorem 4.8.4 are satisfied and we can conclude that the perturbed (coupled) system has transversal homoclinic orbits and hence Smale horseshoes in every energy level $h > 2$, for $\varepsilon$ sufficiently small (depending on $h$).

EXERCISE 4.8.9. Investigate the fates of the resonant tori for the Hamiltonian (4.8.42), given by $F(q, p) = h^z < 2$ and $I = I^0 = (1/\omega)(h - h^z) > 0$.

EXERCISE 4.8.10. Consider the perturbed Hamiltonian of Hénon–Heiles type

$$H(\dot{x}, \dot{y}, x, y) = \frac{\dot{x}^2 + \dot{y}^2}{2} + \frac{\omega^2(x^2 + y^2)}{2} - x^2 y - \frac{y^3}{3}$$

$$+ \varepsilon(\alpha y^2 - \beta y^3), \qquad 0 \leq \varepsilon \ll 1; \quad \alpha, \beta = \mathcal{O}(1).$$

Prove that the system is integrable for $\varepsilon = 0$, and nonintegrable for all small $\varepsilon \neq 0$. (Hint: First use the symplectic transformation

$$q_1 = \frac{1}{\sqrt{2}}(x + y), \qquad q_2 = \frac{1}{\sqrt{2}}(x - y),$$

$$p_1 = \frac{1}{\sqrt{2}}(\dot{x} + \dot{y}), \qquad p_2 = \frac{1}{\sqrt{2}}(\dot{x} - \dot{y}).)$$

(Cf. Holmes [1982b] and Exercise 1.8.14.)

We remark that the theory for two-dimensional area preserving maps discussed in this section applies directly to periodically perturbed Hamiltonian systems of the form

$$H^\varepsilon(q, p, t) = F(q, p) + \varepsilon H^1(q, p, t). \tag{4.8.49}$$

The Duffing equation example considered in Sections 4.5–4.6 takes this form when the damping, $\delta$, is set to zero. Decreasing $\delta$ for fixed $\gamma, \omega > 0$, then, we approach the Hamiltonian limit, and the resulting bifurcations (cf. Figure 4.6.3), in which countable sets of periodic and homoclinic orbits are created, should be seen as steps in the creation of the homoclinic tangles and stochastic layers illustrated in Figure 4.8.3.

In closing, we note that Theorems 4.8.3–4.8.4 have higher-dimensional analogues, applicable to $n$-degree of freedom systems (cf. Holmes and Marsden [1982a, b]) and that Gruendler [1982, 1985] has also obtained multidimensional extensions of the Melnikov theory. In fact, in certain cases the Melnikov theory extends to perturbations of infinite-dimensional Hamiltonian evolution equations arising from PDEs. Holmes and Marsden [1981] provide an analysis for some such situations, with an example from mechanics which is essentially an infinite-dimensional generalization of the Duffing equation modelling the forced, damped vibrations of a continuous beam under axial load considered in Section 2.2. Also Holmes [1981b] has applied these methods to the sine–Gordon partial differential equation

$$\phi_{tt} - \phi_{xx} + \sin \phi = -\varepsilon \phi_t,$$

with time-dependent boundary conditions

$$\phi_x(0, t) = \varepsilon H,$$

$$\phi_x(1, t) = \varepsilon(H + I(t)).$$

# CHAPTER 5

# Hyperbolic Sets, Symbolic Dynamics, and Strange Attractors

## 5.0. Introduction

The solutions of ordinary differential equations can have an erratic time dependence which appears in some ways to be random. We have seen several such examples in Chapter 2. The present chapter is devoted to a discussion of simple, geometrically defined systems in which such chaotic motion occurs. We shall describe both the irregular character of individual solutions and the complicated geometric structures associated with their limiting behavior. The principal technique which we use is called symbolic dynamics and the general approach to the questions we adopt is referred to as dynamical systems theory. We shall not develop this theory systematically but will state some of its major results and provide a brief guide to its literature. Our strategy in solving specific problems will generally involve the use of numerical or perturbation methods, such as those of Chapter 4, to establish the existence of interesting geometrical structure in appropriate Poincaré maps, followed by the use of the methods of this chapter.

Let us begin with some considerations about a truly random process: (ideal) coin tossing experiments. Imagine a coin which is repeatedly tossed an infinite number of times. We would like to summarize the results of such a sequence compactly. Arbitrarily assigning a value 0 to heads and 1 to tails, the outcome of an experiment is given by a sequence $\mathbf{a} = \{a_i\}_{i=1}^{\infty}$, $a_i = 0, 1$. If we regard $\mathbf{a}$ as the binary expansion of a number $x \in [0, 1]$, then we can represent any outcome to the sequences of tosses by a single number. Formally, if $\Sigma$ is the set of all semi-infinite sequences of 0's and 1's, and $I = [0, 1]$, then we have defined a mapping $\phi : \Sigma \to I$. Note that, apart from sequences $\mathbf{a}$ which terminate in an infinite string of 0's or 1's, the mapping $\phi$ is 1-1. Thus, if we start with $x \in I$, we can recover $\phi^{-1}(x)$—the sequence of coin tosses as a string of 0's and 1's—from the binary representation of $x$.

The binary representation of a number $x \in I$ can be defined iteratively by looking at the integer part of $2^n x \pmod 2$ provided that $2^n x$ is never itself an integer. If we define $f : I \to I$ by $f(x) = 2x \pmod 1$, then the $n$th term of the binary representation of $x$ is 0 or 1 as $f^n(x)$ is in $(0, \frac{1}{2})$ or $(\frac{1}{2}, 1)$. If $f^n(x) = \frac{1}{2}$, then $x$ has two representations. The sequence of coin tosses corresponding to $x$ is the sequence of integers in its binary representation, so there is an explicit correspondence between the set of all sequences of coin tosses (a probability sample space) and the points of the interval, which involves the map $f(x)$ defined above. From a probabilistic point of view, the process of picking a random initial condition (with respect to the Lebesgue measure) and observing whether each $f^n(x)$ falls to the left or right of $\frac{1}{2}$ is completely equivalent to an infinite sequence of fair coin tosses. We have demonstrated a 1:1 relationship between realizations of a random process and orbits of a deterministic dynamical system. The reader should compare this map with the Poincaré return maps for the van der Pol and Lorenz systems described in Chapter 2. We return to the Lorenz map later in this chapter.

We want to extend this simple relationship between a mapping and a random process to as large a class of mappings as possible. The technique for doing this is called *symbolic dynamics*. For limit sets with a *hyperbolic structure*, we shall develop a relationship with a (topological) finite state Markov chain in a manner which generalizes the above discussion. Hyperbolic limit sets are an archetype for random-like behavior of solutions of ordinary differential equations. However, nonhyperbolic limit sets are often encountered in examples of practical importance: they occur in the Duffing, van der Pol, and bouncing ball examples of Chapter 2, as we shall see. One would like to study their symbolic dynamics, but this has not been done in a satisfactory or systematic manner, except in the special case of mappings defined on the line. Here there is an extensive general theory, and we refer the reader to the book by Collet and Eckman [1980] for a recent account of this work, various aspects of which we shall summarize, where appropriate, in this and the next chapter.

We begin by describing the symbolic dynamics of a one-dimensional mapping $g : \mathbb{R} \to \mathbb{R}$ defined by $g(x) = ax(1 - x)$ with $a > 2 + \sqrt{5}$, as a prelude to our analysis of the horseshoe in the next section. The mapping $g$ has a fixed point at 0, and $g(1) = 0$. The graph of $g$ on the closed unit interval $I$ is shown in Figure 5.0.1. The requirement that $a > 2 + \sqrt{5}$ implies that there is a $\lambda > 1$ such that $|g'(x)| \geq \lambda$ on $g^{-1}(I) \cap I$.

EXERCISE 5.0.1. Show that the largest possible value of $\lambda$ is $a\sqrt{1 - 4/a}$.

Although the orbits $\{g^i(x)\}_{i=0}^{\infty}$ of almost all points $x \in I$ eventually escape from $I$ and tend towards $-\infty$, there is an invariant set of points $\Lambda$ whose iterates remain in $I$. This set $\Lambda$ can be described by the use of symbolic dynamics in the following way.

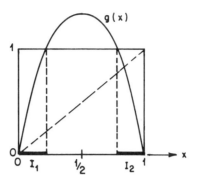

Figure 5.0.1. The graph $g(x)$. Note the subintervals $I_1 = [0, \frac{1}{2} - \frac{1}{2}\sqrt{1 - 4/a}]$ and $I_2 = [\frac{1}{2} + \frac{1}{2}\sqrt{1 - 4/a}, 1]$.

Partition $I \cap g^{-1}(I)$ by writing its two components as $I_1$ and $I_2$. We can associate to each $x \in \Lambda$ a sequence $\{a_i\}_{i=0}^\infty$ of 1's and 2's defined by $a_i = j$ if $g^i(x) \in I_j$. The sequence $\{a_i\}$ labels the iterates of $x$ according to the left–right pattern they follow. The essential observation we make about this process is that if $J \subset I$ is a subinterval, then $g^{-1}(J)$ consists of exactly two subintervals, one contained in $I_1$ and the other contained in $I_2$. Moreover, we note that the length of $J$ is at least $\lambda$ times the length of each component of $g^{-1}(J)$ because $|g'(x)| \geq \lambda$ if $g(x) \in I$.

From the above observation we derive several consequences:

(1) Every symbol sequence is associated to some point of $\Lambda$. To prove this, it suffices (by the finite intersection property for compact sets) to prove that each set of the form

$$I_{a_0, \ldots, a_n} = I_{a_0} \cap g^{-1}(I_{a_1} \cap g^{-1}(I_{a_2} \cap \cdots \cap g^{-1}(I_{a_n}) \cdots))$$

is a closed, nonempty set. This follows inductively from the observation above because $g^{-1}(I_{a_0, \ldots, a_n})$ intersects both $I_1$ and $I_2$.

(2) Distinct points of $\Lambda$ have distinct symbol sequences associated to them. This follows immediately from the fact that the lengths of the sets $I_{a_0, \ldots, a_n}$ are all less than $\lambda^{-(n+1)}$.

(3) $\Lambda$ is a Cantor set. To see this, write $\Lambda_n = \bigcup I_{a_0, \ldots, a_n}$, the union being taken over all $(n + 1)$-tuples of 1's and 2's. Then $\Lambda_n$ consists of $2^{n+1}$ closed subintervals, and each component of $\Lambda_{n-1}$ contains exactly two components of $\Lambda_n$. Since the lengths of the components of $\Lambda_n$ tend to 0 with increasing $n$, $\Lambda = \bigcap_{n \geq 0} \Lambda_n$ is a *Cantor set.* *

(4) The sequence associated to $g(x)$ is obtained from the sequence associated to $x$ by dropping the first term.

Thus we have shown that we can label each point $x \in \Lambda$ uniquely by a semi-infinite sequence $\phi(x) = \{a_i(x)\}_{i=0}^\infty$ where the $a_i$'s are 1's and 2's.

---

* By a Cantor set we mean a closed set which contains no interior points or isolated points.

Moreover, the $a_i$'s are chosen to reflect the dynamics—the orbit structure—of $f$. In this way we have reduced the study of the mapping to an essentially combinatorial problem involving the symbols $\{1, 2\}$.

EXERCISE 5.0.2. Show that there is a countable infinity of periodic orbits and asymptotically periodic orbits for $g$.

EXERCISE 5.0.3. Work out the symbolic dynamics for the cubic map $x \to f(x) = ax - x^3$ and describe the structure of its invariant set for "large" $a$. (Hint: Find $a_c$ such that for $a > a_c$ there is a subinterval $I = [-\sqrt{1 + a}, \sqrt{1 + a}]$ such that $I \cap f^{-1}(I)$ has three components on which $|f'(x)| > 1$.)

## 5.1. The Smale Horseshoe: An Example of a Hyperbolic Limit Set

The one-dimensional mapping $g$ described above is closely related to another example, the *Smale horseshoe*, which is a hyperbolic limit set that has been a principal motivating example for the development of the modern theory of dynamical systems. We shall now describe this example in detail, using its symbolic dynamics. The example is described in terms of an invertible planar map which can be thought of as a Poincaré map arising from a three-dimensional autonomous differential equation or a forced oscillator problem. In Section 5.3, below, we will see how the horseshoe arises whenever one has transverse homoclinic orbits, as in the Duffing equation. In Section 2.4 we have already met an example in which maps and horseshoes arise directly.

We begin with the unit square $S = [0, 1] \times [0, 1]$ in the plane and define a mapping $f : S \to \mathbb{R}^2$ so that $f(S) \cap S$ consists of two components which are mapped rectilinearly by $f$. See Figure 5.1.1. Note the orientation of the image: in particular, the horizontal boundaries $AB$, $DC$ are mapped to horizontal intervals.

One can think of the map as performing a linear vertical expansion and a horizontal contraction of $S$, by factors $\mu$ and $\lambda$, respectively, followed by a folding, the latter being done so that the folded portion falls outside $S$, Figure 5.1.1(b). Thus, restricted to $S \cap f^{-1}(S)$, the map is linear.

Reversing the folding and stretching, one can easily see that the inverse image, $f^{-1}(S \cap f(S)) = S \cap f^{-1}(S)$, is two horizontal bands $H_1 = [0, 1] \times [a, a + \mu^{-1}]$ and $H_2 = [0, 1] \times [b, b + \mu^{-1}]$, on each of which $f$ will have a constant Jacobian

$$\begin{pmatrix} \pm\lambda & 0 \\ 0 & \pm\mu \end{pmatrix}, \qquad (+ \text{ on } H_1, - \text{ on } H_2), \qquad (5.1.1)$$

with $0 < \lambda < \frac{1}{2}$ and $\mu > 2$. On each $H_i$, the map $f$ compresses horizontal segments by a factor of $\lambda$ and stretches vertical segments by a factor of $\mu$.

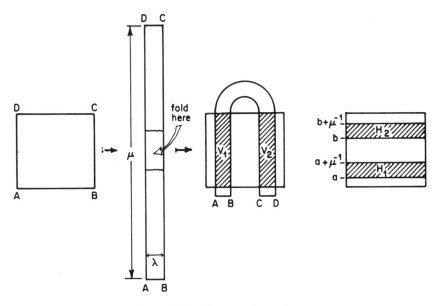

Figure 5.1.1. The Smale horseshoe.

As $f$ is iterated, most points either leave $S$ or are not contained in an image $f^i(S)$. Those points which do remain in $S$ for all time form a set $\Lambda = \{x \mid f^i(x) \in S, -\infty < i < \infty\}$. This set $\Lambda$ has a complicated topological structure that we now describe. Each horizontal band $H_i$ is stretched by $f$ to a rectangle $V_i = f(H_i)$ which intersects both $H_1$ and $H_2$. Since $f$ is rectilinear on $H_i$, those points that end up in $H_i$ after applying $f$ come from thinner horizontal strips in $H_i$. See Figure 5.1.2.

Now $H_1 \cup H_2 = f^{-1}(S \cap f(S))$, so the four thinner strips constitute $f^{-2}(S \cap f(S) \cap f^2(S))$. If we continue this argument inductively, we find that $f^{-n}(S \cap f(S) \cap \cdots \cap f^n(S))$ is the union of $2^n$ horizontal strips. The thickness of each of these is $\mu^{-n}$ since $|\partial f/\partial y| = \mu$ at all points of $H_1 \cup H_2$ and the first $(n-1)$ iterates of the horizontal strips remain inside $H_1 \cup H_2$.

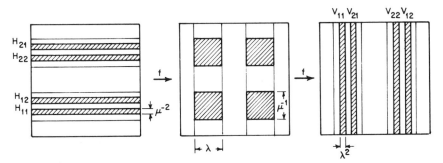

Figure 5.1.2. Iteration of $f$: $V_{ij} = f^2(H_{ij})$.

The intersection of all these horizontal strips (as $n \to \infty$) forms a Cantor set of horizontal segments. We shall discuss this structure below.

Consider now the image under $f^n$ of one of the $2^n$ horizontal strips in $f^{-n}(S \cap f(S) \cap \cdots \cap f^n(S))$. Using the chain rule, we have

$$
Df^n = \begin{pmatrix} \pm \lambda^n & 0 \\ 0 & \pm \mu^n \end{pmatrix}
$$

at these points, so the image is a rectangle of horizontal width $\lambda^n$ which extends vertically from the top to the bottom of the square. The map $f^n$ is 1–1, so the images of the horizontal strips are distinct. We conclude that $S \cap f(S) \cap \cdots \cap f^n(S)$ is the union of $2^n$ vertical strips, each of width $\lambda^n$. The intersection of these sets over all $n \geq 0$ is a Cantor set of vertical segments composed of those points which are in the images of all the $f^n$. To be in $\Lambda$, a point $x$ must be in both a vertical segment and a horizontal segment from the collection described above. Therefore, topologically, $\Lambda$ is itself a Cantor set: its components are each points and each point of $\Lambda$ is an accumulation point for $\Lambda$. In Figure 5.1.3 we show the sixteen components of $f^{-2}(S) \cap f^{-1}(S) \cap S \cap f(S) \cap f^2(S)$, to give an idea of the structure of $\Lambda$.

So far we have essentially repeated the informal description of the horseshoe in Section 2.4. We can, however, achieve a more complete description which contains information about the dynamics of each point by a construction similar to the one we used for the one-dimensional mappings in the previous section. In this construction, we note which horizontal band $H_1$ or $H_2$ each iterate of a point $x \in \Lambda$ visits, and use this information as an actual characterization of the point. Each point $x \in \Lambda$ will be characterized by a bi-infinite sequence, since here the map is invertible, unlike $x \to 2x \pmod 1$.

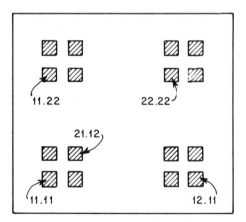

Figure 5.1.3. The small black rectangles of width $\lambda^2$ and height $\mu^{-2}$ are the components of $\bigcap_{n=-2}^{2} f^n(S)$. The four symbol sequences $\{a_{-2}a_{-1} \cdot a_0 a_1\}$ refer to Exercise 5.1.1, below.

A bi-infinite sequence is one whose index set is all of $\mathbb{Z}$: we use the notation $\mathbf{a} = \{a_i\}_{i=-\infty}^{\infty}$.

**Theorem 5.1.1.** *There is a 1–1 correspondence $\phi$ between $\Lambda$ and the set $\Sigma$ of bi-infinite sequences of two symbols such that the sequence $\mathbf{b} = \phi(f(x))$ is obtained from the sequence $\mathbf{a} = \phi(x)$ by shifting indices one place: $b_i = a_{i+1}$. The set $\Sigma$ has a metric defined by*

$$d(a, b) = \sum_{i=-\infty}^{\infty} \delta_i 2^{-|i|}, \qquad \delta_i = \begin{cases} 0 & \text{if } a_i = b_i, \\ 1 & \text{if } a_i \neq b_i. \end{cases} \qquad (5.1.2)$$

*The map $\phi$ is a homeomorphism from $\Lambda$ to $\Sigma$ endowed with this metric.*

PROOF. The proof of this theorem provides a basic illustration of how symbolic dynamics works. Take the two symbols of the theorem to be 1 and 2. The map is defined by the recipe

$$\phi(x) = \{a_i\}_{i=-\infty}^{\infty}, \quad \text{with } f^i(x) \in H_{a_i}. \qquad (5.1.3)$$

In words, $x$ is in $\Lambda$ if and only if $f^i(x)$ is in $H_1 \cup H_2$ for each $i$, and we associate to $x$ the sequence of indices that tells us which of $H_1$ and $H_2$ contains each $f^i(x)$. Unlike the map $f = 2x \pmod 1$, this definition of $\phi$ is unambiguous because $H_1$ and $H_2$ are disjoint. This description of $\phi$ leads immediately to the shift property required by the theorem: since $f^{i+1}(x) = f^i(f(x))$, it follows that $\phi(f(x))$ is obtained from $\phi(x)$ by shifting indices. To see that $\phi$ is both 1–1 and continuous, we look at the set of $x$'s which each possess a given central string of symbols. Specifying $b_{-m}, b_{-m+1}, \ldots b_0, \ldots, b_n$ we denote as $R(b_{-m}, b_{-m+1}, \ldots b_0, \ldots, b_n)$ the set of $x$'s for which $f^i(x) \in H_{b_i}$ for $-m \leq i \leq n$. We observe inductively that $R(b_{-m}, \ldots, b_n)$ is a rectangle of height $\mu^{-(n+1)}$ and width $\lambda^m$, obtained from the intersection of a horizontal and a vertical strip. As one lets $m, n \to \infty$, the diameter of the sets $R(b_{-m}, \ldots, b_n) \to 0$. Consequently, $\phi$ is both 1–1 and continuous.

The final point is that $\phi$ is onto. This is crucial for the applications of symbolic dynamics. The reason that $\phi$ is onto is that for each choice of $b_{-m}, \ldots, b_n$, the set $R(b_{-m}, \ldots, b_n)$ is nonempty. To see this, reference to Figure 5.1.2 is helpful. Note that $R(b_0, \ldots, b_n)$ is a horizontal strip mapped vertically from top to bottom of the square $S$ by $f^{n+1}$. Therefore, $f^{n+1}(R(b_0, \ldots, b_n))$ intersects each $H_i$ and $R(b_0, \ldots, b_n, b_{n+1})$ is a nonempty horizontal strip extending across $S$. Similarly, we have already observed that $S \cap f(S) \cap \cdots \cap f^m(S)$ consists of $2^m$ vertical strips. Each of these is a set of the form $R(b_{-m}, \ldots, b_{-1})$ and all sequences $(b_{-m}, \ldots, b_{-1})$ occur. Finally, $R(b_{-m}, \ldots, b_n)$ is nonempty because every vertical strip $R(b_{-m}, \ldots, b_{-1})$ intersects every horizontal strip $R(b_0, \ldots, b_n)$ and $R(b_{-m}, \ldots, b_n) = R(b_{-m}, \ldots, b_{-1}) \cap R(b_0, \ldots, b_n)$. $\qquad \square$

EXERCISE 5.1.1. Verify that the central parts of the symbol sequences attached to the rectangles of Figure 5.1.3 are correct and label the remaining shaded rectangles by their (finite) symbol sequences.

The correspondence $\phi$ between $\Lambda$ and $\Sigma$ imparts to $\Lambda$ a symbolic description which is an extraordinarily useful tool for understanding the dynamics of $\Lambda$. It is helpful to give a formal name to the process of "shifting indices." Thus

$$\sigma: \Sigma \to \Sigma, \qquad (5.1.4)$$

the *shift map*, is defined by $\sigma(\mathbf{a}) = \mathbf{b}$ with $b_i = a_{i+1}$. The basic property of the theorem is now restated as the equation

$$\phi \circ (f|_\Lambda) = \sigma \circ \phi. \qquad (5.1.5)$$

This equation expresses the topological conjugacy of $f|_\Lambda$ and $\sigma$. Written as $f|_\Lambda = \phi^{-1} \circ \sigma \circ \phi$, it has the immediate consequence that

$$f^n|_\Lambda = \phi^{-1} \circ \sigma^n \circ \phi, \qquad (5.1.6)$$

so that $\phi$ maps orbits of $f$ in $\Lambda$ to orbits of $\sigma$ in $\Sigma$. The description of $\sigma$ is explicit enough that many dynamical properties are readily determined. For example, a periodic orbit of period $n$ for $\sigma$ consists of a sequence which is periodic: $a_i = a_{i+n}$ for all $i$ in the sequence $\mathbf{a}$. Fixing $n$, we readily count the sequences with the property $a_i = a_{i+n}$ and find that $f^n$ has $2^n$ fixed points in $\Lambda$. This set includes all points which are periodic with period $n$ or a divisor of $n$.

EXERCISE 5.1.2. Show that all the periodic orbits in $\Lambda$ are of saddle type. Show that $\Lambda$ contains a countable infinity of heteroclinic and homoclinic orbits. Show that $\Lambda$ contains orbits which are not asymptotically periodic. List the first few periodic orbits (with periods, say $\leq 5$) and locate them on Figure 5.1.3. Show that $\Lambda$ contains an uncountable collection of nonperiodic orbits and describe their symbol sequences.

EXERCISE 5.1.3. Display a point of $\Sigma$ whose $\sigma$-orbit is dense in $\Sigma$. (Hint: Two points of $\Sigma$ are close if they agree in a long "central block": find a sequence which contains all finite strings of 1's and 2's.)

The description of the horseshoe we have just given is "robust" with respect to small changes in the mapping $f$. The reader should try to imagine what changes will take place if the assumption that $f$ is rectilinear on $H_1 \cup H_2$ is dropped. We imagine perturbing $f$ to a mapping $\tilde{f}$ in such a way that the Jacobian derivative of $\tilde{f}$ can be nonconstant but still is close to that of $f$. Qualitatively, nothing changes. The sets $S \cap \tilde{f}(S) \cap \cdots \cap \tilde{f}^n(S)$ will still consist of $2^n$ "vertical" strips (which are, however, no longer exactly rectangles). Similarly, the set $\tilde{f}^{-n}(S) \cap \cdots \cap S$ will consist of $2^n$ "horizontal" strips which are no longer exactly rectangles. Nonetheless, the set of points all of whose $\tilde{f}$-iterates remain in $S$ form a set $\tilde{\Lambda}$ which is topologically conjugate to the shift $\Sigma$. This result of Smale [1963, 1967] is an early example of a *structural stability* theorem.

We can summarize the results of this section as follows:

**Theorem 5.1.2.** *The horseshoe map f has an invariant Cantor set Λ such that*:

(a) Λ *contains a countable set of periodic orbits of arbitrarily long periods.*
(b) Λ *contains an uncountable set of bounded nonperiodic motions.*
(c) Λ *contains a dense orbit.*

*Moreover, any sufficiently $C^1$ close map $\hat{f}$ has an invariant Cantor set $\tilde{\Lambda}$ with $\hat{f}|_{\tilde{\Lambda}}$ topologically equivalent to $f|_\Lambda$.*

We shall take up the question of nonlinear maps possessing horseshoes in the next section.


## 5.2. Invariant Sets and Hyperbolicity

The example described above, the Smale horseshoe, provides a good intuitive basis for the way in which orbits of mappings, and hence of ordinary differential equations, can be chaotic. Later in this chapter, we shall describe a general theory of "Axiom A" dynamical systems which builds upon this example. The concept of structural stability makes this generalization a very natural one, but the class of Axiom A systems is not adequate to encompass the various examples described in Chapter 2. There are many unresolved issues about the details of the dynamics in these examples, so that our discussion becomes more tentative toward the end of the chapter when we apply the theory developed here to these examples. Of particular interest will be the question of when a "typical" trajectory can be expected to have the chaotic dynamical features of the horseshoe. Our aim being to describe what is known, observed, and suspected, we include only details about the proofs of results which we feel are illuminating.

A few topological definitions are necessary at the beginning of our discussion. The horseshoe Λ described in Section 5.1 has a rather complicated topological structure, but it cannot be further split into closed invariant subsets because there are orbits which are dense in Λ. We want to focus on sets like these, which carry most of the interesting dynamical information of a flow. If $\phi$ is a discrete or continuous flow, then a fundamental property of all sets we consider is that they be invariant. (Recall from Section 1.6 that the set $S$ is *invariant* if $\phi_t(S) = S$ for all $t$.) There are various kinds of invariant sets; the ones which interest us the most will be composed of asymptotic limit sets of points. Again let $\phi_t$ be a discrete or continuous flow, and recall from Section 1.6:

**Definition 5.2.1.** The $\alpha$ *limit set of* $x$ *for* $\phi_t$ is the set of accumulation points of $\phi_t(x)$, $t \to -\infty$. The $\omega$ *limit set of* $x$ *for* $\phi_t$ is the set of accumulation points of $\phi_t(x)$, $t \to \infty$. The $\alpha$ and $\omega$ limits of $x$ are its asymptotic limit sets. ($y$ is an accumulation point of $\phi_t(x)$, $t \to \infty$ if there is a sequence $t_i \to \infty$ such that $\phi_{t_i}(x) \to y$.)

EXERCISE 5.2.1. If $\phi_t$ is the flow of the system

$$\dot{r} = r(1 - r^2),$$
$$\dot{\theta} = 1,$$

in polar coordinates, and $x \neq 0$ is inside the unit circle, show that the $\alpha$-limit set of $x$ is 0 and the $\omega$-limit set of $x$ is the unit circle.

EXERCISE 5.2.2. Show that the $\alpha$- and $\omega$-limit sets of $x$ for a flow $\phi_t$ are invariant.

While we certainly want to pay attention to all the $\omega$-limit sets of a system, as they contain the asymptotic behavior of the system as $t \to +\infty$, other ways of distinguishing the limit sets have been more fruitful. Much of the literature of dynamical systems theory has been formulated in terms of the nonwandering set of a flow, which we again recall:

**Definition 5.2.2.** A point $x$ is *nonwandering for* $\phi_t$ if for every neighborhood $U$ of $x$, and $T > 0$, there is $t > T$ such that $\phi_t(U) \cap U \neq \varnothing$. The set of nonwandering points of $\phi_t$ is denoted by $\Omega$.

EXERCISE 5.2.3. Show that $\Omega$ forms a *closed* set.

EXERCISE 5.2.4. If $y$ is in the $\omega$-limit set of $x$ for $\phi_t$, show that $y$ is nonwandering.

The definition of the nonwandering set embodies a very weak concept of recurrence. Nonetheless, there is a still weaker idea of recurrence, called chain recurrence, which is useful in discussions of structural stability.

**Definition 5.2.3.** The point $x$ is *chain recurrent* if for every $\varepsilon > 0$, there are points $x = x_0, x_1, \ldots, x_n = x$, and times $t_1, \ldots, t_n \geq 1$ such that the distance from $\phi_{t_i}(x_{i-1})$ to $x_i$ is smaller than $\varepsilon$. The set of chain recurrent points of $\phi_t$ is denoted by $\Gamma$.

The ideas of nonwandering points and chain recurrent points are indicated in Figure 5.2.1. The utility of the chain recurrent set has been emphasized by Bowen [1978] and Conley [1978].

EXERCISE 5.2.5. Give an example with $\Gamma \neq \Omega$. (Hint: Construct a flow on $S^1$ with exactly one equilibrium point.)

The following definitions allow us to decompose the nonwandering or chain recurrent set into pieces which are indecomposable in dynamical terms, and make the rather vague discussion of attractors in Section 1.6 more precise:

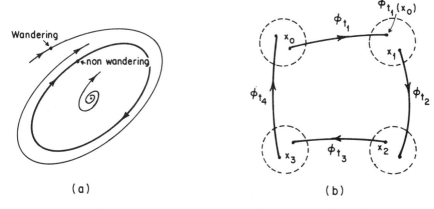

Figure 5.2.1. (a) Wandering and nonwandering points of a flow; (b) a chain recurrent point.

**Definition 5.2.4.** A closed invariant set $\Lambda$ is *topologically transitive* if $\phi_t$ has an orbit which is dense in $\Lambda$.

EXERCISE 5.2.6. Show that the set consisting of the homoclinic orbit and the saddle at $(0, 0)$ for the system

$$\dot{x} = y,$$

$$\dot{y} = x - x^2,$$

is topologically transitive.

A similar definition which allows for an easier decomposition of the chain recurrent set is the following:

**Definition 5.2.5.** The closed invariant set $\Lambda$ is *indecomposable* if for every pair of points $x$, $y$, in $\Lambda$ and $\varepsilon > 0$, there are $x = x_0, x_1, \ldots, x_{n-1}, x_n = y$ and $t_1, \ldots, t_n \geq 1$ such that the distance from $\phi_{t_i}(x_{i-1})$ to $x_i$ is smaller than $\varepsilon$.

EXERCISE 5.2.7. Show that, if $\Omega_1$ and $\Omega_2$ are indecomposable and $\Omega_1 \cap \Omega_2$ is nonempty, then $\Omega_1 \cup \Omega_2$ is indecomposable.

This exercise implies that the maximal indecomposable sets are well defined and disjoint. The chain recurrent set $\Gamma$ can then be written uniquely as a disjoint union of maximal indecomposable sets $\Gamma_i$. In bounded regions, the $\Gamma_i$ will lie at positive distances from one another. Thus these sets are the ones in which all of the recurrence of the dynamics occurs, and each of the $\Gamma_i$'s is indecomposable in such a way that small changes in a trajectory which leave it inside $\Gamma$ cannot carry it outside of $\Gamma_i$. (This definition is *not* standard in the dynamical systems literature, where most authors work with maximal topologically transitive sets, but a choice must be made here. For Axiom A systems, sensible choices are equivalent. For more general systems, the best choices have not been established. We have tried to err on the side of working with the largest invariant sets possible. Technical points such as this have been the bane of efforts to develop a sound mathematical theory which encompasses the examples of Chapter 2.)

We now want to give qualitative analyses of indecomposable invariant sets. The extent to which this is possible depends largely on our ability to characterize the *homoclinic* behavior within the set in terms of *hyperbolic structures*. These efforts can be viewed as attempts to extend as much of the analysis of the horseshoe as is possible to a general indecomposable set. We will focus on discrete systems for the most part. The discussion of the Lorenz attractor in Section 5.7 illustrates additional complications which can occur in continuous time systems.

**Definition 5.2.6.** Let $\Lambda$ be an invariant set for the discrete dynamical system defined by $f : \mathbb{R}^n \to \mathbb{R}^n$. A *hyperbolic structure* for $\Lambda$ is a continuous invariant direct sum decomposition $T_\Lambda \mathbb{R}^n = E_\Lambda^u \oplus E_\Lambda^s$ with the property that there are constants $C > 0, 0 < \lambda < 1$ such that:

(1) if $v \in E_x^u$, then $|Df^{-n}(x)v| \leq C\lambda^n|v|$;
(2) if $v \in E_x^s$, then $|Df^n(x)v| \leq C\lambda^n|v|$.

Several remarks are in order:

(i) In this definition, $T_\Lambda \mathbb{R}^n$ consists of all the tangent vectors to $\mathbb{R}^n$ at all points of $\Lambda$. For each $x \in \Lambda$, $T_x \mathbb{R}^n$ is the tangent space at $x$ and $T_x \mathbb{R}^n = E_x^u \oplus E_x^s$ is a direct sum splitting of this vector space into subspaces of dimensions $u$ and $s$ ($u + s = n$).
(ii) The derivative $Df$ of $f$ maps $T_x \mathbb{R}^n$ to $T_{f(x)} \mathbb{R}^n$. The invariance of the definition requires that $Df(E_x^s) = E_{f(x)}^s$ and $Df(E_x^u) = E_{f(x)}^u$. The continuity of the splitting means that as $x$ varies in $\Lambda$, one can find continuously varying bases for $E_x^u$ and $E_x^s$. In general, the splitting cannot be chosen to be smooth.
(iii) The hyperbolicity conditions (1) and (2) state that, infinitesimally, vectors in $E^s$ (resp. $E^u$) are contracted exponentially in forward (resp. backwards) time at an exponential rate $\lambda$ which is *uniform for all points of $\Lambda$ and all choices of vectors in the invariant subspaces*.

Uniform hyperbolicity is difficult to establish in examples. A set of numerical procedures tested on a collection of examples would represent a significant advance in efforts to make dynamical systems theory a tool for the rigorous analysis of specific systems. Nonetheless, hyperbolicity considerations play a central role in most considerations of chaotic dynamical behavior.

(PARTIALLY) WORKED EXERCISE 5.2.8. Construct a hyperbolic structure for the piecewise linear horseshoe of Section 5.1.

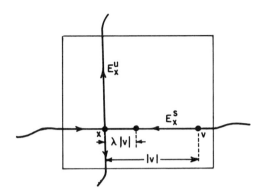

Figure 5.2.2. A splitting for the horseshoe.

SOLUTION. The splitting $E_x^u \oplus E_x^s$ at each point $x \in \Lambda$ is provided by taking appropriate pieces of the sets $\bigcap_{n=0}^{\infty} f^n(S)$ and $\bigcap_{n=0}^{\infty} f^{-n}(S)$, respectively, each of which intersects $S$ in a Cantor set of segments. Thus, for each $x$, $E_x^u$ is a vertical line and $E_x^s$ is a horizontal line; Figure 5.2.2. Moreover, since the map is linear on $H_1 \cup H_2$, and

$$Df = \begin{pmatrix} \pm\lambda & 0 \\ 0 & \pm\mu \end{pmatrix},$$

the estimates of the definition are obtained immediately if $\lambda$ of that definition is taken to be $\max\{\lambda, \mu^{-1}\}$ and $C = 1$. The splitting is constant because the map is linear on each piece with the same eigenvectors.

In this exercise we considered the piecewise linear horseshoe map restricted to $H_1 \cup H_2 \subset S$. The idea of hyperbolicity may be applied to more general nonlinear maps also, and we now give a description of how hyperbolicity can be checked in such cases. We adapt our exposition from Moser [1973], which the reader should consult for more details, and we concentrate on the horseshoe and generalizations of it.

Since we wish to extend our methods to nonlinear maps, we need a looser definition of horizontal and vertical substrips $\{H_i\}$, $\{V_i\}$ of the square $S = \{(x, y) \in \mathbb{R}^2 \mid 0 \le x \le 1, 0 \le y \le 1\}$:

**Definition 5.2.7.** A *vertical curve* $x = v(y)$ is a curve for which

$$0 \le v(y) \le 1, \qquad |v(y_1) - v(y_2)| \le \mu|y_1 - y_2| \quad \text{in } 0 \le y_1 \le y_2 \le 1$$
(5.2.1)

for some $0 < \mu < 1$. Similarly a *horizontal curve* $y = h(x)$ is one for which

$$0 \le h(x) \le 1, \qquad |h(x_1) - h(x_2)| \le \mu|x_1 - x_2| \quad \text{in } 0 \le x_1 \le x_2 \le 1.$$
(5.2.2)

Given two nonintersecting vertical curves $v_1(y) < v_2(y)$, $y \in [0, 1]$, we can define a *vertical strip*

$$V = \{(x, y) \mid x \in [v_1(y), v_2(y)]; y \in [0, 1]\};$$
(5.2.3)

and given two nonintersecting horizontal curves $h_1(x) < h_2(x)$, we have a *horizontal strip*

$$H = \{(x, y) \mid x \in [0, 1]; y \in [h_1(x), h_2(x)]\}. \tag{5.2.4}$$

Finally, the width of vertical and horizontal strips is defined as

$$d(V) = \max_{y \in [0, 1]} |v_2(y) - v_1(y)|, \qquad d(H) = \max_{x \in [0, 1]} |h_2(x) - h_1(x)|. \tag{5.2.5}$$

It is now easy to see the following:

**Lemma 5.2.1.** *If* $V^1 \supset V^2 \supset V^3 \ldots$ *is a sequence of nested vertical (or horizontal) strips and if* $d(V^k) \to 0$ *as* $k \to \infty$ *then* $\bigcap_{k=1}^{\infty} V^k \stackrel{\text{def}}{=} V^{\infty}$ *is a vertical (or horizontal) curve.*

**Lemma 5.2.2.** *A vertical curve* $v(y)$ *and a horizontal curve* $h(x)$ *intersect in precisely one point.*

Thus to each pair of curves (and hence to each pair of sequences of nested strips) there corresponds exactly one point $x \in S$. We can now state our hypotheses on the map $f : S \to \mathbb{R}^2$. Note that we can have any number of horizontal and vertical strips (in fact Moser permits countable sets of strips, $\mathscr{S}$).

H1 Let $\mathscr{S}$ be the set $\{1, 2, \ldots, N\}$ and let $H_i$, $V_i$ for $i \in \mathscr{S}$ be disjoint horizontal and vertical strips and let $f(H_i) = V_i$, $i \in \mathscr{S}$.

H2 $f$ contracts vertical strips and $f^{-1}$ contracts horizontal strips uniformly; i.e. letting $v_1, v_2 \in V_i$ be any two vertical curves bounding a vertical substrip $V_i' \subseteq V_i$, then $f(V_i') \cap V_j$ is a vertical strip and

$$d(f(V_i') \cap V_j) \le vd(V_i')d(V_j)/d(V_i),$$

for some $v \in (0, 1)$ and $i, j \in \mathscr{S}$. Similarly, letting $h_1, h_2 \in H_i$ be any two horizontal curves bounding a horizontal substrip $H_i' \subseteq H_i$, then $f^{-1}(H_i') \cap H_j$ is a horizontal strip and $d(f^{-1}(H_i') \cap H_j) \le vd(H_j)$.

As an alternative to H2, if we have a $C^1$ map $f = (f_1, f_2)$ with linearization $Df$ or

$$\begin{pmatrix} \xi \\ \eta \end{pmatrix} \mapsto \begin{bmatrix} \dfrac{\partial f_1}{\partial x} & \dfrac{\partial f_1}{\partial y} \\[2ex] \dfrac{\partial f_2}{\partial x} & \dfrac{\partial f_2}{\partial y} \end{bmatrix} \begin{pmatrix} \xi \\ \eta \end{pmatrix} \stackrel{\text{def}}{=} \begin{bmatrix} a & b \\ c & d \end{bmatrix} \begin{pmatrix} \xi \\ \eta \end{pmatrix}, \tag{5.2.6}$$

then H2 can be replaced by

H3 There exist sets (sector-bundles) $S^u = \{(\xi, \eta) \mid |\xi| < \mu|\eta|\}$ defined over $\bigcup_{i \in \mathscr{S}} V_i$ and $S^s = \{(\xi, \eta) \mid |\eta| < \mu|\xi|\}$ defined over $\bigcup_{i \in \mathscr{S}} H_i$ with $0 < \mu < 1$ such that $Df(S^u) \subset S^u$ and $Df^{-1}(S^s) \subset S^s$. Moreover, if $Df(\xi_0, \eta_0) = (\xi_1, \eta_1)$ and $Df^{-1}(\xi_0, \eta_0) = (\xi_{-1}, \eta_{-1})$ then $|\eta_1| \ge (1/\mu)|\eta_0|$ and $|\xi_{-1}| \ge (1/\mu)|\xi_0|$ (see Figure 5.2.3).

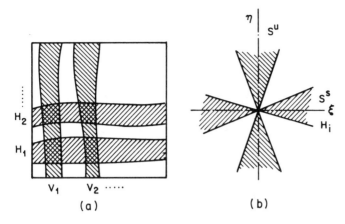

Figure 5.2.3. Nonlinear strips and sectors. (a) The strips $H_i$, $V_i$; (b) the sectors $S^u$, $S^s$.

In fact, since we are ultimately only interested in checking hyperbolicity in the neighborhood of $\Lambda$, H3 need only hold on the $N^2$ "squares" $\bigcup_{i,j \in \mathscr{S}} (V_i \cap H_j)$ rather than on full strips. This relaxation is sometimes useful in applications, as we shall see.

We now have

**Proposition 5.2.3.** H1 and H3, with $0 < \mu < \frac{1}{2}$, imply H2 with $\nu = \mu/(1 - \mu)$.

In the proof of this result one first shows that, as in the piecewise linear case, the preimage under $f$ of any horizontal strip $H_i$ is a set of thinner horizontal strips $H_{ji}$ and the image of any vertical strip $V_k$ is a set of thinner vertical strips $V_{kl}$. One then verifies the contraction estimates that $d(H_{ji}) \le \nu d(H_j)$ and $d(V_{kl}) \le \nu d(V_k)$, with $\nu = \mu/(1 - \mu)$.

**Theorem 5.2.4.** If $f$ is a two-dimensional homeomorphism satisfying H1 and H2 then $f$ possesses an invariant set $\Lambda$, topologically equivalent to a shift $\sigma$ on $\Sigma$: the set of bi-infinite sequences of elements of $\mathscr{S}$. That is, there exists a homeomorphism $h$ of $\Sigma$ onto $\Lambda$ such that $f|_\Lambda = h \circ \sigma \circ h^{-1}$. Furthermore, if $f$ is a $C^r$ diffeomorphism ($r \ge 1$) satisfying H1 and H3 with $0 < \mu < \frac{1}{2}$, and $|\mathrm{Det}(Df)|$, $|\mathrm{Det}(Df)|^{-1} \le \frac{1}{2}\mu^{-2}$, then $\Lambda$ is hyperbolic.

**Remark.** In the area preserving case, $|\det Df| = 1$. and so the final hypothesis is easily satisfied in examples of conservative or weakly dissipative systems.

PROOF.(Moser [1973]). We generalize the arguments of Section 5.1. Define inductively the strips $V_{a_{-1}a_{-2}...a_{-n}} = f(V_{a_{-2}a_{-3}...a_{-n}}) \cap V_{a_{-1}}$ and $H_{a_0a_1a_2...a_n} = f^{-1}(H_{a_1a_2...a_n}) \cap H_{a_0}$. (Note that $V_{ji} = f(V_i) \cap V_j$ and $H_{ji} = f^{-1}(H_i) \cap H_j$ of H2 are special cases and that $V_{kji} = f(V_{ji}) \cap V_k = f^2(V_i) \cap f(V_j) \cap V_k$, etc.) By H2 we have $d(V_{a_{-1}a_{-2}...a_{-n}}) \le \nu d(V_{a_{-2}...a_{-n}}) \le \nu^{n-1} d(V_{a_{-n}})$ and

$d(H_{a_0 a_1 \ldots a_n}) \le v^n(H_{a_n}) \le v$. Thus, taking the limits as $n \to \infty$, by Lemma 5.2.1, for each pair of sequences we have unique vertical and horizontal curves

$$V(a) = \bigcap_{n=1}^{\infty} V_{a_{-1} a_{-2} \ldots a_{-n}}, \qquad H(a) = \bigcap_{n=0}^{\infty} H_{a_0 a_1 \ldots a_n},$$

and so by Lemma 5.2.2 there is a unique point $x = V(a) \cap H(a)$ correspond- ing to the sequence $a(x) = \{\ldots a_{-n} \ldots a_{-2} a_{-1} \cdot a_0 a_1 a_2 \ldots\} \in \Sigma$. Our con- struction implies that the sequence $a(x)$ describes the orbit of $x$ under $f$, since

$$f(x) = \left( \bigcap_{n=0}^{\infty} V_{a_0 a_{-1} \ldots a_{-n}} \right) \cap \left( \bigcap_{n=1}^{\infty} H_{a_1 a_2 a_3 \ldots a_n} \right)$$

corresponds to the shifted sequence

$$\sigma(a(x)) = \{\ldots a_{-n} \ldots a_{-2} a_{-1} a_0 \cdot a_1 a_2 \ldots a_n \ldots\}.$$

Finally, it remains to show that the relationship $x = h(a(x))$ is a homeo- morphism. Continuity of $h$ follows from the fact that if $a_k = a'_k$ for $|k| \le j$, then $x = h(a)$ and $x' = h(a')$ both lie in $V_{a_{-1} \ldots a_{-j}}$ and $H_{a_0 \ldots a_j}$. Thus $|x - x'| \le (1/(1 - \mu))(v^j + v^{j-1})$, since $d(V_{a_{-1} \ldots a_{-j}}) < v^{j-1}$ and $d(H_{a_0 \ldots a_j}) < v^j$. Since the components $H_i$, $V_i$ are disjoint, $h$ is injective. We leave it to the reader to check that $h^{-1}$ is continuous and to verify the last statement using Proposition 5.2.3.                                                                                                  □

**Remark.** Since a diffeomorphism $f$ satisfying the hypotheses of Theorem 5.2.4 can evidently be subjected to small $C^1$ perturbations without violating H1 and H3, this result shows that $f|_\Lambda$ is structurally stable, as claimed in Section 5.1.

As an example of the use of this method we return to the bouncing ball example of Section 2.4 (equation (2.4.4))

$$f_{\alpha, \gamma}(\phi, v) \to (\phi + v, \alpha v - \gamma \cos(\phi + v)). \qquad (5.2.7)$$

We have already demonstrated the existence of topological horseshoes, strips $H_i$, $V_i$ such that $f_{\alpha, \gamma}(H_i) = V_i$, $i = 1, 2$. Here we shall show that these horseshoes are hyperbolic, under somewhat more restrictive conditions. We start by recalling the earlier result. We then obtain the sector bundle estimates H3 for this example.

**Lemma 5.2.5.** *For $\alpha = 1$ and $\gamma \ge 4\pi$ the map $f_{\alpha, \gamma}$ possesses a topological horseshoe; i.e., there are "horizontal" and "vertical" strips $H_i$, $V_i$, such that $f(H_i) = V_i$, $i = 1, 2$.*

Here we relax Moser's conditions (5.2.1)–(5.2.2) in that a "horizontal" strip is understood to be bounded by curves $v = v(\phi)$ with $|v'| < 2$ and a "vertical" strip is bounded by curves $\phi = \phi(v)$ with $|\phi'| < \frac{1}{2}$.

PROOF. We note that, for $\alpha = 1$, $f$ is periodic in both $\phi$ and $v$ with period $2\pi$. Without loss of generality, we pick as basic domain $Q$ the parallelogram

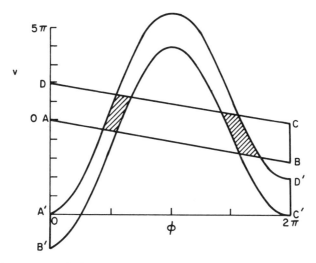

Figure 5.2.4. The horseshoe for $f_{\alpha,\gamma}$ of equation (5.2.7): $\alpha = 0$, $\gamma = 5\pi$.

$ABCD$ bounded by the lines $\phi + v = 0$ $(AB)$, $\phi + v = 2\pi$ $(CD)$, $\phi = 0$ $(AD)$, $\phi = 2\pi$ $(BC)$. We note that $Q$ is foliated by the family of lines $\phi + v = k$, $k \in [0, 2\pi]$ and that the images of such lines under $f$ are vertical lines $\phi = k$, $v \in [k - 2\pi - \gamma \cos k, k - \gamma \cos k]$. Finally, the images of the boundaries $\phi = 0$ and $\phi = 2\pi$ are curves $v = \phi - \gamma \cos \phi$, $v = \phi - 2\pi - \gamma \cos \phi$, see Figure 5.2.4.

To obtain two disjoint strips, we examine the images of the segments in $Q$ along the lines $\phi + v = k$ with $k = 0$, $\pi$, and $2\pi$. Each image is a vertical segment with $\phi$ constant and $v$ varying in the segments $[-2\pi - \gamma, -\gamma]$, $[-\pi + \gamma, \pi + \gamma]$, and $[-\gamma, 2\pi - \gamma]$, respectively. It is easy to check that, for $\gamma > 4\pi$, the images of the segments with $k = 0$ and $2\pi$ both lie below $Q$ while that of the segment with $k = \pi$ lies above $Q$. Moreover, simple computations reveal that, for $\gamma \geq 4\pi$, the slopes of the curves bounding $V_1$ and $V_2$ are greater in magnitude than 2 and those bounding $H_1$ and $H_2$ is less in magnitude than 2. $\qquad\square$

**Remarks.** 1. More delicate estimates reveal that $\gamma$ can be somewhat reduced without destroying the topology of the strips.

2. Since the images $A'D'$ and $B'C'$ of $AD$ and $BC$ are given by $v = \phi - \gamma \cos \phi$ and $v = \phi - 2\pi - \gamma \cos \phi$, respectively, it is easy to compute bounds between which the vertical strips $V_1$ and $V_2$ must lie. They are given by the appropriate roots of the following equations.

$V_1$: roots of $\phi - \gamma \cos \phi = -\phi$ and $\phi - 2\pi - \gamma \cos \phi = 2\pi - \phi$

(between 0 and $\pi$); (5.2.8)

$V_2$: roots of $\phi - 2\pi - \gamma \cos \phi = 2\pi - \phi$ and $\phi - \gamma \cos \phi = -\phi$

(between $\pi$ and $2\pi$). (5.2.9)

It is easy to see that, as $\gamma$ increases, these roots converge on $\pi/2$ and $3\pi/2$, respectively, and thus that the widths of $V_1$ and $V_2$ decrease with increase of $\gamma$. For example, solutions of (5.2.9) show that, for $\gamma = 5\pi$, the $\phi$ coordinates of points in $V_1$ and $V_2$ lie in the intervals (1.39, 2.13) and (4.48, 5.49), respectively. Using the inverse map $f^{-1}$ we find that the corresponding horizontal strips $H_1$, $H_2$ are similarly bounded by $\phi + v = 1.39$, 2.13 and 4.48, 5.49 (all coordinates are given in radians). We then obtain

**Lemma 5.2.6.** *For $\gamma$ sufficiently large ($5\pi$ is sufficient), there are sector bundles $S^u(p)$, $S^s(p)$ based at points $p \in \bigcup_{i,j=1,2} (H_i \cap V_j)$ centered on lines $\phi = $ const. and $\phi + v = $ const., respectively, and each of angular extent $\pi/4$, such that $Df(S^u(p)) \subset S^u(p)$ and $Df^{-1}(S^s(p)) \subset S^s(p)$. Moreover, $Df(p)$ expands vertical distances by a factor of at least 5.5 and $Df^{-1}$ expands horizontal distances by a factor of at least 4.5. See Figure 5.2.5.*

**Remark.** Here, lines $\phi + v = $ const. are termed "horizontal."

PROOF. Linearizing the map, we have

$$Df = \begin{bmatrix} 1 & 1 \\ r & 1+r \end{bmatrix}, \qquad Df^{-1} = \begin{bmatrix} 1+r & -1 \\ -r & 1 \end{bmatrix}, \qquad (5.2.10)$$

where $r = \gamma \sin(\phi + v)$ or $\gamma \sin \phi$, respectively. From the estimates above, for $\gamma \geq 5\pi$ we have $\phi + v \in (1.39, 2.13) \cup (4.48, 5.49)$ for $p \in H_1 \cup H_2$ and similar bounds for $\phi$ for $p \in V_1 \cup V_2$. This shows that $|\sin(\phi + v)|, |\sin \phi| > 0.716$ for points $p \in \bigcup_{i,j=1,2} (H_i \cap V_j)$, or $|\gamma \sin(\phi + v)|, |\gamma \sin \phi| > 11.24$.

We shall only consider the estimate for sector $S^u$ here, since that for $S^s$ is obtained similarly. Consider the image of the sector $S^u$ shown in Figure 5.2.5 under the map $Df$. Let the "edges" of $S^u$ be defined by the points ($\pm 0.414$, 1) (tan $\pi/8 \approx 0.414$). Taking $r = 11.24$ we obtain the images (1.414, 16.89), (0.586, 7.587), respectively, and taking $r = -11.24$ we obtain (1.414, $-14.893$),

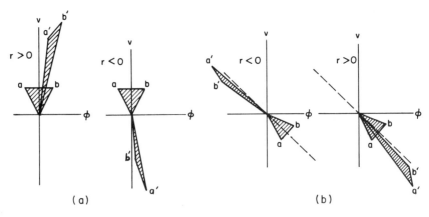

Figure 5.2.5. The sectors $S^u(p)$ and $S^s(p)$. (a) $S^u(p)$; (b) $S^s(p)$.

(0.586, −5.587), respectively. Since $|r| = |\gamma \sin(\phi + v)| > 11.24$ for points $p \in H_i \cap V_j$ and increases of $|r|$ make the images of the sectors even larger and thinner, our estimate is obtained.

The computation of "horizontal" sectors, bounded by lines of angle $-\pi/8, -3\pi/8$, is slightly more awkward but proceeds in a similar manner. We end by noting that, for $\gamma > 5\pi, |\gamma \sin(\phi + v)|, |\gamma \sin \phi| > 11.24$ and thus that our estimates hold for all $\gamma > 5\pi$.                                              $\square$

These two lemmas establish H1 and H3. In our choice of sectors we have chosen the parameter $\mu$ as $\mu = \tan \pi/8 \approx 0.414$. The sector bundle estimates guarantee that $f$ maps vertical strips to vertical strips with a contraction of at least $v = \mu/(1 - \mu) \approx 0.706$ and that $f^{-1}$ maps horizontal strips with a similar contraction.

Using Theorem 5.2.4, these facts, together with the fact that, for $\alpha = 1$, $\det(Df) = \det(Df^{-1}) = 1 < 1/2\mu^2 \approx 2.917$, imply the following:

**Theorem 5.2.7.** *For $\gamma \geq 5\pi$ and $\alpha = 1$ the map (5.2.7) possesses an invariant, hyperbolic, Cantor set $\Lambda$. The map $f$ restricted to $\Lambda$ is homeomorphic to the shift on two symbols.*

Thus we have proved that the bouncing ball problem has a horseshoe $\Lambda$ for $\alpha = 1$ and $\gamma$ sufficiently large. Moreover, the structural stability of $\Lambda$ implies that we can reduce $\alpha$ slightly without losing the horseshoe. Alternatively, fixing $\alpha < 1$ we can repeat the estimates performed above to obtain a lower bound on values $\gamma$ for which horseshoes exist:

EXERCISE 5.2.9. Show that, for $\gamma$ sufficiently large and $\alpha = \frac{1}{2}$, the bouncing ball map $f_{\alpha, \gamma}$ has a hyperbolic horseshoe. Estimate $\gamma$.

EXERCISE 5.2.10 (cf. Devaney and Nitecki [1979]). Show that the Hénon map $(x, y) \rightarrow (\alpha + \beta y - x^2, x)$ has a hyperbolic horseshoe for $\alpha > (5 + 2\sqrt{5})(1 + |\beta|)^2/4$ and $\beta \neq 0$.

Although horseshoes are not attractors, their presence in discrete dynamical systems has important consequences for physically observable motions. (If the maps in questions are Poincaré maps for flows, then similar observations apply to the solutions of the associated differential equations.) We have already seen, in Chapter 2, numerical evidence that solutions of differential equations and maps can depend sensitively on initial conditions when horseshoes are present. This dependence can be understood in terms of the structure of the sets $E_\Lambda^s$ and $E_\Lambda^u$ of Definition 5.2.6 (cf. Figure 5.2.2). For the (linear) horseshoe, each of these sets is the product of an interval with a Cantor set. This implies that, if $x$ lies on an orbit asymptotic to an orbit $\gamma_1$ in $\Lambda$, then in any neighborhood of $x$ there are points $y$ whose orbits are asymptotic to orbits $\gamma_j \subset \Lambda$ with symbol sequences which completely differ from that of $\gamma_1$ after sufficiently many places. Moreover, there are open sets of orbits starting near $x$ which eventually escape from the neighborhood of $\Lambda$, leaving "parallel to" different members of the unstable set $E^u$. Thus orbits starting close together can meet quite different fates.

The orbits asymptotic to $\Lambda$ form a stable manifold, $W^s(\Lambda)$, which can provide very complicated boundaries to the domains of attraction of different attractors: in effect the horseshoe behaves like a chaotic saddle point. Thus, when horseshoes are present, one expects to see (long) chaotic transients before the orbits "settle down" to equilibria or periodic behavior. While almost all orbits (with respect to Lesbesgue measure) passing near $\Lambda$ eventually leave its neighborhood, the presence of $\Lambda$ dramatically affects their behavior. The stable manifold theorem for hyperbolic sets expresses concisely the way in which the points of a hyperbolic invariant set behave like saddles, and also leads to a topological characterization of the chaotic nature of the dynamics within the invariant set:

**Theorem 5.2.8** (Hirsch and Pugh [1970]). *Let $\Lambda$ be a compact invariant set for a $C^r$ diffeomorphism $f: \mathbb{R}^n \to \mathbb{R}^n$ with a hyperbolic structure $E^s_\Lambda \oplus E^u_\Lambda$. Then there is an $\varepsilon > 0$ and there are two collections of $C^r$ manifolds $W^s_\varepsilon(x)$, $W^u_\varepsilon(x)$, $x \in \Lambda$ which have the following properties:*

(1) $y \in W^s(x)$ *if and only if* $d(f^n(x), f^n(y)) \le \varepsilon$ *for all* $n \ge 0$.
   $y \in W^u_\varepsilon(x)$ *if and only if* $d(f^n(x), f^n(y)) \le \varepsilon$ *for all* $n \le 0$.
(2) *The tangent spaces of $W^s_\varepsilon(x)$ and $W^u_\varepsilon(x)$ at $x$ are $E^s_x$ and $E^u_x$, respectively.*
(3) *There are constants $C > 0$, $0 < \lambda < 1$ such that if $y \in W^s_\varepsilon(x)$, then $d(f^n(x), f^n(y)) \le C\lambda^n$ for $n \ge 0$, and if $y \in W^u_\varepsilon(x)$, then $d(f^{-n}(x), f^{-n}(y)) \le C\lambda^n$ for $n \ge 0$.*
(4) $W^s_\varepsilon(x)$ *and $W^u_\varepsilon(x)$ are embedded disks in $\mathbb{R}^n$. The mappings from $\Lambda$ to the function space of $C^r$ embeddings of disks into $\mathbb{R}^n$ given by $x \to W^s_\varepsilon(x)$ and $x \to W^u_\varepsilon(x)$ are continuous.*

The full meaning of (4) and the proof of the stable manifold theorem require more discussion of function spaces and their topologies than is appropriate here. The one aspect of (4) which we shall use is the fact that the tangent spaces and the diameters of the $W^s_\varepsilon(x)$ and $W^u_\varepsilon(x)$ vary continuously with $x$. Together with the transversality of the subspaces in a hyperbolic structure, this leads to the following corollary:

**Corollary 5.2.9.** *Let $\Lambda$ be a compact hyperbolic invariant set for a diffeomorphism $f: \mathbb{R}^n \to \mathbb{R}^n$. Then there are $\delta > 0$ and $\varepsilon > 0$ such that if $x, y \in \Lambda$ and $d(x, y) < \delta$, then $W^s_\varepsilon(x) \cap W^u_\varepsilon(y)$ contains exactly one point.*

PROOF. When $x = y$, we know that the tangent spaces of $W^s_\varepsilon(x)$ and $W^u_\varepsilon(x)$ are $E^s_x$ and $E^u_x$. These are complementary, so $W^s_\varepsilon(x)$ and $W^u_\varepsilon(x)$ intersect transversally at $x$. Thus for $\varepsilon$ sufficiently small, $W^s_\varepsilon(x) \cap W^u_\varepsilon(x)$ contains just $x$. Using the continuity of the invariant manifolds and the compactness

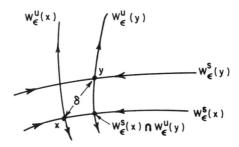

Figure 5.2.6. Corollary 5.2.9.

of $\Lambda$, $\varepsilon$ may be chosen independent of $x$, and $\delta > 0$ can be found so that the conclusion of the corollary holds.                                                                 □

We illustrate the idea in Figure 5.2.6.

For the linear horseshoe of Section 5.1, the local stable and unstable bundles $\bigcup_x W_\varepsilon^s(x)$ and $\bigcup_x W_\varepsilon^u(x)$ are contained in the sets $E_\Lambda^s$ and $E_\Lambda^u$ discussed above (cf. Figure 5.2.2).

Throughout this section, and in general in our study of the global dynamics of nonlinear systems, we have made and will make extensive use of local linearization. For example, in the proof of the Smale–Birkhoff homoclinic theorem in the next section, and in the analysis of homoclinic orbits for maps and flows in Chapter 6, while the qualitative knowledge of global properties of the dynamics is necessary to ensure that orbits return close to their starting points, the quantitative estimates which actually enable hard results to be obtained rely on local analysis of linearized systems. We end this section with a result which uses these ideas, the proof of which we leave as an exercise:

**Theorem 5.2.10** (The Lambda Lemma, Palis [1969]). *Let $f$ be a $C^1$ diffeomorphism of $\mathbb{R}^n$ with a hyperbolic fixed or periodic point $p$ having $s$ and $u$ dimensional stable and unstable manifolds ($s + u = n$), and let $D$ be a $u$-disk in $W^u(p)$. Let $\Delta$ be a $u$-disk meeting $W^s(p)$ transversely at some point $q$. Then $\bigcup_{n \geq 0} f^n(\Delta)$ contains $u$-disks arbitrarily $C^1$ close to $D$.*

EXERCISE 5.2.11. Prove Theorem 5.2.10. (Hint: Suppose that $p$ is fixed, since if it is $k$-periodic, we can replace $f$ by $f^k$. Introduce local coordinates $(x, y) \in \mathbb{R}^s \times \mathbb{R}^u$ in a neighborhood $U$ of $p$, so that the stable and unstable manifolds are locally given by $y = 0$ and $x = 0$, respectively. Consider the fate of points in $f^k(\Delta)$, and tangent vectors to $f^k(\Delta)$, once they lie in $U$; show that their behavior is dominated by the linearized map $Df(p)$ (cf. Palis [1969], Newhouse [1980]).)

We close with the remark that, if there is a transversal homoclinic point $q \in W^u(p) \cap W^s(p)$, then $\Delta$ of Theorem 5.2.10 can be chosen to be a $u$-disc in $W^u(p)$. In this case the lambda lemma implies that $W^u(p)$ accumulates on

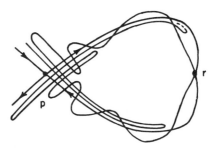

Figure 5.2.7. The lambda lemma implies homoclinic tangles.

itself, and, similarly $s = u = 1$ $(n = 2)$, this leads to the homoclinic tangle which we have already met; cf. Figure 5.2.7. We note that it is *not* necessary that the map be area (or volume) preserving for this structure to occur.


## 5.3. Markov Partitions and Symbolic Dynamics

The stable manifold theorem (Theorem 5.2.8) and its corollary provide the foundation for endowing hyperbolic invariant sets with a good symbolic representation of their dynamics. Throughout this section we let $\Lambda$ be a compact hyperbolic maximal invariant set for a $C^r$ diffeomorphism $f: \mathbb{R}^n \to \mathbb{R}^n$. Furthermore, we fix $\varepsilon$ and $\delta$ as in the stable manifold theorem and its corollary: $W_\varepsilon^s(x)$ and $W_\varepsilon^u(x)$ are $C^r$ embedded disks and $d(x, y) < \delta$ implies that $W_\varepsilon^s(x) \cap W_\varepsilon^u(y)$ is exactly one point.

**Proposition 5.3.1.** *If $\Lambda$ is indecomposable and maximal and $x$, $y \in \Lambda$ with $d(x, y) < \delta$, then $W_\varepsilon^s(x) \cap W_\varepsilon^u(y) \in \Lambda$ (and $W_\varepsilon^u(x) \cap W_\varepsilon^s(y) \in \Lambda$).*

**Proof.** Let $z \in W_\varepsilon^s(x) \cap W_\varepsilon^u(y)$ and $\rho > 0$. Pick $n_1, n_2 > 0$ such that $d(f^{n_1}(z), f^{n_1}(x)) < \rho/2$ and $d(f^{-n_2}(z), f^{-n_2}(y)) < \rho/2$. The fact that $f^{n_1}(x)$, $f^{-n_2}(y) \in \Lambda$, and $\Lambda$ is indecomposable, implies that there exists a $\rho/2$ chain recurrence $\{u_1, \ldots, u_k\}$ from $f^{n_1}(x)$ to $f^{-n_2}(y)$ (cf. p. 236). It follows that $\{z, \ldots, f^{n_1}(z), u_2, \ldots, u_{k-1}, f^{-n_2}(z), \ldots, z\}$ is a $\rho$ chain recurrence for $z$, and hence that $z$ is in the same maximal, indecomposable part of the chain recurrent set as $x$ and $y$.  $\square$

EXERCISE 5.3.1. Verify that this proposition holds for the invariant set $\Lambda$ of the horseshoe.

This proposition is an existence result which implies that the forward and backwards asymptotic behavior of a point have a substantial degree of statistical independence. This concept will be incorporated into the definition of Markov partitions. To begin the construction of Markov partitions, we want to regard the stable and unstable manifold as establishing a "coordinate grid" on $\Lambda$ which is adapted to its dynamics. Assume for the rest of this section that $\Lambda$ is indecomposable and maximal.

**Definition 5.3.1.** A *rectangle R for* $\Lambda$ is a closed subset of $\Lambda$ with the property that $x, y \in R$ implies that $W^s_\varepsilon(x) \cap W^u_\varepsilon(y)$ is exactly one point, and this point is in $R$.

EXERCISE 5.3.2. In the horseshoe example with $\varepsilon = 1$, show that the rectangles are given by the intersection of $\Lambda$ with rectangles having horizontal and vertical sides.

EXERCISE 5.3.3. If $x \in \Lambda$, show that there are $\delta, \varepsilon > 0$ such that $R = \{y |$ there exist $w, z$ with $y = W^u_\varepsilon(w) \cap W^s_\varepsilon(z), w \in W^s_\delta(x), z \in W^u_\delta(x)\}$ is a rectangle (see Figure 5.2.6).

This exercise implies that $R$ can be identified with the Cartesian product of $W^s_\varepsilon(x) \cap R$ and $W^u_\varepsilon(x) \cap R$.

EXERCISE 5.3.4. Show that the intersection of two rectangles is a rectangle.

We note that the rectangles defined here are not rectangles in the usual sense, but may be disconnected sets (typically Cantor sets), obtained by intersecting "usual" rectangles with pieces of $\Lambda$. For the horseshoe, for example, we have the natural candidates for two rectangles $R_1 = H_1 \cap \Lambda$ and $R_2 = H_2 \cap \Lambda$. It is, however, conventional to draw these rectangles as if they were rectangles in the usual sense, cf. Figure 5.3.1. The reason for not using usual rectangles is that, while $f$ may be defined on such rectangles, iterates may not be well behaved except on points of $\Lambda$. For instance, in constructing the horseshoe we only specified the action of $f$ on $S$, and hence cannot follow the orbits of points once they leave $S$.

The rectangles of $\Lambda$ will be candidates for the "states" which we use in a symbolic description of $\Lambda$. They are the counterpart (and generalization) of the sets $H_i$ in the horseshoe example. Insofar as possible, we partition $\Lambda$ into rectangles so that a point of $\Lambda$ can be characterized by symbol sequences which specify the rectangles visited by its trajectory. It is not reasonable to hope that there will be a 1–1 correspondence between $\Lambda$ and *all* symbol sequences, as was the case with the horseshoe. Nonetheless, a good symbolic description of $\Lambda$ can be obtained from a Markov partition for $\Lambda$. If $R$ is a rectangle and $x \in R$, we set $W^s_\varepsilon(x, R) = W^s_\varepsilon(x) \cap R$, and $W^u_\varepsilon(x, R) = W^u_\varepsilon(x) \cap R$. The *interior* of a rectangle $R$, denoted int $R$, is defined to be $\{x \in R |$ there exists $\delta > 0$ with $W^u_\delta(x) \cap \Lambda \subset R$ and $W^s_\delta(x) \cap \Lambda \subset R\}$. The *boundary* of $R$ is $R - $ int $R$.

**Definition 5.3.2.** A *Markov partition for* $\Lambda$ is a finite collection of rectangles $\{R_1, \ldots, R_m\} = \mathcal{R}$ such that:

(1) $\Lambda = \bigcup_{i=1}^m R_i$;
(2) int $R_i \cap$ int $R_j = \emptyset$ if $i \neq j$;
(3) $f(W^u(x, R_i)) \supset W^u(f(x), R_j)$ and $f(W^s(x, R_i)) \subset W^s(f(x), R_j)$, whenever $x \in$ int $R_i$, $f(x) \in$ int $R_j$. See Figure 5.3.1.

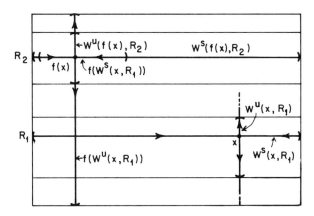

Figure 5.3.1. For the horseshoe $f(W^u(x, R_1)) \supset W^u(f(x), R_2)$ and $f(W^s(x, R_1)) \subset W^s(f(x), R_2))$.

We now have:

**Theorem 5.3.2** (Bowen [1978]). *Compact, maximal, indecomposable hyperbolic invariant sets have Markov partitions.*

The reader should consult Bowen [1978] or Shub [1978] for the construction of Markov partitions for a hyperbolic set $\Lambda$. The key step in the proof of the theorem relies upon the *shadowing* property of *pseudo-orbits* in $\Lambda$. One says that a sequence $\mathbf{x} = \{x_i\}_{i=a}^b$ is an $\alpha$-*pseudo-orbit* for $f$ if $d(x_{i+1}, f(x_i)) < \alpha$ for all $a \le i < b$. The point $y$ $\beta$-*shadows* $\mathbf{x}$ if $d(f^i(y), x_i) < \beta$ for $a \le i \le b$. One can think of a pseudo-orbit as a trajectory in a random perturbation of $f$ which allows one to perturb $f$ independently each successive iterate by an amount smaller than $\alpha$ before proceeding. Such orbits are essentially like those realized in numerical iterations of maps by computer, or in physical experiments. The following proposition states that all of the pseudo-orbits are still approximated by trajectories of $f$ itself. See Figure 5.3.2.

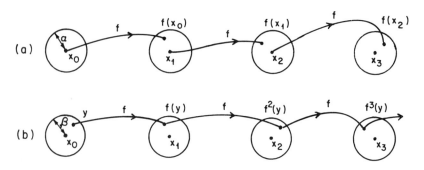

Figure 5.3.2. (a) an $\alpha$-pseudo-orbit; (b) the point $y$ $\beta$-shadows $\mathbf{x}$.

**Proposition 5.3.3** (Shadowing Lemma, Bowen [1970, 1978]). *Let $\Lambda$ be a hyperbolic invariant set. Then for every $\beta > 0$, there is an $\alpha > 0$ such that every $\alpha$-pseudo-orbit $\{x_i\}_{i=a}^b$ in $\Lambda$ is $\beta$-shadowed by a point $y \in \Lambda$.*

For a proof see Bowen [1978] or Newhouse [1980].

This proposition implies that, while a computer may not calculate the orbit which you hope for, what it does find is nonetheless an approximation to some true orbit of the system.

Markov partitions lead directly to good symbolic dynamics for hyperbolic sets $\Lambda$. The symbol space is a set of bi-infinite sequences on a finite set (the rectangles of the partition) and the dynamics of $f$ is reproduced by the shift map on sequences. We now describe formally the resulting object, a *subshift of finite type*.

Let $\mathscr{R} = (R_1, \ldots, R_m)$ be a Markov partition for $\Lambda$ consisting of closed rectangles. Define the $m \times m$ matrix $A = (A_{ij})$ by $A_{ij} = 0$ if

$$\text{int } R_i \cap f^{-1}(\text{int } R_j) = \varnothing$$

and $A_{ij} = 1$ if int $R_i \cap f^{-1}(\text{int } R_j) \neq \varnothing$. Define $\Sigma_A$ to be the set of bi-infinite sequences $\mathbf{a} = \{a_i\}_{i=-\infty}^{\infty}$, $a_i \in \{1, \ldots, m\}$, satisfying the property $A_{a_i a_{i+1}} = 1$ for all $i \in \mathbb{Z}$. The shift map $\sigma$ of such bi-infinite sequences is given by $\sigma(\mathbf{a}) = \mathbf{b}$ if $b_i = a_{i+1}$. Clearly $\sigma(\Sigma_A) = \Sigma_A$. The set $\Sigma_A$ together with the shift map $\sigma$ is called the *subshift of finite type* with *transition matrix $A$*. We note that $\Sigma_A$ is a closed subset of the set of bi-infinite sequences of the symbols $\{1, \ldots, m\}$ and is therefore a compact metric space with metric coming from $\Sigma$. The symbolic dynamics of $\Lambda$ associated with $\mathscr{R}$ is given by the following proposition:

**Proposition 5.3.4.** *For each $\mathbf{a} \in \Sigma_A$, the set $\bigcap_{j \in \mathbb{Z}} f^{-j}(R_{a_j})$ consists of exactly one point, denoted $\pi(\mathbf{a})$. The map $\pi: \Sigma_A \to \Lambda$ is a continuous surjection, $\pi \circ \sigma = f \circ \pi$, and $\pi$ is 1–1 over the set $\bigcap_{i \in \mathbb{Z}} (f^i(\bigcup_{i=j}^m \text{int } R_j)) \subset \Lambda$.*

For a proof, see Bowen [1970].

Apart from the set of $\Lambda$ consisting of points whose trajectories intersect the boundary of a rectangle, this proposition states that the subshift of finite type gives a faithful topological representation of the hyperbolic set $\Lambda$ together with its dynamics. If $\Lambda$ is zero-dimensional (a Cantor set like the invariant set of horseshoe), then a Markov partition for $\Lambda$ may be chosen so that its rectangles are pairwise disjoint. Thus zero-dimensional hyperbolic sets are topologically conjugate to subshifts of finite type. Higher-dimensional $\Lambda$ are obtained from subshifts of finite type by making identifications of sequences in much the same way that certain real numbers have two different decimal representation (cf. the introduction to this chapter).

EXERCISE 5.3.5. Define $f: \mathbb{C} \to \mathbb{C}$ (the complex numbers) by $f(z) = z^2$. Show that the circle $S^1 = \{e^{2\pi i\theta} | 0 \leq \theta < 1\}$ is invariant for $f$. Let $\Sigma$ be the one-sided symbol sequence space $\Sigma = \{\{a_i\}_{i=0}^{\infty} | a_i = 0 \text{ or } 1\}$ and $\sigma: \Sigma \to \Sigma$ be the shift map. Show that the map $\pi: \Sigma \to S^1$ defined by $\pi(\mathbf{a}) = e^{2\pi i\theta}$ with $\mathbf{a}$ being a binary representation of $\theta$, satisfies the equation $f \circ \pi = \pi \circ \sigma$. Prove that there is a $z \in S^1$ such that $\{z^{2^n}\}_{n \geq 0}$ is dense in $S^1$.

EXERCISE 5.3.6. Return to the van der Pol map of Section 2.1 in the light of the above discussion, and study its dynamics.

Virtually all analyses of chaotic behavior in specific dynamical systems involve the identification of hyperbolic invariant sets. Often it is unclear if the sets which have been identified are parts of larger indecomposable invariant sets. This is an important issue to which we return later in this chapter. As an application of symbolic dynamics, we shall prove a theorem of Smale which establishes a criterion for the existence of hyperbolic invariant sets in a flow. The construction in the proof of this result is typical of most arguments yielding chaotic invariant sets.

**Theorem 5.3.5** (The Smale–Birkhoff Homoclinic Theorem). *Let $f: \mathbb{R}^n \to \mathbb{R}^n$ be a diffeomorphism such that $p$ is a hyperbolic fixed point and there exists a point $q \neq p$ of transversal intersection between $W^s(p)$ and $W^u(p)$. Then $f$ has a hyperbolic invariant set $\Lambda$ on which $f$ is topologically equivalent to a subshift of finite type.*

PROOF. The idea of the proof is to find a picture which looks like the horseshoe example for some iterate of the map $f$; Figure 5.3.3. We do this by first picking a small neighborhood $U$ of $p$ whose properties will be determined during the course of the proof. Since $f^i(q) \to p$ as $i \to +\infty$ and as $i \to -\infty$, there are $m, n > 0$ such that $q \in f^l(U)$ for $l \geq m$ and $q \in f^{-k}(U)$ for $k \geq n$ (Figure 5.3.4). By carefully selecting $U$, we shall obtain a Markov partition of disjoint sets consisting of $U$, $V \subset f^{-k}(U) \cap f^l(U)$, and $f^i(V)$ for $-l < i < k$. The process for doing this requires close examination of trajectories passing near $q$.

At $q$, let $W_s$ and $W_u$ be small neighborhoods of $q$ in its stable and unstable manifolds, respectively. We examine the iterates $f^{-l}(W_s)$ and $f^{+k}(W_u)$.

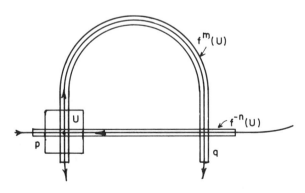

Figure 5.3.3. Theorem 5.3.5.

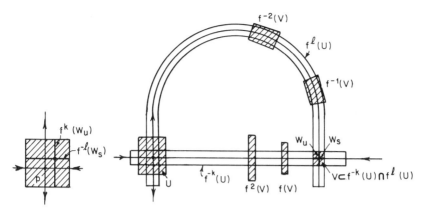

Figure 5.3.4. More on Theorem 5.3.5.

Since $f^k(q) \in U$ when $k \geq n$, $f^k(W_u) \cap U$ is nonempty. Moreover, $f^k(W_u)$ intersects $W^s$ transversally at $f^k(q)$ since $W_u$ intersects $W^s$ transversally at $q$. Local analysis embodied in the $\lambda$-lemma of Palis [1969] (cf. Newhouse [1980]) implies that $f^k(W_u) \cap U$ approaches $W^u_\varepsilon(p)$ as $k \to \infty$. Similarly, $f^{-l}(W_s) \cap U$ approaches $W^s_\varepsilon(p)$ as $l \to \infty$. Therefore, for $k$, $l$ sufficiently large and $W_u$, $W_s$ chosen of the proper size, $f^k(W_u)$ and $f^{-l}(W_s)$ have exactly one point of intersection and this lies in $U$. See Figure 5.3.4. Finally, we shrink $U$ and pick $V$ to be the component of $f^{-k}(U) \cap f^l(U)$ which contains $q$. This produces sets $U$ and $V$ so that the sets $U$ and $f^i(V)$, $-l < i < k$, are disjoint.

With the choices we have made, we end with a collection of disjoint sets $\{U; f^i(V), -l < i < k\}$ which forms a Markov partition $\mathscr{R}$ for a zero-dimensional invariant set. The transition matrix $A$ of $\mathscr{R}$ is the $(k + l) \times (k + l)$ matrix:

$$
\begin{bmatrix}
1 & 1 & 0 & \cdots & 0 \\
0 & 0 & 1 & \cdots & 0 \\
\vdots & \vdots & \vdots & & \vdots \\
0 & 0 & & \cdots & 1 \\
1 & 0 & & \cdots & 0
\end{bmatrix}.
$$

In particular, for $f^{k+l}$, the set $U$ intersects itself in vertical strips containing $W^u_\varepsilon(p)$ and $f^k(W_u)$ and we obtain a shift on two symbols for $f^{k+l}$ which gives $f^{k+l}$ a horseshoe.  $\square$

EXERCISE 5.3.7. Compare our version of the Smale–Birkhoff theorem with those of Smale [1963] and Moser [1973, pp. 181–188].

We end this section with an outline of the analysis of the attracting set for the van der Pol map of Levi discussed in Section 2.1 and illustrated in Figure 2.1.5(b). We refer the reader to Levi [1981] for more details. Here we

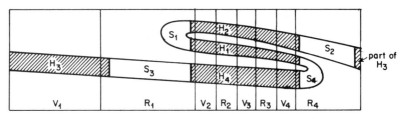

Figure 5.3.5. The van der Pol map of the annulus $F: R^+ \to R^+$. For definitions of $R^+$ see Section 2.1, p. 76.

are more specific than in Section 2.1, and define the map of the annulus geometrically as follows (Figure 5.3.5). We assume that the annulus $D$ is divided into eight vertical* strips $V_i$, $R_i$, $i = 1, \ldots, 4$ as indicated, such that the four images

$$H_i = F(V_i)$$

are horizontal* strips and the four images $S_i = F(R_i)$ lie in $R_1 \cup R_4$ as indicated. We define the attracting set $\mathscr{A}$ as usual:

$$\mathscr{A} = \bigcap_{n \geq 0} F^n(D). \tag{5.3.1}$$

We further assume that $F|_{V_i}$ contracts vertically and expands horizontally on the strips $V_i$, $H_i$ in a controlled manner so that analogues of Lemmas 5.2.1 and 5.2.2 hold. Note that uniform horizontal expansion cannot occur everywhere, and in particular that in $R_1$ and $R_4$ the map contracts in both directions.

Levi [1981] proves that $\mathscr{A}$ contains two attracting fixed points, corresponding to the two solutions of periods $(2k \pm 1)T$ discussed in Section 2.1. Moreover, he shows that $\mathscr{A}$ also contains an invariant hyperbolic Cantor set $\Lambda$, together with its unstable manifolds $W^u(\Lambda)$, and that $F|_\Lambda$ is conjugate to a subshift on four symbols. We now sketch the analysis.

$\Lambda$ is the set of points whose orbits never leave $\bigcup_i V_i$, or

$$\Lambda = \bigcap_{-\infty < n < \infty} F^n \left( \bigcup_{i=1}^4 V_i \right). \tag{5.3.2}$$

Clearly $\Lambda$ contains a standard horseshoe, since the pair of strips indexed 2 and 4 obey $F(V_i) = H_i$ with the $H_i$ intersecting the $V_i$ as required (up to a change of "direction"). However, $\Lambda$ contains infinitely many other points, since we can construct a transition matrix $A = [A_{ij}]$ with $A_{ij} = 1$ if $F(V_i) \cap V_j$ is nonempty and zero otherwise:

$$A = \begin{bmatrix} 0 & 1 & 1 & 1 \\ 0 & 1 & 1 & 1 \\ 1 & 0 & 0 & 0 \\ 0 & 1 & 1 & 1 \end{bmatrix}. \tag{5.3.3}$$

* In the sense of definition 5.2.5.

Equipped with $A$, we can define a subspace $\Sigma_A \subset \Sigma$, the space of bi-infinite sequences of four symbols such that $A_{a_i a_{i+1}} = 1$ for all sequences $\{a_i\} \in \Sigma_A$ and $i \in \mathbb{Z}$. Proposition 5.3.4 then implies that $F|_\Lambda$ is conjugate to the subshift of finite type with transition matrix $A$, and hence that there is a 1:1 correspondence between orbits of $F|_\Lambda$ and all sequences which do not contain the "forbidden" pairs 11, 21, 32, 33, 34, or 41. We note that the subshift contains the full shift on the two symbols 2, 4, since there are no forbidden pairs of this form.

We next consider the set of points in $D$ which do not lie in $W^s(\Lambda)$. It should be clear that

$$W_{\text{loc}}^s(\Lambda) = \left( \bigcap_{n \geq 0} F^{-n}\left( \bigcup_i V_i \right) \right) \cap D \tag{5.3.4}$$

is a Cantor set of vertical segments, separated by the preimages

$$\left( \bigcup_{n \geq 0} F^{-n}\left( \bigcup_i R_i \right) \right) \cap D \tag{5.3.5}$$

of the four strips $R_i$. It is also not too hard to see that any point which is mapped into $\bigcup_i R_i$ is attracted to one or the other of the fixed points which lie in $R_1$ and $R_4$, since

$$F(R_1), F(R_3) \subset R_1$$

and

$$F(R_2), F(R_4) \subset R_4. \tag{5.3.6}$$

Since $W^u(\Lambda)$ is invariant and the local unstable manifolds $W_\varepsilon^u(x)$ of points $x \in \Lambda$ are clearly contained in $D$, we have $\text{Cl}(W^u(\Lambda)) \subseteq \mathscr{A}$. To prove that $\mathscr{A} = \text{Cl}(W^u(\Lambda))$ we note that any point in $D$ lies either on $W^s(\Lambda)$ or in the stable manifold of one or the other of the sinks. In particular, $D$ can be covered by open sets each of which contain a piece of $W^s(\Lambda)$, and the lambda lemma then implies that there are images of these sets which lie arbitrarily close to points in $W^u(\Lambda)$, showing that $\mathscr{A} \subseteq \text{Cl}(W^u(\Lambda))$ and hence that $\mathscr{A} = \text{Cl}(W^u(\Lambda))$, as claimed.

# 5.4. Strange Attractors and the Stability Dogma

Theorem 5.3.5, proved above, gives the existence of hyperbolic limit sets under very general hypotheses. It does not, however, broach the subject of "strange attractors." Having encountered this issue several times already, it is time to give it a thorough discussion. For concreteness, let us pick the Duffing equation as an example to consider in these terms and ask whether it does have strange attractors.

Numerical integration of the forced Duffing equation certainly *appears* to yield trajectories that are not asymptotically periodic. Moreover, the analytic methods of Chapter 4 allow us to prove the existence of transversal homoclinic points in this system, and hence to deduce that there are hyperbolic invariant sets. We do not know, however, that the computed trajectory tends to the hyperbolic invariant set located analytically. Indeed, the hyperbolic invariant set located by Smale's theorem is *not* attracting, and the set of points asymptotic to it will have zero measure for $C^2$ maps (the Duffing equation is analytic). Thus the results are complementary, but they *do not imply* that the typical trajectory will be asymptotically chaotic. In fact in some cases we observe transient periods of chaos followed by asymptotically periodic motions. The *attracting* invariant sets remain to be determined.

In trying to piece together a coherent picture of this situation, we enter a realm in which the theory remains in an unsatisfactory state. There are paradoxes in which different theorems appear to be steering us toward opposite conclusions. The first paradox involves the definition of *attractor*. The most naive definition of an attractor for a flow $\phi_t$ is that it is a closed indecomposable invariant set $\Lambda$ with the property that $\Lambda$ has a neighborhood $U$ with $\phi_t(x) \in U$ for $t \geq 0$ and $\phi_t(x) \to A$ for all $x \in U$ (cf. Section 1.6). There are difficulties with this definition, however. For a planar vector field, a saddle loop at which the saddle trace is negative appears to be attracting from the interior of the loop but not the exterior. This example is not structurally stable, but it is also unlikely that structurally stable strange attractors always appear in systems such as the Duffing equation. It is not even evident that the forced Duffing equation will have invariant sets larger than a periodic orbit which do attract a whole neighborhood. Therefore, we shall adopt a less restrictive definition.

**Definition 5.4.1.** An *attractor* is an indecomposable closed invariant set $\Lambda$ with the property that, given $\varepsilon > 0$, there is a set $U$ of positive Lebesgue measure in the $\varepsilon$-neighborhood of $\Lambda$ such that $x \in U$ implies that the $\omega$-limit set of $x$ is contained in $\Lambda$, and the forward orbit of $x$ is contained in $U$. We shall call an attractor *strange* if it contains a transversal homoclinic orbit.

One reason for picking this definition is that numerical results with dynamical systems appear to give results compatible with the ideal experiment of picking a point randomly with respect to Lebesgue measure (Bennetin

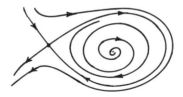

Figure 5.4.1. An attracting homoclinic loop.

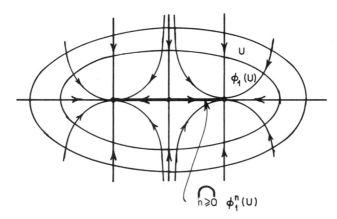

Figure 5.4.2. An attracting set need not be an attractor.

*et al.* [1978, 1979]). Thus our definition of attractor should be a good idealization of the statement that there is a positive probability that computed trajectories will tend to $\Lambda$.*

The requirement of indecomposability may be illustrated by the following example (cf. Exercise 1.6.4). Consider the time 1 map $\phi_1$ for the flow generated by

$$\dot{x} = x - x^3,$$

$$\dot{y} = -y,$$

the phase portrait of which is shown in Figure 5.4.2. There are two sinks at $(x, y) = (\pm 1, 0)$ and a saddle at the origin. Clearly a suitably chosen open set $U$ containing these points is mapped into itself by $\phi_1$ and the limit set $\bigcap_{n=0}^{\infty} \phi_1^n(U)$ is the closed interval $[-1, 1]$. If we drop the indecomposability requirement, then $[-1, 1]$ is an attractor, but almost all points in $U$ are asymptotic to one of the sinks at $(\pm 1, 0)$, and thus the set $\bigcap_{n=0}^{\infty} \phi_1^n(U)$ is not of much direct interest. However, in some problems, such as the Duffing example, while the set $\bigcap_{n=0}^{\infty} \phi_1^n(U)$ does contain periodic sinks and hence is decomposable, the basins of attraction of the sinks are so narrow and convoluted that small (numerical or physical) perturbations prevent typical orbits from achieving periodic asymptotic behavior, although they do remain arbitrarily close to $\bigcap_{n=0}^{\infty} \phi_1^n(U)$. Here it might be sensible to drop the indecomposability requirement.

Suppose that we have found an attractor in a dynamical system for a specific parameter value. We next wish to know if it persists for nearby values and if (similar) attractors exist for "most" parameter values. There are alternative points of view as to what kinds of sets should be regarded as important and which sets are negligible in evaluating "typical" behavior.

* For a recent study of the definition of attractors with physical examples in mind, see Ruelle [1981], who makes use of the idea of chain recurrence.

Here we have adopted the position that phenomena which occur on sets of positive Lebesgue measure are *not* negligible. This viewpoint is at least partially in conflict with the attitude that one need only consider *generic sets*: countable intersections of open dense sets. The conflict here is that there are generic sets whose complements have positive measure. As an illustration, we describe one such set which has relevance for the analysis of one-dimensional mappings (see Section 5.6, below).

Choose numbers $0 < \alpha < 1$ and $0 < \beta < 1$. We will construct a Cantor set $C$ by an inductive construction which defines sets $C_0 \supset C_1 \supset C_2 \supset \cdots$ with $C = \bigcap_{i=0}^{\infty} C_i$, where $C_i$ is the union of $2^i$ closed intervals of equal length. If the length of each interval in $C_i$ is $\gamma_i$, we obtain $C_{i+1}$ from $C_i$ by removing from each interval $I$ of $C_i$ the open interval $J$ of length $\alpha\beta^i\gamma_i$ centered at the midpoint of $I$. Then $I - J$ consists of two intervals, each of length $(\gamma_i/2)(1 - \alpha\beta^i)$. Letting $l_i = 2^i\gamma_i$, the total length of all intervals in $C_i$, we have $l_{i+1} = l_i(1 - \alpha\beta^i)$. It follows that $l_i = l_0 \prod_{j=0}^{i-1} (1 - \alpha\beta^j)$, so that the measure of $C$ is $l_0 \prod_{j=0}^{\infty} (1 - \alpha\beta^j)$. Since $\sum_{j=0}^{\infty} [1 - (1 - \alpha\beta^j)] = \sum_{j=0}^{\infty} \alpha\beta^j = \alpha/(1 - \beta) < \infty$, $\prod_{j=0}^{\infty} (1 - \alpha\beta^j)$ is positive. Thus $C$ has positive measure even though its complement is an open-dense set of the line.

There are one-parameter families of one-dimensional mappings in which strange attractors are present for sets of parameter values much like $C$. This phenomenon is described more fully in Chapter 6. Based on the premise that one-dimensional mappings are good idealizations of diffeomorphisms of the plane that contract area, one can *conjecture* that there are sets of parameter values of positive measure for the Duffing, van der Pol, and bouncing ball examples for which the systems have strange attractors. We note that this is only conjecture, and the conjecture contrasts with theorems of Newhouse which we state in Chapter 6. The answer to this conjecture is perhaps *the* outstanding theoretical problem in our current attempts to bring dynamical systems theory to bear upon these examples. From a practical point of view, one can argue that the issue is not so important because the presence of noise in real systems and numerical errors in computations will make long period stable periodic orbits with convoluted basins of attraction indistinguishable from strange attractors. However, the issue is one which highlights the incompleteness of our understanding of the dynamics of even the geometrical models of our various examples.

In discussing the "prevalence of strange attractors" from a practical point of view, we should consider questions of *structural stability*. We recall from Section 1.7 that a dynamical system is structurally stable when small $C^1$ perturbations yield topologically equivalent systems. Historically, structurally stable strange attractors occupy an important place in the development of dynamical systems theory in spite of the fact that all known examples of such attractors are defined geometrically rather than through explicit equations motivated by physical models. Indeed, experience with planar vector fields suggested that all vector fields have structurally stable perturbations, and that one could therefore usually ignore systems which were not structurally stable (cf. Peixoto's theorem). This principle was embodied in a *stability dogma*, in which structurally unstable systems were regarded as somehow suspect. This dogma stated that, due to measurement uncertainties, etc., a model of a physical system was valuable only if its qualitative properties did not change with perturbations. In the case of dynamical systems, this meant that structural stability was *imposed* as an *a priori* restriction on "good" models of physical phenomena.

The logic which supports the stability dogma is faulty. It is true that some dynamical systems (such as the Lotka–Volterra equations of population biology and the undamped harmonic oscillator) are not good models for the phenomena they are supposed to represent because perturbations give rise to different qualitative features. However, the presumed strange attractors of our examples are not structurally stable, and we are confident that these systems *are* realistic models for the chaotic behavior of the corresponding (deterministic) physical systems. *But, since the systems are not structurally stable, details of their dynamics which do not persist in perturbations may not correspond to verifiable physical properties of the system.* The situation is analogous to the question of whether the length of a rod is rational or irrational (a qualitative property). Limitations on measurement processes do not allow a sensible answer.

Thus the stability dogma might be reformulated to state that the only properties of a dynamical system (or a family of dynamical systems) which are *physically relevant* are those which are preserved under perturbations of the system. The definition of physical relevance will clearly depend upon the specific problem. This is quite different from the original statement that the only good systems are ones with *all* of their qualitative properties preserved by perturbations.

In this discussion of the stability dogma, we have not focused on the question of which perturbations are to be allowed for a given system. Physical problems often have a symmetry or satisfy constraints that one wants to preserve in the perturbations being considered. For example, both the van der Pol and Duffing equations are preserved by rotation of the plane by $\pi$ and this symmetry reflects symmetry properties of the underlying physical systems giving rise to these equations. Therefore, discussions of structural stability in the context of individual examples requires that one specify the allowable perturbations to a given system. This issue will play a prominent role in our treatment of multiple bifurcations in Chapter 7.

## 5.5. Structurally Stable Attractors

Although structurally stable hyperbolic attractors are hard to find in physical examples, they still serve as an archetype and guide to thinking about the dynamics of other "strange attractors." In this section we shall review certain aspects of the theory of hyperbolic attractors and describe why the Hénon transformation of the plane, the bouncing ball map, and the forced Duffing equation should not be expected to have hyperbolic strange attractors. The theory of structural stability is generally presented in terms of diffeomorphisms of compact manifolds. We shall assume that all systems we work with are defined on a compact set $D$ of $\mathbb{R}^n$ with smooth boundary such that $f(D) \subset \text{int } D$. The "compact" theory will then work for such systems.

The problem of characterizing structurally stable attractors has not been fully solved unless one adopts a very strong definition of structural stability. See Robbin [1972] for a review of these questions. If $f$ is a diffeomorphism, then a more general type of perturbation of the dynamical system generated by $f$ than merely allowing a fixed perturbation of the generator $f$ is to allow a different perturbation on each application of $f$. More specifically, we can study the time-dependent dynamical process formed by selecting a collection of perturbations $f_i$, $i \in \mathbb{Z}$ of $f$ and forming the process $\tau = \{f_i\}_{i=-\infty}^{\infty}$ which sends $x$ at time $n$ to $f_n \circ \cdots \circ f_1(x)$ for $n > 0$ and $f_{-n}^{-1} \circ \cdots \circ f_{-1}^{-1}(x)$ for $n < 0$. Here $\tau$ can be regarded as a *time-dependent* perturbation of $f$. Franks [1974] defined $f$ to be *time-dependent stable* if, for each choice of a small time-dependent perturbations $\tau$, there is a homeomorphism $h$ such that $h$ maps trajectories of $f$ to trajectories of $\tau$.

The advantage of this strong definition of structural stability is that it readily allows one to prove that a time-dependent stable attractor has a hyperbolic structure. This result is closely related to the ideas of shadowing and pseudo-orbits which we introduced in Section 5.3. In particular, it is immediately clear that every pseudo-orbit of a dynamical system is a trajectory for a time-dependent perturbation. Therefore a time-dependent stable diffeomorphism has the shadowing property. As we saw earlier with the construction of Markov partitions, this is tantamount to the imposition on an indecomposable invariant set $\Lambda$ of the rectangular canonical coordinates associated with a hyperbolic structure. The converse result that a hyperbolic attractor $\Lambda$ is time-dependent stable is implied by Proposition 5.3.3: that $\Lambda$ does have the shadowing property for pseudo-orbits.

The question as to whether structural stability implies time-dependent stability is still open. However, assuming that this is the case, we shall focus our attention on hyperbolic attractors in the remainder of the section. These attractors have Markov partitions which provide a good symbolic description of them and yield strong conclusions about topological aspects of their dynamics, such as the existence of countably many periodic orbits of large periods and the existence of dense orbits. The hyperbolic structure can also be used to gain a better understanding of the attractor.

The topological study of hyperbolic invariant sets is a subject which has grown rapidly in the past few years, and the reader should consult Franks [1982] for an extensive review. Here we shall restrict our attention to attractors of three-dimensional flows and two-dimensional diffeomorphisms, to minimize the topological technicalities and make the most direct connections with our examples. For discrete systems, these attractors were studied very early in the development of the Smale program by Williams [1967]. His work on one-dimensional attractors is the prototype of work on the topology of Axiom A systems. It gives us a geometric picture of necessary conditions which must be met if a three-dimensional flow is to have a hyperbolic attractor.

If a two-dimensional diffeomorphism $f$ has a hyperbolic attractor $A$, then the hyperbolic structure of $A$ has one-dimensional stable and unstable manifolds. Transversal homoclinic points only exist for periodic orbits whose stable and unstable manifolds both have positive dimension. Moreover, the unstable manifolds of points of $A$ must belong entirely to $A$ because hyperbolic attractors have neighborhoods consisting of points which approach the attractor. Thus the only hyperbolic attractors which are possible for a diffeomorphism $f: \mathbb{R}^2 \to \mathbb{R}^2$ have topological* dimension one and are unions of the curves forming the unstable manifolds of individual points.

EXERCISE 5.5.1. Let $U$ be a neighborhood of a hyperbolic attractor $A$ and $\bigcup_{x_i \in A} W^u(x_i)$ be the union of the unstable manifolds of the points in $A$. Prove that $A = \mathrm{Cl}(\bigcup_{x_i \in A} W^u(x_i))$, as claimed. (Hint: First show that $\mathrm{Cl}(\bigcup_{x_i \in A} W^u(x_i)) \subseteq A$. Then consider the fate of points $y \in U$ under $f$.)

The above discussion suggests that, while points are mapped under $f$ towards the attractor, once they get close to it, their orbits separate locally at an exponential rate along the expanding direction. Since a hyperbolic attractor can contain no stable periodic orbits (or it would be decomposable), almost all pairs of orbits asymptotic to it are torn apart and ultimately display the statistical independence we have already met in the horseshoe and other examples. In this case, however, almost all orbits *never* settle down to asymptotically periodic behavior, though they do approach the closure of the set of unstable manifolds. The typical local structure in a hyperbolic attractor for a two-dimensional map is the product of a curve and a Cantor set.

We remark that on the torus $T^2$, there are *Anosov* diffeomorphisms $f: T^2 \to T^2$ defined by the requirement that *all* of $T^2$ constitutes a hyperbolic attractor. To study these, $T^2$ is regarded as the quotient of $\mathbb{R}^2$ by the integer lattice. The simplest such examples are defined via the projection of diffeomorphisms of $\mathbb{R}^2$ defined by $2 \times 2$ matrices $A$ with integer coefficients and determinant 1; e.g., $A = \begin{pmatrix} 2 & 1 \\ 1 & 1 \end{pmatrix}$. For these "linear" Anosov diffeomorphisms, the stable and unstable manifolds are the projections of lines parallel to the eigenvectors of $A$. All points of $T^2$ with rational coordinates are periodic because $A$ preserves the set of points whose coordinates are rational numbers with denominators dividing a given integer $k$ (cf. Section 1.4,

---

* Different notions of dimension, such as "capacity" and Hausdorff dimension, have been suggested in connection with the description of strange attractors. See Section 5.8 below.

Figure 1.4.3). All Anosov diffeomorphisms of $T^2$ are equivalent to one of these linear diffeomorphisms (Franks [1970]).

The theory of Williams gives a topological classification of one-dimensional hyperbolic attractors and a construction which is equivalent to following the symbolic dynamics of the attractor in the expanding direction. We now outline his ideas.

A neighborhood $U$ of a hyperbolic attractor $A$ is the union of pieces of stable manifolds of points of $A$. For a two-dimensional mapping, each stable manifold is one dimensional, so each component is an interval. If $f(U) \subset U$, then each component of a stable manifold in $U$ is mapped into another such component. We conclude that $f$ induces a well-defined map $\tilde{f}$ on the quotient space $B = B(U)$ defined by identifying $x$, $y \in U$ if $y$ is in the component of $W^s(x)$ containing $x$; that is, loosely speaking, we project along members of the stable foliation: cf. Figure 5.5.1.

The map $\tilde{f}: B \to B$ keeps track of the way in which components of $W^s(x) \cap U$ are mapped to components of $W^s(f(x)) \cap U$. Clearly $\tilde{f}$ is never $1:1$ and the quotient space $B = B(U)$ can have a certain amount of pathology, but it will be one dimensional because each small segment of an unstable manifold projects into $B$ in a $1$–$1$ way (Figure 5.5.1). By making a good choice of $U$ (e.g., insisting $U$ have smooth boundary whose points of tangency with stable manifolds are quadratic) and using closed neighborhoods, one can arrange that $B$ is a *branched manifold*. We do not need the full strength of the definition of branched manifolds; all we use here is the consequence

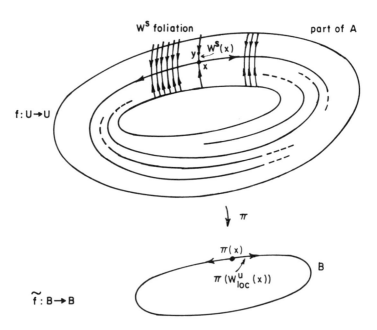

Figure 5.5.1 A stable foliation of $U$.

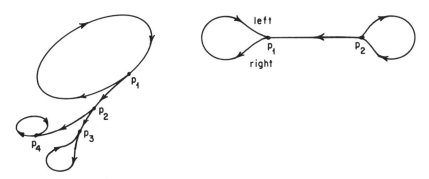

Figure 5.5.2. Some branched 1-manifolds with vertices $p_i$. Note that edges meet tangentially at a vertex (see the remark following Theorem 5.5.1.)

that $B$ is homeomorphic to a graph formed by joining a finite number of points with segments called edges. In addition, the edges meeting a vertex $p$ are partitioned into two classes which we shall call the right and left edges of $p$, cf. Figure 5.5.2.

If $A$ is indecomposable and connected, then the map $\tilde{f}: B \to B$ has the property that each interval $I$ contained in $B$ eventually maps onto $B$; i.e., there is an $n$ (depending on $I$) such that $\tilde{f}^n(I) = B$. Moreover, if the neighborhood $U$ of $A$ is chosen as the union of rectangles in a Markov partition, then $\tilde{f}$ maps vertices to vertices. There is a final condition on $\tilde{f}$ that it maps edges on opposite sides of a vertex $p$ into edges on opposite sides of $\tilde{f}(p)$.

Williams proved that the conditions above are sufficient to characterize a one-dimensional attractor (Williams [1967]):

**Theorem 5.5.1.** *Let $G$ be a graph whose edges at a vertex $p$ are partitioned into left and right so that both classes are nonempty and let $g: G \to G$ be a mapping satisfying:*

(1) *$g$ maps vertices to vertices, and $g$ maps edges on opposite sides of the vertex $p$ to edges on the opposite sides of $g(p)$.*

(2) *$g$ is locally 1–1 at points which are not vertices. If $g(p)$ is a vertex and $V$ a neighborhood of $p$ in $G$, then $g(V)$ contains points on both sides of $g(p)$.*

(3) *If $I$ is an interval in $G$, then there is an $n$ with $g^n(I) \supset G$.*

*Then there is a two-dimensional diffeomorphism $f$ such that $f$ has an indecomposable attractor $A$ and there is a neighborhood $U$ of $A$ with $G$ homeomorphic to $B(U)$ and $g$ topologically equivalent to $\tilde{f}: B \to B$.*

**Remark.** Not every graph $G$ can be mapped into the plane. If $G$ can be mapped into the plane with each edge a smooth curve and such that edges on opposite sides of a vertex meet the vertex tangentially from opposite sides, then $f$ can be chosen to be a planar diffeomorphism. We will give some examples below.

The proof of Williams' theorem involves an *inverse limit construction* which explicitly gives a model for $A$. Starting with the map $g: G \to G$, one considers the set $A_s$ of bi-infinite sequences $(\ldots x_{-n}, \ldots, x_0, \ldots, x_n)$ with the property that $g(x_i) = x_{i+1}$. The map $g$ induces a map $\bar{g}: A_s \to A_s$ defined by $\bar{g}(\mathbf{x}) = \mathbf{y}$ with $y_i = g(x_i) = x_{i+1}$. (The set $A_s$ can be given a topological structure in a natural way as an inverse limit of the sequence $\ldots G \underset{g}{\to} G \underset{g}{\to} G$.) Knowing $x_i$ determines $x_j$ for $j > i$, but not $x_j$ for all $j < i$ because $g$ is not 1–1. In fact there is an entire Cantor set of points of $A_s$ which correspond to each $x_i$. Williams [1974] proves that, if there is a hyperbolic attractor $A$ of $f$ with a neighborhood $U$ such that $\hat{f}: B \to B$ is topologically equivalent to $g: G \to G$, then $f|_A$ is topologically equivalent to the map $\bar{g}$ on $A_s$. He also gives a topological construction of an $f$ with this property.

The construction above is closely related to the symbolic dynamics for a hyperbolic attractor $A$. If $g: G \to G$ is a mapping of a graph which sends vertices to vertices, and is 1–1 on edges then we can label the edges with symbols $\{1, \ldots, n\}$ and form a transition matrix $A = [A_{ij}]$ by setting $A_{ij} = \{\# \text{ pts } x \text{ of edge } i \text{ with } g(x) = y \text{ for each } y \text{ of edge } j\}$. The *one-sided subshift of finite type* $\Sigma_r$ with transition matrix $[A_{ij}]$ is the set of (ordinary) sequences $\mathbf{x} = \{x_i\}_{i=0}^{\infty}$ with the property that $A_{x_i x_{i+1}} \neq 0$ for all $i \geq 0$. The shift map $\sigma$ is defined as before, but it is not 1–1 here. There is a symbol map $\phi: \Sigma_r \to G$ (which is 1–1 except on the inverse image of trajectories of vertices) which satisfies $\phi \circ \sigma = f \circ \phi$ as before. If we apply the inverse limit construction described above to $\Sigma_r$ and to $G$, then we recover the symbolic dynamics of $A_s$.

Examples of hyperbolic attractors for three-dimensional maps have been known for some time. In particular, an analysis of the *solenoid*, defined on a three-torus, can be found in Smale [1967]; also see Ruelle and Takens' [1971] paper on turbulence, in which they embedded a solenoid in the Poincaré map for a quadruply periodic flow on a four-torus. However, the simplest known example of a planar hyperbolic attractor is due to Plykin [1974]. A slight modification (cf. Newhouse [1980]) is illustrated in Figure 5.5.3. We take a compact subset $D \subset \mathbb{R}^2$ with three holes, as illustrated. In each hole we place a source and we define $f: D \to \mathbb{R}^2$ geometrically so that $f(D)$ lies inside the interior of $D$ as shown. The attractor is $A = \bigcap_{n \geq 0} f^n(D)$. Moreover, we can choose a smooth stable foliation $\mathscr{F}$ of $D$ so that, if $l \in \mathscr{F}$, then $f(l) \in \mathscr{F}$, as indicated. The leaves $l$ of $\mathscr{F}$ are just pieces of the stable manifolds $W^s(x_i)$ of points $x_i$ in $A$, as required. The branched manifold $B$ and graph $G$ associated with the problem is illustrated in Figure 5.5.4, and the map $g: G \to G$ acts as follows:

$$g: A \to B$$
$$g: B \to B + D + C - D - B,$$
$$g: C \to A, \tag{5.5.1}$$
$$g: D \to D - C - D.$$

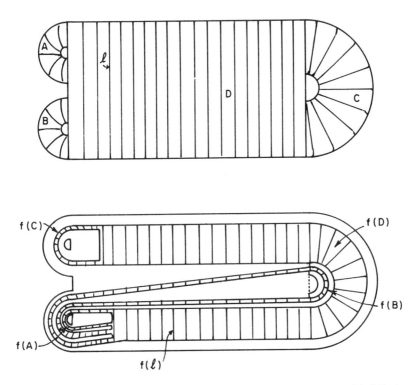

Figure 5.5.3. The Plykin attractor. (a) The disc with three holes $D$ and the stable foliation; (b) $f(D) \subset D$.

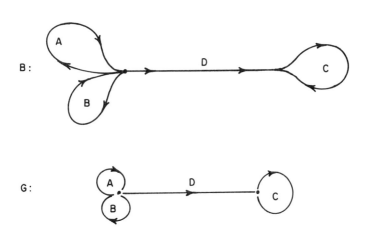

Figure 5.5.4. The branched manifold $B$ and graph $G$ for the Plykin attractor.

(Note that $g$ takes vertices to vertices.) Moreover, we have the transition matrix

$$(A_{ij}) = \begin{array}{c} \\ A \\ B \\ C \\ D \end{array} \begin{array}{cccc} A & B & C & D \\ \begin{bmatrix} 0 & 1 & 0 & 0 \\ 0 & 2 & 1 & 2 \\ 1 & 0 & 0 & 0 \\ 0 & 0 & 1 & 2 \end{bmatrix} \end{array}, \tag{5.5.2}$$

where, for example, $A_{44} = 2$, since every point in edge number 4 ($D$) has two images in edge $D$, and $A_{24} = 2$, since every point in edge 2 ($B$) has two images in edge $D$.

$f$ can be chosen such that, in addition to the topological properties above, $Df$ contracts uniformly on the tangent spaces $E^s_{x_i}$ to $W^s(x_i)$ and expands uniformly on the unstable spaces $E^u_{x_i}$ tangent to $W^u(x_i)$, as in Definition 5.2.6. Thus $A$ is a hyperbolic attractor, and since it contains horseshoes (directly from the construction) it meets our definition of a strange attractor. Locally $A$ is the product of a curve and a Cantor set.

The Plykin attractor has been incorporated into a map of the two-torus, corresponding to a Poincaré map for a flow on the three-torus, to show that coupling of three limit cycle oscillators can lead to chaotic motions (Newhouse, et al. [1978], cf. Newhouse [1980]).

EXERCISE 5.5.2. Try to construct a hyperbolic attractor with only two holes, starting with the graph $G$ and map of Figure 5.5.5. What goes wrong?

$$g: A \to A,$$
$$g: B \to A,$$
$$g: C \to C + B - C.$$

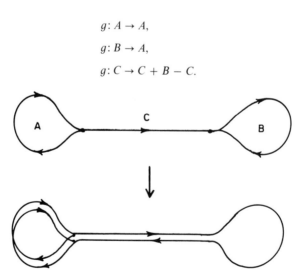

Figure 5.5.5. Exercise 5.5.1.

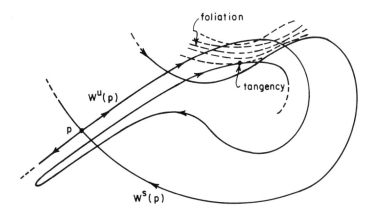

Figure 5.5.6. Nonexistence of a stable foliation in the Duffing system.

The topology of the Plykin example is already fairly complicated, but it does provide an "ideal" prototype for hyperbolic attractors which might occur in the Duffing and van der Pol equations. However, it seems unlikely that there is a foliation of stable manifolds in either of these cases which would always intersect an unstable manifold transversally. For example, consider Figure 5.5.6 (cf. Figure 2.2.8), and suppose that the attractor is $Cl(W^u(p))$. There should be a stable foliation (including $W^s(p)$), transverse to $W^u(p)$ everywhere, but, since $W^u(p)$ folds upon itself without apparently going around several holes in the Plykin attractor, any attempt to foliate $U$ by stable manifolds necessarily involves some members of the foliation which are tangent to $W^u(p)$. Thus the topological prerequisites for the existence of a hyperbolic attractor do not seem to be met in these examples.

Analyses of the bouncing ball map (2.4.4) and the Hénon map (Hénon [1976]) encounter similar problems: in both cases an attracting set $\bigcap_{n \geq 0} f^n(D)$ can be found, but a continuous stable foliation of $D$ cannot. However, Misiurewicz [1980] has shown that a piecewise linear version of the Hénon map due to Lozi [1978] *does* have a hyperbolic strange attractor. The map is

$$(x, y) \rightarrow (1 + y - a|x|, bx). \qquad (5.5.3)$$

Since the invariant manifolds for this map are piecewise linear, the tangencies of stable and unstable manifolds are replaced by corners. Apart from a point set of measure zero, all stable and unstable manifolds intersect transversally and hence hyperbolicity estimates can be carried through almost everywhere.

In the following section we discuss in more detail evidence concerning the existence of nonhyperbolic "strange" attractors in the Duffing and similar systems. In contrast, Section 5.7 is devoted to the Lorenz equations, where use is made of some aspects of continuous time systems which have been neglected in this chapter thus far.

## 5.6. One-Dimensional Evidence for Strange Attractors

Connected hyperbolic attractors have the property that the unstable manifold of a single point is dense in the attractor. For the forced Duffing and van der Pol equations, we can conjecture that there are parameter values for which the unstable manifold of a periodic orbit lies in an indecomposable invariant set. Such an invariant set would be an outstanding candidate for the strange attractor we seek. To discuss the state of affairs relative to this conjecture, we look at a family of diffeomorphisms of the plane which are defined directly, rather than through the solutions of a forced oscillator, and which share some of the features of the bouncing ball problem.

Hénon [1976] conducted a numerical study of the diffeomorphisms $F_{a,b} \colon \mathbb{R}^2 \to \mathbb{R}^2$ defined by

$$F_{a,b}(x, y) = (y, 1 + bx - ay^2). \tag{5.6.1}$$

Figure 5.6.1 reproduces from his study portions of the orbit of a single point* (cf. Figures 2.4.5 and 2.4.7 for the bouncing ball problem). To the extent that the points of this trajectory appear to fill out a one-dimensional curve, indistinguishable from the unstable manifold of the saddle point $\bar{x} = \bar{y} = (1/2a)(-(1 - b) + \sqrt{(1 - b)^2 + 4a})$, Hénon's computation represents numerical evidence for the existence of a strange attractor which contains the unstable manifolds of its periodic orbits. In particular, the iterates of Figure 5.6.1 appear to lie on a set which is locally the product of a curve and a Cantor set, and it has been proved that certain hyperbolic attractors, such as the Plykin example of Section 5.5, have precisely this local structure (Williams [1967, 1974]). However, as in the Duffing example, these unstable manifolds fold back upon themselves (rather than going around holes as in the Plykin example) and there is thus no hope that a foliation of stable manifolds transverse to these unstable curves can exist. So the attractor, if it indeed exists, is not hyperbolic.

There is an analogue of the problem concerning the existence of an attractor for Hénon's map which has been rigorously analyzed. This question involves one-dimensional mappings and can be interpreted as the singular limit of Hénon's mapping as $b \to 0$. The Jacobian derivative of the Hénon mapping $F_{a,b}$ has determinant $-b$. Thus $F_{a,b}$ contracts areas by a factor $|b|$. The limit $b = 0$ is singular and represents mappings which contract the entire plane onto a curve. The value of $F_{a,0}$ is independent of the $x$-coordinate of $(x, y)$, so that we can follow trajectories of $F_{a,0}$ in terms of the one-dimensional mappings

$$f_a(y) = 1 - ay^2. \tag{5.6.2}$$

---

* Hénon used the different, but equivalent form $(x, y) \to (1 + y - ax^2, bx)$.

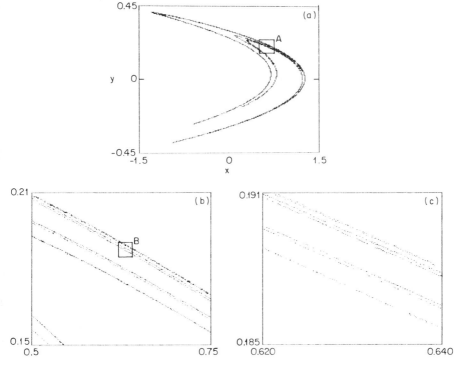

Figure 5.6.1. Orbits of the Hénon map $(x, y) \rightarrow (1 + y - ax^2, bx)$, for $a = 1.4, b = 0.3$. (a) 10,000 iterates of $(x_0, y_0) = (0.631, 0.189)$ (near a saddle point); (b) enlargement of the region $A$ in (a): $10^5$ iterates; (c) enlargement of the region $B$ in (b): $10^6$ iterates. (From Hénon [1976].)

For $0 < a \leq 2$, $f_a$ maps the interval

$$I = \left[ -\frac{1}{2a}(1 + \sqrt{1 + 4a}), \frac{1}{2a}(1 + \sqrt{1 + 4a}) \right]$$

into itself. The theory of one-dimensional mappings has been extensively developed in the last few years, enough so that the existence of nonhyperbolic strange attractors has been established in that context. In this section we shall describe the relevant theorems of Jakobson [1978, 1981] and Collet and Eckmann [1980], and speculate on their relationship to the Hénon mapping and forced oscillation examples. Going through this section, the reader might find it helpful to refer forward to Section 6.3, in which parametrized families of one-dimensional maps are studied.

Work on iteration of maps of the interval has concentrated on studying one-parameter families of maps $f : I \rightarrow I$, $I = [0, 1]$ which have a single critical point. The family of quadratic functions;

$$f_\mu(x) = \mu x(1 - x), \qquad 0 < \mu \leq 4, \tag{5.6.3}$$

is the prototypical example of the families which have been considered. The singular Hénon maps $f_a(y)$ of (5.6.2) may be transformed into this family easily. It turns out that there is a very rich structure for such families, and an extensive theory has developed since 1975. Some aspects of the theory are illuminating for the kinds of behavior which one might expect from a non-hyperbolic attractor. The recent monograph of Collet and Eckmann [1980] is a good introduction to the general area. Here we shall focus only upon those issues which are relevant to the strange attractor problem.

Consider a continuous map $f : I \to I$ on the unit interval $I = [0, 1]$, with $f(0) = f(1) = 0$, and having a single critical point $c$; i.e., $f$ is strictly increasing on $[0, c)$ and strictly decreasing on $(c, 1]$. If $f$ satisfies certain further restrictions: in particular if $f$ is three times differentiable, $f'(0) > 1$, and the *Schwarzian derivative*,

$$S(f) = \left[ \frac{f'''}{f'} - \frac{3}{2} \left( \frac{f''}{f'} \right)^2 \right]$$  (5.6.4)

is negative on $I - \{c\}$, one can conclude that almost all points of $I$ (with respect to Lesbesgue measure) have the same asymptotic behavior. We shall call such a map an NS map for short. Thus there may be a set of points of measure zero which can be decomposed into expanding, hyperbolic invariant sets, but the attracting portion of the nonwandering set will be indecomposable. The attractor might simply be a periodic orbit, but it is also possible that this attracting set contains expanding (=unstable) periodic orbits. In the context of one-dimensional mappings, such orbits play the rôle of the transversal homoclinic points of diffeomorphisms.

The three types of asymptotic behavior possible for an NS map can be labelled as *stable, critical,* and *strange attractor*. In the stable case, the only attractor for $f$ is a stable periodic orbit. By a theorem of Singer [1978] (cf. Collet and Eckmann [1980] and Section 6.3), at most one such orbit can exist and moreover, the trajectory of $c$ will be asymptotic to this stable orbit.

In the critical case, the attractor is a Cantor set $\Lambda$ (whose measure has not been determined) and the critical point $c$ lies in $\Lambda$. While there is an infinite attractor in the critical case, $\Lambda$ does not have the sensitivity to initial conditions, and positive Liapunov exponents,* which occur for a strange attractor. One way (but not the only way) in which the critical case occurs is as the limit of a sequence of period doubling flip bifurcations, where there is additional "universal" structure described in Section 6.8.

The strange attractor case occurs when the attractor $\Lambda$ itself is a finite union of closed intervals. One simple example of a strange attractor for the NS map occurs in the quadratic map.

$$f(x) = 4x(1 - x).$$  (5.6.5)

* See Section 5.8.

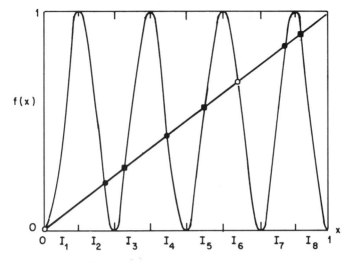

Figure 5.6.2. The map $f^3$, with the eight subintervals $I_j$ and the fixed points shown. Note that two of the latter (at $x = 0, x = \frac{3}{4}$) are fixed for $f$ and the remaining six form two orbits of period 3 for $f$. Note that $f^3(I_j) = I$ for all $j$.

This mapping sends $I = [0, 1]$ *onto* itself, so the trajectory of $c$ is given by $f(c) = 1$ and $f''(c) = 0, n \geq 2$. A recursive argument implies that there are $2^n$ subintervals of $I$ each of which are mapped onto $I$ by $f^n$. Each of these subintervals contains a fixed point of $f^n$. In Figure 5.6.2 we show $f^3$, which has $2^3 - 1 = 7$ critical points, and we indicate the eight intervals $I_1, \ldots, I_8$.

One can prove additionally that these periodic points are dense in $I$ and that the derivative of $f^n$ at each of its nonzero fixed points is $\pm 2^n$. Moreover, the measure with distribution $dx/(\pi\sqrt{x(1 - x)})$ is invariant and ergodic for the mapping $f$. These statement are easy consequences of the fact that $f$ is topologically conjugate to the piecewise linear map

$$\tilde{f}(\bar{x}) = \begin{cases} 2\bar{x}, & \bar{x} \in [0, \frac{1}{2}], \\ 2 - 2\bar{x}, & \bar{x} \in [\frac{1}{2}, 1], \end{cases} \tag{5.6.6}$$

via the homeomorphism

$$h(x) = \frac{2}{\pi} \arcsin \sqrt{x}, \tag{5.6.7}$$

(cf. Ulam and von Neumann [1947], also see Ruelle [1977]). The map $\tilde{f}$ of (5.6.6) leaves the measure $d\bar{x}$ invariant and one obtains the measure for $f$ from the transformation $h^{-1}$. Thus the entire interval $I$ behaves like a strange attractor for this particular example.

There is a simple topological criterion which, if satisfied, guarantees that an NS map will have a connected strange attractor $\Lambda \subset I$. The idea behind this criterion is that any subinterval of $\Lambda$ should have some iterate which

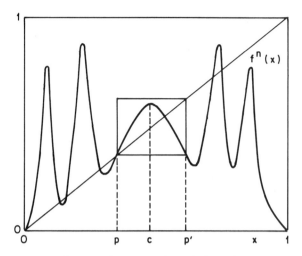

Figure 5.6.3. A central restrictive point.

eventually covers all of $\Lambda$. One can reduce the criterion to checking that certain special intervals have iterates which eventually contain all of $\Lambda$. A *central point p* for $f$ is defined to be a fixed point of some $f^n$ with the property that $f^n$ is monotone on the interval $[p, c]$. If $p' \neq p$ is the point with $f(p') = f(p)$, then the central point $p$ is *restrictive* if $f^n([p, p']) \subset [p, p']$. Geometrically, this means that the graph of $f^n$ does not leave the box depicted in Figure 5.6.3. In this situation iterates of this interval, which contains $c$, cannot expand to cover $I$. Guckenheimer [1979] proves that an NS map $f$ has a connected strange attractor if and only if there are no restrictive central points.

Return now to consider a one-parameter family of NS maps $f_\mu : I \to I$. If there are values $(\mu_0, \mu_1)$ of $\mu$ for which $f_{\mu_0}(c) = c$ and $f_{\mu_1}(c) = 1$, then we shall say that $f_\mu$ is a *full family* (cf. Section 6.3). Denote by $M_p$, $M_c$, and $M_s$ the sets of parameter values in $M = (\mu_0, \mu_1)$ giving rise to periodic, critical, and strange attractors, respectively. The set $M_p$ contains intervals, and we conjecture that it is dense in $M$ for many families $f$. The question of immediate concern is the size of $M_s$. Jakobson [1981] has proved a theorem which includes the result that the set $M_s$ has positive Lebesgue measure. Thus, if one picks a parameter value at random from $M$, there is a positive probability that the resulting map will have a strange attractor which is a finite union of intervals. (A similar result holds for diffeomorphisms of the circle with irrational rotation number; cf. Section 6.2.)

The arguments of Jakobson [1981] and Collet and Eckmann [1980] are long and intricate. In terms of the criterion we have stated above, one wants to study the set $M_s \subset M$ of parameter values for which the mappings $f_\mu$ have no restrictive central points. Examining $f_\mu^n$ as $\mu$ varies, those values of $\mu$ for which there is a restrictive central point of period $n$ are to be discarded.

Proceeding inductively, one hopes to eliminate a smaller *proportion* of the remaining parameter values with larger $n$, so that $M_s$ will eventually be described as a Cantor set of positive measure. (Recall the Cantor set constructed in Section 5.4, and see Section 6.2 below.) There are several different types of estimates which are required to make the argument work. These involve three different aspects of the mapping:

(1) The size of $(f_\mu^n)''(c)$;
(2) the variation of $f_\mu^n(c)$ with $\mu$; and
(3) the extent to which $f_\mu^n$ differs from a quadratic mapping on a subinterval $[p, p']$ where $p$ is a central point of period $n$.

One would like to apply this theory to the Hénon mapping when $|b| \in (0, 1)$ is fixed, and thus obtain the existence of strange attractors in this case. One could argue that the one-dimensional mapping gives a good approximation to the behavior of the Hénon mapping along its unstable manifolds and that there should be a large set of $a$'s which yield strange attractors. Unfortunately, things are not so simple, and the one-dimensional theory does not carry over directly to this new situation, so that the existence of strange attractors for the Hénon mapping remains an open and difficult question. In Chapter 6 we shall see some qualitative differences between the bifurcations of the one-dimensional mappings and the Hénon mapping. Until the issue is resolved, we are left with only faith to support us in the belief that the orbits which appear in Hénon's computations are asymptotic to an attractor and not perhaps merely to an attracting set containing stable periodic motions of arbitrarily long periods with vanishingly small domains of attraction.

## 5.7. The Geometric Lorenz Attractor

The issues discussed in Section 5.5 involving the tangency of stable and unstable manifolds in an attractor do not arise for a class of attractors modelled after numerical studies of the Lorenz system (2.3.1). In this section, we shall consider these geometrically defined examples using symbolic dynamics and the inverse limit construction of Williams [1967, 1974]. These attractors are unusual in that they contain a (saddle) equilibrium point of the flow which introduces a discontinuity into the Poincaré map used to pass from a continuous to a discrete time system. This discontinuity allows separation of nearby trajectories to take place without the bending and folding present in the Hénon map or the complicated topology of the Plykin example.

Recall our description of the Lorenz system in Section 2.3. For the parameter values $\sigma = 10$, $\beta = \frac{8}{3}$, and $\rho \cong 28$, one observes numerically the strange attractor depicted in Figure 2.3.2. We shall give a geometric description of a flow that is based upon an examination of the return map $F$ of the (nonlinear) "rectangle" $\Sigma$ contained in the plane $z = \rho - 1$ of Figure 2.3.2. The rectangle $\Sigma$ has opposite sides which pass through the equilibrium points $q^-, q^+$; and $\dot{z} < 0$ at all points of the interior of $\Sigma$, so that $\Sigma$ is a cross section for the flow. The equilibria $q^-$ and $q^+$ are each saddle points with one-dimensional stable manifolds $W^s(q^{\pm})$, parts of which form opposite sides of the cross-section $\Sigma$ in our geometric model. The unstable eigenvalues of $q^-$ and $q^+$ are complex. Since $\dot{z} < 0$ on the interior of $\Sigma$, all of the trajectories pass down through $\Sigma$. Most of them spiral around either $q^-$ or $q^+$ and return to $\Sigma$, but there must be a separating boundary, $D$, between those which spiral around $q^-$ and those which spiral around $q^+$. This boundary is assumed to be in the stable manifold of a third equilibrium point $p$ which lies below $\Sigma$; see Figure 5.7.1.

To describe the geometric Lorenz attractor from the flow depicted in Figure 5.7.1 one makes four additional assumptions about the flow. The first of these assumptions is that the eigenvalues $\lambda_1, \lambda_2, \lambda_3$ of $p$ satisfy the condition $0 < -\lambda_1 < \lambda_2 < -\lambda_3$ where $\lambda_1$ is the eigenvalue of the $z$-axis, which is assumed to be invariant under the flow. The second assumption is that there is a family $\mathscr{F}$ of curves in $\Sigma$ which contains $D$ and has the property that $\mathscr{F}$ is invariant under the return map $F$ of $\Sigma$. This means that if $\gamma \subset \mathscr{F}$, then $F(\gamma)$ is also contained in an element of $\mathscr{F}$, provided that $F$ is defined on $\gamma$ ($F$ is not defined on $D$). The family $\mathscr{F}$ is part of a *strong stable foliation* for the flow which is defined in a neighborhood of the attractor (Robinson [1981a]).

The third assumption one makes is that all points of the interior of $\Sigma - D$ do return to $\Sigma$, and that the return map $F$ is "sufficiently" expanding (cf. property (3) below) in the direction transverse to the family of curves $\mathscr{F}$.

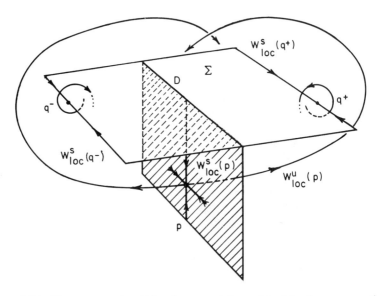

Figure 5.7.1. The cross section $\Sigma$ for the geometric model of the Lorenz equations.

Finally, one assumes that the flow is symmetric with respect to rotation by $\pi$ around the $z$-axis.

These "geometric" assumptions have not all been verified for the differential equations (2.3.1), but in principle they can be verified by numerical methods. Some progress has been made in this direction (cf. Sinai and Vul [1981], Shub [1982]).

Expressed analytically, these assumptions mean that there is a system of coordinates $(u, v)$ on $\Sigma$ such that $F$ has the following properties:

(1) The curves in the family $\mathscr{F}$ are given by $u = $ constant; $D$ is the set with $u = 0$.
(2) There are functions $f$ and $g$ such that $F$ has the form $F(u, v) = (f(u), g(u, v))$ for $u \neq 0$ and $F(-u, -v) = -F(u, v)$.
(3) $f'(u) > \sqrt{2}$ for $u \neq 0$ and $f'(u) \to \infty$ as $u \to 0$.
(4) $0 < \partial g / \partial v < c < 1$ for $u \neq 0$ and $\partial g / \partial v \to 0$ as $u \to 0$.

The image of $F$ in these coordinates is depicted in Figure 5.7.2. Here $\Sigma_+ = \{(u, v) | u > 0\}$ and $\Sigma_- = \{(u, v) | u < 0\}$.

Apart from the fact that $F$ is not defined on $D$, conditions (3) and (4) on $F$ imply that there is a hyperbolic structure. In particular, for sufficiently small angles $\alpha$, the $\alpha$-sectors centered on lines parallel to the $v$-axis around each point will satisfy Moser's conditions (cf. Section 5.2). Moreover, only a countable union of vertical curves in $\Sigma$ have $F$ trajectories which terminate on $D$, while all the other $F$ trajectories remain inside $\Sigma$. Therefore, any invariant set of $F$ will have a suitably defined hyperbolic structure and be attracting. The points of such an invariant set can be characterized by symbol sequences relative to the partition $\{\Sigma^-, \Sigma^+\}$ of $\Sigma$. Nonetheless, the topology of the geometric Lorenz attractor is considerably more complicated than the topology of the horseshoe, as we describe below.

The limit $F(u, v)$ is independent of $v$ as $u \to 0^+$ or $u \to 0^-$ because $\partial F / \partial v \to 0$. We denote $\lim_{u \to 0^-} F(u, v) = (r^+, t^+)$ and $\lim_{u \to 0^+} F(u, v) = (r^-, t^-)$ and call attention to the reversal of $+$ and $-$ in these formulae. In view of property (4), the vertical strip $V$ of $\Sigma$ defined by $r^- \leq u \leq r^+$ is mapped into itself,

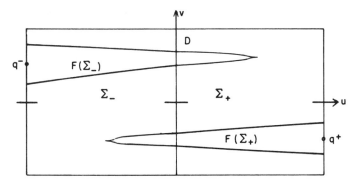

Figure 5.7.2. $\Sigma_\pm$ and $F(\Sigma_\pm)$.

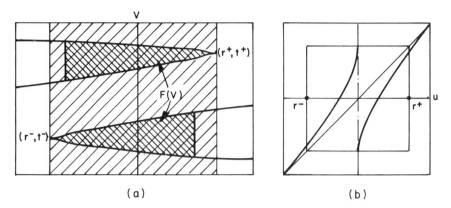

Figure 5.7.3. More on the maps $F$ and $f$. (a) $F(V) \subset V$; (b) The graph $f(u)$.

apart from $D$, where $F$ is not defined. Thus all points of the interior of $\Sigma$ have trajectories which eventually enter $V$ and then remain there. This is readily seen from the graph of the function $f$. See Figure 5.7.3. Henceforth, we restrict our attention to $V$. We assert that $A = \bigcap_{n \geq 0} F^n(V)$ is an attractor for the map $F$. Clearly, all points of $V$ either tend to $A$ or have trajectories that end on $D$, where $F$ is not defined. Consider now a point $x \in A$ and a rectangular neighborhood $U$ of $x$. We shall prove that the images of $U$ are dense in $A$. This is done in two steps.

We first show that there is an $n > 0$ such that one component of $F^n(U)$ extends horizontally across $V$. It is at this point that we use the hypothesis that $f' > \sqrt{2}$. Let $l(I)$ denote the length of the interval $I$ and let $I \subset (r_-, r_+)$ be an interval with $0 \notin I$, and consider $f^2(I)$. Since $0 \notin I$, $f(I)$ is connected. If $0 \notin f(I)$, then replace $I$ by $f(I)$ which satisfies $l(f(I))/l(I) > \sqrt{2}$ and continue to iterate. If $0 \in f(I)$, then $f^2(I)$ has two components. One of these components must be longer than $I$ because $(f^2)' > 2$ from property (3) and the chain rule. Moreover, if $0 \in f(I)$, then the components of $f^2(I)$ have endpoints at $r_-$ and $r_+$. Thus, $0 \in f(I)$ implies that either $0 \notin f^2(I)$ or $f^2(I)$ contains one of the intervals $(r_-, 0)$ or $(0, r_+)$. If $(r_-, 0)$ or $(0, r_+)$ is in $f^2(I)$, then $f^4(I) = (r_-, r_+)$. (Note that $f' > \sqrt{2}$ and $f(-u) = -f(u)$ imply that $f^2(r_-) < 0$ and $f^2(r_+) > 0$.) If $0 \in f(I)$ and $0 \notin f^2(I)$, then we replace $I$ by the longer component of $f^2(I)$ and continue the argument. Since our interval has finite length, the process of iterating the longest component in the image of an interval must eventually produce an integer $n$ with $f^n(I) = (r_-, r_+)$.

EXERCISE 5.7.1. Construct a piecewise linear map $f$ on $I = [-1, 1]$ with $f' \leq \sqrt{2}$ such that there are subintervals $J \in [-1, 1]$, for which $f^n(J)$ never covers $I$ (suggested by J. Sandefur).

The second step in our argument is to pick any point $s \in \bigcap_{n \geq 0} F^n(V)$. Then, given $\varepsilon > 0$, we shall locate a point of $U$ whose trajectory passes within distance $\varepsilon$ of $s$. Properties (2) and (4) of the map $F$ imply that the distance from $F^n(x, y_1)$ to $F^n(x, y_2)$ decreases exponentially, with

$$d(F^n(x, y_1), F^n(x, y_2)) < c^n|y_1 - y_2|. \qquad (5.7.1)$$

Given any $\varepsilon > 0$, we can therefore find an $m$ such that

$$d(F^m(x, y_1), F^m(x, y_2)) < \varepsilon \qquad (5.7.2)$$

for any $(x, y_1)$, $(x, y_2) \in \Sigma$. Since $s \in \bigcap_{n \geq 0} F^n(V)$, there is a point $(u, v) \in \Sigma$ such that $F^m(u, v) = s$. Now the first (expansion) part of the argument above gives an $n$, a $w$, and a point $(x, y)$ of $U$ such that $F^n(x, y) = (u, w)$. We then have that $F^{m+n}(x, y)$ is within distance $\varepsilon$ of the point $F^m(u, v) = s$. This proves that the trajectories originating in $U$ form a dense set in $A$.

EXERCISE 5.7.2. Show that $A$ has a dense orbit.

We conclude that $A$ is an attractor. All points of $V$ approach $A$ at an exponential rate (with uniform bounds). Just as for the horseshoe, the points of $A$ whose trajectories do not intersect $D$ can be characterized by their symbol sequences $\mathbf{a} = \{a_i\}_{i=-\infty}^{\infty}$ with respect to the partition of $V - D$ into its two components. The horizontal coordinates of points are determined by the $a_i$ for $i \geq 0$ and the vertical coordinates are determined by the $a_i$ for $i < 0$. Unlike the horseshoe, however, not all symbol sequences $\mathbf{a}$ are associated to points of $A$. For example, since $f^k(r_-) > 0$ for some $k$, there will be an integer $k$ such that no point of $\mathbf{a}$ has a string of $k$ successive 0's in its symbol sequence. It may even happen that the set of symbol sequences which occur for points of $A$ do not form a subshift of finite type. We postpone further consideration of these issues to Section 6.4 where the bifurcations of the Lorenz system are discussed.

To describe the topology of $A$ we still have to consider those trajectories which end in $D$, and especially those which originate at the points $(r^{\mp}, t^{\mp}) = \lim_{u \to 0^{\pm}} F(u, v)$ on the boundary of $V$. The principal effect of the trajectories which end in $D$ is to provide the "glue" which connects points of $A$ whose symbol sequences agree for $i < 0$. This can be more easily seen by first considering a somewhat different map $G: V - D \to V$ which is piecewise linear in $v$. Define $G$ by requiring that

$$G(u, v) = (f(u), h_u(v)),$$

with

$$
\begin{aligned}
h_u(v) &= \alpha v - \beta, \qquad u > 0, \\
h_u(v) &= \alpha v + \beta, \qquad u < 0,
\end{aligned}
\qquad (5.7.3)
$$

for suitable $0 < \alpha < \frac{1}{2}$ and $\beta$, and $\beta$ chosen so that $G$ is 1–1. See Figure 5.7.4.

Partitioning $V$ into $V_0 = \{(u, v)|r^- \leq u < 0\}$ and $V_1 = \{(u, v)|0 < v \leq r^+\}$ and exploiting the properties of $f$ established above, we deduce that $G^n(V)$

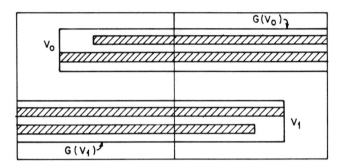

Figure 5.7.4. The piecewise linear map $G$. The four shaded strips are $G^2(V)$.

consists of a certain number of rectangles (which need not each extend across $V_0$ and $V_1$). Moreover, $\Gamma = \bigcap_{n \geq 0} G^n(V)$ will be an attractor of $V$ and will consist of horizontal segments.

EXERCISE 5.7.3. Prove inductively that each horizontal line which intersects $G^n(V)$, intersects $G^n(V)$ in a segment.

Relative to the partition of $V$ by $V_0$, $V_1$, the points which do not have unique symbol sequences are those whose trajectories terminate on $D$. If $(u_1, v)$ and $(u_2, v)$ have symbol sequences $\mathbf{a}$ and $\mathbf{b}$ for $G$ such that $a_i = b_i$ for $i < n$ but $a_n \neq b_n$, then there is a point $(u_3, v)$ on the segment joining $(u_1, v)$ and $(u_2, v)$ such that $G^n(u_3, v) \in D$.

EXERCISE 5.7.4. State and prove properties of $F$ analogous to those deduced above for $G$.

The principal qualitative difference between the attractor $\Gamma$ of $G$ and the attractor $A$ of $F$ comes from their vertical "ends." Topologically, $A$ can be obtained from $\Gamma$ by pinching together vertically all of the points which lie in the image of a vertical segment $\{G^n(u, v) \mid u = r^{\pm}\}$. Thus a rough picture of $A$ is that it consists of an uncountable number of segments, each transverse to the vertical direction in $V$. These segments of $\Lambda$ are tied together in bundles at their ends in a way which is determined by the trajectories of the points $r^{\pm}$ for the one-dimensional mapping $f$.

Finally, one wants to obtain a picture of the attractor of the geometric Lorenz flow from $A$. The procedure is to build a suspended flow which takes into account the presence of the singular point $p$ which is contained in the attractor. Consider the solid volume depicted in Figure 5.7.5, whose upper surface is $V$. We define a linear flow on this solid with the appropriate eigenvalues at $p$ and with $D$ contained in the stable manifold of $p$. The flow is tangent to the curved surfaces of the solid and to the bottom segment. On the front and back surfaces, the flow is into the solid while trajectories emerge from the vertical ends. These emergent trajectories are continued around so that $F$ describes the return map for $V$. With this description of the flow, the geometric Lorenz attractor is the union of trajectories passing through

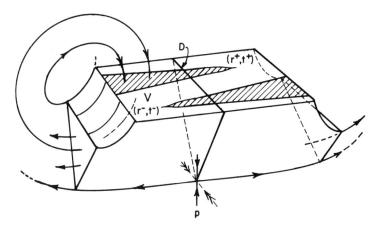

Figure 5.7.5. The geometric Lorenz flow.

$A \subset V$ together with the point $p$. Locally, the attractor appears to be a *Cantor book*: a family of surfaces parametrized by a Cantor set which are sewn together along a curve. This curve (the *spine* of the book) is the unstable manifold of $p$, see Figure 5.7.6.

We end this section by making the observations at the end of Section 2.3, concerning structural stability, more precise. We note that the spine of the geometric Lorenz attractor must be preserved by homeomorphisms. Since the preimages $f^{-n}(0)$ are dense in $(r^-, r^+)$, by arbitrarily small adjustments in the mapping $f$, we can arrange that the $f$ trajectories of the points $r^{\pm}$ terminate at 0. When this happens, there are homoclinic trajectories inside the attractor, and the spine $W^u(p)$ is homeomorphic to a figure eight. In contrast, if the $f$ trajectories of $r^{\pm}$ do not terminate at 0, then the spine is not compact and is simply connected. Thus the geometric Lorenz attractor is not structurally stable. Those attractors which have compact spines and those which do not each occur densely in some open set of vector fields defining these flows.

We will return to the Lorenz equation to consider the sequence of bifurcations in which the attractor is created in Section 6.4 below.

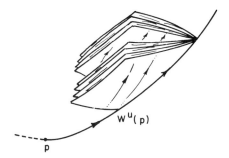

Figure 5.7.6. The local structure of the geometric Lorenz attractor.

## 5.8. Statistical Properties: Dimension, Entropy, and Liapunov Exponents

The previous sections leave one with the impression that a complete qualitative description of the dynamics of many nonlinear oscillators is an overwhelming task beyond our capabilities. In this final section we adopt a probabilistic point of view which asks for less and therefore is applicable in very general settings. We allow ourselves to discard exceptional trajectories in looking for typical properties within a system. A quantitative description of what is exceptional and what is typical forces us to rely heavily on the concept of a *measure*. This section is therefore more abstract than the remainder of the book and requires more mathematical background. Moreover, the subject introduced here, *smooth ergodic theory*, has sufficient scope that only a brief introduction is possible. It is undergoing rapid development and attempts to apply it to specific examples of the sort considered in this book are promising. We remark that Chapters 6 and 7 may be read independently of this section. For background information on ergodic theory from the view point of dynamical systems, see the book of Cornfeld *et al.* [1982] or the shorter notes of Sinai [1976].

Let $f: X \to X$, be a mapping of the metric space $X$. An *invariant measure* $\mu$ for $f$ is a (Borel) measure with the property that $\mu(f^{-1}(A)) = \mu(A)$ for all (Borel) sets $A$. Invariant measures are the mathematical objects which describe quantitative properties of the dynamical system generated by $f$. A *probability measure* $\mu$ is one for which $\mu(X) = 1$.

Here are some examples of invariant probability measures:

(a) If $p$ is a periodic point of period $n$ for $f$, the measure which assigns mass $1/n$ to each point of the orbit of $p$ is an invariant probability measure.
(b) If $f: [0, 1] \to [0, 1]$ is the map $f(x) = 2x$ (mod 1), then Lebesgue measure $dx$ is an invariant probability measure.
(c) If $f: [0, 1] \to [0, 1]$ is defined by $f(x) = 4x(1 - x)$, then the measure $dx/(\pi\sqrt{x(1 - x)})$ is invariant (cf. Section 5.6).

EXERCISE 5.8.1. Show that the "tent" map

$$f(x) = \begin{cases} 2x, & x \in [0, \tfrac{1}{2}], \\ 2 - 2x, & x \in [\tfrac{1}{2}, 1], \end{cases}$$

defined on $I = [0, 1]$ preserves Lebesgue measure.

EXERCISE 5.8.2. Show that if $f: \mathbb{R}^n \to \mathbb{R}^n$ is smooth and finitely many to one and $\mu = h(x)dx$ is a smooth invariant measure, then $\mu$ is invariant if and only if $h(x) = \sum_{f(y)=x} h(y)/|\det[Df(y)]|$. (In particular, Lebesgue measure is invariant for a diffeomorphism $f$ if and only if $|\det(Df)| \equiv 1$ (area preservation).)

We next examine the horseshoe in some detail to illustrate how invariant measures can be used to describe quantitative features of a dynamical system. For the purposes of computation, it is convenient to use the symbolic description of the horseshoe as a shift and to describe invariant measures for the shift map $\sigma: \Sigma \to \Sigma$, with $\Sigma$ the symbol space of bi-infinite sequences of $\{1, 2\}$. One class of invariant measures on $\Sigma$ is that associated with viewing $\Sigma$ as an infinite sequence of *unfair* coin tosses. Suppose that 1 has probability $p$ and 2 has probability $q = 1 - p$. There is an invariant measure $\mu = \mu_p$ on $\Sigma$ which is determined by defining $\mu$ on the sets $\Sigma_1 = \{\mathbf{a} \mid a_0 = 1\}$ and $\Sigma_2 = \{\mathbf{a} \mid a_0 = 2\}$ and requiring probabilitistic independence of the coin tosses. Thus $\Sigma_i$ corresponds to the 0th outcome being $i$ and we set $\mu(\Sigma_1) = p$ and $\mu(\Sigma_2) = q$. Invariance of $\mu$ implies that $\mu(\sigma^{-j}(\Sigma_i)) = \mu(\Sigma_i)$ for all $j \in \mathbb{Z}$. Here $\sigma^j(\Sigma_i)$ is the set of sequences in which the $j$th outcome is $i$. *Independence* is the statement that, for any sequence $b \in \Sigma$ and $k, l \in \mathbb{Z}$

$$\mu\left( \bigcap_{j=k}^{l} \sigma^j(\Sigma_{b_j}) \right) = \prod_{j=k}^{l} \mu(\sigma^j(\Sigma_{b_j})) = p^\alpha (1 - p)^\beta, \tag{5.8.1}$$

where $\alpha$ is the number of occurrences of 1 in $\{b_j\}_{j=k}^{l}$ and $\beta$ is the number of occurrences of 2 in $\{b_j\}_{j=k}^{l}$. Probabilistically, the joint probability of several outcomes of different tosses is the product of the probabilities of each individual toss.

The measures $\mu_p$ have properties which are quite different from one another. The central limit theorem implies that almost all sequences for $\mu_p$ have the property that, as the length of a finite block increases, the proportions of 0's and 1's in the finite block tend to $p$ and $(1 - p)$, respectively. Consequently, the measures $\mu_p$ are mutually singular: sets of full measure for $\mu_p$ have zero measures for $\mu_{p'}$ if $p' \neq p$. Nonetheless, these measures are *ergodic*: there are no sets $E$ such that both $E$ and its complement are invariant for $\sigma$ and have positive measure. The *entropies* of the $\mu_p$ also differ. The intuitive definition of the entropy of $\mu$ is the maximum expected gain of information (relative to a partition) from knowing one additional point in a typical trajectory. The precise definition is given below. Using the probabilistic interpretation of $\mu_p$ as independent coin tosses, one obtains the same amount of information from each additional point in a trajectory, and this amount is $-(p \log p + (1 - p) \log(1 - p))$.

To see that all of these measures $\mu_p$ have some relevance for dynamics, consider the following piecewise linear modification of the horseshoe mapping $f: S \to \mathbb{R}^2$. In the modified mapping, we assume that the vertical expansion factor $\gamma$ is different on the two components of $f^{-1}(S) \cap S$ (cf. Section 5.1); the horizontal contraction factor $\lambda$ remains the same. If the two vertical factors are $\gamma_1$ and $\gamma_2$, then the heights of the corresponding horizontal strips $H_1$ and $H_2$ are $\gamma_1^{-1}$ and $\gamma_2^{-1}$. See Figure 5.8.1. If we specify a finite symbol sequence $a_{-m} \ldots a_{-1} a_0 a_1 \ldots a_n$, then the area of the rectangle

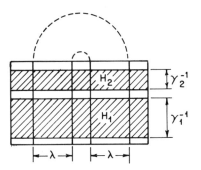

Figure 5.8.1. The uneven horseshoe.

$R_\mathbf{a} = \bigcap_{i=-m}^{n} f^{-i}(H_{a_i})$ is $\lambda^{+m}\gamma_1^{-k}\gamma_2^{-(n+1-k)}$ where $k$ is the number of occurrences of 1 among $a_0, \ldots, a_n$. The total area contained in these rectangles is $\lambda^m(\gamma_1^{-1} + \gamma_2^{-1})^{n+1}$ and the proportion in the rectangle $R_\mathbf{a}$ is

$$\frac{\gamma_1^{-k}\gamma_2^{-(n+1-k)}}{(\gamma_1^{-1} + \gamma_2^{-1})^{n+1}} = \frac{\gamma_1^{n+1-k}\gamma_2^{k}}{(\gamma_1 + \gamma_2)^{n+1}} = \left(\frac{\gamma_1}{\gamma_1 + \gamma_2}\right)^{n+1-k} \left(\frac{\gamma_2}{\gamma_1 + \gamma_2}\right)^{k}.$$

The measure described above with $p = \gamma_2/(\gamma_1 + \gamma_2)$ therefore assigns a mass to those sequences in $R_\mathbf{a}$ which is the proportion of the area of $R_\mathbf{a}$ to the area contained in $\bigcup R_\mathbf{b}$, the union being taken over all finite sequences $\mathbf{b} = \{b_i\}_{i=-m}^{n}$.

Using Markov partitions, the invariant measures on any hyperbolic invariant set can be studied in terms of the invariant measures on subshifts of finite type. As we have just seen for the full shift, subshifts of finite-type support many invariant measures which differ markedly from one another. However, there is one special invariant measure for a hyperbolic attractor $A$ which is singled out by its relationship with Lebesgue measure through the dynamics of the system. To describe it, we begin with the idea of the typical behavior of trajectories which approach $A$. Here, "typical" is interpreted as meaning "for almost all initial points with respect to Lebesgue measure." One way to measure the asymptotic behavior of the typical trajectory is to look at its time averages.

**Definition 5.8.1.** Let $f: \mathbb{R}^n \to \mathbb{R}^n$ define a discrete dynamical system and let $g: \mathbb{R}^n \to \mathbb{R}$ be a real valued function. The *time average* of $g$ on the (forward) trajectory of $x$ is $\lim_{N \to \infty} (1/N) \sum_{i=0}^{N-1} g(f^i(x))$, if the limit exists. For a continuous flow $\phi_t: \mathbb{R}^n \to \mathbb{R}^n$, the time average of $g$ on the (forward) trajectory of $x$ is $\lim_{T \to \infty} (1/T) \int_0^T g(\phi_t(x)) \, dt$.

We then have (cf. Bowen and Ruelle [1975]):

**Theorem 5.8.1** (Sinai–Ruelle–Bowen). *Let $\phi_t: \mathbb{R}^n \to \mathbb{R}^n$ be a $C^2$ flow with a hyperbolic attractor $A$. Then there is a unique measure $\mu$ supported on $A$ with* *the following property: If U is a neighborhood of A such that $\phi_t(U) \subset U$ for all $t \geq t_0$, $A = \bigcap_{t \geq 0} \phi_t(U)$, and g is a continuous function, then for almost all $x \in U$ (with respect to Lebesgue measure),*

$$\lim_{T \to \infty} \frac{1}{T} \int_0^T g(\phi_t(x)) \, dt = \int_A g \, d\mu. \tag{5.8.2}$$

If $f: \mathbb{R}^n \to \mathbb{R}^n$ defines a discrete system, then the analogous theorem holds with

$$\lim_{N \to \infty} \frac{1}{N} \sum_{i=0}^{N-1} g(f^i(x)) = \int_A g \, d\mu. \tag{5.8.3}$$

A measure satisfying (5.8.2) is called an *asymptotic* measure.

The measure $\mu$ of Theorem 5.8.1 has properties analogous to the modified horseshoe example described above. Fix an $\varepsilon > 0$ small and a Markov partition for the attractor $A$ of a discrete system $f: \mathbb{R}^n \to \mathbb{R}^n$. For a set $R \subset A$ define $V(R)$ to be the $n$-dimensional Lebesgue measure of $(\bigcup_{x \in R} W_\varepsilon^s(x))$. The measure assigned to the set $R_a$ consisting of points with a given finite symbol sequence $a_{-m}, \ldots, a_{-1}a_0, \ldots, a_n$ is approximately proportional to the relative volume $V(R_a)$ divided by $V(A)$.

The volumes $V(R_a)$ are closely related to the *Liapunov exponents* of the flow, which we define below. The relative volume $V(R_a)/V(A)$ is determined by the rate of expansion of tangent vectors along the unstable manifold of $A$. Very roughly, $f^{+n}(R_a)$ is mapped across the $a_n$ element of the Markov partition in the unstable directions. The relative size $V(R_a)/V(R_{a_n})$ is therefore approximated by $\det(Df^n | W^u)$, the derivative being evaluated at a point of $R_a$. As $n \to \infty$, the maps $Df^n$ on the unstable subspaces are growing exponentially. The rate of stretching may be different in different directions. The definition of Liapunov exponents formalizes the concept of these stretching rates:

**Definition 5.8.2.** Let $f: \mathbb{R}^n \to \mathbb{R}^n$ define a discrete dynamical system and select a point $x \in \mathbb{R}^n$. Suppose that there are subspaces $V_i^{(1)} \supset V_i^{(2)} \supset \cdots \supset V_i^{(n)}$ in the tangent space at $f^i(x)$ and numbers $\mu_1 \geq \mu_2 \geq \cdots \geq \mu_n$ with the properties that

(1) $Df(V_i^{(j)}) = V_{i+1}^{(j)}$.
(2) $\dim V_i^{(j)} = n + 1 - j$.
(3) $\lim_{N \to \infty} (1/N) \ln \| \sqrt{(Df^N)^*(Df^N)} \cdot v \| = \mu_j$ for all $v \in V_0^{(j)} - V_0^{(j+1)}$,

where $(Df^N)^*$ is the transpose of $Df^N$.
Then the $\mu_j$ are called the *Liapunov exponents* of $f$ at $x$.

If $x = \bar{x}$ is a fixed point, then the subspaces $V_i^{(j)} = V^{(j)}$ do not depend upon $i$ and are simply the eigenspaces associated with (sets of) eigenvalues of

$Df(\bar{x})$. The Liapunov exponents are the logarithms of the moduli of these eigenvalues.

For example, if $Df(\bar{x}) = \begin{bmatrix} \lambda_1 & 0 \\ 0 & \lambda_2 \end{bmatrix}$ with $\lambda_1 > \lambda_2$, then we pick $V^{(1)} =$ span$\{(1, 0), (0, 1)\} = \mathbb{R}^2$ and $V^{(2)} = $ span$\{(0, 1)\}$. Properties (1) and (2) are immediately satisfied and we have

$$\mu_1 = \lim_{N \to \infty} \frac{1}{N} \left\{ \ln \left\| \begin{bmatrix} \lambda_1^N & 0 \\ 0 & \lambda_2^N \end{bmatrix} \begin{pmatrix} v_1 \\ v_2 \end{pmatrix} \right\| \right\} \quad \text{for } v \in V^{(1)} - V^{(2)}.$$

$$= \lim_{N \to \infty} \frac{1}{N} \ln |\lambda_1^N| = \ln |\lambda_1|,$$

$$\mu_2 = \lim_{N \to \infty} \frac{1}{N} \left\{ \ln \left\| \begin{bmatrix} \lambda_1^N & 0 \\ 0 & \lambda_2^N \end{bmatrix} \begin{pmatrix} v_1 \\ v_2 \end{pmatrix} \right\| \right\} \quad \text{for } v \in V^{(2)}$$

$$= \lim_{N \to \infty} \frac{1}{N} \ln |\lambda_2^N| = \ln |\lambda_2|. \tag{5.8.4}$$

Just as the idea of stable and unstable manifolds was generalized to orbits other than fixed points and periodic cycles, this definition generalizes the idea of eigenvalues to give average linearized contraction and expansion rates on an orbit.

Note that the subspace $V_0^{(1)} - V_0^{(2)}$ consists of all vectors in $T_x \mathbb{R}^n$ which grow at the faster possible rate, $V_0^{(2)} - V_0^{(3)}$ consists of those which grow at the next fastest rate, and so forth.

The *multiplicative ergodic theorem* of Oseledec [1968] (cf. Ruelle [1979]) implies that the Liapunov exponents of $f$ exist in great generality if $f$ is $C^1$ and $Df$ is Hölder continuous for some exponent $\theta$. For any $f$ invariant measure $\mu$, almost all points with respect to $\mu$ have Liapunov exponents. Ruelle [1979] proves the striking result that there are smooth invariant stable submanifolds in $\mathbb{R}^n$ tangent to the subspaces involved in Definition 5.8.2 for negative Liapunov exponents. These theorems of Oseledec and Ruelle (cf. also Pesin [1977]) lend confidence to the hope that statistical approaches to the study of general dynamical systems can deal with broader classes of systems than those to which the geometric methods described in this book can be applied. Within the physics community, especially, there has been intense activity directed toward the implementation of algorithms for computing such quantities as Liapunov exponents, invariant measures, and entropies. There is also hope that the geometric and statistical points of view can be amalgamated to a large degree. Katok [1980, 1981] provides a good beginning for two-dimensional diffeomorphisms.

EXERCISE 5.8.3. Use a microcomputer or programmable calculator to compute Liapunov exponents for the map $x \to ax(1 - x)$ for various $a \in (3, 4]$ and for the Lorenz map of equation (2.3.6). Compare your results with the value $\mu = \log 2$ obtained for the tent map and for $x \to 2x \pmod 1$.

In the remainder of this section we discuss two types of quantities associated with an invariant set: entropy and dimension. We begin with a statement of the topological versions of these ideas following Takens [1980], and then discuss the appropriate definitions in a measure theoretic context. Let $\Lambda$ be a compact invariant set for a diffeomorphism $f: \mathbb{R}^m \to \mathbb{R}^m$. For an integer $n > 0$ and a number $\varepsilon > 0$, and $(n, \varepsilon)$ *separated set* $S \subset \Lambda$ is a set which has the property that $x, y \in S$ and $x \neq y$ implies that there is $0 \leq i < n$ such that $d(f^i(x), f^i(y)) > \varepsilon$.

**Definition 5.8.3.** Let $s(n, \varepsilon)$ be the maximum cardinality of an $(n, \varepsilon)$ separated subset of $\Lambda$. Define

$$h(f, \varepsilon) = \limsup_{n \to \infty} \frac{1}{n} \ln s(n, \varepsilon), \qquad (5.8.5)$$

and

$$h(f) = \lim_{\varepsilon \to 0} h(f, \varepsilon). \qquad (5.8.6)$$

Then $h(f)$ is called the *topological entropy* of $f$. The *capacity* of $\Lambda$ is

$$\liminf_{\varepsilon \to 0} \frac{s(1, \varepsilon)}{-\ln \varepsilon}. \qquad (5.8.7)$$

The capacity of $\Lambda$ does not depend on $f$. It is a measure of the growth rate in the number of $\varepsilon$-balls which are required to cover $\Lambda$ as $\varepsilon \to 0$. Fixing $\varepsilon > 0$ as a criterion for separation, $h(f)$ is a measure of the growth rate of the number of distinct trajectories in $\Lambda$ as a function of the length of the trajectory. The concept of Hausdorff dimension is closely related to that of capacity:

**Definition 5.8.4.** The *Hausdorff dimension* of a metric space $X$ is the infimimum of the numbers $\alpha$ with the following property: for any $\varepsilon > 0$, there is a $\delta > 0$ and a cover $\mathcal{U}$ of $X$ such that the sets $B \in \mathcal{U}$ all have diameter smaller than $\delta$ and $\sum_{B \in \mathcal{U}} (\text{diam } B)^\alpha < \varepsilon$.

Note that Hausdorff dimension agrees with the usual notion of integer dimension for smooth manifolds and Euclidean spaces.

As an example we consider the middle third Cantor set. The uniform construction of the set implies that we can cover the set with $2^n$ closed intervals each of length $((1 - \frac{1}{3})/2)^n = 3^{-n}$; here $3^{-n}$ plays the role of $\delta$. To determine the Hausdorff dimension we consider the final inequality in the definition, which is in this case

$$2^n (3^{-n})^\alpha < \varepsilon. \qquad (5.8.8)$$

Letting $n \to \infty$, we require the infimum of the $\alpha$'s such that $\lim_{n\to\infty} (2^n 3^{-n\alpha}) \to 0$ or, equivalently, $n(\ln 2 - \alpha \ln 3) \to -\infty$. We obtain

$$\alpha = \frac{\ln 2}{\ln 3}. \tag{5.8.9}$$

as an upper-bound for the Hausdorff dimension of the middle third Cantor set. The estimate is, in fact, sharp, but we do not prove this here

EXERCISE 5.8.4. Show that the capacity of the middle third Cantor set is also $\ln 2/\ln 3$. Find the capacity and Hausdorff dimension for the middle $\beta$ Cantor set, $\beta \in (0, 1)$, and for the Cantor set produced by removing a proportion $\beta\gamma^n$ from each closed interval at the $n$th stage, where $\gamma, \beta \in (0, 1)$ (cf. Section 5.4).

These topological definitions have the drawback that a typical trajectory may not approach an attractor $\Lambda$ in such a way that the relevant limits are realized. Measure theoretic concepts are available which modify Definitions 5.8.3 and 5.8.4. However, the "standard" definitions are more cumbersome to state, and the translation between the definitions given here and the standard ones is not trivial (cf. Brin and Katok [1981], Young [1982]).

**Definition 5.8.5.** Let $\mu$ be an invariant, ergodic, probability measure for $f : \mathbb{R}^n \to \mathbb{R}^n$ with compact support. Set

$$V(x, \varepsilon, n) = \{y \in \mathbb{R}^n \mid d(f^i(x), f^i(y)) < \varepsilon, 0 \leq i < n\}.$$

Then for almost all $x$ with respect to $\mu$, the number

$$\lim_{\varepsilon \to 0} \liminf_{n \to \infty} \left\{ -\frac{1}{n} \ln \mu(V(x, \varepsilon, n)) \right\} = h_\mu(f)$$

is independent of $x$. The number $h_\mu(f)$ is the $\mu$-entropy of $f$. The *Hausdorff dimension of* $\mu$ (denoted $\mathrm{HD}(\mu)$) is the infimum of the Hausdorff dimension of subsets $Y$ of $\mathbb{R}^n$ such that $\mu(Y) = 1$.

We end with the remark that there are apparent relationships between the concepts of Liapunov exponents, entropy, and Hausdorff dimension:

**Theorem 5.8.2** (Young [1982]). *Let* $f : M \to M$ *be a* $C^2$ *diffeomorphism of a compact surface* $M$ *and let* $\mu$ *be an ergodic Borel probability measure with Liapunov exponents* $\lambda_1 \geq \lambda_2$. *Then* $\mathrm{HD}(\mu) = h_\mu(f)(1/\lambda_1 - 1/\lambda_2)$ *whenever the right-hand side of this equation is not* $0/0$.

Higher-dimensional analogues of this theorem remain unproved. Notable here are the conjectures of Yorke (Frederickson *et al.* [1982]) about the dimension of attractors.

We close with an example similar to those proposed by Kaplan and Yorke [1979a], which illustrates the formula of Theorem 5.8.2. This example is related to the geometric Lorenz model discussed earlier. Consider the map

$$F(x, y) = (f(x), g(x, y)),$$  (5.8.10)

with

$$f(x) = \begin{cases} 2x + 1, & x \in [-1, 0), \\ 2x - 1, & x \in (0, 1], \end{cases}$$

$$g(x, y) = \begin{cases} \mu y + \frac{1}{2}(1 - \mu), & x \in [-1, 0), \\ \mu y - \frac{1}{2}(1 - \mu), & x \in [0, 1], \end{cases} \quad 0 < \mu < \tfrac{1}{3},$$

defined on the set $\Sigma = \Sigma_+ \cup \Sigma_-$, with

$$\Sigma_- = \{(x, y) \mid -1 < x < 0, -1 < y < 1\},$$  
$$\Sigma_+ = \{(x, y) \mid 0 < x < 1, \quad -1 < y < 1\}.$$  (5.8.11)

We illustrate the map in Figure 5.8.2. Note that there are unstable (saddle) points at $(x, y) = (\pm 1, \mp \tfrac{1}{2})$. Although $\Sigma$ is not mapped into its interior, we define the attracting set

$$A = \overline{\bigcap_{k \geq 0} F^k(\Sigma)},$$

as in previous examples, noting that points on the boundaries $x = \pm 1$ are asymptotic to the two saddles, and the orbits of points which land on $x = 0$ terminate. As in Section 5.7, we can prove that $A$ has a dense orbit, and, by our construction, $A$ is the product of a Cantor set and the horizontal interval $[-1, 1]$.

At each stage in the construction of the Cantor set, we remove a proportion $1 - 2\mu$ of each vertical interval (two "ends" and a middle). Thus, at the $n$th stage, the set $\bigcap_{0 \leq k \leq n} F^n(\Sigma)$ consists of $2^n$ "horizontal" rectangles

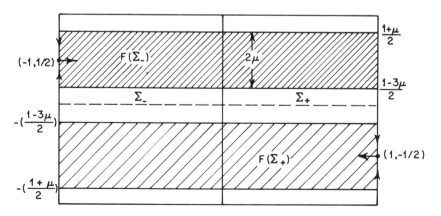

Figure 5.8.2. A piecewise linear map of Lorenz type.

each of length 2 and height $2 \cdot \mu^n$. Thus we can cover the attractor with $N \approx 2^n \cdot (2/2 \cdot \mu^n) = (2/\mu)^n$ squares each of diameter $2 \cdot \mu^n$. Using the Hausdorff dimension inequality of Definition 5.8.4, we therefore require the infimum of the $\alpha$'s such that

$$\lim_{n \to \infty} \left[ \left( \frac{2}{\mu} \right)^n \cdot (2\mu^n)^\alpha \right] = 0,$$

or

$$\lim_{n \to \infty} [2^n \cdot \mu^{n(\alpha-1)} \cdot 2^\alpha] = 0$$

$$\Rightarrow \lim_{n \to \infty} [n \ln 2 + n(\alpha - 1) \ln \mu + \alpha \ln 2] = -\infty.$$

This suggests

$$\alpha = 1 + \ln 2/\ln\left(\frac{1}{\mu}\right), \tag{5.8.12}$$

which is also the capacity in this case. We note that this is $1 + d$, where $d$ is the Hausdorff dimension of the middle $(1 - 2\mu)$ Cantor set (cf. Exercise 5.8.4).

For this example, the derivative

$$DF = \begin{bmatrix} 2 & 0 \\ 0 & \mu \end{bmatrix} \tag{5.8.13}$$

is constant, giving Liapunov exponents of $\ln 2 > 0 > \ln \mu$, respectively, and area contraction by a factor $\det(DF) = 2\mu < 1$, on each application of the map. Since the expansion rate is constant in the $x$ direction, the subsets $V(x, \varepsilon, n)$ of Definition 5.8.5 can be chosen to be disks (or squares) of diameter $\varepsilon/2^n$, so that

$$h_\mu(f) = \lim_{\varepsilon \to 0} \lim_{n \to \infty} \inf\left\{ -\frac{1}{n} \ln(\varepsilon \cdot 2^{-n}) \right\}$$

$$= \lim_{\varepsilon \to 0} \lim_{n \to \infty} \inf\left\{ \frac{n}{n} \ln 2 - \frac{1}{n} \ln \varepsilon \right\} \tag{5.8.14}$$

$$= \ln 2.$$

The formula of Theorem 5.8.2 therefore yields

$$HD(\mu) = \ln 2 \left( \frac{1}{\ln 2} - \frac{1}{\ln \mu} \right)$$

$$= \ln 2 \left( \frac{\ln(1/\mu) + \ln(2)}{\ln 2 \ln(1/\mu)} \right)$$

$$= 1 + \ln 2/\ln\left(\frac{1}{\mu}\right), \tag{5.8.15}$$

in agreement with (5.8.12).

# CHAPTER 6

# Global Bifurcations

In Chapter 3 we dealt with the local bifurcation properties of equilibrium points and periodic orbits. The theory developed there relied upon coordinate transformations which bring general systems into normal forms, from which dynamical information can be deduced from the Taylor series of a vector field or map at a single point.

In this chapter, we shall consider dynamical properties which cannot be deduced from local information. Rather, the situations we describe here involve *global* aspects of flows. The simplest of these involves the occurrence of homoclinic orbits for planar vector fields. We have already met several examples of such homoclinic orbits and also of heteroclinic orbits. These orbits provided the starting point for the use of perturbation methods to locate transversal homoclinic points in the forced Duffing equation in Chapter 4, and in Chapter 5 we described some of the rich dynamical behavior which can result from perturbations of them.

In the first section we return to planar homoclinic and heteroclinic orbits and develop a bifurcation theory which describes the typical ways in which they break under small perturbations and new invariant sets are created thereby. We go on to discuss diffeomorphisms of the circle, which we have already met in connection with flows on tori, and we describe some of the properties of rotation numbers in parametrized families of such maps. In the third section we take up the discussion of Section 5.6, on noninvertible maps of the interval, and describe the bifurcations of such maps. Section 6.4 concerns the Lorenz equations once more, and we use a one-parameter family of one-dimensional maps of the type introduced in Section 2.3 to describe the global bifurcations in which attractors of Lorenz type are created.

In Section 6.5 we then return to homoclinic orbits, and describe the dynamics near a homoclinic orbit to a fixed point in a class of three-dimensional

flows. The final two sections deal with the creation of transverse homoclinic orbits in two-dimensional diffeomorphisms. In discussing the van der
Pol, Duffing, and bouncing ball models, we have seen that such orbits
typically arise after the stable and unstable manifolds of a fixed or periodic
point meet in a tangency. In these sections we describe some aspects of the
infinite sequences of bifurcations in which horseshoes are created as the
tangential intersection breaks to yield transverse homoclinic points.

   In the final section we return to one-dimensional maps to outline the
application of renormalization group methods to the detection and analysis
of universal scaling behavior in iterated maps. The use of such ideas, originally proposed in condensed matter physics, was suggested in the present
context by Feigenbaum [1978]. Since then a sizable "scaling industry"
has arisen, particularly in the physics community.

## 6.1. Saddle Connections

The simplest global bifurcations occur for planar vector fields when there is a
trajectory joining two saddle points or forming a loop containing a single
saddle point. Consider the system

$$\dot{x} = \mu + x^2 - xy,$$
$$\dot{y} = y^2 - x^2 - 1, \tag{6.1.1}$$

(cf. Exercise 1.9.1). When $\mu = 0$, this system has two equilibrium points
$(0, \pm 1)$ and the eigenvalues at each point have opposite sign. Moreover, the
$y$-axis is invariant for the flow because $x = 0$ implies $\dot{x} = 0$. Thus the segment
$(-1, 1)$ of the $y$-axis is a trajectory connecting the two saddle points at $(0, \pm 1)$.
This is not a structurally stable situation, and we want to describe what happens when $\mu$ is perturbed from 0. A simple perturbation calculation shows
that the saddle points are perturbed to $(\pm \mu, \pm 1) + O(\mu^2)$. Along the segment
of the $y$-axis from $(-1, 1)$, the flow now has a nonzero horizontal component
$\dot{x} = \mu$, and we conclude that the trajectory which joined the two saddle
points for $\mu = 0$ is perturbed into two saddle separatrices neither of which
crosses the $y$-axis. The phase portraits for the flow are illustrated in Figure
6.1.1.

EXERCISE 6.1.1. Verify the results of the perturbation calculations outlined above using
the Melnikov method. (Since the unperturbed problem is not Hamiltonian, you will
have to do Exercise 4.5.1 first.)

   There is a qualitative difference between the pictures for $\mu < 0$ and $\mu > 0$.
When $\mu < 0$, the separatrix of the upper equilibrium $p_1$ lies to the left of the
separatrix of $p_2$ and trajectories can go from $x = +\infty$ to $x = -\infty$ as $t$
increases. When $\mu > 0$, the separatrix of $p_1$ is to the right of the separatrix

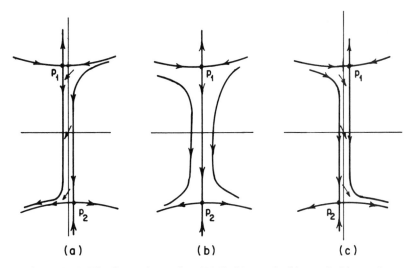

Figure 6.1.1. The flows of equation (6.1.1). (a) $\mu < 0$; (b) $\mu = 0$; (c) $\mu > 0$.

of $p$ and trajectories can go from $x = -\infty$ to $x = +\infty$ as $t$ increases. The limiting behavior of some of the orbits has clearly changed.

The crossing of saddle separatrices may be associated with other changes in the qualitative features of a flow. The simplest situation occurs with planar flows, where the existence of saddle loops is associated with the appearance and disappearance of periodic orbits.

Consider the following example:

$$\dot{x} = y,$$
$$\dot{y} = x - x^2 + \mu y. \tag{6.1.2}$$

When $\mu = 0$, the system is divergence-free and has the first integral

$$H(x, y) = \frac{y^2}{2} - \frac{x^2}{2} + \frac{x^3}{3}. \tag{6.1.3}$$

The origin is a saddle point, two of whose separatrices coincide and form a loop $\gamma$. The interior of $\gamma$ is filled by a family of closed orbits. See Figure 6.1.2.

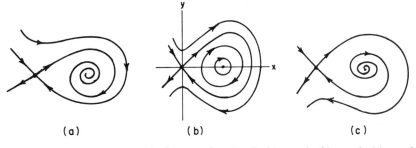

Figure 6.1.2. The phase portraits for equation (6.1.2). (a) $\mu < 0$; (b) $\mu = 0$; (c) $\mu > 0$.

EXERCISE 6.1.2. Verify the phase portraits above for $\mu \neq 0$. (Hint: Look at the divergence; use Melnikov's method.)

This is not a structurally stable situation, since the most degenerate periodic orbits one expects to find in a one-parameter family are isolated saddle-nodes. Nonetheless, the example does illustrate that saddle loops are associated with periodic orbits.

The degeneracy of equation (6.1.2) (divergence zero) implies that the saddle point contained in the separatrix loop has trace zero. (In fact the system is Hamiltonian in this case.) If we modify the example to the system

$$\begin{aligned} \dot{x} &= y, \\ \dot{y} &= x - x^2 + \mu y + axy, \end{aligned} \tag{6.1.4}$$

with $a \neq 0$, then $\mu = 0$ no longer gives the saddle loop. For $a > 0$, the saddle loop occurs when $\mu < 0$ and the saddle point has negative trace.

EXERCISE 6.1.3. Verify this statement and use Melnikov's method to find the line through the origin in the $(\mu, a)$ plane which is a tangent to the homoclinic bifurcation set of equation (6.1.4). (Hint: Let $\mu = \varepsilon \bar{\mu}, a = \varepsilon \bar{a}$ and perturb from $\varepsilon = 0$.)

The following theorem then provides us with a characterization of the periodic orbits which will be found near such nondegenerate codimension one bifurcations.

**Theorem 6.1.1.** *Consider a system of differential equations $\dot{x} = f(x, \mu)$; $x \in \mathbb{R}^2, \mu \in \mathbb{R}$, such that the following conditions hold:*

(1) *When $\mu = \mu_0$, there is a hyperbolic saddle point $p_0$ and a homoclinic orbit (saddle loop) $\gamma_0 \subset W^u(p_0) \cap W^s(p_0)$. Let $q \neq p_0$ be a point on $\gamma_0$.*

(2) *Let $M$ be a one-dimensional transversal section to $\gamma_0$ at $q$. Let $I = -[\mu_0 - \varepsilon, \mu_0 + \varepsilon]$ be an interval in parameter space. As $\mu$ varies, let $p_\mu = p(\mu)$ be the curve of saddle points with $p(\mu_0) = p_0$ and $u(\mu), s(\mu)$ be smooth curves in $\mathbb{R}^2 \times \mathbb{R}$ contained in $(M \times I) \cap W^u(p_\mu)$ and $(M \times I) \cap W^s(p_\mu)$. Assume $(d/d\mu)(u(\mu) - s(\mu)) \neq 0$ at $\mu = \mu_0$. If, in addition,*

(3) *when $\mu = \mu_0$, $\text{tr}[Df(p_0)] < 0$ (resp. $> 0$), then there is a family $\Gamma$ of stable (resp. unstable) periodic orbits in $(x, \mu)$ space for the systems $\dot{x} = f(x, \mu)$, whose closure contains $\gamma_0 \times \{\mu_0\}$. The periods of these periodic orbits are unbounded as $\mu \to \mu_0$. There is an $\varepsilon$ close to 0 ($\varepsilon$ may be negative) such that if $\mu$ lies in the interval between $\mu_0$ and $\mu_0 + \varepsilon$, then $\dot{x} = f(x, \mu)$ has exactly one periodic orbit in the family $\Gamma$. See Figure 6.1.3.*

PROOF. The proof of this theorem is formulated in terms of the return map $P_\mu$ for the cross section $M$. We only consider the stable case, since the unstable one is proved similarly.

Since $M$ is transversal to the vector $f(x_0, \mu_0)$ at $q = M \cap \gamma_0$, by restricting attention to a small enough neighborhood of $(q, \mu_0)$, we may assume that

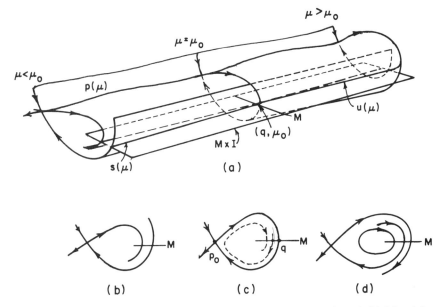

Figure 6.1.3. A saddle connection bifurcation (positive trace case shown). (a) $M$ and the curves $u(\mu)$, $s(\mu)$; (b) $\mu < \mu_0$; (c) $\mu = \mu_0$; (d) $\mu > \mu_0$.

$f(x, \mu)$ is always transversal to $M$ there. The loop $\overline{\gamma_0} = \gamma_0 \cup \{p_0\}$ has a corner at $p_0$ with tangents pointing along stable and unstable eigenvectors of $p_0$. Some of the orbits near $\gamma_0$ can escape from its neighborhood as $t \to \pm\infty$, as Figure 6.1.3(c) shows. To eliminate such orbits, we define the return map $P_{\mu_0}$ only on that portion of $M$ which lies inside $\overline{\gamma_0}$. Orbits starting near $\gamma_0$ on this side flow past $p_0$ and remain close to $\gamma_0$ (cf. Figure 6.1.3).

When $\mu = \mu_0$, we argue that, as trajectories on the inside of $\gamma_0$ flow past $p_0$, they are strongly attracted to $\gamma_0$. We introduce coordinates $(x, y)$ at $p_0$ so that the local stable and unstable manifolds are the coordinate axes. Assume that the eigenvalues of $Df(p_0)$ are $-\alpha$ and $\beta$, with $\alpha > \beta > 0$. Choose $\delta$ such that $1 > \delta > \beta/\alpha$. Then, if $(x, y)$ is sufficiently small, we have $|dy/dx| = |y(\beta + \cdots)/x(-\alpha + \cdots)| < \delta|y/x|$. This means that the trajectories of our flow are less steep than $\delta|y/x|$ near the origin. Consequently, Gronwall estimates (cf. Lemma 4.1.2) imply that, when $\varepsilon > 0$ is sufficiently small, the solution with initial conditions $(\varepsilon, y_0)$ reaches the horizontal line $y = \varepsilon$ at a point $(x_1, \varepsilon)$ with $|x_1| < |y_0|^{1/\delta}(\varepsilon)^{1 - 1/\delta}$. Since $\delta < 1$, $|x_1|/|y_0| \to 0^+$ as $y_0 \to 0^+$, as was to be proved. See Figure 6.1.4.

It follows from this argument that the derivative of $P_{\mu_0}$ approaches zero as $x \to q$ in $M$. The return maps $P_\mu$ therefore all have slopes less than one in small neighborhoods of $W^s(p_\mu) \cap M$. Observe also that the graph of $P_{\mu_0}$ approaches the diagonal in the plane as one approaches $q = \gamma_0 \cap M$, because $\gamma_0$ is a homoclinic orbit. Hypothesis (2) now implies that, as $\mu$ varies, the endpoint of the graph of $P_\mu$ moves across the diagonal with nonzero

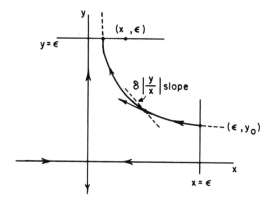

Figure 6.1.4. The flow near a contracting saddle point.

speed. (Note that the map $P_\mu$ is only defined for points on $M$ "inside" $W^s(p_\mu) \cap M$, cf. Figure 6.1.5.) We conclude that the graphs of the functions $P_\mu$ look qualitatively like those of Figure 6.1.5, up to reflection in the $\mu$-axis. This diagram contains the remainder of the proof of the theorem. In particular, since the slopes of the $P_{\mu_0}$ are smaller than one, each vector field has at most one periodic orbit near $\gamma_0$ and those periodic orbits which do occur are stable. Moreover, for $\mu$ to one side of $\mu_0$, periodic orbits necessarily occur.                                                                    □

   In the next chapter we shall encounter loops formed from several saddle separatrices. Theorem 6.1.1 can be generalized to deal with this situation,

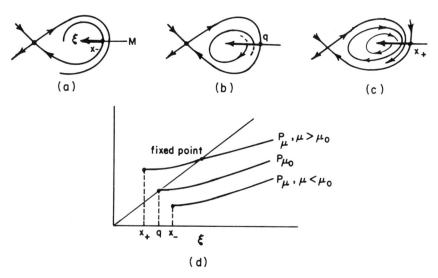

Figure 6.1.5. The Poincaré map $P_\mu$ and the associated vector fields. The domain of $P_\mu$ in $M \times \{\mu\}$ is shown as a heavy line. (a) $\mu < \mu_0$; (b) $\mu = \mu_0$; (c) $\mu > \mu_0$; (d) the map.

and the stability of the periodic orbits which occur is then determined by the quantity $\log\left(\prod\limits_{i=1}^{n} \dfrac{\lambda_i}{\mu_i}\right)$, where $-\lambda_i < 0 < \mu_i$ are the eigenvalues of the $i$th saddle in the loop; cf. Reyn [1979].

EXERCISE 6.1.4. Carry out the generalization of Theorem 6.1.1 referred to above.

EXERCISE 6.1.5. Show that, when $1 + \gamma < \beta - 1$, the following system has a loop (a homoclinic cycle) containing three saddle points:

$$\dot{x} = x\left(1 - x^2 + \beta\left(\frac{1 + \gamma}{1 - \beta}\right)y^2\right),$$

$$\dot{y} = y\left(\left(\frac{1 + \gamma}{\beta - 1}\right) + \gamma x^2 + \left(\frac{1 + \gamma}{1 - \beta}\right)y^2\right).$$

Can you determine from the trace formula whether the loop forms an $\alpha$ or an $\omega$ limit set for nearby points? (cf. Sections 7.4–7.5). (Hint: First show that the circle $x^2 + y^2 = 1$ is invariant for the flow.)

Roughly speaking, Theorem 6.1.1 describes a bifurcation of a periodic orbit to "infinite period." It does not resolve other global features which may be associated with bifurcations involving homoclinic orbits to a fixed point. This is especially true in dimensions larger than two. One example of the possible complications associated with such bifurcations occurs in the Lorenz system. There, a whole set of trajectories whose cross section is conjugate to a shift on two symbols is created when the origin acquires a pair of homoclinic orbits. The details of this situation are described in Section 6.4. We also refer the reader to Section 6.5 where the work of Silnikov on another homoclinic bifurcation for three-dimensional systems is discussed.

## 6.2. Rotation Numbers

The next global bifurcation phenomenon we discuss involves diffeomorphisms of the circle. In the weakly nonlinear van der Pol oscillator and in the analysis of Hopf bifurcations for periodic orbits, we have encountered diffeomorphisms of the plane which map a (smooth) closed curve into itself. Thus far we have postponed a full discussion of the dynamics which take place on such an invariant curve, but it is now time to consider this matter in more detail. We denote the circle by $S^1$ and consider diffeomorphisms $f: S^1 \to S^1$. We shall regard $S^1$ as $\{e^{2\pi i\theta} | \theta \in \mathbb{R}\}$, where $\theta$ is a coordinate defined modulo 1.

The starting point for the theory of (orientation preserving) diffeomorphisms on $S^1$ is that there is a cyclic order on $S^1$ which is preserved. If $f: S^1 \to S^1$ is an orientation preserving diffeomorphism and $x < y < z$ in the cyclic order, then $f(x) < f(y) < f(z)$. This order preserving property greatly constrains the dynamics—enough so that any $f$ has *almost* the same dynamics as a rigid rotation $R_\alpha$ defined by $R_\alpha = \theta + \alpha$. We will explore these topological properties in a way that should make clear the correspondence between the theory and the theory of noninvertible one-dimensional mappings.

Let $f: S^1 \to S^1$ be a orientation preserving difffeomorphism and choose an orientation on $S^1$. We pick an arbitrary point $x \in S^1$ and partition $S^1$ into the two arcs $I_0 = [x, f(x))$ and $I_1 = [f(x), x)$. For any point $y \in S^1$, we define the number

$$\rho_y(f) = \lim_{n \to \infty} \frac{1}{n} (\text{cardinality}\{f^i(y) | 0 \le i < n \text{ and } f^i(y) \in I_0\}). \quad (6.2.1)$$

Intuitively, $\rho_y(f)$ is the asymptotic proportion of the points on the trajectory of $y$ which lie in $I_0$.

We now state some important properties of $\rho_y(f)$.

**Proposition 6.2.1.** $\rho_y(f)$ *exists and is independent of* $y$.

PROOF. Denote by $N(y, k)$ the cardinality of $\{f^i(y) | 0 \le i < k \text{ and } f^i(y) \in I_0\}$. Note that $\rho_y(f) = \lim_{n \to \infty} (1/n)N(y, n)$. We first make two observations about the function $N$:

(1) $$N(y, k + l) = N(y, k) + N(f^k(y), l); \quad (6.2.2)$$

(2) For any $y, z \in S^1$ and $k \in \mathbb{Z}, |N(y, k) - N(z, k)| \le 1. \quad (6.2.3)$

The first observation comes immediately from the definition. The second observation depends upon an analysis of the points of discontinuity of $N(y, k)$. Since $x, f(x)$ form the boundaries of $I_0$ and $I_1$, points of discontinuity of $N(y, k)$ must be of the form $f^{-i}(x)$ for $k > i \ge -1$. But if $y < z$ and $y$ and $z$ are close to $f^{-i}(x)$, with $y < f^{-i}(x) < z$, then we have $f^i(y) < x < f^i(z) < f(x)$ and $f^{i+1}(y) < f(x) < f^{i+1}(z)$. Thus the only possible discontinuities are $f^{-k+1}(x)$ and $f(x)$. If $f^{-k}(x) \ne x$, then the points $f^{-k+1}(x)$ and $f(x)$ divide the circle into two arcs on which $N(y, k)$ is constant with values that differ by 1.

From these observations, we obtain the inequality

$$|N(y, nk + l) - (nN(y, k) + N(y, l))| \le n. \quad (6.2.4)$$

Letting $n \to \infty$, we find that $\rho_y(f)$ exists and that

$$\left| \rho_y(f) - \frac{1}{k} N(y, k) \right| \le \frac{1}{k}. \quad (6.2.5)$$

Now (6.2.3) implies immediately that $\rho_y(f)$ is independent of $y$, so we write $\rho(f) = \rho_y(f)$. The number $\rho(f)$ is the *rotation number* of $f$. $\qquad \square$

EXERCISE 6.2.1. Show that $\rho(f)$ is also independent of the choice of $x$ if $\rho(f) \ne 0$ or 1. What happens if $f$ has a fixed point?

**Remark**. The standard definition of $\rho(f)$ involves studying lifts of $f$ from $S^1$ to $\mathbb{R}$ (cf. Coddington and Levinson [1955], Chapter 16, or Hale [1969], pp. 64–76).

As an example, consider the rigid rotation $R_\alpha(\theta) \equiv \theta + \alpha$. We claim that $\rho(R_\alpha) = \alpha$. For $\alpha$ rational, this follows immediately from the definition (6.2.1). For $\alpha$ irrational, we apply the following proposition:

**Proposition 6.2.2.** *Let $f$, $g$ be orientation preserving homeomorphisms of $S^1$. If $\varepsilon > 0$ and $\rho(f) \neq 0$ or $1$, there is a $\delta > 0$ such that $\sup|f - g| < \delta$ implies $|\rho(f) - \rho(g)| < \varepsilon$.*

PROOF. We use the inequality (6.2.5). Given $\varepsilon > 0$, pick $k$ such that $1/k < \varepsilon/2$ and $y \in S^1$ such that $y \neq f(x)$ or $f^{-k+1}(x)$. There is then a $\delta > 0$ such that $N_g(y, k) = N_f(y, k)$ for all $g$ with $\sup|f - g| < \delta$. Then $|\rho(g) - \rho(f)| < \varepsilon$, as claimed. $\quad\square$

**Proposition 6.2.3.** *Let $f_t$ be an orientation preserving family of homeomorphisms of $S^1$ such that:*

*(1) $t_0 < t_1$ implies that $f_{t_0} < f_{t_1}$; and*
*(2) no $f_t$ has a fixed point.*

*Then $t_0 < t_1$ implies $\rho(f_{t_0}) \leq \rho(f_{t_1})$.*

PROOF. Each $f_t$ is a homeomorphism, and $t_0 < t_1$ implies $f_{t_0} < f_{t_1}$. Consider a point of discontinuity of $N_{f_t}(y, k)$ as a function of $t$. This occurs for a value of $t$ for which $f_t^l(y) = x$ for some $l \leq k$. We have $f_{t_0}^l(y) < x < f_{t_1}^l(y)$ and $N_{f_{t_0}}(y, l) < N_{f_{t_1}}(y, l)$ for $t_0 < t < t_1$ and $t_1 - t_0$ small. Note that $f_{t_0}^l(y) < x < f_{t_1}^l(y)$ implies $f_{t_0}^{l+1}(y) < f_{t_0}^l(x) < f_{t_1}^l(x) < f_{t_1}^{l+1}(y)$, so that $l < k$ implies $N_{f_{t_0}}(y, l + 1) = N_{f_{t_1}}(y, l + 1)$. Hence $l = k$ and each $(1/k)(N_{f_t}(x, k))$ is a nondecreasing function of $t$. Thus, the limit $\rho(f_t)$ is also nondecreasing. $\quad\square$

EXERCISE 6.2.2. Prove that $\rho(f^n) = n\rho(f) \pmod 1$.

**Proposition 6.2.4.** *$\rho(f)$ is rational if and only if $f$ has a periodic orbit.*

PROOF. If $f$ has a periodic orbit of period $n$ containing $y$, then $N(y, n \cdot k) = kN(y, n)$ and hence $\rho(f) = \lim_{n \to \infty} N(y, n)/n$ is rational. If $\rho(f) = m/n$ is rational, then $\rho(f^n) = 0$ (or $1$) and the proposition will be proved if we show that $f^n$ has a fixed point. If $\rho(g) = 0$, we assert that $\{g^i(y)\}$ is monotone with either $y < g(y) < g^2(y) < \cdots < y$ or $y > g(y) > g^2(y) > \cdots > y$. Were this not the case, then there would be a $k$ with

$$y < g(y) < g^2(y) < \cdots < g^{k-1}(y) < y < g^k(y)$$

or

$$y > g(y) > g^2(y) > \cdots > g^{k-1}(y) > y > g^k(y).$$

In the first case $\rho(g) > 1/k$ and in the second case $\rho(g) < 1 - 1/k$; in both there is a contradiction. If $\{g^i(y)\}$ is monotone, then it clearly converges and the limit is a fixed point of $g$. □

To illustrate how some of these ideas apply to examples which we have already discussed, we consider the flow on the two-torus generated by

$$\dot{\theta}_1 = 1 - \gamma \sin 2\pi(\theta_1 - \theta_2),$$

$$\dot{\theta}_2 = \omega + \gamma \sin 2\pi(\theta_1 - \theta_2), \tag{6.2.6}$$

where $\theta_1, \theta_2 \in S^1 = [0, 1]$ mod 1, $\omega$ is fixed in the range $[0, 1]$ and $\gamma \geq 0$ is a variable parameter. Such systems arise in studies of linearly coupled limit cycle oscillators; cf. Cohen and Neu [1979], Neu [1979], Rand and Holmes [1980], and Cohen *et al.* [1982]; also see Section 1.8.

Let $\Sigma \subset T^2 = S^1 \times S^1$ be the cross section

$$\Sigma = \{(\theta_1, \theta_2) | \theta_1 = 0\}, \tag{6.2.7}$$

and consider the Poincaré map $P_\gamma : \Sigma \to \Sigma$ induced by the flow of (6.2.6). $P_\gamma$ is a diffeomorphism of the circle.

EXERCISE 6.2.3. Consider the scalar equation in $\phi = \theta_1 - \theta_2$:

$$\dot{\phi} = (1 - \omega) - 2\gamma \sin 2\pi\phi,$$

obtained by subtracting the two components of (6.2.6). Show that this equation can be solved exactly and use the solution to show that, for $2\gamma > (1 - \omega)$, the Poincaré map $P_\gamma$ has two hyperbolic fixed points. Hence show that the rotation number $\rho(\gamma) = 1$ for $2\gamma \geq (1 - \omega)$. Describe the bifurcation occurring at $2\gamma = (1 - \omega)$.

We now suppose that $2\gamma < (1 - \omega)$, in which case the solution of equations (6.2.6) based at $\theta_1 = 0, \theta_2 = \theta$ can be written

$$\theta_1(t) = \frac{\theta}{2} + \frac{1}{2\pi} \left( \left( \frac{1 + \omega}{2} \right) t + \arctan u(t) \right),$$

$$\theta_2(t) = \frac{\theta}{2} + \frac{1}{2\pi} \left( \left( \frac{1 + \omega}{2} \right) t - \arctan u(t) \right), \tag{6.2.8}$$

where

$$u(t) = \frac{1}{1 - \omega} \left\{ 2\gamma + \sigma \tan\left( \frac{\sigma t}{2} + c \right) \right\},$$

$$\sigma = \sqrt{(1 - \omega)^2 + 4\gamma^2},$$

and where $c$ is defined via

$$\sigma \tan c = -\{2\gamma + (1 - \omega) \tan(\pi\theta)\},$$

so that

$$u(0) = -\tan\left(\frac{\theta}{2}\right).$$

EXERCISE 6.2.4. Verify that equations (6.2.8) are correct.

The exact solution of (6.2.8) enables us to write the Poincaré map as

$$P_\gamma(\theta) = \frac{\theta}{2} + \frac{1}{2\pi}\left(\left(\frac{1+\omega}{2}\right)\tau - \arctan u(\tau)\right), \qquad (6.2.9)$$

where $\tau$ is the time required for $\theta_1$ to go from 0 to 1:

$$1 = \frac{\theta}{2} + \frac{1}{2\pi}\left(\left(\frac{1+\omega}{2}\right)\tau + \arctan u(\tau)\right). \qquad (6.2.10)$$

EXERCISE 6.2.5. Prove that, for $2\gamma < (1 - \omega)$, $P_\gamma$ is a rigid rotation. (Hint: Use the invariance of the vector field of (6.2.6) under translations along $\theta_1 = \theta_2$.)

We now compute the rotation number $\rho(P_\gamma) = \rho(\gamma)$ for $2\gamma < (1 - \omega)$. From (6.2.9)–(6.2.10) (and using the result of Exercise 6.2.1), we have

$$\rho(P_\gamma) = \rho(\gamma) = \frac{1}{n}\left\{\frac{(1+\omega)\tau_n}{2\pi} - n\right\} = \frac{(1+\omega)\tau_n}{2\pi n} - 1, \qquad (6.2.11)$$

where $\tau_n$ solves

$$2\pi n = \left(\frac{1+\omega}{2}\right)\tau_n + \arctan u(\tau_n). \qquad (6.2.12)$$

If we let $n \to \infty$, then $\tau_n$ may be calculated by taking the average slope of the oscillating function $\arctan u(\tau_n)$. Now $\arctan u(\tau)$ changes by $\pi$ in the time $\tau^*$ taken for $u(\tau)$ to go from $-\infty$ to $+\infty$, or for $(\sigma t/2 + c)$ to go from $-\pi/2$ to $\pi/2$. Thus $\tau^* = 2\pi/\sigma$. We conclude that the average slope of $\arctan u(\tau)$ is $\pi/(2\pi/\sigma) = \sigma/2$, so that, for large $n$ (and hence large $\tau_n$) (6.2.12) yields $\tau_n = 2\pi n(2/(1 + \omega + \sigma))$. Using this in (6.2.11) we obtain

$$\rho(\gamma) = \frac{1 + \omega - \sigma}{1 + \omega + \sigma}, \qquad (6.2.13)$$

where $\sigma$ was given in (6.2.8). A graph of the rotation number appears in Figure 6.2.1. Note that, while it is continuous at $\gamma = 1 - \omega$, it is nondifferentiable with limit $\rho'(\gamma) \to \infty$, as $\gamma \to (1 - \omega)/2$ from below. Thus, after 1:1 phase locking is lost when $\gamma$ passes through $1 - \omega$, decreasing, the rotation number passes rapidly through infinitely many rational and irrational values and approaches $\omega$ as $\gamma \to 0$.

The properties embodied in Propositions 6.2.1–6.2.4 summarize the topological properties of homeomorphisms of the circle. In a sense, the rigid rotations provide models for the asymptotic behavior of any trajectory.

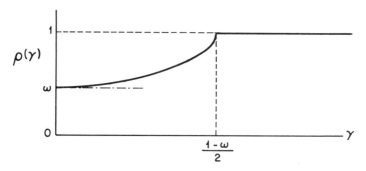

Figure 6.2.1. The rotation number for $P_y$.

If $\rho(f)$ is rational, all trajectories of $f$ are asymptotically periodic, while if $\rho(f)$ is irrational, the trajectories occur with the same order as those for the mapping $R_\alpha$; $\alpha = \rho(f)$. However, we would like to go beyond these general properties for homeomorphisms to obtain information which depends upon the smoothness of a map $f$. We focus first upon the properties of a single diffeomorphism with irrational rotation number.

Let $f$ be a diffeomorphism with $\rho(f) = \alpha$ irrational and let $x \in S^1$. In the discussion above, we have shown that a function $h$ defined on the trajectory of $x$ by $h(f^n(x)) = n\alpha$ (mod 1) preserves the ordering of points on the circle. Moreover, $\{n\alpha \text{ (mod 1)}\}$ is dense in the circle because $\alpha$ is irrational. Consequently, $h$ extends uniquely to a nondecreasing function on the circle. Explicitly, we have

$$h(y) = \lim_{n \to \infty} \frac{1}{n} (\text{cardinality}\{k \,|\, 0 \le k < n \text{ and } f^k(x) \in [x, y)\}). \quad (6.2.14)$$

EXERCISE 6.2.6. Check that $h(y) = n\alpha$ (mod 1) if $y = f^n(x)$ with this definition, by showing that $h(f(y)) = h(y) + \alpha$ (mod 1).

We now wish to know whether the map $h$, above, is a homeomorphism, since if it is, it is then a topological conjugacy from $f$ to the rigid rotations $R_\alpha$. Whether $h$ is a homeomorphism or not is determined by whether the $f$ trajectories of points are dense in $S^1$. Since $n_1\alpha \ne n_2\alpha$ mod(1) for $n_1 \ne n_2$, $h(f^{n_1}(x)) \ne h(f^{n_2}(x))$ for all $n_1 \ne n_2$. If every pair of points in $S^1$ is separated by points in the trajectory of $x$, then $h$ is a homeomorphism.

If $h$ is not a homeomorphism, there will be some values $x$ for which $h^{-1}(x)$ is a closed arc $I$. If this is so, then $h$ is constant on $f^n(I)$ and takes distinct values on $f^{n_1}(I)$ and $f^{n_2}(I)$ if $n_1 \ne n_2$. Consequently, the arcs $f^n(I)$ are all disjoint. Since the circle $S^1$ has finite length, the intervals $f^n(I)$ must be very short when $|n|$ is large. This implies that for large $n$, the derivative of $f^n$ is very large at some points of $f^{-n}(I)$ and the derivative of $f^n$ is very small at some points of $I$. In fact the function $\log(f^n)'$ will have a variation which is not uniformly bounded with $n$.

Denjoy [1932] observed that the oscillation of $\log(f^n)'$ cannot grow indefinitely on an arc such as $I$ if $\log f'$ itself has bounded variation. The argument is simple but fundamental. If $x, y \in I$, and $J = (x, y)$, then we know that the arcs $f^i(J)$ are disjoint. Using the chain rule to express $(f^n)'$ as a product, we have

$$\log(f^n)'(y) - \log(f^n)'(x) = \sum_{i=0}^{n-1} (\log f'(f^i(y)) - \log f'(f^i(x))). \quad (6.2.15)$$

The right-hand side of this equation is smaller than the total variation of $\log f'$ because the arcs $f^i(J)$ are disjoint. (If $f$ is twice differentiable, then we obtain

$$\log(f^n)'(y) - \log(f^n)'(x) = \sum_{i=0}^{n-1} \int_{f^i(x)}^{f^i(y)} \frac{f''(\xi)}{f'(\xi)} d\xi \le \int_{S^1} \frac{|f''(\xi)|}{f'(\xi)} d\xi. \Bigg) \quad (6.2.16)$$

This estimate of the "nonlinear distortion" of $f^n$ is closely related to estimates which appear in the proofs of the statistical properties of the asymptotic measures of hyperbolic attractors. Here it leads to Denjoy's theorem:

**Theorem 6.2.5** (Denjoy [1932]). *If $f$ is a diffeomorphism of $S^1$ such that $\rho(f) = \alpha$ is irrational and $\log f'$ has bounded variation, then $f$ and $R_\alpha$ are topologically conjugate.*

**PROOF.** Define $R_\alpha(\theta) = \theta + \alpha$. Pick a large integer $n$ with the property that $|n\alpha \pmod 1| < |k\alpha \pmod 1|$ for $k < n$. If $K$ is the (shorter) arc from 0 to $n\alpha$, then we assert that $R_\alpha^k(K) \cap R_\alpha^l(K) = \varnothing$ for $0 \le k, l < n$ and $k \ne l$. (Otherwise there are $x, y$ with $|x - y| < |n\alpha \pmod 1|$ and

$$|x - y| = |(k - l)\alpha \pmod 1|,$$

contradicting the choice of $n$.) Now consider $f$ and assume that there is an interval $J$, all of whose images are disjoint from each other. Since the ordering of points in the orbits of $f$ is the same as the ordering for $R_\alpha$, there is an arc $I$ containing $J$ and $f^n(J)$ with the property that the arcs $f^i(I), 0 \le i < n$, are all disjoint. We conclude that if $y, z \in I$, then

$$\log(f^n)'(z) - \log(f^n)'(y) = \sum_{i=0}^{n-1} (\log f'(f^i(z)) - \log f'(f^i(y)))$$

$$\le \text{variation}(\log f'). \quad (6.2.17)$$

Using (6.2.17) we can find an $M > 0$, independent of $n$, such that

$$\frac{1}{M} < \frac{(f^n)'(y)}{(f^n)'(z)} < M \quad (6.2.18)$$

for all $y, z \in I$. Denoting the length of an interval $K$ by $l(K)$, this gives the estimate

$$\frac{1}{M} < \frac{l(J)/l(f^{-n}(J))}{l(f^n(J))/l(J)} < M, \quad (6.2.19)$$

since the mean value theorem implies that there are points $y \in f^{-n}(J)$ and $z \in J$ such that $(f^n)'(y) = l(J)/l(f^{-n}(J))$ and $(f^n)'(z) = l(f^n(J))/l(J)$. But we have already observed that $l(f^{-n}(J)) \cdot l(f^n(J)) \to 0$ as $n \to \infty$ (since the images of $J$ are disjoint), and this contradicts the inequality. Therefore the map $h$ defined above is 1–1 and a topological conjugacy from $f$ to $R_\alpha$. ☐

Denjoy's theorem is only the beginning of an extensive and deep mathematical story. The conjugacy $h$ from $f$ to $R_\alpha$ of Denjoy's theorem is almost unique. If $h_1$ and $h_2$ are two such conjugacies, then $h_1 h_2^{-1}$ is a conjugacy from $R_\alpha$ to itself. A self-conjugacy $h$ of $R_\alpha$ satisfies the equation $h(x) + n\alpha = h(x + n\alpha)$ (mod 1) for all integers $n$. Since the points $\{n\alpha \pmod 1\}$ are dense, we conclude that $h(0) + y = h(y)$ (mod 1) for all $y$ and fixed $x = 0$. In words, $h$ is itself a rotation, and a pair of conjugacies from $f$ to $R_\alpha$ satisfies $h_1(h_2^{-1}(x)) = x + \beta$ for some constant $\beta$. It therefore makes sense to discuss the smoothness of the conjugacy relating a diffeomorphism $f$ with irrational rotation number $\alpha$ to the rigid rotation $R_\alpha$.

There are two major theorems about this smoothness problem. The first involves perturbations of rigid rotations. We want to solve an equation $h(f(x)) = h(x) + \alpha$ for $h$ with $f$ given and $\alpha$ equal to $\rho(f)$. If $f$ is $R_\alpha + \hat{f}$ with $\hat{f}$ small and $h = $ identity $+ \hat{h}$ with $\hat{h}$ small, then the equation can be linearized by writing $h(f(x)) \simeq f(x) + \hat{h}(R_\alpha(x))$. The linearized equation is

$$\hat{h}(x + \alpha) - \hat{h}(x) = \hat{f}(x). \tag{6.2.20}$$

If $\hat{h}$ and $\hat{f}$ are expanded in Fourier series:

$$\hat{f}(x) = \sum a_m e^{2\pi i m x}; \qquad \hat{h}(x) = \sum b_m e^{2\pi i m x}, \tag{6.2.21}$$

then we have

$$\hat{h}(x + \alpha) - \hat{h}(x) = \sum b_m (e^{2\pi i m \alpha} - 1)e^{2\pi i m x} = \sum a_m e^{2\pi i m x}. \tag{6.2.22}$$

Thus the linearized equation is solved *formally* by equating the Fourier series of $\hat{f}(x)$ and $\hat{h}(x + \alpha) - \hat{h}(x)$ term by term and taking

$$b_m = \frac{a_m}{e^{2\pi i m \alpha} - 1} \quad \text{(if } a_0 = 0\text{)}. \tag{6.2.23}$$

There is a substantial convergence difficulty here, called a *small divisor problem*. The denominators $e^{2\pi i m \alpha} - 1$ cannot be bounded away from 0, thus making the convergence of the Fourier series of $\hat{h}$ problematic. It is certainly necessary to put arithmetic diophantine conditions on $\alpha$ which give estimates for just how small $(e^{2\pi i m \alpha} - 1)$ can be. The strongest inequalities which are satisfied by almost all $\alpha$ have the form that there are constants $c, \varepsilon > 0$

depending on $\alpha$ such that

$$|e^{2\pi i m \alpha} - 1| > \frac{c}{m^{(1+\varepsilon)}} \quad \text{for all } m. \qquad (6.2.24)$$

The $\alpha$'s satisfying inequality (6.2.24) for some $c$, $\varepsilon > 0$ are said to be *badly approximated* by rational numbers. We note that if an integer $n$ (depending on $m$) is selected such that $|\alpha m - n|$ is small, then

$$|e^{2\pi i m} - 1| > |\alpha m - n|, \qquad (6.2.25)$$

because $1 = e^{2\pi i n}$ and $|(d/dx)(e^{2\pi i x})| > 1$. Thus (6.2.24) can be replaced by

$$\left| \alpha - \frac{n}{m} \right| \geq \frac{c}{m^{(2+\varepsilon)}} \quad \text{for all } m, \text{ and } n > 0. \qquad (6.2.26)$$

If such conditions are met, then the Fourier series for $\hat{h}$ can be shown to converge. Moreover, the set of rotation numbers $\alpha$ for which (6.2.26) holds has Lebesgue measure one. For details, see Arnold [1965] and Herman [1977].

For rotation numbers which are sufficiently irrational in this sense, one would like to solve the conjugacy problem by using the above solution to the linearized problem. This requires a "hard" implicit function theorem (Hamilton [1982]). The development of the necessary techniques for solving this problem was carried through by Arnold [1963b, 1965] and Moser [1962] and further pursued by Rüssmann [1970]. There is a great deal of delicate analysis in this KAM theory (cf. Section 4.8) which we shall not pursue here. The final result of the KAM analysis is that a smooth transformation $h$ can be found, taking maps with irrational rotation numbers, $\alpha$, sufficiently close to rigid rotations, into rigid rotations, provided that $\alpha$ satisfies (6.2.26).

To illustrate, we consider a two-parameter family of the form

$$F_{\delta, \varepsilon} : x \rightarrow x + \mu + \delta + f(x, \varepsilon), \qquad (6.2.27)$$

with $f$ one-periodic in $x$ and $|f(x, \varepsilon)| < K|\varepsilon|$; cf. Arnold [1965]. Clearly, for $\varepsilon = 0$, we have a rigid rotation with rotation number $\alpha = \mu + \delta$ and when $\mu + \delta$ is rational, there are circles of periodic points. Typically, there are open sets in $(\varepsilon, \delta)$-space (small "wedges") for which $\rho(F_{\delta, \varepsilon})$ is rational and constant and $F_{\delta, \varepsilon}$ has hyperbolic periodic points. However, if $\mu + \delta$ satisfies (6.2.26), Arnold shows that there is a curve $\delta(\varepsilon)$ with $\delta(0) = \delta$ such that, for sufficiently small $\varepsilon$, the map $F_{\delta(\varepsilon)\,\varepsilon}$ is topologically equivalent to the rigid rotation $R_{\mu+\delta}$. Thus each little resonant wedge in $(\varepsilon, \delta)$ space is separated from its neighboring wedges by a curve of maps with irrational rotation number $\mu$.

EXERCISE 6.2.7 (Arnold [1965]). Consider the family

$$f_{\mu, \varepsilon} : x \rightarrow x + \mu + \varepsilon \cos(2\pi x), \qquad \varepsilon \geq 0.$$

Show that the resonant wedge with $\rho(f_{\mu, \alpha}) = 1$ is bounded by the curves $\mu = \pm|\varepsilon|$. Iterate the map twice and then three times, taking $\mu = \frac{1}{2} + \mathcal{O}(\varepsilon^2)$ and $\mu = \frac{1}{3} + \mathcal{O}(\varepsilon^2)$, respectively, and show that the resonant wedges in which $\rho(f_{\mu, \varepsilon}) = \frac{1}{2}$ and $\frac{1}{3}$ are bounded by the curves

$$\mu = \frac{1}{2} \pm \varepsilon^2 \frac{\pi}{2} + \mathcal{O}(\varepsilon^3),$$

and

$$\mu = \frac{1}{3} + \varepsilon^2 \frac{\sqrt{3}\pi}{6} \pm \varepsilon^3 \frac{\sqrt{7}\pi^2}{6} + \mathcal{O}(\varepsilon^4),$$

respectively. (Arnold [1965] plots graphs of the boundaries of these resonant wedges.)

Successive applications of perturbation theory to the family considered in this exercise show that a resonant wedge containing systems with periodic orbits of period $m$ is of width $\mathcal{O}(\varepsilon^m)$. Thus, fixing $\varepsilon > 0$, small, if one removes the values of $\mu$ in each wedge of order $m$ in sequence, then at any stage one is left with a collection of intervals corresponding to maps which have either irrational rotation numbers or periodic orbits of period $> m$. Carrying this process to the limit $m \to \infty$, we are left with a Cantor set of parameter values corresponding to systems with irrational rotation numbers. Since the amount removed at each stage decreases like $m\varepsilon^m$, and $m\varepsilon^m \to 0$ as $m \to \infty$, the resulting Cantor set of maps with irrational $\rho(F_{\delta, \varepsilon})$ can be expected to have positive Lebesgue measure (cf. the construction in Section 5.4). That this is indeed the case was proved by Arnold [1965] and more generally by Herman [1977].

We shall now sketch the analysis leading to a generalization of this special example, which includes the second major result on the smoothness of conjugacies. This result is due to Herman [1976, 1977]. He greatly extends the work of Arnold and Moser by removing the restrictions that one must deal with diffeomorphisms which are small perturbations of a rigid rotation.

The smoothness results are very technical, but they play an essential role in the subtle aspects of the dynamics of a one-parameter family of diffeomorphisms of a circle. If $f_\lambda : S^1 \to S^1$ is a one-parameter family of diffeomorphisms, then the function $\rho(\lambda) = \rho(f_\lambda)$ tells us a great deal about how the dynamics changes. For the sake of exposition, let us confine our considerations to families which are increasing in the sense that $\lambda_1 < \lambda_2$ implies $f_{\lambda_1}(x) < f_{\lambda_2}(x)$ for all $x \in S^1$. Herman obtains a number of properties of the function $\rho(\lambda)$ for an increasing family that we now state.

(1) $\rho(\lambda)$ is a nondecreasing function of $\lambda$.
(2) If $\rho(\lambda_0) = n/m$ is rational and $f_{\lambda_0}^m(x) < x$ for some $x \in S^1$, then there is a $\lambda_1 > \lambda_0$ with $\rho(\lambda_1) = \rho(\lambda_0)$.
(3) The Lebesgue measure of $\{\lambda \,|\, \rho(\lambda)$ is irrational$\}$ is positive, provided that $\rho(\lambda)$ is not a constant function.

The first property follows from topological arguments of the kind presented earlier in this section. The second property follows from the fact that $f_{\lambda_0}^m$ must have a fixed point $p$, with nearby values $x$ at which $f_{\lambda_0}^m(x) < x$ and the graph of $f^m_\lambda$ lies above the graph of $f^m_{\lambda_0}$ for $\lambda > \lambda_0$, and will therefore cross the diagonal near $p$ if $\lambda - \lambda_0$ is small, giving a fixed point for $f^m$. The proof of the third property invokes the full strength of Herman's analytical results. Properties (2)–(3) yield a situation analogous to the one described in Section 5.6 for Jakobson's theorem. *In a generic family, an open-dense set of parameter values yield rational rotation numbers, but the complementary set for which there are irrational rotation numbers has positive measure.*

The application of this theory to Hopf bifurcations of maps and periodic orbits of flows is immediate. It gives a description of the dynamics one expects to find on the invariant closed curves in this problem. It may also be applied directly to the weakly forced van der Pol equation of Section 2.1.

We recall from Section 3.5 that the normal form of a two-dimensional map under-going Hopf bifurcation is given, through terms of third order, by

$$(r, \theta) \to (r(1 + d(\mu - \mu_0) + ar^2), \theta + c + br^2), \qquad (6.2.28)$$

where $a$, $b$, $c$, and $d$ are constants and $\mu$ is the bifurcation parameter. For this truncated map, we therefore obtain invariant closed curves given by

$$r = \sqrt{\frac{d}{a}(\mu_0 - \mu)}, \qquad (6.2.29)$$

either above or below $\mu_0$, depending upon the signs of $a$ and $d$. Using (6.2.29) in (6.2.28), the rigid rotation on the (invariant) circle is given by

$$\theta \to \theta + \alpha(\mu),$$

where $\qquad\qquad\qquad\qquad\qquad\qquad\qquad\qquad\qquad\qquad\qquad\qquad$ (6.2.30)

$$\alpha(\mu) = c + \frac{bd}{a}(\mu_0 - \mu).$$

We conclude that, to *leading order*, if $b$, $d$, and $a$ are nonzero, then the rotation number changes linearly with $\mu$.

However, since higher-order terms were neglected in (6.2.28), the rigid rotation (6.2.30) is merely the leading part of a one-parameter family of diffeomorphisms of the circle, and the theory sketched above implies that there are perturbations of the family (6.2.28) for which the rotation number will be constant and rational on open sets of parameter values. This leads to the phenomenon of *entrainment*. In the case of flows, the periodic points correspond to periodic orbits on the torus which are *phase locked*. For open sets of parameter values, stable and unstable hyperbolic orbits alternate, and at the ends of such parameter intervals saddle-node bifurcations occur. The length of these intervals decreases rapidly as the period of the orbits increases. In contrast, Herman's work shows that the parameter values which separate such intervals of structurally stable behavior and yield irrational flows, form a set of positive measure. Thus the theory of Arnold and Herman provides a satisfactory explanation for the ephemeral nature of entrainment for orbits of high period *and* for the abundance of flows which appear to be genuinely quasiperiodic.


# 6.3. Bifurcations of One-Dimensional Maps

We have already discussed one-dimensional mappings in Section 5.6 in relation to the problem of the existence of strange attractors. Here we shall give more details on the *kneading theory* of mappings with one critical point and the implications for bifurcation theory. As with the other sections in this chapter, we try to present the essential ideas involved while avoiding most details. A much more complete exposition of the theory we discuss here may be found in the recent monograph of Collet and Eckmann [1980] and the literature cited therein.

Throughout this section we shall assume that $f : I \to I$ is a map of the unit interval into itself which has a *negative Schwarzian derivative*: i.e., $f$ is three times differentiable, and

$$S(f) = \frac{f'''(x)}{f'(x)} - \frac{3}{2}\left(\frac{f''(x)}{f'(x)}\right)^2 < 0, \qquad (6.3.1)$$

for all $x$ at which $f'(x) \neq 0$. This is a technical assumption which is motivated by the following theorem of Singer [1978] (cf. Misiurewicz [1981], Collet and Eckmann [1980]):

**Theorem 6.3.1.** *Let $f : I \to I$ be a $C^3$ map with negative Schwarzian derivative. If $\gamma$ is a stable periodic orbit, then there is a critical point\* of $f$ or an endpoint of $I$ whose trajectory approaches $\gamma$.*

EXERCISE 6.3.1. Prove Theorem 6.3.1. (Hint: First show that if $S(f) < 0$, $S(g) < 0$ then $S(f \circ g) < 0$, so that the N.S. property holds for iterates $f^n$ of $f$. Use this to conclude that $|(f^n)'|$ can have no positive minima and hence that a neighborhood of any stable periodic point necessarily contains either a preimage of a critical point or an endpoint of I.)

We shall assume further that $f(0) = f(1) = 0$ and that $f$ has a single critical point $c$ in the interior of $I$. If 0 is a stable fixed point, we assume that $f^n(c) \to 0$ as $n \to \infty$. With these assumptions, the mapping $f$ has at most one stable periodic orbit. In Section 5.6 we discussed the kinds of asymptotic behavior most points can have if $f$ does not have a stable periodic orbit. Here we shall examine one-parameter families of maps satisfying the above hypotheses for the order in which stable periodic orbits appear as a parameter is varied.

---

\* A critical point of $f$ is a point with $f'(c) = 0$.

Let $f_\mu: I \to I$ be a one-parameter family of mappings with $\mu \in [\mu_0, \mu_1]$ which satisfies the hypotheses above and has the further properties that:

(1) $f_{\mu_0}(c) = c$;
(2) $f_{\mu_1}(c) = 1$.

(In general the critical point $c = c(\mu)$ is allowed to depend on $\mu$.) We call such a family a *full family*. The prototype of such families is the quadratic logistic equation:

$$f_\mu(x) = \mu x(1 - x); \qquad \mu \in [2, 4]. \qquad (6.3.2)$$

Here $c$ is identically $\frac{1}{2}$. There must be many bifurcations of periodic orbits in a full family. Indeed, $f_{\mu_0}$ has only two fixed points as its periodic orbits, while $f_{\mu_1}^n$ has at least $2^n$ fixed points for all $n > 0$. Thus, as $\mu$ increases in $[\mu_0, \mu_1]$, new periodic orbits frequently appear. The order in which they do so is tightly constrained by the one dimensionality of the maps, and it is this phenomenon that we now describe.

The way in which periodic orbits appear in a full family can be described using symbolic dynamics. The framework and terminology introduced by Milnor and Thurston [1977] for this situation is called the *kneading calculus* or *kneading theory*. It describes much of the structure of an individual map and determines much of the bifurcation structure in a family. We start with a few necessary definitions. Our symbolic description of a map $f$ will always be phrased in terms of the partition of the interval $I$ into the two subintervals $I_0 = [0, c)$, $I_1 = (c, 1]$ and the point $C = \{c\}$. The intervals $I_0$ and $I_1$ are the *laps* of $f$. If $x \in I$, the $n$th *address* $A_n(x)$, $n \geq 0$, is $I_0$, $C$, or $I_1$ as $f^n(x) \in I_0$, $f^n(x) = c$, or $f^n(x) \in I_1$. The *itinerary* of $x$, $\mathbf{A}(x)$, is the sequence $\{A_n(x)\}$ of its successive addresses. The itinerary of $f(c)$ we shall call the *kneading sequence* of $f$. (Previous usage has been to call the itinerary of $c$ the kneading sequence.)

Defining the signs $\varepsilon(I_0)$, $\varepsilon(C)$, and $\varepsilon(I_1)$ as $+1$, $0$, and $-1$, respectively, the *invariant coordinate* $\boldsymbol{\theta}(\mathbf{a}) = \{\theta_n(\mathbf{a})\}_{n \geq 0}$ is defined by

$$\theta_n(x) = \left[\prod_{i=0}^{n-1} \varepsilon(a_i)\right] a_n(x)$$

for any symbol sequence $\mathbf{a}$. We write $\boldsymbol{\theta}(x) = \boldsymbol{\theta}(\mathbf{A}(x))$ for $x \in I$.

The (invariant) coordinates can be ordered by assigning the order $-I_1 < -C < -I_0 < 0 < I_0 < C < I_1$ to the signed symbols which appear in the definition of $\boldsymbol{\theta}(x)$ and then using a lexicographic ordering for the sequences. This means that $\boldsymbol{\theta}(x) < \boldsymbol{\theta}(y)$ if there is an $n$ with $\theta_i(x) = \theta_i(y)$ for $i < n$ and $\theta_n(x) < \theta_n(y)$. This ordering of invariant coordinates reflects the ordering of points of the line.

**Proposition 6.3.2** (Milnor and Thurston [1977]). *If $x < y$, then $\theta(x) \le \theta(y)$.*

PROOF. Let $x < y$ and suppose $\theta(x) \ne \theta(y)$. Let $n$ be the smallest integer with $\theta_n(x) \ne \theta_n(y)$. Then $f^i$ is monotone on the interval $J = [x, y]$ for $i \le n$ because $c$ does not lie in $f^i(J)$ for $i < n$. We next observe that $f^n$ is decreasing or increasing on $J$ as the sign of $\theta_n(x)$ is $-1$ or $+1$. In either case, one checks easily that $\theta_n(x) < \theta_n(y)$.                                                                $\square$

This proposition, the *monotonicity of the invariant coordinate*, allows one to make detailed symbolic calculations. Since $f(c)$ is the maximum value of a mapping $f$, $\theta(f(c))$ is larger than the invariant coordinate of $f(x)$ for any other point $x$. One can use a criterion based on this observation to develop an existence theorem for trajectories with a given itinerary. We note that $A(f(x)) = \sigma(A(x))$, where $\sigma$ is the shift map on sequences. Therefore, any $x \in I$ satisfies the criterion that $\theta(\sigma^n(A(x))) \le \theta(f(c))$ for all $n \ge 0$.

We now have:

**Theorem 6.3.3.** *If $\mathbf{a}$ is a symbol sequence such that $\theta(\sigma^n(\mathbf{a})) < \theta(f(c))$ for all $n \ge 0$, then there is a point $x$ such that $\mathbf{a} = A(x)$.*

The kneading sequence of $f$ essentially determines all of the symbol sequences which appear as itineraries for $f$. See Guckenheimer [1979] for complete details.

This theorem leads directly to the structure of the bifurcation set which we seek. We can define an ordering of symbol sequences by using the criterion of the theorem. Given two symbol sequences $\mathbf{a}$ and $\mathbf{b}$, we say that $\mathbf{a} \prec \mathbf{b}$ if there is an integer $m \ge 0$ such that $\theta(\sigma^n(\mathbf{a})) < \theta(\sigma^m(\mathbf{b}))$ for all $n \ge 0$. We then conclude that the presence of a point $x$ with $A(x) = \mathbf{b}$ for a mapping $f$ implies that there is a $y$ with $A(y) = \mathbf{a}$. Changes in the set of itineraries occur only by changes in the kneading sequence and the set of all possible kneading sequences is ordered by $\prec$. For maps which are differentiable (so that $f'(c) = 0$), then all of the possible kneading invariants are realized in any full family. We conclude that between parameters $\mu_0$ and $\mu_1$, there are bifurcations which represent transitions to all kneading sequences between those of $f_{\mu_0}$ and $f_{\mu_1}$. As a special case, we have the following:

**Theorem 6.3.4.** *There is an ordering of all periodic sequences of the symbols $I_0$ and $I_1$ with the following properties:*

(1) *Two sequences $\mathbf{a}$ and $\mathbf{b}$ are equivalent if and only if $\sigma^n(\mathbf{a}) = \mathbf{b}$ for some $n$.*

(2) *If $f_\mu$, $\mu \in [\mu_0, \mu_1]$, is a one-parameter family of mappings such that $f_{\mu_0}$ and $f_{\mu_1}$ have periodic kneading sequences with the kneading sequence of $f_{\mu_0}$ smaller than the kneading sequence of $f_{\mu_1}$, then there is a set of bifurcations in the family $f_\mu$ corresponding to the creation of each periodic orbit intermediate between the kneading sequences of $f_{\mu_0}$ and $f_{\mu_1}$.*

EXERCISE 6.3.2 (Šarkovskii [1964], Li and Yorke [1975], Stefan [1977]). Let $f$ be a continuous map of the unit interval $I$ such that two closed intervals $L = [a, b]$, $R = [c, d]$ with $b \leq c$ are mapped under $f$ with $L \cup R \subseteq f(R)$ and $R \subseteq f(L)$. Suppose further that *either* $b < c$ or $f^2(b) \notin R$. Prove that $f$ has points of period $n$ for all $n$. Hence prove that if a map $g$ has a point of period 3, then it has points of all periods. (Hint: Look for orbits $\{f^k(x)\}$ with $f^k(x) \in R, 0 \leq k \leq n - 2, f^{n-1}(x) \in L$ and $f^n(x) = x \in R$.)

Rather than pursue the analytic details further as in Guckenheimer [1977, 1979] or Collet and Eckmann [1980], we shall illustrate a symbolic computation with kneading sequences which recovers the theorem of Šarkovskii [1964] (cf. Stefan [1977]) for the class of mappings we consider. The computation involves finding the smallest periodic orbit of each period relative to the order we have introduced here. It is a good exercise in using the symbolic calculus. We shall consider only itineraries which do not contain a $C$, since the ones which do are not the smallest itineraries of a periodic orbit with given period.

We begin with the observation that the constant itinerary **a** with $a_n = I_0$ for all $n$ represents only points which are asymptotic to a fixed point (0) in $I_0$. Thus all periodic orbits with periods larger than 1 contain the symbol $I_1$ in their itineraries. The itinerary with the largest invariant coordinate for a point of the orbit therefore begins with an $I_1$. Next observe that a sequence which begins $I_1 I_0$ is larger than one which begins $I_1 I_1$. Similarly, a sequence which begins $I_1 I_0 I_0$ is larger than one which begins $I_1 I_0 I_1$.

**Lemma 6.3.5.** *Consider the sequences* **a** *and* **b** *with* $a_0 = b_0 = I_1, a_1 = b_1 = I_0, a_j = b_j = I_1$ *for* $2 \leq j \leq l, a_l = I_0, b_l = I_1$. *If* $l$ *is even, then* $\theta(\mathbf{b}) < \theta(\mathbf{a})$. *If* $l$ *is odd, then* $\theta(\mathbf{a}) < \theta(\mathbf{b})$.

PROOF. Easy exercise, using the ordering on p. 307. □

**Lemma 6.3.6.** *Let* **a** *be an itinerary with* $a_i = I_1$ *when* $i$ *is even and* **b** *be an itinerary with* $b_0 = I_1$ *and* $b_i = I_0$ *for some even* $i$. *Then there is an integer* $k$ *such that* $\theta(\sigma^l(\mathbf{a})) < \theta(\sigma^k(\mathbf{b}))$ *for all* $l \geq 0$.

PROOF. Let $i$ be the smallest even integer with $b_i = I_0$. Choose $k$ even so that $b_{k+1} = I_0$ and $b_j = I_1$ for $k + 1 < j < i$. Then $\sigma^k(\mathbf{b}) = \mathbf{c}$ has $c_1 = c_{i-k} = I_0$ and $c_j = I_1$ for $1 < j < i - k$ and $j = 0$. Consider now $\sigma^l(\mathbf{a}) = \mathbf{d}$ for any $l \geq 0$. If $l$ is odd, then $d_1 = I_1$, and this implies that $\theta(\mathbf{d}) < \theta(\mathbf{c})$. If $l$ is even, then $d_j = I_1$ for all even $j$. If $d_1 = I_1$, then $\theta(\mathbf{d}) < \theta(\mathbf{c})$. If $d_1 = I_0$, pick $m$ to be the smallest integer larger than 1 with $d_m = I_0$. Now compare $\theta(\mathbf{c})$ and $\theta(\mathbf{d})$. If $m < i - k$, then $j = m$ is the smallest integer with $c_j \neq d_j$. We have $c_m = I_1$, $d_m = I_0$ and an even number of $I_1$'s in each sequence prior to $m$. Therefore $\theta(\mathbf{d}) < \theta(\mathbf{c})$. If $m > i - k$, then $j = i - k$ is the smallest integer with $c_j \neq d_j$. We have $c_{i-k} = I_0$, $d_{i-k} = I_1$ and an odd number of $I_1$'s in each sequence prior to $(i - k)$. Therefore $\theta(\mathbf{d}) < \theta(\mathbf{c})$. This concludes the proof because $m$

and $(i - k)$ have opposite parity. We have determined that $\theta(\sigma^l(\mathbf{a})) < \theta(\sigma^k(\mathbf{b}))$ for all $l \geq 0$.                                                          □

**Lemma 6.3.7.** *The largest point in the smallest periodic orbit of odd period $n$ has itinerary $\mathbf{a}$ with $a_i = I_0$ if $i \equiv 1$ (mod $n$) and $a_i = I_1$ otherwise.*

PROOF. If there is a $k$ with $a_k = I_1$ and $a_{k+1} = a_{k+2} = I_0$, then $\sigma^k(\mathbf{a})$ has larger invariant coordinates than any sequence with isolated occurrences of $I_0$. Therefore, the sequence $\mathbf{a}$ of this lemma has the property that $a_k = I_0$ implies $a_{k-1} = a_{k+1} = I_1$. Next we observe that there is a block of symbols of even length in the sequence, each of which is $I_1$ except the first and last. Lemma 6.3.5 implies that the longer these blocks, the smaller the itinerary. The itinerary specified in the present lemma has the block of this type of maximum length $n + 1$ for a periodic sequence of period $n$.                    □

**Corollary 6.3.8.** *If $n > 1$ is odd, there is a periodic orbit of period $n + 2$ which is smaller than all periodic orbits of period $n$.*

Using Lemma 6.3.6, we can construct periodic orbits of any even period which are smaller than all periodic orbits of odd period. These orbits have the property that every even indexed term in their itineraries is $I_1$. We observe that the graph of $f^2$ for a map $f$ whose kneading invariant is of this form looks like Figure 6.3.1. There is a subinterval $J$ of $[0, 1]$ around $c$ which is mapped into itself. The itinerary for $f^2$ of a point in $J$ will be subject to the same considerations as those which we have applied to $f$ above, but with the rôle of the symbols $I_0$ and $I_1$ reversed (since the map is "upside down"). Thus sequences $\mathbf{a}$ with $a_i = I_1$ for $i$ even and $a_i = I_0$ for $i \equiv 1$ (mod 4) are smaller than sequences $\mathbf{b}$ with $b_i = I_1$ for $i$ odd and $b_i = I_1$ for some $i \equiv 1$ (mod 4). This leads to:

**Lemma 6.3.9.** *If $k, l$ are odd and $m > n$, then there are periodic sequences of period $k \cdot 2^m$ which are smaller than all periodic sequences of period $l \cdot 2^n$.*

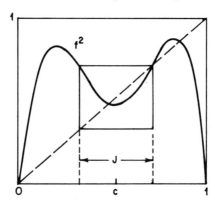

Figure 6.3.1. The graph of $f^2$. Note the invariant subinterval $J$.

PROOF. Exercise. (Use the above discussion.)    □

**Lemma 6.3.10.** *If $k$, $l$ are both odd and $1 < k < l$, there are periodic sequences of period $l \cdot 2^n$ which are smaller than all periodic sequences of period $k \cdot 2^n$.*

PROOF. Exercise.    □

This discussion of periodic orbits ends with the following theorem of Šarkovskii, which is true in a very general context.

**Theorem 6.3.11** (Šarkovskii [1964]; Stefan [1977]; Block *et al.* [1979]). *Consider the following ordering of all positive integers:*

$$1 \vartriangleleft 2 \vartriangleleft 4 \vartriangleleft \cdots \vartriangleleft 2^k \vartriangleleft 2^{k+1} \vartriangleleft \cdots$$

. . . . . . . . . .

$$\cdots \vartriangleleft 2^{k+1} \cdot (2l+1) \vartriangleleft 2^{k+1} \cdot (2l-1) \vartriangleleft \cdots \vartriangleleft 2^{k+1} \cdot 5 \vartriangleleft 2^{k+1} \cdot 3 \vartriangleleft \cdots$$
$$\cdots \vartriangleleft 2^k \cdot (2^l+1) \vartriangleleft 2^k \cdot (2l-1) \vartriangleleft \cdots \vartriangleleft 2^k \cdot 5 \vartriangleleft 2^k \cdot 3 \vartriangleleft \cdots$$

. . . . . . . . . .

$$\cdots \vartriangleleft 2(2l+1) \vartriangleleft 2(2l-1) \vartriangleleft \cdots \vartriangleleft 2 \cdot 5 \vartriangleleft 2 \cdot 3 \vartriangleleft \cdots$$
$$\cdots \vartriangleleft (2l+1) \vartriangleleft (2l-1) \vartriangleleft \cdots \vartriangleleft 5 \vartriangleleft 3.$$

*If $f$ is a continuous map of an interval into itself with a periodic point of period $p$ and $q \vartriangleleft p$ in this ordering, then $f$ has a periodic point of period $q$.*

In particular, this result contains the statement "period 3 implies chaos" of Li and Yorke [1975].

Note that the first sequence of periods in the Šarkovskii ordering consists of the powers of 2 in increasing order. This corresponds to a sequence of period doubling bifurcations in a one-parameter family as it moves from parameter values at which it has only fixed points to parameter values at which there are an infinite number of periodic orbits. These cascades of period doubling bifurcations have a rich structure which was first observed by Feigenbaum [1978]. There are properties associated with these cascades which are universal in the sense that they are independent of the particular map within large classes of mappings (cf. Collet and Eckmann [1980], Collet *et al.* [1980]); we discuss these in Section 6.8. Also, bifurcation to "chaos" through period doubling bifurcations appears to be a common feature of systems which are approaching a homoclinic tangency between stable and unstable manifolds of a periodic orbit. It has been numerically observed by Huberman and Crutchfield [1979] and Ueda [1981] in the forced Duffing equation. It has also been observed experimentally in fluid experiments (Libchaber and Maurer [1982]), thereby indicating the relevance of bifurcation theory for understanding the transition to turbulence in fluid systems (cf. Gollub and Benson [1980]).

## 6.4. The Lorenz Bifurcations

We can use the theory of one-dimensional mappings outlined in Section 6.3 to give a geometric description of the bifurcations which produce the Lorenz attractor. The process is quite subtle and involves both the symmetry of the system and two homoclinic transitions of different kinds. The main features of the bifurcation sequence were discovered by Kaplan and Yorke [1979b], while the characterization of the geometric Lorenz attractors themselves is due to Guckenheimer and Williams [1979] (cf. Section 5.7).

Let us recall the information about equilibrium points of the Lorenz equations (2.3.1) described in Section 2.3. For $\rho \le 1$, the Lorenz system has an equilibrium at the origin which is globally stable. At $\rho = 1$, a pitchfork bifurcation occurs, leaving the origin with one positive eigenvalue and creating two nonzero stable equilibria. These equilibria become unstable at a subcritical Hopf bifurcation when unstable periodic orbits collapse onto them. This bifurcation occurs at $\rho = \frac{470}{19} \approx 24.74$ (with $\sigma = 10$, $\beta = \frac{8}{3}$). In the parameter range $1 < \rho < 24.74$ between these bifurcations of equilibria, there are a variety of striking global changes in the dynamics of the Lorenz system. These changes can be described in terms of one-dimensional mappings, but the rigorous reduction of the system (2.3.1) to the pictures we present has not yet been accomplished. As in Section 5.7, we offer a family of geometrically defined models which remains an unsubstantiated interpretation of the true behavior of the Lorenz system, but whose behavior appears to agree with the results of numerical studies of the Lorenz system.

The nonzero equilibria $q^{\pm} = (\pm\sqrt{\beta(\rho - 1)}, \pm\sqrt{\beta(\rho - 1)}, \rho - 1)$ of the Lorenz system with $\rho > 1$ lie on the line obtained by intersecting the plane $z = \rho - 1$ with the plane $x = y$. We can study the further bifurcations of the system in the parameter range of interest by using the cross section $\Sigma$ (cf. Sections 2.3 and 5.7) in the plane $z = \rho - 1$ and its Poincaré return map $F$. We can also proceed by introducing a *strong stable foliation* (Robinson [1981a]) consisting of curves of points which asymptotically converge to one another exponentially as the flow evolves, as described in Section 5.7. The flow induces on this family of curves the semiflow depicted in Figure 2.3.3. The return map of this semiflow is a discontinuous map of the interval $I$. We shall work primarily with this one-dimensional mapping in this section, although reference to $\Sigma$ and its return map $F$ will be useful at some points of the discussion.

We abstract from the Lorenz system a family of one-dimensional mappings which correspond to the return map of the strong stable foliation for the parameter range which is approximately $10 < \rho < 30$ (we keep $\sigma = 10$ and $\beta = \frac{8}{3}$) fixed throughout). At the beginning of this parameter range, the flow is still simple, with a nonwandering set consisting solely of the three equilibria. The nonzero equilibria $q^{\pm}$ are stable, but one pair of eigenvalues is complex at each of them. The flow is illustrated in Figure 6.4.1(a) and the correspond-

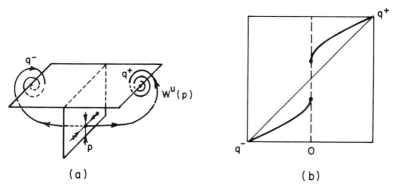

Figure 6.4.1. The Lorenz system for $\rho \approx 10, \sigma = 10, \beta = \frac{8}{3}$. (a) The flow; (b) the map $f_\rho$.

ing return map $f_\rho$ of the strong stable foliation in Figure 6.4.1(b). As $\rho$ increases, a bifurcation occurs in which the unstable manifold of $p$ becomes a pair of homoclinic trajectories. From numerical computations this occurs at $\rho = \rho_t \approx 13.296$. See Figure 6.4.2.

Recall that the slope of the map $f_\rho$ becomes infinite as one approaches its point of discontinuity, due to the fact that the unstable eigenvalue of $p$ is larger than the magnitude of one of the stable eigenvalues. This fact leads to the existence of a hyperbolic invariant set topologically equivalent to the suspension of a horseshoe, immediately following the homoclinic bifurcation illustrated in Figure 6.4.2. In terms of the return map $f_\rho$, the invariant set is created in the following way. The homoclinic bifurcation is followed by flows for which each branch of the unstable manifold of $p$ crosses to the opposite side of the stable manifold of $p$ as it descends for the first time: Figure 6.4.3. This forces the graph of $f_\rho$ to intersect the diagonal in a pair of fixed points $r^-$, $r^+$. These fixed points of $f$ correspond to unstable periodic orbits of the flow generated in much the same way that periodic orbits are generated by the saddle loop bifurcations in the plane described in Section 6.1.

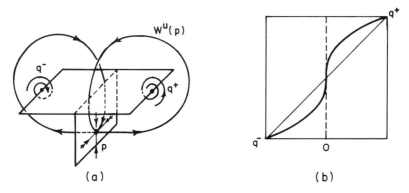

Figure 6.4.2. The Lorenz system for $\rho = \rho_t \approx 13.926$. (a) The flow; (b) the map $f_\rho$.

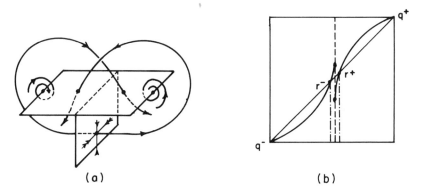

(a)                                          (b)

Figure 6.4.3. The Lorenz system for $\rho > \rho_t$. (a) The flow; (b) the map $f_\rho$.

Now we focus our attention on the graph of $f_\rho$ in the interval $[r^-, r^+]$. An expanded picture appears in Figure 6.4.4. The slope of $f$ is large in the interval $[r^-, r^+]$; consequently, each branch of the graph of $f$ passes vertically across the entire square which has opposite vertices on the diagonal above $r^-$ and $r^+$. Denoting by $d$ the point of discontinuity for $f_\rho$, $f_\rho$ has the property that the intervals $[r^-, d)$ and $(d, r^+]$ are each mapped onto $[r^-, r^+]$ with derivative larger than 1. Using $[r^-, d)$ and $(d, r^+]$ as a Markov partition of $[r^-, r^+]$, we identify a hyperbolic invariant set $\Lambda$ whose symbolic dynamics are a one-sided (full) shift on two symbols. A similar construction for the return map $F$ of the cross section $\Sigma$ produces a horseshoe as an invariant set of $F$. In Figure 6.4.5 we show two strips $V_i$ in the cross section $\Sigma$ and their images $H_i$ under $f$. Note that $F$ has a shift on two symbols with transition matrix

$$A = \begin{bmatrix} 1 & 1 \\ 1 & 1 \end{bmatrix}. \tag{6.4.1}$$

EXERCISE 6.4.1. Convince yourself that the above statement is correct. Use the methods of Section 5.2 (cf. Section 5.7).

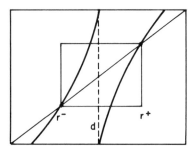

Figure 6.4.4. The map $f_\rho$, expanded in $[r^-, r^+]$.

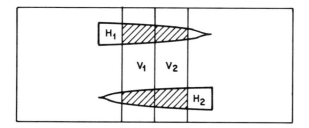

Figure 6.4.5. A shift for the Lorenz map $F$.

EXERCISE 6.4.2. Construct a Markov partition with *four* elements and show that the invariant set corresponding to the two-shift found above can be described as a subshift on four symbols with transition matrix

$$A = \begin{bmatrix} 1 & 1 & 0 & 0 \\ 0 & 0 & 1 & 1 \\ 0 & 0 & 1 & 1 \\ 1 & 1 & 0 & 0 \end{bmatrix}. \tag{6.4.2}$$

Hence conclude that the $F^{2n}$ has $4^n$ fixed points ($F^2$ has a full shift on four symbols). (Hint: Cf. the construction due to Levi described at the end of Section 5.3, or see Kaplan and Yorke [1979b].)

From (6.4.1) we can conclude that the map $F^n$ has $2^n$ fixed points. The horizontal expansion and vertical contraction implies that they are all of saddle type. For more details, see Kaplan and Yorke [1979b].

Kaplan and Yorke call this parameter regime one of *preturbulence*. While almost all trajectories ultimately tend to $q^-$ or $q^+$, there are long transients whose behavior has been studied statistically by Piangiani and Yorke [1979]. The hyperbolic invariant set separates regions of phase space which have different patterns of revolutions about $q^-$ or $q^+$ before they reach a small neighborhood of one of these points and "settle down." However, we note that the measure of the Cantor set $\Lambda$ is zero, and almost all orbits starting in $[r^-, r^+]$ eventually leave this interval and approach $q^-$ or $q^+$.

The bifurcation which makes the invariant set $\Lambda$ of the preturbulent Lorenz flow an *attractor* is subtle. As $\rho$ increases, the fixed points $r^-$ and $r^+$ of $f_\rho$ move toward $q^-$ and $q^+$, while the values $f_\rho(d)$ move toward $q^-$ and $q^+$ *more slowly*. One reaches a parameter value $\rho = \rho_a$ at which $f_{\rho_a}(d^-) = r^+$ and $f_{\rho_a}(d^+) = r^-$. Numerical investigations suggest $\rho_a \cong 24.06$ (Kaplan and Yorke [1979b]). For this parameter value the graph of $f$ is illustrated in Figure 6.4.6.

The interval $(r^-, r^+)$ is now mapped *into itself* by $f_{\rho_a}$ and the points of this interval can no longer approach the stable fixed points $q^-$ and $q^+$. The invariant set $\Lambda$ of the flow has now become an attractor (although its domain of attraction is not a neighborhood of $\Lambda$ because points near the periodic

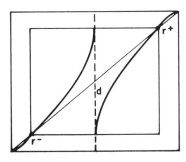

Figure 6.4.6. Another bifurcation for $f_\rho$: $\Lambda$ becomes an attractor.

orbits corresponding to $r^\pm$ can tend to $q^\pm$). The topology of $\Lambda$ has changed from the suspension of a Cantor set to an object which contains *two-dimensional surfaces*, as in Section 5.7. We have here the birth of the (geometric) Lorenz attractor occurring at a parameter value for which there are heteroclinic trajectories from the equilibrium $p$ to the periodic orbits corresponding to $r^\pm$.

For values of $\rho$ following the heteroclinic bifurcation which we have just described, we obtain geometric Lorenz attractors of the type described in Section 5.7. These attractors first appear before the points $q^\pm$ have lost their stability. The periodic orbits corresponding to $r^\pm$ are *not* part of the attractor and they are now isolated in the nonwandering set. The stable manifolds of these periodic orbits separate the trajectories which approach $q^\pm$ from the trajectories which approach the attractor $\Lambda$. As $\rho$ increases further, the periodic orbits corresponding to $r^\pm$ collapse onto $q^\pm$ in the subcritical Hopf bifurcation at $\rho = \rho_h \approx 24.74$ discussed at the beginning of the section and in Section 2.3. Thus the Hopf bifurcation is almost completely irrelevant to the dynamics which concern the attractor itself. The Hopf bifurcation merely marks the end of a parameter regime for which there are multiple attractors for the system.

The final issue which we discuss concerns the structural stability of the Lorenz attractors. In Section 5.7, we noted that these attractors fail to be structurally stable in a severe way. Nonetheless, one can reconstruct a geometric Lorenz attractor from the return map $f$ on its strong stable foliation. We saw in the previous sections that there are strong limitations on the topological equivalence classes and bifurcations of one-dimensional mappings. Using these one-dimensional techniques, it is possible to classify the different possible topological equivalence classes of Lorenz attractors. We now outline the procedure for classifying the topological equivalence classes of the one-dimensional return maps, referring the reader to Section 5.7, Guckenheimer and Williams [1979], and Robinson [1981a] for more details concerning the reconstruction of the attractors of the flow from $f_\rho$. Henceforth we drop explicit reference to the parameter $\rho$.

Figure 2.3.4 illustrates the type of one-dimensional maps whose bifurcations we wish to study. Suitably normalized, these are maps $f:[-1, 1] \rightarrow [-1, 1]$ which have the following properties:

(a)  $f$ has a single discontinuity at 0 and $\lim_{x \rightarrow 0^\pm} f(x) = \mp 1;$

(b)  $f'(x) > 1$  for all $x \neq 0$;  (6.4.3)

(c)  $f(-x) = -f(x)$.

The techniques used in Section 6.3 to study bifurcations of one-dimensional mappings can be used here to classify the maps satisfying (6.4.3). The interval $[-1, 1]$ is partitioned into $I_0 = [-1, 0)$ and $I_1 = (0, 1]$ and the symbolic dynamics of the map is studied for this partition.

**Theorem 6.4.1.** *The topological equivalence class of a map $f:[-1, 1] \rightarrow [-1, 1]$ satisfying (6.4.3) is determined by the symbol sequence of the point $(-1)$.*

PROOF. The symbol sequence $\mathbf{a}^+$ of $+1$ is obtained from the symbol sequence $\mathbf{a}^-$ of $-1$ by interchanging 0's and 1's in the symbol sequence because $f^n(+1) = -f^n(-1)$ for all $n$. If $\mathbf{b}^\pm$ are the limits of the symbol sequences obtained from $x$ as $x \rightarrow 0^\pm$, then $\sigma(\mathbf{b}^+) = \mathbf{a}^-$ and $\sigma(\mathbf{b}^-) = \mathbf{a}^+$ from (6.4.3). (Here $\sigma$ is the shift map on sequences.) We make the convention that both sequences $\mathbf{b}^\pm$ are associated to 0 and that $f(0)$ is the *two* points $\pm 1$. Using these facts, we can characterize the symbol sequences which actually occur for points of $[-1, 1]$. We shall use the lexicographic ordering on symbols to compare them. The following lemma is a major step in our proof:

**Lemma 6.4.2.** *Let $\mathbf{c} = \{c_i\}_{i=0}^{\infty}$ be a sequence of the symbols 0 and 1. There is a point $x \in [-1, 1]$ with symbol sequence $\{c_i\}$ if and only if $c_n = 0$ implies $\sigma^n(c) \geq \mathbf{a}^-$ and $c_n = 1$ implies $\sigma^n(c) \leq \mathbf{a}^+$.*

PROOF. Suppose $\mathbf{c}$ is the symbol sequence of $x$. We show that the inequalities on symbol sequences hold. Suppose $f^n(x) \in I_0$, so that $c_n = 0$. Then $x > c$ implies that $f(x) > f(c)$. If $\sigma^n(\mathbf{c}) \neq \mathbf{a}$ and $k$ is the smallest integer for which $c_{n+k} \neq a_k$, then $f$ is increasing on each interval $[f^i(-1), f^{n+i}(x))$ for $i \leq k$. Thus, we must have $c_{n+k} = 1$ and $a_k = 0$. The inequality of symbol sequences for $c_n = 1$ is similar.

Suppose now that $\mathbf{c}$ is a symbol sequence satisfying the above inequalities. We shall prove that $\bigcap_{n \geq 0} f^{-n}(I_{c_n}) \neq \varnothing$. The finite intersection property implies that this will be true if $\bigcap_{n=0}^{N} f^{-n}(I_{c_n}) \neq \varnothing$ for all $N \geq 0$. Clearly $I_{c_0}$ is nonempty. We now prove that if the intersections $\bigcap_{n=0}^{N-1} f^{-n}(I_{c_n})$ are nonempty for all sequences $\mathbf{c}$ satisfying the inequalities, then the intersections $\bigcap_{i=0}^{N} f^{-n}(I_{c_n})$ are nonempty. Consider a sequence $\mathbf{c}$ with $c_0 = 0$. Then $\bigcap_{n=0}^{N-1} f^{-n}(I_{c_{n+1}})$ is nonempty, since $\sigma(\mathbf{c})$ satisfies our inequalities if $\mathbf{c}$ does. We assert that

$$f^{-1}\left(\bigcap_{n=0}^{N-1} f^{-n}(I_{c_{n+1}})\right) = \bigcap_{n=0}^{N-1} f^{-(n+1)}(I_{c_{n+1}})$$

intersects $[-1, 0)$. This is true unless $\bigcap_{n=0}^{N-1} f^{-n}(I_{c_{n+1}})$ lies in the interval $[-1, f(-1)]$ consisting of points $x$ for which $f^{-1}(x)$ is a single point in $[0, 1]$. If $\bigcap_{n=0}^{N-1} f^{-n}(I_{c_{n+1}}) \subset [-1, f(-1))$, then the symbol sequences of these points are strictly *smaller* than the symbol sequence $\sigma(\mathbf{a}^-)$ of $f(-1)$. (The expansiveness of $f$ implies that distinct points have different symbol sequences.) Since $c_0 = 0 = a_0$, we conclude that $\mathbf{c} < \mathbf{a}$, contradicting the inequality assumed for $\mathbf{c}$. Thus, $\bigcap_{n=0}^{N-1} f^{-n}(I_{c_{n+1}})$ contains points in $[f^{-1}(-1), 1]$, and $\bigcap_{n=0}^{N} f^{-n}(I_{c_n})$ is nonempty. The lemma is proved.

In the course of proving the lemma, we remarked that (6.4.3(b)) implies that distinct points have distinct symbol sequences. Therefore, apart from the minor complication that points whose trajectory includes 0 have two symbol sequences associated to them,* the correspondence between symbol sequences and points is 1–1 and order preserving. If $f$ and $g$ are two maps satisfying (6.4.3) with the same symbol sequences for $(-1)$, then we can use the lemma to define an order preserving map $\phi$ from $f$ to $g$ which associates points with the same symbol sequence. We leave it to the reader to check that $\phi$ will be a topological equivalence and hence complete the proof of the theorem.                                                                                      □

EXERCISE 6.4.3. The possibilities for the symbol sequence of $(-1)$ for a map satisfying (6.4.3) are limited. Show that if $f$ satisfies (6.4.3) and the slope of $f$ is greater than $\sqrt{2}$, then there is an $\alpha$ with $\sqrt{2} < \alpha \leq 2$ such that $f$ is topologically equivalent to the piecewise linear map $g: [-1, 1]$ defined by

$$g(x) = \begin{cases} 1 + \alpha x, & x \leq 0, \\ -1 + \alpha x, & x \geq 0. \end{cases}$$

Glendinning, "The structure of $\beta$ transformations and conjacies to Lorenz maps" (preprint, University of Warwick) provides an analysis of what happens when the slope of $f$ is between 1 and $\sqrt{2}$.

We end this section by remarking that the symmetry properties of the Lorenz system play an important rôle in the bifurcations we have described. It is an open-ended research project to determine how these bifurcations are modified if the Lorenz system is perturbed so as to destroy the rotational symmetry about the $z$-axis. Guckenheimer [1980b] contains a few remarks about this problem.

## 6.5. Homoclinic Orbits in Three-Dimensional Flows: Šilnikov's Example

In this section we consider a three-dimensional flow in which there is a homoclinic trajectory to an saddle point with complex eigenvalues. Šilnikov [1965] showed that if the real eigenvalue has larger magnitude than the real part of the complex eigenvalues, then there are horseshoes present in return

---

* If $f''(\pm 1) = 0$ for some $n$, then we can still associate just *two* symbol sequences to 0, taking these to be the limits of symbol sequences associated to $x \neq 0$ as $x \to 0$.

maps defined near the homoclinic orbit. The sequence of Lorenz bifurcations described in Section 6.4 showed that a homoclinic orbit can be associated with the appearance of a (suspended) horseshoe to one side of the bifurcation in the parameter space. The Šilnikov example is different in that the horseshoes exist at and in a full neighborhood of the parameter value for which a homoclinic orbit occurs. The Lorenz bifurcation also required a symmetry in the system which the Šilnikov example does not. Here we shall outline only one of the cases considered by Šilnikov. The thesis of Tresser (cf. Tresser [1982]), contains much more information about systems of the Šilnikov type.

Consider a flow $\phi_t$ in $\mathbb{R}^3$ which has an equilibrium point at the origin with a real positive eigenvalue $\lambda$ and pair of complex eigenvalues $\omega, \bar{\omega}$ which have negative real parts (the case of $\lambda$ negative and $\text{Re}(\omega)$ positive is dealt with similarly). The stable manifold theorem allows us to introduce coordinates so that the local unstable manifold is contained in the $z$-axis and the local stable manifold is contained in the $(x, y)$ plane. We assume further that the trajectory $\gamma$ in $W^u(0)$ which points upward near $0$ is a homoclinic orbit which enters the $(x, y)$ coordinate plane and spirals toward the origin as $t \to \infty$; see Figure 6.5.1.

We shall study a return map for orbits near $\gamma$ in this section and prove the following theorem:

**Theorem 6.5.1** (Šilnikov [1965]). *If $|\text{Re } \omega| < \lambda$, then the flow $\phi_t$ can be perturbed to $\phi'_t$ such that $\phi'_t$ has a homoclinic orbit $\gamma'$ near $\gamma$ and the return map of $\gamma'$ for $\phi'_t$ has a countable set of horseshoes.*

The proof of this theorem requires a careful analysis of the flow of trajectories passing near the origin. To simplify the analysis we make an initial perturbation of $\phi_t$ which makes its vector field linear in a neighborhood $U$ of the origin. We assume throughout the remainder of the proof that our

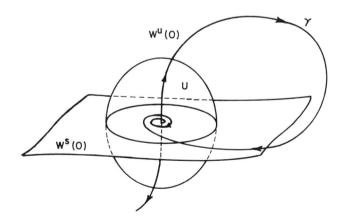

Figure 6.5.1. A homoclinic orbit to a spiral saddle.

original vector field has this property. In this case, the solution of the system of equations

$$\begin{pmatrix} \dot{x} \\ \dot{y} \\ \dot{z} \end{pmatrix} = \begin{bmatrix} \alpha & -\beta & 0 \\ \beta & \alpha & 0 \\ 0 & 0 & \lambda \end{bmatrix} \begin{pmatrix} x \\ y \\ z \end{pmatrix}; \qquad \omega = \alpha + i\beta, \qquad (6.5.1)$$

which governs the flow near the origin, is given by

$$\begin{pmatrix} x \\ y \\ z \end{pmatrix}(t) = \begin{pmatrix} e^{\alpha t}((\cos \beta t)x(0) - (\sin \beta t)y(0)) \\ e^{\alpha t}((\sin \beta t)x(0) + (\cos \beta t)y(0)) \\ e^{\lambda t}z(0) \end{pmatrix}. \qquad (6.5.2)$$

As trajectories flow past the origin, the "radial" coordinate $r = \sqrt{x^2 + y^2}$ decreases while $|z|$ increases.

We define two surfaces $\Sigma_0$ and $\Sigma_1$ by

$$\begin{aligned} \Sigma_0 &= \{(x, y, z) | x^2 + y^2 = r_0^2 \text{ and } 0 < z < z_1\}, \\ \Sigma_1 &= \{(x, y, z) | x^2 + y^2 < r_0^2 \text{ and } z = z_1 > 0\}. \end{aligned} \qquad (6.5.3)$$

We assume that $\Sigma_0$ and $\Sigma_1$ are in the neighborhood $U$ where the flow is linear. Trajectories flow from $\Sigma_0$ to $\Sigma_1$ according to (6.5.2). We want to compute the mapping $\psi: \Sigma_0 \to \Sigma_1$ which associates to each point $a \in \Sigma_0$, the first intersection with $\Sigma_1$ of the trajectory starting at $a$. See Figure 6.5.2.

The formula for $\psi$ is given by solving $z_1 = e^{\lambda t}z(0)$ for $t$ to obtain the time of flight $t = \lambda^{-1} \ln(z_1/z(0))$, and substituting the result into (6.5.2). We obtain

$$\phi_t(x, y, z) = \left(\left(\frac{z_1}{z}\right)^{\alpha/\lambda} [(\cos \gamma)x - (\sin \gamma)y], \left(\frac{z_1}{z}\right)^{\alpha/\lambda} [(\sin \gamma)x + (\cos \gamma)y], z_1\right),$$

$$(6.5.4)$$

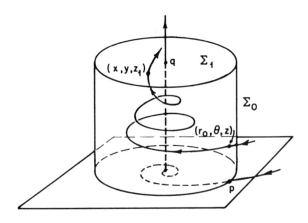

Figure 6.5.2. The sections $\Sigma_0$ and $\Sigma_1$.

where $\gamma = \beta \lambda^{-1} \ln(z_1/z)$. Setting $x = r_0 \cos \theta$, $y = r_0 \sin \theta$, we may then express $\psi : \Sigma_0 \to \Sigma_1$ as a two-dimensional diffeomorphism whose domain has coordinates $\theta = \tan^{-1}(y/x)$ and $z$, and whose range has coordinates $x, y$:

$$\psi(\theta, z) = \left( r_0 \left( \frac{z_1}{z} \right)^{\alpha/\lambda} \cos(\theta + \gamma), \; r_0 \left( \frac{z_1}{z} \right)^{\alpha/\lambda} \sin(\theta + \gamma) \right) \tag{6.5.5}$$

$$\stackrel{\text{def}}{=} (\psi_1(\theta, z), \psi_2(\theta, z)).$$

We note that $\psi$ maps a vertical segment $\theta = $ const. of $\Sigma_0$ to a logarithmic spiral which encircles the $z$-axis and lies in $\Sigma_1$. In order to see that the vertical segment is stretched as it wraps around the spiral, we shall need the Jacobian derivative of $\psi$. We have, from (6.5.5)

$$D\psi(\theta, z) = \begin{bmatrix} \dfrac{\partial \psi_1}{\partial \theta} & \dfrac{\partial \psi_1}{\partial z} \\[2ex] \dfrac{\partial \psi_2}{\partial \theta} & \dfrac{\partial \psi_2}{\partial z} \end{bmatrix},$$

and the derivative may be written as the product of two matrices

$$D\psi(\theta, z) = r_0 \left( \frac{z_1}{z} \right)^{\alpha/\lambda} \begin{bmatrix} \cos \gamma & -\sin \gamma \\ \sin \gamma & \cos \gamma \end{bmatrix} \begin{bmatrix} -\sin \theta & \dfrac{-\alpha \cos \theta + \beta \sin \theta}{\lambda z} \\[2ex] \cos \theta & \dfrac{-\alpha \sin \theta - \beta \cos \theta}{\lambda z} \end{bmatrix} \tag{6.5.6}$$

(such a decomposition will be useful later). Thus we find that

$$\det(D\psi) = \left( \frac{\alpha r_0^2 z_1^{2\alpha/\lambda}}{\lambda} \right) z^{-(1 + 2\alpha/\lambda)}. \tag{6.5.7}$$

This shows that for $z$ sufficiently small, $\psi$ contracts (resp. expands) areas if $2\alpha < -\lambda$ (resp. $2\alpha > -\lambda$), consistent with the fact that the vector field has divergence $2\alpha + \lambda$ at the saddle point. However, since $-1 < \alpha/\lambda < 0$, even when $\psi$ contracts areas, vertical segments $\{\theta = $ const.; $z \in (0, z_0)$ in $\Sigma_0$ are mapped into logarithmic spirals of radius $r_0(z_1/z)^{\alpha/\lambda}$, and thus their lengths are stretched by an amount which becomes unbounded as $z \to 0$.

EXERCISE 6.5.1. Verify that equations (6.5.6) and (6.5.7) are correct.

Denote now by $p$ the intersection of $W^u(0)$ with $\Sigma_0$ and by $q$ the point $(0, 0, z_1) = W^u(0) \cap \Sigma_1$ (see Figure 6.5.2). The flow from $q$ to $p$ along $W^u(0)$ is nonsingular, so there is a diffeomorphism $\phi$ from a neighborhood of $q$ in $\Sigma_1$ to a neighborhood of $p$ in the surface $\tilde{\Sigma}_0 = \{(x, y, z) | x^2 + y^2 = r_0, |z| < z_0\}$ which sends a point to the first intersection of its trajectory with $\tilde{\Sigma}_0$. The return map of $\Sigma_0$ near $p$ is given by $\phi \circ \psi$ for those points $r$ with $\phi(\psi(r)) \in \Sigma_0$. See Figure 6.5.3. Changing coordinates by a rotation around

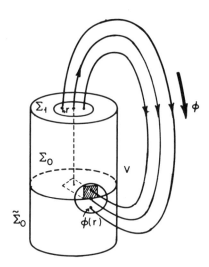

Figure 6.5.3. The return map $\phi$.

the $z$-axis, we may assume that $p$ lies on the $x$-axis ($\theta = 0$). Pick a small domain $V$ in $\Sigma_0$ defined by

$$V = \{(r, \theta, z) \mid r = r_0, |\theta| < \delta, 0 < z < \varepsilon\}, \qquad (6.5.8)$$

where $\varepsilon$ is chosen such that $\psi(V)$ is contained in the domain of $\phi$ and $\delta$ will be specified later (Figure 6.5.3). The image of $\phi \circ \psi(V)$ is illustrated in Figure 6.5.4.

Each vertical segment of $V$ is mapped to a spiral around $q$ in $\Sigma_1$, and $\phi$ maps this spiral diffeomorphically into $\tilde{\Sigma}_0$ with $\phi(q) = p$. If $\delta$ is chosen sufficiently large, relative to $\varepsilon$, then each spiral in $\phi \circ \psi(V)$ cuts through $V$ vertically many times until one reaches a portion of the spiral which does not cut through the top of $V$.

We want to locate subsets of $V$ in which $\phi \circ \psi$ has a horseshoe. Note that (6.5.5) implies that if $z$ varies along a vertical segment $J \subset V$ with $z \in (z', z'')$,

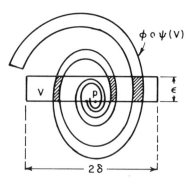

Figure 6.5.4. $V$ and $\phi \circ \psi(V)$.

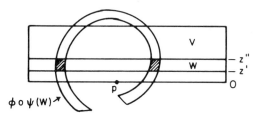

Figure 6.5.5.  $\phi \circ \psi|_W$ has a horseshoe: $W \cap \phi \circ \psi(W)$ shown shaded.

where $(\beta/\lambda)(\log(z_1/z') - \log(z_1/z'')) = 2\pi$ or $z''/z' = \exp(2\pi\lambda/\beta)$, then $\psi(J)$ makes one full turn of a logarithmic spiral. The distance of $\psi(J)$ from $q$ is of the order of $(z_1/z)^{\alpha/\lambda}$, $z \in J$. The ratio $(z_1/z)^{\alpha/\lambda}/z \to \infty$ as $z \to 0$. Fix $\delta > 0$ small in the definition of $V$ and pick $(z', z'')$ with the properties

(i) $z''/z' = \exp(2\pi\lambda/\beta)$,

(ii) $z'$, $z''$ are sufficiently small that $(z_1/z)^{\alpha/\lambda} < \delta$ for all $z \in (z', z'')$.  (6.5.9)

(iii) if $|\theta| < \delta$, then the images $\phi \circ \psi(r_0, \theta, z')$ and $\phi \circ \psi(r_0, \theta, z'')$ do not lie in $\Sigma_0$.

Define $W = \{(r_0, \theta, z) \in V \,|\, z \in (z', z'')\}$ and observe that (6.5.9) implies that the image $\phi \circ \psi(W)$ looks like a horseshoe mapping; see Figure 6.5.5.

To demonstrate that $W \cap \phi \circ \psi(W)$ contains a horseshoe, we need to show that $D(\phi \circ \psi)$ satisfies the sectorial conditions H1 and H3 of Theorem 5.2.4. Satisfaction of the first is immediate. Moreover, since $\phi$ is a diffeomorphism on $\Sigma_1$, from (6.5.6) we have

$$D(\phi \circ \psi)(r_0, \theta, z)$$

$$= r_0 \left(\frac{z_1}{z}\right)^{\alpha/\lambda} A \begin{bmatrix} \cos\gamma & -\sin\gamma \\ \sin\gamma & \cos\gamma \end{bmatrix} \begin{bmatrix} -\sin\theta & \dfrac{-\alpha\cos\theta + \beta\sin\theta}{\lambda z} \\ \cos\theta & \dfrac{-\alpha\sin\theta - \beta\cos\theta}{\lambda z} \end{bmatrix},$$

(6.5.10)

where $A = D\phi(\psi(r_0, \theta, z))$. We can rewrite (6.5.10) in the form

$$z^{-(\alpha/\lambda + 1)} BC \begin{bmatrix} z & 0 \\ 0 & 1 \end{bmatrix},$$

where

$$B = r_0 z_1^{\alpha/\lambda} \cdot A \begin{bmatrix} \cos\gamma & -\sin\gamma \\ \sin\gamma & \cos\gamma \end{bmatrix},$$

and the matrix

$$
C = \begin{bmatrix} -\sin\theta & \dfrac{-\alpha\cos\theta + \beta\sin\theta}{\lambda} \\[2ex] \cos\theta & \dfrac{-\alpha\sin\theta - \beta\cos\theta}{\lambda} \end{bmatrix} \tag{6.5.11}
$$

is nonsingular and approximately constant in each component of $W \cap \phi \circ \psi(W)$, since $|\theta| < \delta$. Moreover, since the points in $W \cap \phi \circ \psi(W)$ have images under $\psi$ whose arguments in $\Sigma_1$ differ by approximately $\pi$, the values of $B$ in the two components of $W \cap \phi \circ \psi(W)$ differ approximately by multiplication by

$$
-I = \begin{bmatrix} -1 & 0 \\ 0 & -1 \end{bmatrix}.
$$

When $z$ is small, $z^{-(\alpha/\lambda + 1)}$ is large and $z^{-\alpha/\lambda}$ is small. We observe that the eigenvalues of the product

$$
\mu \begin{bmatrix} b_{11} & b_{12} \\ b_{21} & b_{22} \end{bmatrix} \begin{bmatrix} c_{11} & c_{12} \\ c_{21} & c_{22} \end{bmatrix} \begin{bmatrix} 0 & 0 \\ 0 & 1 \end{bmatrix} = \mu \begin{bmatrix} 0 & b_{11}c_{12} + b_{12}c_{22} \\ 0 & b_{21}c_{12} + b_{22}c_{22} \end{bmatrix} \tag{6.5.12}
$$

are 0 and $\mu(b_{21}c_{12} + b_{22}c_{22})$, with eigenvectors $(1, 0)$ and $((b_{11}c_{12} + b_{12}c_{22}), (b_{21}c_{12} + b_{22}c_{22}))$, respectively. Since $\mu = z^{-(\alpha/\lambda + 1)}$, as $z \to 0$ the second eigenvalue becomes large as we choose strips with $z', z'' \to 0$ and, provided the ratio $|(b_{21}c_{12} + b_{22}c_{22})/(b_{11}c_{12} + b_{12}c_{22})|$ remains bounded away from zero, then the eigenvectors of 0 and $\mu(b_{21}c_{12} + b_{22}c_{22})$ remain in disjoint cones. Moreover, the eigenvectors and eigenvalues of (6.5.10) are close to those of singular matrix (6.5.12) for, being distinct, they vary smoothly with the matrix coefficients. If $(b_{21}c_{12} + b_{22}c_{22})$ is not bounded away from 0 for points of $V \cap \phi \circ \psi(V)$ as $z \to 0$, then we must perturb the flow $\phi_t$ to accomplish this. A perturbation which has the effect of rotating trajectories around $\gamma$ so that the perturbed return map is the composition (on the left) with a rigid rotation does the trick; see Figure 6.5.6.

The sectorial structure for (6.5.11) is now obtained from the limiting values of the eigenvectors of this product as $z', z'' \to 0$. (The limit values do exist, because $\theta \to 0$ and $\gamma \pmod{\pi}$ approaches the value which determines the line in $\Sigma_1$ which $D\phi(q)$ maps to the horizontal line in the tangent space to $\Sigma_0$ at $p$.) Any small sectors around these limiting values satisfy (H3) when $z'$ and $z''$ are sufficiently small. Choosing a sequence of rectangles $W$ in $V$ whose heights decrease geometrically, we obtain a sequence of horseshoes for our perturbed flow, which still has a homoclinic orbit for the equilibrium at the origin.  □

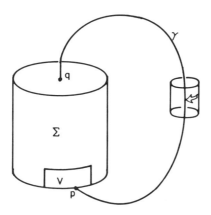

Figure 6.5.6. A small perturbation of $\phi_t$.

**Corollary 6.5.2** (Šilnikov [1965]). *Let $X$ be a $C^r$ vector field on $\mathbb{R}^3$ which has an equilibrium $p$ such that:*

(i) *the eigenvalues of $p$ are $\alpha \pm i\beta$, $\lambda$ with $|\alpha| < |\lambda|$ and $\beta \neq 0$;*
(ii) *there is a homoclinic orbit for $p$.*

*Then there is a perturbation $Y$ of $X$ such that $Y$ has invariant sets containing transversal homoclinic orbits.*

EXERCISE 6.5.2. Analyze the bifurcation behavior associated with a homoclinic trajectory to an equilibrium with eigenvalues $\alpha \pm i\beta$, $\lambda$ with $0 < \lambda < -\alpha$ and $\beta \neq 0$. (Hint: See Holmes [1980c] or Tresser [1982].)

Arnéodo *et al.* [1981, 1982] give examples of three-dimensional systems for which the hypotheses of these theorems can be checked. Flows containing such homoclinic trajectories have also been found in phase portrait analysis for nerve axon equations (cf. Evans *et al.* [1982], Feroe [1982], Hastings [1982]). For more examples in which these ideas have been used, see Šilnikov [1967, 1970] and Holmes [1980c].

# 6.6. Homoclinic Bifurcations of Periodic Orbits

The discrete version of the saddle loop bifurcations discussed in Section 6.1 is much more complicated than in the former case. When the stable and unstable manifolds of a saddle fixed point of a diffeomorphism $f: \mathbb{R}^2 \to \mathbb{R}^2$ intersect transversally, as we have proved in Chapter 5, $f$ has an infinite number of periodic orbits. Thus, bifurcations must occur in the process of deforming a diffeomorphism with a finite number of periodic orbits so that transversal homoclinic orbits appear. This process is one that was first observed in the highly nonlinear forced Van der Pol equation by Cartwright and Littlewood [1945], cf. Section 2.1. We have also met some of these bifurcations in the Melnikov theory of Chapter 4 (Section 4.6). In this section, we shall examine some bifurcations of periodic orbits which are associated with a periodic orbit having tangential stable and unstable manifolds. The results we describe are due to Gavrilov and Silnikov [1972, 1973] and to Newhouse [1979, 1980]. We do not strive for generality, but examine the simplest example which displays phenomena of interest.

Let $f_\mu$ be a one-parameter family of diffeomorphisms of $\mathbb{R}^2$ which all have a fixed point at the origin with eigenvalues $\rho$ and $\lambda$ satisfying $\rho(\mu) < 1 < \lambda(\mu) < \rho^{-1}(\mu)$. Such a saddle point, with $|\rho\lambda| < 1$, is called *dissipative*. Assume that, when $\mu > 0$, $W^s(0) \cap W^u(0) = \{\varnothing\}$, and that when $\mu = 0$, there is a point $p_0$ at which $W^s(0) \cap W^u(0)$ have a *quadratic tangency* for $f_0$. By this we mean that the curvatures of $W^s(0)$ and $W^u(0)$ are different. When $\mu < 0$ is small, $W^s(0)$ and $W^u(0)$ should have two points of transversal intersection near $p_0$. We assume further that when $\mu = 0$, the segments of $W^s(0)$ and $W^u(0)$ from 0 to $p_0$ bound a region which has a convex angle at 0 and contains points of $W^u(0)$ in its interior. See Figure 6.6.1. Clearly many other cases of tangency can occur (cf. Gavrilov and Silnikov [1973]).

The first result we describe, due to Gavrilov and Silnikov, shows that there are sequences of saddle-node bifurcations accumulating at $\mu = 0$ from both above and below.

We treat the situation in a fashion similar to that employed in describing Smale's [1963, 1967] theorem on the existence of horseshoes. Here we shall describe invariant sets that change with $\mu$, their periodic points appearing, disappearing, or changing their stability in the process. The bifurcations of these periodic orbits will necessarily be saddle-nodes or flips, because the area contraction of $f_\mu$ near 0 dominates any expansion of $f_\mu$ elsewhere for the sets we consider, and we obtain one-dimensional center manifolds associated with the passage of simple eigenvalues of $Df_\mu^n$ through $+1$ or $-1$. To begin the construction of the invariant sets, we use the stable manifold theorem to pick coordinates in which the local stable manifold of 0 is a segment of the $x$-axis and the local unstable manifold of 0 is a segment of the $y$-axis. This

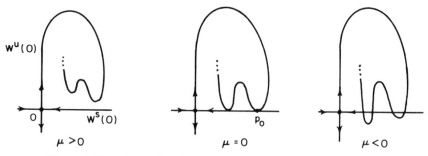

Figure 6.6.1. A homoclinic bifurcation for a planar diffeomorphism.

can be done for all $\mu$ simultaneously because the local stable and unstable manifolds of 0 vary continuously with $\mu$. Let us suppose then, that $W^u(0)$ and $W^s(0)$ are "flat" in some neighborhood $V$ of the origin.

First consider the system when $\mu = 0$. Let $p_0 = (0, y_0)$ in our coordinate neighborhood $V$ be a point of tangency between $W^s(0)$ and $W^u(0)$. Let $U \subset V$ be a small rectangular neighborhood of $p_0$. We want to follow the iterates of $U$ until they intersect $U$ again, and then examine the hyperbolicity of this intersection. Recall that we have assumed that the stable and unstable manifolds intersect in the manner depicted in Figure 6.6.1. There is clearly an integer $k$ such that $f^k(p_0)$ lies on the local stable manifold of 0. Moreover, the curve $W^u(0)$ intersects the $x$-axis tangentially at $f_0^k(p_0)$, and we have assumed that it can be represented near this point as the graph of a function with positive second derivative. See Figure 6.6.2.

We next follow further iterates of $p_0$ and $W^u(0)$ from $f_0^k(p_0)$. To simplify our estimates and focus upon the essential features of this iteration, we shall examine a particular example and leave the generalization of the arguments as an exercise. In particular, we assume that $f_0$ is linear in its coordinate neighborhood about 0, so that

$$f_0(x, y) = Df_0(x, y) = (\rho x, \lambda y) \tag{6.6.1}$$

for all $(x, y)$ in the neighborhood. We assume further that $f_0^k$ is a quadratic mapping at $p_0$ of the form

$$f_0^k(x, y - y_0) = (x_0 - \beta(y - y_0), \gamma x + \delta(y - y_0)^2), \tag{6.6.2}$$

in our specified neighborhood $V$ of $p_0$. Here $p_0 = (0, y_0), f_0^k(p_0) = (x_0, 0)$, and $\beta$, $\gamma$, and $\delta$ are positive constants ($\delta\beta^{-2}$ is the curvature of $W^u(0)$ at

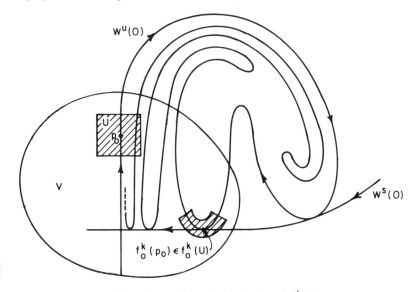

Figure 6.6.2. The neighborhoods $V$, $U$, and $f^k(U)$.

$f^k_0(p_0)$ in our coordinate system). This particular nonlinear map carries vertical lines $x = c$ in $U$ into parabolas $y = \gamma c + (\delta/\beta^2)(x - x_0)^2$ near $f^k(p_0) \in W^s(0)$.

For all $n \geq 0$, $f^{n+k}_0$ is a quadratic mapping provided $f^{k+i}_0(x, y)$ is in our coordinate neighborhood $V$ for $i \leq n$: explicitly, we have

$$f^{n+k}_0(x, y) = (\rho^n(x_0 - \beta(y - y_0)), \lambda^n(\gamma x + \delta(y - y_0)^2)). \qquad (6.6.3)$$

We want to show that, if $n$ is large, then there is a rectangle with $p_0$ on its boundary for which $f^{n+k}_0$ has a horseshoe. Pick $\varepsilon > 0$, $\rho < v < \lambda^{-1}$, and $n$ so that the rectangle $S_n$ of width $v^n$, whose left edge is the segment $(y_0 - \varepsilon, y_0 + \varepsilon)$ of the $y$-axis, lies in $V$ and its image $f^k(S_n)$ also lies in $V$. We will show that $f^{k+n}_0$ has horseshoes in $S_n$ for $n$ large.

First observe that the abscissas of points in $f^{n+k}_0(S_n)$ are all smaller than $\rho^n(x_0 + \beta\varepsilon)$ and this is smaller than $v^n$ when $n$ is large. Next observe that the ordinate of $f^{n+k}_0(x, y) = \lambda^n(\gamma x + \delta(y - y_0)^2)$ is approximately $y_0$ when $|y - y_0| \simeq (y_0\lambda^{-n}/\delta)^{1/2}$ since $\lambda^n\gamma x < \lambda^n\gamma v^n \to 0$ as $n \to \infty$. Thus the image $f^{n+k}_0(S_n)$ intersects $S_n$ in two vertical strips whose inverse images are approximately horizontal strips near values of $y$ which satisfy $|y - y_0| = (y_0\lambda^{-n}/\delta)^{1/2}$, cf. Figure 6.3.3. The derivative of $f^{n+k}$ in $V$ is

$$Df^{n+k}_0 = \begin{bmatrix} 0 & -\beta\rho^n \\ \gamma\lambda^n & 2\delta\lambda^n(y - y_0) \end{bmatrix}. \qquad (6.6.4)$$

When $|y - y_0| = (y_0\lambda^{-n}/\delta)^{1/2}$ and $n$ is large, $Df^{n+k}_0$ is approximated by the singular map

$$\begin{bmatrix} 0 & 0 \\ \gamma\lambda^n & 2(y_0\delta\lambda^n)^{1/2} \end{bmatrix}, \qquad (6.6.5)$$

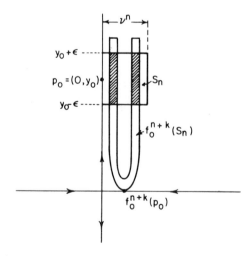

Figure 6.6.3. Quadratic tangencies imply horseshoes.

which has eigenvectors $(1, - \gamma\lambda^{n/2}/2\sqrt{y_0}\delta)$ and $(0, 1)$. While the vertical component of $(1, - \gamma\lambda^{n/2}/2\sqrt{y_0}\delta)$ is large, it is of a smaller order of magnitude than $\lambda^{n/2}\rho^{-n}$, which is the magnitude of the slope of any vector which was originally near vertical. One checks easily that there is a constant $c$ and a sector at each point of $f^{-(n+k)}(S_n) \cap S_n$ bounded by vectors of slope $\pm c\rho^{-n}\lambda^{+n/2}$ which is mapped into itself by $Df_0^{n+k}$ (cf. the analysis of Section 5.2). The vectors in this sector are stretched by an amount which is bounded below and therefore contain the expanding vectors of a hyperbolic structure. For this example, we have therefore proved the following theorem:

**Theorem 6.6.1.** *Let $f_0: \mathbb{R}^2 \to \mathbb{R}^2$ be a diffeomorphism having a fixed point $p$ at which the eigenvalues $\rho$ and $\lambda$ of $Df(p)$ satisfy $\rho < 1 < \lambda < \rho^{-1}$. Assume further that $W^u(p)$ and $W^s(p)$ have a point $p_0$ of tangential intersection, and that the region $D$ bounded by the segments $W^u(p)$ and $W^s(p)$ from $p$ to $p_0$ is mapped into itself by $f_0$. If $V$ is a neighborhood of $p$, then there is an integer $M$ such that $m \geq M$ implies that the mapping $f^m$ has an invariant set $\Lambda_m$ in $V$ topologically conjugate to a shift on two symbols (i.e., a horseshoe).*

**Remark.** The structural stability of $\Lambda_m$ guarantees that we can perturb the map slightly (for example, by adding higher-order terms, or small non-linearities in $V$) without destroying $\Lambda_m$.

We also remark that this analysis yields a *countable* set of horseshoes $\Lambda_m$ for all $m(=n+k) \geq M$. In the papers of Gavrilov and Silnikov [1972, 1973] other cases of tangencies are considered for both orientation preserving and reversing diffeomorphisms. In some of these situations horseshoes do not occur in the neighborhood of $p_0$ until the tangency is perturbed to yield transverse homoclinic points.

Having established that near a homoclinic bifurcation complicated dynamics may already be present, we next explore what happens when the homoclinic tangency is perturbed in a one-parameter family. We continue the discussion in terms of the quadratic example which we introduced above, but we incorporate it in a one-parameter family of maps $f_\mu$. Working in a fixed neighborhood of the origin, we assume that $f_\mu$ is the linear map defined by $f_\mu(x, y) = (\rho x, \lambda y)$ with $\rho < 1 < \lambda < \rho^{-1}$ as before, for all $\mu$, and assume that there is an integer $k > 0$ and a point $p_0 = (0, y_0)$ near which $f_\mu^k$ is given by

$$f_\mu^k(x, y) = (x_0 - \beta(y - y_0), \mu + \gamma x + \delta(y - y_0)^2). \quad (6.6.6)$$

In this case the map $f_\mu^{n+k}$ becomes

$$f_\mu^{n+k}(x, y) = (\rho^n(x_0 - \beta(y - y_0)), \lambda^n(\mu + \gamma x + \delta(y - y_0)^2)). \quad (6.6.7)$$

Variation of $\mu$ then causes the bifurcation sequence of Figure 6.6.1. When $\mu > 0$, points to the right of the $y$-axis near $p_0$ have a minimum vertical coordinate $\mu$. Therefore, an additional $n$ iterates carries all of these points above $p_0$ if $n > \log(y_0/\mu)/\log \lambda$. This contrasts with the situation we described when $\mu = 0$. Hence, as $\mu$ increases from 0, many of the horseshoes which we located for $\mu = 0$ disappear. One can in fact solve our model equations explicitly for saddle-node bifurcations of periodic orbits of period $(n + k)$ and even for the subsequent flip bifurcations which occur as $\mu$ decreases towards 0 (cf. Gavrilov and Silnikov [1973]):

EXERCISE 6.6.1. Show that an infinite sequence $\mu_{n+k}$ of saddle-node bifurcations of periodic orbits of successively higher periods $n + k$, accumulates on $\mu = 0$ from above for the family $f_\mu^{n+k}$ of (6.6.7). Show that the sinks created in these bifurcations subsequently undergo flip bifurcations at a second sequence $\mu'_{n+k}$ of values also accumulating on $\mu = 0$ from above.

Qualitatively, the picture of what happens when $\mu > 0$ is similar to the picture one has for the Hénon mapping or for an appropriate iterate of the strongly nonlinear van der Pol and Duffing return maps. The parabolic sections of $W^u(0)$ rise at the exponential rate $\lambda^n$ starting at $(x_0, \mu)$. Thus, increasing $\mu$ pulls the images of $f^{n+k}(S_n)$ up through $S_n$, destroying an entire horseshoe $\Lambda^{n+k}$ of $f^{n+k}(S_n)$ in the process, cf. Figure 6.6.4.

All of the issues concerned with the presence of strange attractors discussed in Section 5.6 arise in the analysis of perturbations of a diffeomorphism with a homoclinic tangency. We also note that the derivative of $f_\mu^{n+k}$ has determinant which approaches 0 exponentially with $n$, so that the bifurcations become better approximated by those of one-dimensional mappings as $\mu$ approaches 0 from above (but see the comments at the end of Section 6.7).

To see what happens when $\mu$ approaches 0 from below, note that $\mu < 0$ implies that $f_\mu$ has a transversal homoclinic point near $f_\mu^k(p_0)$. Therefore, there is an iterate $f_\mu^l$ of $f_\mu$ which has a horseshoe contained in a small neighborhood of the segment of $W^s(0)$ from 0 to $f_\mu^k(p_0)$. In particular, there are

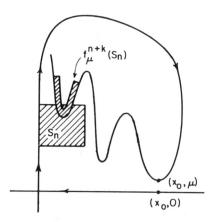

Figure 6.6.4. Disappearance of the horseshoe $\Lambda^{k+n}$ for $f_\mu^{k+n}$.

orbits of $f_\mu^l$ whose symbol sequences for this horseshoe are periodic with blocks of length $m + n$ of the form $\underbrace{10\ldots0}_{m}\ \underbrace{10\ldots0}_{n}$ for arbitrary $m$, $n$, where 1 and 0 are symbols corresponding to a Markov partition, as in Sections 5.1–5.2. We shall argue that there is no such periodic orbit for $f_0^l$, and hence that saddle-node bifurcations of periodic orbits have occurred between 0 and $\mu = \mu_1 < 0$.

Because $f_\mu$ is linear in the coordinate neighborhood $V$, a point $(x, y)$ which starts near $f_\mu^k(p_0)$ and remains at a level below $y_0$ for exactly $r$ iterates must satisfy $\lambda^{-r+1}y_0 \le y < \lambda^{-r}y_0$. Now $f_\mu^r(x, y)$ has an $x$ coordinate which is approximately $x_0\rho^r$. This implies that there is a lower bound on the $y$ coordinate of $f_\mu^{k+r}(x, y)$ which is proportional to $\rho^r$. Thus if $\lambda^l\rho^r$ is large, $f_\mu^{k+r+s}(x, y)$ will have large $y$ coordinate for some $s \le t$. We conclude that, if $n$ is large relative to $m$, there is no periodic point for $f_0^l$ which starts near $f_0^k(p_0)$, then remains close to 0 for $m$ iterates, then maps to the region around $f_0^k(p_0)$ again and remains close to 0 for $n$ iterates, and finally returns to its starting point.

This implies that there is a sequence of bifurcations of periodic orbits which accumulate at $\mu = 0$ from below. In the terminology of Gavrilov and Silnikov [1973], we have proved that the type of homoclinic tangency considered here is *inaccessible* from both sides. (Gavrilov and Silnikov [1973] explicitly exhibit saddle-node bifurcations for a family of maps of the form $f_\mu^{k+m} \circ f_\mu^{k+n}$, with $f_\mu^{k+i}$ as in (6.6.7)). The homoclinic (tangency) bifurcation is itself embedded in an infinite set of bifurcations and cannot be encountered as the initial change one sees for the birth of a strange attractor from a structurally stable (Morse–Smale) diffeomorphism. Nonetheless, one would like to be able to start with a simple diffeomorphism (say with a single globally stable fixed point) and follow a sequence of bifurcations which lead to the development of increasingly complicated and finally chaotic dynamics. Such a process would be analogous to many experiments in which the transition to chaos has been observed. The theory we have described above indicates the homoclinic tangency of the type we have considered is not a reasonable candidate for a bifurcation which lies at the transition. However, such bifurcations can certainly be expected to occur within chaotic regimes, and we have already seen the important rôle played by transverse homoclinic points. We therefore go on to consider the implications of homoclinic tangencies in more detail in the next section.

## 6.7. Wild Hyperbolic Sets

Newhouse [1974, 1979, 1980] has proved that, in addition to the bifurcations described by Gavrilov and Silnikov, there is much more subtle and complicated dynamical behavior associated with a homoclinic tangency. In this section, we outline the essential features of these theorems of Newhouse and point out how they cast doubt on the existence of strange attractors for maps of the plane which are close to those with a homoclinic tangency. The proofs are technical and we only outline them here. However, the important products of these results are examples of diffeomorphisms having infinite sets of stable periodic orbits of large period. Moreover, such diffeomorphisms are residual in open sets of the space of $C^2$ diffeomorphisms which are near diffeomorphisms with a homoclinic tangency and can thus be expected to occur in typical families (such as the one considered in the last section).

Our first step is to show that certain types of tangency are persistent. Let $\Lambda$ be a zero-dimensional hyperbolic invariant set in the plane. The stable and unstable manifolds, $W^s(\Lambda)$ and $W^u(\Lambda)$, are each locally the product of a Cantor set with an interval (think of the horseshoe example of Section 5.1). We are interested in conditions which guarantee that there will be a point of tangency between two of the segments in these sets. See Figure 6.7.1. Note that this implies the existence of a heteroclinic orbit, containing points outside $\Lambda$, and connecting two orbits $\{f^n(x)\}$, $\{f^n(y)\}$ (not necessarily periodic) of $\Lambda$, but not necessarily a homoclinic orbit.

If we regard $W^u(\Lambda)$ as a Cantor set of horizontal lines, then the question is whether some curve in $W^s(\Lambda)$ has a point with a horizontal tangency which is contained in $W^u(\Lambda)$. Naively, it would appear that if this is the case, then a slight vertical shift of $W^s(\Lambda)$ would separate the Cantor set at which $W^s(\Lambda)$ has a horizontal tangent from $W^u(\Lambda)$, as happens when $\Lambda$ is a point or a periodic orbit, as in the last section. However, for general hyperbolic sets, Newhouse observed that this is not always possible, for reasons which we now explain in a setting divorced from the dynamics.

Let $\Gamma_1$ and $\Gamma_2$ be two Cantor sets contained in the unit interval. We ask for conditions on $\Gamma_1$ and $\Gamma_2$ which imply that they intersect. One such condition is that $\Gamma_1$ and $\Gamma_2$ be Cantor sets with overlapping support and

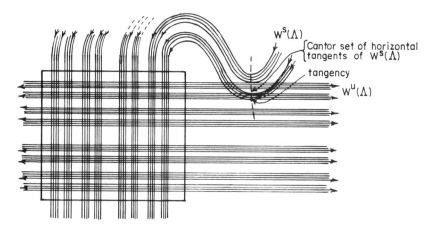

Figure 6.7.1. Tangency of stable and unstable manifolds of a hyperbolic set.

positive Lebesgue measure (see Section 5.4) and that the sum of their measures be greater than 1. Then $\mu(\Gamma_1 \cap \Gamma_2) = \mu(\Gamma_1) + \mu(\Gamma_2) - \mu(\Gamma_1 \cup \Gamma_2) > 0$ and $\Gamma_1 \cap \Gamma_2 \neq \varnothing$. However, the Cantor sets which arise in hyperbolic sets have Lebesgue measure zero, so that this condition is not appropriate here. For example, in the piecewise linear horseshoe, the Cantor sets are ones defined by a "middle $\alpha$" construction in which one inductively removes the open segment which is the middle $\alpha$ proportion of a closed interval. The Cantor sets defined in this fashion have zero Lebesgue measure, but nonzero *Hausdorff dimension* (cf. Section 5.8). If this dimension has resonable properties, then it should satisfy the inequality $\dim(\Gamma_1 \cap \Gamma_2) \geq \dim(\Gamma_1) + \dim(\Gamma_2) - 1$ (with suitable hypotheses). Newhouse introduces a related concept, the *thickness* of a Cantor set, which leads naturally to the hypotheses we need.

We represent the Cantor set $\Gamma \subset \mathbb{R}$ by writing $\mathbb{R} - \Gamma = \bigcup_{i=-2}^{\infty} U_i$ with $U_{-2}$ and $U_{-1}$ the unbounded components of $\mathbb{R} - \Gamma$ and the remaining $U_i$ disjoint open intervals. The $U_i$'s are the *gaps* of $\Gamma$ and the sets $C_j = \mathbb{R} - \bigcup_{-2 \leq i < j} U_i$ $(j \geq 0)$ form a *defining sequence* for $\Gamma$. The components of $C_j$ are closed intervals, which we call *bridges*. (The interval $C_0 = \mathbb{R} - (U_1 \cup U_2)$ contains the Cantor set.) Each $U_j$ is a subinterval of a bridge $B_j$ of $C_j$ which divides $B_j$ into two bridges $B_j^l$ (left) and $B_j^r$ (right) of $C_{j+1}$. Denoting the length of an interval $J$ by $l(J)$, we define

$$\tau(\{C_j\}) = \inf_{j>0} \left\{ \frac{l(B_j^l)}{l(U_j)}, \frac{l(B_j^r)}{l(U_j)} \right\}. \tag{6.7.1}$$

The thickness $\tau(\Gamma)$ is

$$\tau(\Gamma) = \sup\{\tau(\{C_i\}) | \{C_i\} \text{ is defining sequence for } \Gamma\}. \tag{6.7.2}$$

For example, if $\Gamma$ is the Cantor set defined by the middle $\alpha$-proportion construction on I, then for any defining sequence $\{C_i\}$ which removes at step $i$ a set $U_j$ which is as long as possible, we have

$$\frac{l(B_j^l)}{l(U_j)} = \frac{l(B_j^r)}{l(U_j)} = \left(\frac{1-\alpha}{2}\right) l(B_j) / \alpha\, l(B_j) = \frac{1-\alpha}{2\alpha}, \tag{6.7.3}$$

cf. Figure 6.7.2. Thus $\tau(\Gamma) = (1 - \alpha)/2\alpha$.

**Proposition 6.7.1** (Newhouse [1970]). *If $\Gamma_1$ and $\Gamma_2$ are two Cantor sets in $\mathbb{R}$ such that $\tau(\Gamma_1) \cdot \tau(\Gamma_2) > 1$, $\Gamma_1$ is not contained in a gap of $\Gamma_2$ and $\Gamma_2$ is not contained in a gap of $\Gamma_1$, then $\Gamma_1 \cap \Gamma_2 \neq \varnothing$.*

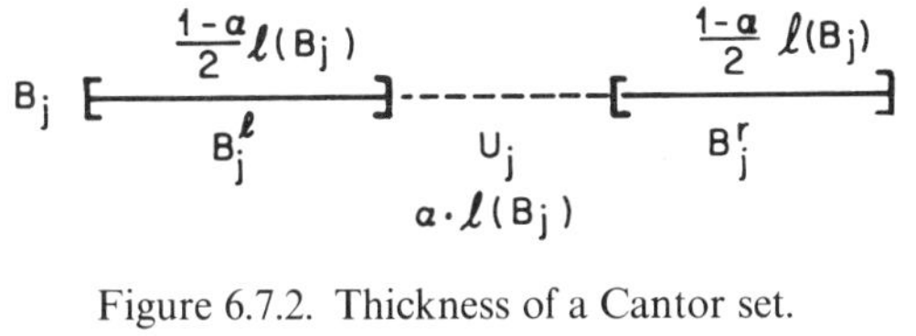

Figure 6.7.2. Thickness of a Cantor set.

PROOF. Choose defining sequences $\{C_i\}$ from $\Gamma_1$ and $\{D_i\}$ for $\Gamma_2$ so that $\tau(\{C_i\}) \cdot \tau(\{D_i\}) > 1$. If $C_0$ and $D_0$ were disjoint, then $\Gamma_1$ is contained in an unbounded gap of $\Gamma_2$ and vice versa. Thus $C_0 \cap D_0 \neq \varnothing$. We show inductively that $C_i \cap D_i \neq \varnothing$. Since these sets are compact and $C_i \cap D_i \supset C_{i+1} \cap D_{i+1}$, the finite intersection property then implies that $\Gamma_1 \cap \Gamma_2 \neq \varnothing$.

Suppose now that $C_i \cap D_i \neq \varnothing$. Let $B^1$ and $B^2$ be bridges of $C_i$ and $D_i$ such that $B^1 \cap B^2 \neq \varnothing$. Let $B^1 - U_i = B^1 \cap C_{i+1}$ and $B^2 - V_i = B^2 \cap D_{i+1}$ ($U_i$ or $V_i$ may be empty). We assert that $(B^1 - U_i) \cap (B^2 - V_i) \neq \varnothing$. There are two cases to consider: the case in which $B^1 \subset B^2$ or vice versa and the case in which $B^1 - (B^1 \cap B^2)$ and $B^2 - (B^1 \cap B^2)$ are both nonempty. See Figure 6.7.3.

If $B^1 \subset B^2$ and $(B^1 - U_i) \cap (B^2 - V_i) = \varnothing$, then $B^1 \subset V_i$. One of the gaps $W$ of $C_i$ bordering $B^1$ (say the one on the left) satisfies $l(B^1)/l(W) \geq \tau(\{C_i\})$, and the left component $B^{2l}$ of $B^2 - V_i$ satisfies $l(B^{2l})/l(V_i) \geq \tau(\{D_i\})$. Therefore, $1 < (l(B^1)/l(V_i)) \cdot (l(B^{2l})/l(W)) < l(B^{2l})/l(W)$ or $l(W) < l(B^{2l})$. This implies that $(B^2 - V_i) \cap (C_i - B^1) \neq \varnothing$ so that $D_{i+1} \cap C_{i+1} \neq \varnothing$. This completes the case $B^1 \subset B^2$. If $B^1 - (B^1 \cap B^2)$ and $B^2 - (B^1 \cap B^2)$ are both nonempty and $B_1$ contains a point to the left of $B^2$, then the right end of $B^1$ lies in $B^2$ and the left end of $B^2$ lies in $B^1$. Denote by $B^{1r}$ the right component of $B^1 - U_i$ and by $B^{2l}$ the left component of $B^2 - V_i$. Then $l(B^{1r})/l(U_i) \geq \tau(\{C_i\})$ and $l(B^{2l})/l(V_i) \geq \tau(\{D_i\})$ implies that $(l(B^{1r})/l(V_i)) \cdot (l(B^{2l})/l(U_i)) \geq 1$. Consequently, it cannot happen that $B^{1r} \subset V_i$ and that $B^{2l} \subset U_i$. We conclude that $(B^1 - U_i) \cap (B^2 - V_i) \neq \varnothing$, and the proposition is proved. $\qquad\square$

From this proposition, Newhouse concludes that there are hyperbolic invariant sets $\Lambda$ in $\mathbb{R}^2$ whose stable and unstable manifolds have tangencies which cannot be destroyed by small perturbations. We will sketch his arguments.

There are thicknesses $\tau^s(\Lambda)$ and $\tau^u(\Lambda)$ associated with $W^s(\Lambda)$ and $W^u(\Lambda)$. The definition of $\tau^u(\Lambda)$ is as follows. Pick $y \in \Lambda$ and a curve $\gamma$ transverse to $W^u(y)$ with $\gamma(0) = y$. Define $\tau^u(y, \gamma, \Lambda) = \inf_{\varepsilon > 0} \{\sup \tau(\Gamma) | \Gamma \text{ is a Cantor set in Image } \gamma|_{(-\varepsilon, \varepsilon)} \cap W^u(\Lambda, f)\}$. Newhouse [1979] shows that $\tau^u(y, \gamma, \Lambda)$ is independent of $y$ and $\gamma$ and may therefore be denoted by $\tau^u(\Lambda)$. Moreover, he proves that $\tau^u(\Lambda)$ is positive and varies continuously with $C^2$ perturbations of the diffeomorphism defining $\Lambda$. Similar definitions and results hold for $\tau^s(\Lambda)$. Consequently, if a diffeomorphism $f$ of $\mathbb{R}^2$ has a hyperbolic invariant

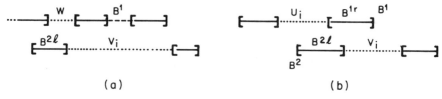

(a)                                                    (b)

Figure 6.7.3. The thickness proposition. (a) $B^1 \subset V_i \subset B^2$; (b) $B^1$ and $B^2$ overlap.

set $\Lambda$ with $\tau^s(\Lambda)\tau^u(\Lambda) > 1$ and if $W^s(\Lambda)$ has a point of tangential intersection with $W^u(\Lambda)$, then there is an $\varepsilon > 0$ such that all $C^2$, $\varepsilon$ perturbations of $f$ have hyperbolic invariant sets near $\Lambda$, which have tangential intersections. This observation is the basis for the early work of Newhouse in his thesis [1970, 1974]. Such sets $\Lambda$ are called *wild hyperbolic sets*.

The second observation made by Newhouse is that a diffeomorphism $f$ which has a homoclinic tangency to a saddle point $p$ can be perturbed to one which has a hyperbolic invariant set $\Lambda$ with $\tau^s(\Lambda)\tau^u(\Lambda) > 1$ and having a tangency between $W^s(\Lambda)$ and $W^u(\Lambda)$. The illustration of a homoclinic tangency which we consider here is the following: Figure 6.7.4, which is the "last" tangency before a horseshoe containing the point $p$ is created.

Here $f$ is a diffeomorphism with a saddle $p$ having eigenvalues $\rho$, $\lambda$ with $\rho < 1 < \lambda < \rho^{-1}$, as in the last section. We observe that, since there is a point $q$ of transversal intersection between $W^u(p)$ and $W^s(p)$, there is a hyperbolic invariant set $\Lambda_1$ near the orbit of $q$ which is described by Smale's theorem [1963]. Moreover, this invariant set has an unstable thickness $\tau^u(\Lambda_1)$ which is approximately constant as $f$ is perturbed.

By perturbing $f$ near its homoclinic tangency, we want to create a new hyperbolic invariant set $\Lambda_2$ with a large stable thickness $\tau^s(\Lambda_2)$, so that $\tau^u(\Lambda_1) \cdot \tau^s(\Lambda_2) > 1$. To do this, we consider the image of a small rectangle $R$ with base on $W^s(p)$ and centered at the point $t$ of homoclinic tangency. For large $n$, $f^n(R)$ will lie close to $W^u(p)$ and extend far along $W^u(p)$ to come close to $t$, Figure 6.7.5.

With increasing $n$, we want to study $f^n(R_n)$, where $R_n \subset R$ is a rectangle consisting of points whose images lie close to $W^u(p) \subset R$. The rectangle $R_n$ extends horizontally across $R$ and has a vertical height (thickness) proportional to $\lambda^{-n}$. The width of $f^n(R_n)$ in the direction normal to $W^u(p)$, and its distance from $W^u(p)$, is proportional to $\rho^n$. Thus the width of $f^n(R_n)$ relative to the height of $R_n$ tends to zero.

Newhouse [1979] performs careful estimates which show that, for a quadratic tangency, one can select a number $n$ and a rectangle $R_n \subset R$ such

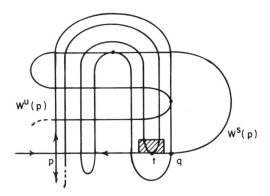

Figure 6.7.4. Homoclinic tangencies lead to wild hyperbolic sets.

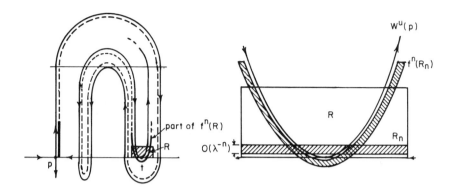

Figure 6.7.5. Creating a set with large stable thickness.

that $f^{-n}(R_n \cap f^n(R_n)) = H_n^1 \cup H_n^2$ is a pair of horizontal strips, the sum of whose heights (normal to $W^s(p)$) is almost that of $R_n$. Thus $f^n$ has an invariant set $\Lambda^n \subset \mathbb{R}^n$ of large stable thickness. To show that $\Lambda^n$ is hyperbolic, sector estimates of the type discussed in Section 5.2 must be carried out, and here these are very delicate, since the "vertical" strips $R^n \cap f^n(R_n) = V_n^1 \cup V_n^2$ have boundaries which are almost tangent to $W^s(p)$ and hence to the horizontal boundaries of $R_n$. See Figure 6.7.6. (Cf. Robinson [1982] for another derivation of this result, and for additional information.)

Here we give an alternative construction of an invariant set with large stable thickness, based on the fact that, for large $n$, $f^n$ is close to a one-dimensional map. If we rescale the vertical coordinate near $t$ by $\lambda^{-n}$, then the mappings $f^n(R_n)$ approach a limiting map $h$ which has rank 1 and has a graph like that shown in Figure 6.7.7.

The map $h$ can be regarded as one which has infinitely strong contraction along its level sets and is *hyperbolic* if there is an expanding direction at each point which yields the usual hyperbolic estimates. The map $h$ is not hyperbolic in this sense, but there are small perturbations (which raise or lower the graph) which make it hyperbolic if $h$ satisfies an appropriate hypothesis (specifically, if $h$ has a negative Schwarzian derivative). The stable thickness of the limit sets for these perturbations of $h$ can be made as large

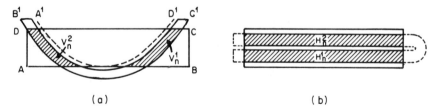

Figure 6.7.6. The construction of an invariant set with large unstable thickness. (a) $R_n \cap f^n(R_n)$; (b) $f^{-n}(R_n) \cap R_n$.

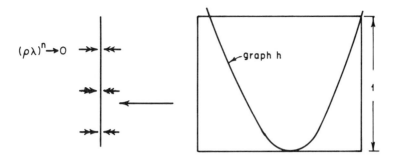

Figure 6.7.7. The singular map $h$.

as one wants, as we show in the lemmas in the Appendix to this section. Hence, taking $n$ sufficiently large, we can find a perturbation of $f$ so that $f^n(R_n)$ has a hyperbolic set $\Lambda_2 \subset R_n$, of large stable thickness. We can also arrange that $W^s(\Lambda_2)$ has a point of tangential intersection with $W^u(\Lambda_1)$.

The construction of $\Lambda_2$ for a perturbation of $f$ with the following properties is a major step towards finding diffeomorphisms with an infinite number of sinks:

(1) $\tau^s(\Lambda_2) \cdot \tau^u(\Lambda_1) > 1$.
(2) $W^s(\Lambda_2) \cap W^u(\Lambda_1)$ and $W^u(\Lambda_2) \cap W^s(\Lambda_1)$ both have points of transversal intersection which are not in $\Lambda_1 \cup \Lambda_2$.
(3) $W^s(\Lambda_2)$ and $W^u(\Lambda_2)$ have a point of tangency.

We have already demonstrated property (1) (recall that $\tau^u(\Lambda_1)$ varies continuously with perturbations of $f$). Property (2) is a consequence of the fact that the curves in $W^u(\Lambda_2)$ and $W^s(\Lambda_2)$ lie close to $W^u(p)$ and $W^s(p)$, respectively, near $t$. Since $p \in \Lambda_1$, and $\Lambda_2$ lies on the same side of $W^s(p)$ as $\Lambda_1$, it follows that we obtain the desired intersections (Figure 6.7.8).

The next step in the construction is a generalization of Smale's theorem [1963, 1967], cf. Section 5.2. Using the transverse intersections (2), we find a hyperbolic invariant set $\Lambda_3 \supset \Lambda_1 \cup \Lambda_2$ which satisfies $\tau^s(\Lambda_3) \cdot \tau^u(\Lambda_3) > 1$. Thus, further perturbations of $f$ will have hyperbolic invariant sets near $\Lambda_3$ which have persistent homoclinic tangencies.

Referring to Figure 6.7.8, we can envisage the creation of tangencies between $W^s(\Lambda_2)$ and $W^u(\Lambda_2)$ by a perturbation which "pulls up" the image $f^n(R)$. In this way we obtain a wild hyperbolic set.

In the previous section, we proved that if $f_0$ has a hyperbolic fixed point $p$ with eigenvalues $\rho < 1 < \lambda < \rho^{-1}$ and a point $p_0$ of tangential intersection between $W^s(p)$ and $W^u(p)$, then there are perturbations of $f_0$ with attracting periodic orbits, which lie close to the orbit of $p_0$. Combining this construction iteratively with the construction of wild hyperbolic sets described above, one can produce diffeomorphisms with a countable infinity of attracting periodic orbits.

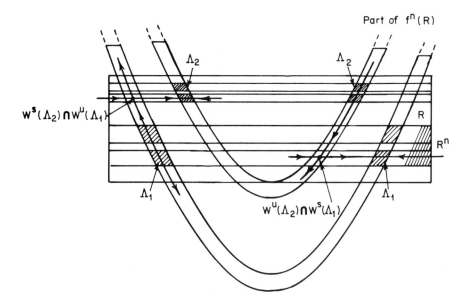

Figure 6.7.8. The invariant sets $\Lambda_1$ and $\Lambda_2$ and the intersections of their manifolds.

The argument can be summarized as follows: Starting with $f_0$, one perturbs to a nearby $f_1$ with a wild hyperbolic set. Newhouse [1980] shows that $f_1$ can be perturbed to $f_2$ with a tangency between the stable and unstable manifolds of a single periodic orbit. His argument uses the density of periodic orbits in $\Lambda_3$ to find one which contains two points near those whose stable and unstable manifolds intersect tangentially; he then perturbs $f_1$ to $f_2$ so that the stable and unstable manifolds of the periodic orbit for $f_2$ intersect tangentially. Finally, he perturbs $f_2$ to a nearby $f_3$ with a stable periodic orbit using the arguments of the preceding section. Throughout, the perturbation from $f_1$ to $f_3$ does not destroy the wild hyperbolic set $\Lambda_3$. Thus there is an $\varepsilon > 0$ such that $C^2$, $\varepsilon$ perturbations of $f_3$ have both a stable periodic orbit and a wild hyperbolic set. Iterating this procedure with decreasing $\varepsilon$ produces diffeomorphisms which have infinitely many attracting periodic orbits. For more details see Newhouse [1979, 1980]. The final result is as follows:

**Theorem 6.7.2.** *Let $p$ be a hyperbolic saddle point of a $C^r$ diffeomorphism $f$ of $\mathbb{R}^2$ with $\det(Df(p)) < 1$. Suppose $W^u(p)$ and $W^s(p)$ are tangent at some point $p_0$. Then arbitrarily $C^r$ near $f$ there is a diffeomorphism $\tilde{f}$ having a wild hyperbolic set near the orbit of $p_0$ and an infinite number of stable periodic orbits.*

These theorems imply that families of two-dimensional diffeomorphisms $f_\mu$ having homoclinic tangencies between the stable and unstable manifolds of a dissipative saddle point (or periodic orbit) for some parameter value $\mu = \mu_0$, can be expected to have infinite sets of periodic sinks for parameter values nearby. Indeed, *finite* sets of periodic sinks persist over *intervals* of parameter values. However, rather as in the cases of maps of the interval and diffeomorphisms of $S^1$, while the systems possessing sinks persist on open sets of parameter values, there may also be relatively large (measurable) sets of systems possessing no stable periodic orbits but having genuine strange attractors instead. The existence of parameter values with no sinks is still undetermined. We do have several examples of systems for which tangencies can be proven to occur, such as the Duffing equation studied by Melnikov theory in Sections 4.5–4.6. The Newhouse construction produces sinks of very high periods in these examples, but the domains of attraction are necessarily very small. Such sinks are therefore probably unobservable in experimental work, and even high precision numerical work is likely to detect only a few such sinks. Newhouse's [1979] conjecture that Hénon's attractor is merely a long-period sink, remains open.

In closing this section we mention the relevance of the one-dimensional results of Section 6.3 to the study of homoclinic bifurcations. In the discussions of this and the preceding section, we saw how, near a point of homoclinic tangency $p_0$ for a family of diffeomorphisms $f_\mu$ possessing a dissipative saddle point $p$, some high iterate of the map, $f_\mu^{n+k}$, can be approximated by a one-dimensional map which is the singular limit of a family of diffeomorphisms of the form (6.6.7) (or a form similar to it). We notice that, for such a family, as $n \to \infty$, the determinant

$$\det(Df^{n+k}) \sim \beta\gamma(\rho\lambda)^n \to 0, \tag{6.7.4}$$

since $\rho\lambda < 1$. Thus for large $n$ it is reasonable to consider families of maps of the form

$$F_{\varepsilon, \mu}(x, y) = (-y, \varepsilon x + g_\mu(y)), \tag{6.7.5}$$

where $g_\mu(y)$ is a quadratic family of one-dimensional maps of the type concon-considered in Section 6.3. $F_{\varepsilon, \mu}$ is essentially a rescaled version of (6.6.7), in which the origin has been translated to $(\rho^n x_0, y_0)$ and $\varepsilon$ plays the rôle of the determinant $\beta\gamma(\rho\lambda)^n$. Thus, as $\varepsilon \to 0$, (6.7.5) models the behavior of successively higher iterates $f_\mu^{n+k}$ of $f_\mu$.

It can be shown that some, *but not all*, of the properties of the one-dimensional family $g$ persist for the diffeomorphisms $F_{\varepsilon, \mu}$, for small $|\varepsilon|$ (Guckenheimer [1977], van Strien [1981]). In particular, accumulating sequences of period doubling bifurcations for $g_\mu$ also occur for $F_{\varepsilon, \mu}$ as $\mu$ varies with $|\varepsilon| < 1$ fixed and sufficiently small (Collet *et al.* [1981]). (In Exercise 6.6.1 we asked to reader to find the first such flip bifurcation for $f_\mu^{n+k}$ explicitly.)

There are also other important parameter values $\mu(l)$ for which the orbit of the critical point $\{g_{\mu(l)}^m(c)\}$ falls on an unstable periodic point of period $l$ for the one-dimensional map. In such cases it is known that $g_{\mu(l)}^m$ (restricted to a finite collection of subintervals) is conjugate to a piecewise linear map (Guckenheimer [1979]). For example, as we saw in Section 5.6, the family

$$g_\mu: y \to \mu y(1 - y) \tag{6.7.6}$$

for $\mu = 4$ has $g_4^2(\tfrac{1}{2}) = 0$ and is conjugate on $[0, 1]$ to

$$f(x) = \begin{cases} 2x, & x \in [0, \tfrac{1}{2}], \\ 2 - 2x, & x \in [\tfrac{1}{2}, 1], \end{cases} \tag{6.7.7}$$

(cf. Ulam and von Neumann [1947] and Section 5.6).

It can be shown that, for $\mu$ near such values $\mu(l)$ and small $|\varepsilon|$, $F_{\varepsilon, \mu}$ has a point of homoclinic tangency to a saddle point of period $l$, and hence that $F_{\varepsilon, \mu}$ has a wild hyperbolic set (van Strien [1981], cf. Holmes and Whitley [1983b]). It follows that there are strongly contracting diffeomorphisms, arbitrarily close to quadratic maps, which have *infinitely many* stable periodic orbits, while the theorem of Singer (Theorem 6.3.1) shows that a given quadratic map has at most *one* stable periodic orbit.

In fact no analogue of a wild hyperbolic set can exist for the singular map $F_{0, \mu}$, since, although hyperbolic sets of large (finite) stable thickness can be created (cf. p. 336), the unstable thickness of all sets is necessarily zero. Thus the product of the stable and unstable thicknesses is always zero and no analogue of persistent tangencies can occur. In this respect, the behavior of the diffeomorphism differs dramatically from that of the one-dimensional map, but the sequence of period doubling and saddle-node bifurcations occurring for $g_\mu$ in the order determined by the kneading theory does nonetheless provide a guide to the behavior of strongly contracting diffeomorphisms, and hence is helpful in the study of homoclinic bifurcations.

## Appendix to Section 6.7: Thickness Lemmas for One-Dimensional Maps

**Lemma 6.7.3.** *Let $h_\mu: I \to \mathbb{R}$ be a one-parameter family of maps with negative Schwarzian derivative such that:*

(1) $h_\mu(0) = h_\mu(1) = 0$ and $h_\mu'(0) > 1$ for all $\mu$;
(2) $h_\mu$ has a single nondegenerate critical point $c$ and $h_\mu(c) \geq 1$, with equality when $\mu = 0$. Then the thickness of the nonwandering sets of the $h_\mu$ is unbounded as $\mu \to 0$.

PROOF. For $\mu > 0$, the nonwandering set $\Lambda_\mu$ of $h_\mu$ is hyperbolic and topologically equivalent to a one-sided shift on two symbols (Guckenheimer [1979]). We shall study the thickness of the $\Lambda_\mu$ by introducing an auxiliary, discontinuous map $g_\mu$ for each $h_\mu$ so that $g_\mu$ still has $\Lambda_\mu$ as an invariant set. To define these maps $g_\mu$, we need a bit more notation. Dropping the subscript $\mu$, we denote by $q$ the fixed point of $h$ in $(0, 1)$ and by $p$ the other point in $I$

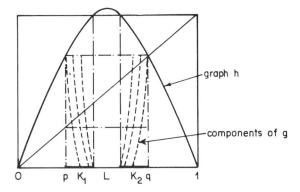

Figure 6.7.9. The map $h$. The two components of $h^{-1}(I) \cap [p, q]$ are shown as heavy lines.

with $h(p) = q$. Write $I = K_1 \cup L \cup K_2$ where $K_1$ and $K_2$ are the components of $h^{-1}(0, 1)$ and $L = h^{-1}(1, \infty)$. For those points in $h^{-1}(I) \cap [p, q]$, we define $g(x) = h^n(x)$ where $n$ is chosen to be the smallest positive integer with $h^n(x) \in [p, q]$. The map $g$ is called the *induced* map of $h$. It has a countable set of discontinuities at points $x$ with $h^n(x) = p$ and $h^k(x) < p$ for $1 < k < n$.

We now assert that the distortion of $g$ is uniformly bounded.

**Definition 6.7.1.** If $g$ is a piecewise smooth one-dimensional mapping which is strictly monotone on each component of continuity, then the *distortion* of $g$ is defined to be $\sup(g'(y)/g'(x))$, the supremum being taken over pairs $(x, y)$ such that $g$ is continuous on $[x, y]$.

Note that $\log(g'(y)/g'(x)) = \int_x^y (g''(\xi)/g'(\xi)) \, d\xi$. If $g(z) = h^n(z)$ on an interval $J = [x, y]$, then

$$\log \frac{g'(y)}{g'(x)} = \sum_{k=0}^{n=1} (\log h'(h^k(y)) - \log h'(h^k(x))) = \sum_{k=0}^{n-1} \int_{h^k(x)}^{h^k(y)} \frac{h''(\xi)}{h'(\xi)} \, d\xi.$$

We shall obtain a uniform bound on the last quantity which is independent of $\mu$ and $n$. The crucial part of this estimate involves determining a bound on the length of $J$ which depends on $n$. If $n$ is large, then most of the iterates $h^k(J), k < n$, lie very close to 0, as follows from the definition of $g$. Denote by $\lambda > 1$ a number such that $h'(0) > \lambda$. Then there is a constant $c$ such that $|(h^k)'(h(x))| > c\lambda^k$ for $0 < k < n - 1$. If $\mu$ were to remain bounded away from 0, then this would be sufficient to obtain a bound of the total length in $\bigcup_{k=0}^{n-1} h^k(J)$ and our estimate would be complete since $h'(\xi)$ would not approach 0. However, when $\mu$ is 0, $h'(\xi) \to 0$ and a more delicate estimate of the lengths are required. Suppose $\mu = 0, x < y < c, h^n(x) = p$ and $h^n(y) = q$. We need to estimate $\int_x^y (h''(\xi)/h'(\xi)) \, d\xi$. Up to bounded factors, the equations which determine $J$ are $h^n(x) = p = \lambda^n(c - x)^2, h^n(y) = q = \lambda^n(c - y^2)$. We conclude that $\lambda^n$ is proportional to $(c - x)^{-2}$ and $(c - y)^{-2}$ and that

$(y - x)(2c - (x + y))$ is proportional to $\lambda^{-2}$ or $(c - y)^2$. Since $(2c - (x + y)) > 2(c - y)$, we conclude $(y - x)/(c - y)$ is bounded, with a bound independent of $n$, and that the length of $J$ is $\mathcal{O}(h'(\xi))$, $\xi \in J$. Therefore, $\sum_{k=0}^{n-1} \int_{h^k(x)}^{h^k(y)} (h''(\xi)/h'(\xi)) \, d\xi$ is uniformly bounded with a bound independent $k$ and $n$.

We now apply the arguments of Newhouse [1979] on thickness, to the mapping $g$. Even when $\mu = 0$, $g$ is hyperbolic where it is defined. A distortion estimate like the one we have obtained above can be applied to the iterates of $g$ to conclude that there is a factor $D > 0$ independent of $n$ such that in a component of $g^{-n}(I)$, $l(g^{-n}(K_i))/l(g^{-n}(L)) > Dl(K_i)/l(L)$. Since $l(K_i)/l(L) \to \infty$ as $\mu \to 0$, the thickness of the invariant set $\Gamma_\mu$ of $g_\mu \to \infty$ as $\mu \to 0$. The invariant set $\Lambda_\mu$ of $h_\mu$ consists of $\Gamma_\mu \cup h_\mu^{-1}(\Gamma_\mu)$. Since $h_\mu^{-1}$ is a (two-valued) function with derivative bounded away from zero, $\tau(\Lambda_\mu) \to \infty$ as $\mu \to 0$. This proves the lemma.  $\square$

**Lemma 6.7.4.** *Let* $h: I \to I$ *be a mapping with negative Schwarzian derivative such that* $h(0) = h(1)$, $h$ *has a single nondegenerate critical point* $c$ *with* $h(c) = 1$, *and* $h'(0) > 1$. *Then there are hyperbolic invariant sets* $\Lambda_n$ *of* $h$ *whose thickness* $\tau(\Lambda_n) \to \infty$ *as* $n \to \infty$.

PROOF. Use the same construction as in Lemma 6.7.3. Take $\Lambda_n$ to be the subshift of finite type in which all sequences occur except those with a block of $n$ consecutive zeros. The induced map $g$ is hyperbolic, and the $\Lambda_n$ are constructed as the invariant sets of $g$ restricted to the complements of a fundamental system of neighborhoods of $c$.  $\square$

# 6.8. Renormalization and Universality

In this section we introduce the technique of *renormalization* which has been used to deduce certain *universal* features associated with the global bifurcations found at the transition to chaotic behavior in some systems. The motivation for the techniques arises in the use of renormalization for studying problems of critical phenomena in condensed matter physics (Wilson [1971a, b]). One of the striking aspects of this theory is that it produces numbers which can be regarded as quantitative predictions about transition to chaos in physical systems. Some of these predictions have been verified with surprising accuracy in physical experiments such as Rayleigh–Benard convection (Libchaber and Mauer [1982]). It is a remarkable occurrence that the theoretical analysis of period doubling bifurcations described in this section led to those wholly unsuspected predictions, which were then substantiated experimentally in systems having an infinite number of degrees of freedom.

We shall describe three problems involving renormalization methods, all of which can be analyzed in terms of one-dimensional mappings. We proceed in the order of increasing mathematical complexity, rather than historical order. The first example, introduced by Pomeau and Manneville [1980] deals with a type of "intermittency" found in the dynamics of one-dimensional mappings. The second example treats the sequences of period doubling bifurcations discovered by Feigenbaum [1978] and already described briefly in Section 6.3. The final example involves the breakdown of quasiperiodicity in a family of maps of the circle with a fixed rotation number. This last example has been studied by a number of different groups, notable among them Feigenbaum *et al.* [1982] and Rand *et al.* [1982]. The motivation for this last problem comes from the study of Hamiltonian systems, in which the issue is the breakdown of KAM surfaces in a system with parameters (cf. Section 4.8). The three problems discussed here are far from an exhaustive list of those which have been treated with renormalization methods; but we have chosen ones that appear most natural within the context of nonlinear oscillations.

The philosophy underlying the renormalization argument is that there are phenomena in dynamical systems theory which appear repetitively on different scales. By systematically adjusting the scales involved so that the observed phenomena remain constant, one can obtain interesting systems which manifest exactly the *self-similarity* which is being studied. In some cases the self-similarity is of a rather simple sort associated with a continuous system whose flow interpolates between the structures being studied, while in the other cases the origin of the self-similarity still has an aura of mathematical mystery. The rigorous arguments dealing with period doubling sequences involve high-precision arithmetic and do not make the phenomena appear inevitable in the way that a satisfying mathematical proof should (Lanford [1982]). The sketches of the three examples presented here are far from complete, but should provide the reader with a starting point from which to approach the literature cited in this section.

# 1. Intermittency

Consider a one-dimensional map $f: I \to I$ with negative Schwarzian derivative and a single critical point $c$. If there is a fixed point $p$ of $f^n$ with $(f^n)'(p) = 1$, then the orbit of $c$ approaches $p$. Moreover, if $(f^n)''(p) \neq 0$, then the orbit of $p$ is stable from one side and unstable from the other. The map $f$ can then be embedded in a one-parameter family $f_\mu$, with $f_0 = f$, which has a saddle-node bifurcation at $\mu = 0$ along the orbit of $p$. Now vary $\mu$ to obtain a mapping $g = f_{\mu_1}$ which has no periodic orbit close to the orbit of $p$. The idea of Pomeau and Manneville [1980] is that the orbits of $g$ will spend large amounts of time approximately following the periodic trajectory of $p$ for $f$. The length of these episodes of almost periodic behavior can be related to

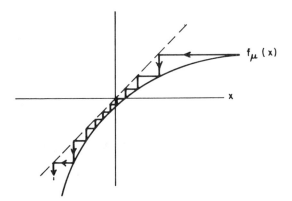

Figure 6.8.1. Near a saddle-node.

$|\mu_1|$. Pomeau and Manneville call these episodes of almost periodic behavior
*intermittent* because they represent regular behavior which is present for
varying periods of time, separated by (possibly) chaotic portions of an orbit.

To study the scaling behavior in more detail, we consider the normal form
for a saddle-node bifurcation in a discrete map: $f_\mu(x) = \mu + x - x^2$. We are
then interested in studying the dynamics of $f_\mu$ near 0 for $\mu < 0$ with small
magnitude. See Figure 6.8.1. An orbit which begins at $x > 0$ will be forced to
spend many iterates near $x = 0$ before it "escapes" to the region where $x < 0$
and $f(x) - x$ is no longer small. We wish to establish that the number of
iterates spent near $x = 0$ is asymptotically proportional to $|\mu|^{-1/2}$ as $\mu \to 0$.
This can be done in two different ways for this problem, and we outline both.

The first method involves replacing the discrete equation $f_\mu(x) =
\mu + x - x^2$ with the continuous version of this problem, the differential
equation

$$\dot{x} = \mu - x^2. \tag{6.8.1}$$

Taking $\mu < 0$, so that (6.8.1) has no fixed points, and solving the initial
value problem (6.8.1) with $x(0) = x_0$, we have

$$\tan^{-1}\left(\frac{x_0}{\sqrt{-\mu}}\right) - \tan^{-1}\left(\frac{x(t)}{\sqrt{-\mu}}\right) = t\sqrt{-\mu}. \tag{6.8.2}$$

It follows that the length of time spent by a trajectory in the region $[-1, 1]$
is given by

$$2\tan^{-1}\left(\frac{1}{\sqrt{-\mu}}\right) = t\sqrt{-\mu},$$

and thus we have

$$t \sim \pi/\sqrt{-\mu}, \tag{6.8.3}$$

as $\mu \to 0^-$.

EXERCISE 6.8.1. Use the results of the above analysis of the continuous system $\dot{x} = \mu - x^2$ to estimate the time spent by trajectories of the discrete system $x \to f_\mu(x) = \mu + x - x^2$ near the origin for small $\mu < 0$.

The second technique for analyzing the trajectories of $f_\mu(x) = \mu + x - x^2$ passing near the origin introduces the idea of renormalization. Consider the second iterate of $f_\mu$:

$$f_\mu \circ f_\mu(x) = \mu + (\mu + x - x^2) - (\mu + x - x^2)^2$$

$$= (2 - \mu)\mu + (1 - 2\mu)x - (2 - 2\mu)x^2 + 2x^3 - x^4. \quad (6.8.4)$$

If $N(\mu)$ iterates we are required for a trajectory of $f_\mu$ to pass near 0, then approximately $N(\mu)/2$ iterates are required for trajectories of $f_\mu^2$ to pass near zero. Retaining only the dominant lowest-order terms, we have

$$f_\mu^2(x) \approx 2\mu + x - 2x^2. \quad (6.8.5)$$

We now rescale $x$ and $\mu$ to make the coefficients agree with the earlier normal form, by setting $X = 2x$ and $M = 4\mu$. We thus obtain the transformed equation

$$g_M(X) = 2\left[ f_{M/4}^2\left(\frac{X}{2}\right) \right] \approx M + X - X^2. \quad (6.8.6)$$

This leads to the recursive relation $N(\mu)/2 \approx N(4\mu)$, which gives the estimate $N(4^{-n}) \sim 2^n$ or $N(\mu) \sim \mu^{-1/2}$, consistent with our earlier analysis.

One can go farther with this type of analysis by looking for an *exact* solution of the functional equation

$$f_\mu^2(x) = R_\alpha f_{\delta\mu}(R_{\alpha^{-1}}x), \quad (6.8.7)$$

where $R_\alpha$ is multiplication by $\alpha$ and gives the scale change of $x$ associated with the "doubling" operator $\mathscr{D}f = f^2$, while $\delta$ gives the scale change for the parameter $\mu$.

EXERCISE 6.8.2. Show that the solutions of the continuous normal form $\dot{x} = \mu + x^2$ do lead to exact solutions of (6.8.7) of the form $f_\mu(x) = \mu^{1/2} \tan(\mu^{1/2}t + \tan^{-1}(\mu^{-1/2}x))$. (Note that, for convenience, we have changed the sign of $x^2$ in equation (6.8.1).)

There is also a "half-way" approach to the exact solution of Exercise 6.8.2. When $\mu = 0$, the equation (6.8.7) has solutions that are given by fractional linear transformations: $g(x) = x/(1 + ax)$. To see this, we simply compute

$$g^2(x) = \frac{x}{1 + 2ax} = \tfrac{1}{2}g(2x). \quad (6.8.8)$$

Now perturbing away from this critical $f$ with a perturbation of the form $\mu + g(x)$ and noting that $g'(0) = 1$, we obtain

$$\mu + g(\mu + g(x)) \approx \mu(1 + g'(g(x))) + g^2(x) \approx \tfrac{1}{2}(4\mu + g(2x)), \quad (6.8.9)$$

where $g(x)$ is small. Once more we conclude that the number of iterates required to pass by zero is of order $(-\mu)^{-1/2}$.

Returning to the problem of intermittency for a one-dimensional mapping $f_\mu: I \to I$ near a saddle-node bifurcation occurring at $\mu = 0$, we have seen that the length of time required to traverse the "bottleneck" region of Figure 6.8.1 scales as $|\mu|^{-1/2}$. Since the mapping is not 1–1, it is possible for a trajectory to land directly in the middle of the bottleneck region without making its way in from the entrance. When this happens, the length of time spent in the bottleneck will be smaller. The result of this effect is a substantial variation in the length of time a trajectory takes to pass through the bottleneck on successive visits. This variation can be modeled with assumptions on the distribution of points as they land in the bottleneck, but the invariant measures of one-dimensional mappings can be very complicated and may not be adequately approximated by a simple model. This makes it difficult to give a complete analysis of the distribution of time intervals for which the system is approximately periodic.

## 2. Period Doubling Sequences

The first (and still the most striking) application of the renormalization methods involves the sequences of period doubling bifurcations by means of which a one-dimensional mapping becomes chaotic. We shall describe the geometry associated with the critical map which divides chaotic and non-chaotic maps in a family and then describe the analysis of Feigenbaum [1978] which leads to universal behavior in this transition. Throughout this section $I = [0, 1] \subset \mathbb{R}$.

A map $f: I \to I$ with a single critical point $c$ which lies between chaotic and nonchaotic maps has a particular kneading sequence which reflects the property that $f^i(c)$ and $f^{i+2^n}(c)$ lie on the same side of $c$, unless $i = 2^k$ for any $k \in \mathbb{Z}$, in which case they lie on opposite sides of $c$. If the mapping has negative Schwarzian derivative, then the nonwandering set of $f$ consists of one unstable periodic orbit of period $2^k$ for each $k$ together with a Cantor set $\Lambda$ depicted in Figure 6.8.2, with the numbers $i$ representing $f^i(c)$.

The Cantor set $\Lambda$ is contained in the union of the $2^n$ intervals $[f^i(c), f^{i+2^n}(c)]$, $1 \le i \le 2^n$. The intersection of these sets over all $n \ge 0$ is $\Lambda$. At each step of the construction, the intervals $[f^i(c), f^{i+2^n}(c)]$, $1 \le i \le 2^n$ are permuted among themselves and there is no substantial spreading of trajectories which start close to one another near $\Lambda$. The kneading sequence

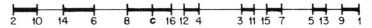

Figure 6.8.2. Eight subintervals in the construction of the Cantor set $\Lambda$; the endpoints are the first 16 iterates: $\{f^i(c)\}_{i=1}^{16}$.

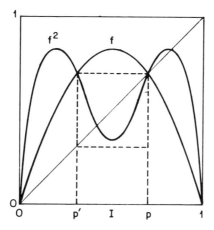

Figure 6.8.3. The critical map $f$ and $f^2$.

**a** of $f$ has the remarkable property that if one defines $b_i = a_{2i}$, then the sequence **b** is the "complement" of **a**, obtained from **a** by changing each symbol. (The sequence is a *Morse sequence*.)

Further insight can be gained by examining Figure 6.8.3, which shows the critical mapping $f$ and its second iterate $f^2$. If $p$ and $p' \neq p$ satisfy $f(p') = f(p) = p$, then the observations about the kneading sequence of $f$ imply that $f^2|_{[p,\, p']}$ is topologically equivalent to $f|_I$. Thus it makes sense to inquire whether $f^2$ and $f$ are related by a linear rescaling. One of Feigenbaum's fundamental observations is that there is a unique even, real analytic mapping $g: \mathbb{R} \to \mathbb{R}$ and real number $\alpha \approx -2.5$ for which $g(0) = 1$, $g''(0) < 0$, and $\alpha g^2(\alpha^{-1}x) = g(x)$. The mapping $g$ can be approximately determined numerically by solving for $g$ as a polynomial of high degree with unknown coefficients.

The mapping $g$ above is a fixed point of the *renormalized doubling operator* $\mathscr{T}$ defined by the relationship

$$\mathscr{T}(f)(x) = \alpha f^2(\alpha^{-1}x); \, x \in [p, p'], \tag{6.8.10}$$

on even functions $f$ with $f(0) = 1$ and $\alpha = (f^2(0))^{-1} = 1/f(1)$. Feigenbaum studied the operator $\mathscr{T}$ numerically and concluded that $g$ was an isolated fixed point and that the linearization $\mathscr{D}\mathscr{T}(g)$ of $\mathscr{T}$ at $g$ in the space of even functions $h$ with $h(0) = 1$ had a single unstable eigenvalue $\delta \approx 4.67$. This analysis leads to the following picture in function space: Figure 6.8.4.

There is a surface $\Sigma$ (of codimension one) consisting of functions with the same kneading sequence as $g$. A neighborhood of $g$ in $\Sigma$ lies in the stable manifold of $g$ for the doubling operator $\mathscr{T}$. Approximately parallel to $\Sigma$ are surfaces $\Sigma_n$ representing mappings for which the critical point is periodic with period $2^n$. The doubling operator $\mathscr{T}$ maps $\Sigma_{n+1}$ to $\Sigma_n$. The distances from

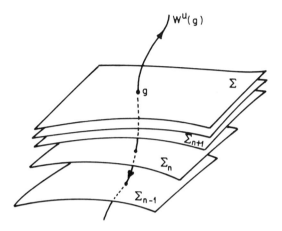

Figure 6.8.4. The structure near the universal map $g$ in function space.

$\Sigma_n$ to $\Sigma$ approach zero like $\delta^{-n}$: $d(\Sigma_n, \Sigma)/d(\Sigma_{n+1}, \Sigma) \to \delta$ as $n \to 0$. For a one-parameter family $f_\mu$ passing transversely through $\Sigma$ close enough to $g$, we can apply a renormalized doubling operator $\tilde{\mathcal{T}}$ for families which adjusts the parameter $\mu$:

$$\tilde{\mathcal{T}}(f_\mu) = (f_{\delta^{-1}\mu}). \tag{6.8.11}$$

The operator $\tilde{\mathcal{T}}$ on families will have a stable fixed point which represents the universal behavior for the limit of a sequence of period doubling bifurcations in a family of mappings with a nondegenerate critical point. The ratio $\delta$ for the distances between successive bifurcations is independent of the family and has been measured experimentally in a number of physical systems. For specific families $f_\mu$ one has

$$\delta^{-1} = \lim_{n \to \infty} \frac{\mu_{n+1} - \mu_n}{\mu_n - \mu_{n-1}}, \tag{6.8.12}$$

where $\mu_n$ is the parameter value at which the orbit of period $2^n$ undergoes a flip bifurcation, giving birth to an orbit of period $2^{n+1}$.

EXERCISE 6.8.3. Find numerically a sequence of period doubling bifurcations in the forced Duffing equation $x + \alpha \dot{x} - \beta x + x^3 = \gamma \cos \omega t$, both for $\beta = 1$ and $\beta = 0$. Compute the number $\delta$ in this example. (Hint: See Feigenbaum [1980].)

On the chaotic side of $\Sigma$, there is another sequence of surfaces $B_n$ which represent *band mergings*: if $f \in B_n$, there is a neighborhood of $c$ for which $f^{2^n}$ has a graph like that depicted in Figure 6.8.5, in which an invariant sub-interval $J \subset I$ is mapped so that $f^{2^n}(J)$ covers $J$ exactly twice. In the function space, the surface $B_n$ is characterized as the set of mappings whose topological entropy is $(1/2^n) \log 2$. The doubling operator maps $B_{n+1}$ to $B_n$, so the surfaces

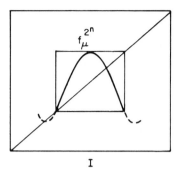

Figure 6.8.5. $f_\mu^{2^n}$, $\mu \in B_n$

$B_n$ also accumulate on $\Sigma$ at the rate $\delta^{-n}$, but from "above" in the convention of Figure 6.8.4.

In subsequent work, Feigenbaum [1979] noted that there is a characteristic structure of the *power spectrum* associated with the period doubling sequences. (See also Nauenberg and Rudnick [1981].) Both the band mergings and the spectral properties can be determined from experimental data. The doubling operator $\mathcal{T}$ can also be allowed to act on dissipative multidimensional mappings without changing its fixed point $g$ or the fact that its linearization has a single unstable eigenvalue.

A rigorous analysis of the doubling operators $\mathcal{T}$ and $\tilde{\mathcal{T}}$ is much more difficult than the analysis of the intermittency associated with saddle-node bifurcations. Lanford [1982] and Campanino and Epstein [1981] have given proofs of the existence of the fixed point function $g$ for $\mathcal{T}$. Lanford also determines that there is a single unstable eigenvalue. Both arguments are closely based on numerical computations and require careful arithmetic estimates. Indeed, Lanford's proof is a "proof by computer" in that by using interval arithmetic and numerical analysis, he provides estimates which substantiate the numerical calculations of Feigenbaum. However, in an earlier paper Collet *et al.* [1980] were able to establish analytically the existence of $g$ and the spectral properties of $\mathcal{D}\mathcal{T}(g)$ for maps of the form $x \to \mu - |x|^{1+\varepsilon}$ with $\varepsilon$ small.

## 3. Quasiperiodic Breakdown

The third example of renormalization techniques which we examine involves a transition from quasiperiodic to chaotic dynamics in a discrete system. We first discuss the phenomenon in the framework of the forced Van der Pol equation to illustrate its nature. For certain parameter ranges, the Poincaré return maps of the weakly nonlinear Van der Pol equation have a family of invariant closed curves whose rotation numbers vary as one changes the frequency of the forcing term (cf. Section 2.1). The analysis of Cartwright and Littlewood [1945] shows that such invariant curves no longer exist when the nonlinearity is sufficiently strong and the equation becomes a relaxation oscillator (cf. Levi [1981] and Section 2.1). Taking the magnitude of the nonlinear term and the forcing frequency as two parameters, there is evidently a curve $\gamma$ in the parameter plane which corresponds to return mappings with a closed invariant curve on which the map has a fixed irrational rotation number. We want to study the mapping which has parameter values lying at the endpoint of the curve: the point at which the closed curve "breaks up."

This mapping $F: \mathbb{R}^2 \to \mathbb{R}^2$ can be studied using an approach similar to that employed in the KAM theory. The goal is to characterize the closed invariant curve $\Lambda$ of $F$ (if it exists) by finding a coordinate change which transforms $F|_\Lambda$ to a rigid rotation on a circle. In the critical case considered here, the KAM theory implies that no *smooth* coordinate change can exist. However, it is possible that a continuous coordinate change which is not differentiable might transform $F|_\Lambda$ to a rotation. A simpler problem which preserves the essential difficulties of this one can be posed. Suppose $f: S^1 \to S^1$ is a smooth map of the circle onto itself which is 1–1, but has an inflection point $p$; i.e., $f'(p) = f''(p) = 0 \neq f'''(p)$. If the rotation number of $f$ is irrational, is there a continuous coordinate change of $S^1$ which transforms $f$ to a rotation? The general answer to this question is far from settled, but renormalization methods have been used to study special rotation numbers.

Suppose that $f: S^1 \to S^1$ has an inflection point $p$ and is topologically equivalent to rotation by $\alpha$. This means that there is a homeomorphism $h$ such that $h \circ f(x) = h(x) + \alpha$ (cf. Section 6.2). Since $f'(p) = 0$, either $h$ or $h^{-1}$ must have singularities and these singularities will be present along the trajectory of $p$. If the rotation number $\alpha$ satisfies special properties, then it is reasonable to search for some kind of self-similarity in the structure of $h$. Numerically based studies of this kind have revealed universal features in these situations, but the structure does not appear to be as rigid as that described above for the period doubling sequences of bifurcations.

In particular, the two-parameter family of maps

$$\theta \to f_{\beta,\gamma}(\theta) = \theta + \beta + \gamma \sin 2\pi\theta; \qquad \beta, \gamma \geq 0. \qquad (6.8.13)$$

has been used as an example. For $\gamma < 1$ this map is a diffeomorphism of the circle, while it is noninvertible for $\gamma > 1$. The critical case occurs at $\gamma = 1$. The second parameter, $\beta$, allows one to adjust the rotation number. We recall that this is the map used by Arnold [1965] as an example of a perturbation of a rigid (irrational) rotation (cf. Section 6.2). Here, however, $\gamma$ is not kept small. We also remark that Glass and Perez [1982] have studied this map for $\gamma > 1$.

The rotation number which has been studied most intensively is $\alpha = (\sqrt{5} - 1)/2 \approx 0.618034\ldots$, the reason being that it has the simplest continued fraction expansion among the irrationals:

$$\frac{\sqrt{5} - 1}{2} = \cfrac{1}{1 + \cfrac{1}{1 + \cfrac{1}{1 + \cfrac{1}{1. \ldots .}}}}$$

Henceforth we refer to this rotation number as $\omega$. Related to this continued fraction expansion are arithmetic relationships which lead to scaling behavior. In particular, we note that $(-\omega)$ is an eigenvalue of the matrix

$$T = \begin{pmatrix} 1 & 1 \\ 1 & 0 \end{pmatrix}, \tag{6.8.14}$$

with eigenvector $(-\omega, 1)^T$. Note also that $T^n$ is the matrix

$$\begin{pmatrix} \phi_{n+1} & \phi_n \\ \phi_n & \phi_{n-1} \end{pmatrix}, \tag{6.8.15}$$

where $\phi_i$ is the $i$th Fibonacci number (indexed so that $\phi_0 = 0$, $\phi_1 = 1$). This yields the identity

$$\phi_{n-1} - \phi_n \omega = (-\omega)^n, \tag{6.8.16}$$

because $(-\omega)^n$ is an eigenvalue of $T^n$ with eigenvector $(-\omega, 1)^T$.

Denoting by $R_\omega$ a rotation of the circle by $\omega$, we have

$$(R_\omega)^{\phi_n}(\theta) - \theta = \phi_n \omega, \quad \text{mod } 1,$$

and so, from (6.8.16) and since $|\omega| < 1$, we obtain

$$|(R_\omega)^{\phi_n}(\theta) - \theta| \equiv \omega^n. \tag{6.8.17}$$

We conclude that the $\phi_n$ iterates of $R_\omega$ form a sequence of mappings which converge geometrically to the identity. Moreover, since $\phi_{n+1} = \phi_n + \phi_{n-1}$ for the Fibonacci numbers, the $(R_\omega)^{\phi_n}$ can be generated by successive compositions of the form:

$$(R_\omega)^{\phi_{n+1}} = (R_\omega)^{\phi_n} \circ (R_\omega)^{\phi_{n-1}}. \tag{6.8.18}$$

These observations about $\omega$ set the stage for attempts to find scaling behavior for the mapping $f$.

In searching for scaling behavior of $f$, there is a difficulty arising from the fact that the circle has bounded length. There must be a largest length scale on the circle to which the arguments can be applied, so the search for a fixed point of a renormalization operator is moved from the circle to the line. Lifting $f$ to a function $\tilde{f}: \mathbb{R} \to \mathbb{R}$ satisfying $e^{2\pi i \tilde{f}(\theta)} = f(e^{2\pi i \theta})$, the equation (6.8.16) leads one to look for behavior of $\tilde{f}^{\phi_n} - R_{\phi_{n-1}}$ which scales geometrically.

Starting with $u = v = \tilde{f}$ and a number $\alpha$ one can study renormalization operators defined by

$$\mathcal{R}_1 \begin{bmatrix} u(\theta) \\ v(\theta) \end{bmatrix} = \begin{bmatrix} \alpha u(\alpha v(\alpha^{-2}\theta)) \\ u(\theta) \end{bmatrix}, \tag{6.8.19}$$

or

$$\mathcal{R}_2 \begin{bmatrix} u(\theta) \\ v(\theta) \end{bmatrix} = \begin{bmatrix} \alpha^2 v(\alpha^{-1}u(\alpha^{-1}\theta)) \\ u(\theta) \end{bmatrix}, \tag{6.8.20}$$

which correspond the passage from $(\tilde{f}^{\phi_n}, \tilde{f}^{\phi_{n-1}})$ to $(\tilde{f}^{\phi_{n+1}}, \tilde{f}^{\phi_n})$ with rescaling by $\alpha$. Numerical calculations produce a function $g$ so that $\begin{bmatrix} g \\ g \end{bmatrix}$ is a fixed point for both $\mathcal{R}_1$ and $\mathcal{R}_2$ with $\alpha \approx -1.29$: Feigenbaum *et al* [1982]; Shenker [1982]. Moreover, the linearization of each $\mathcal{R}_i$ at its fixed point has one unstable eigenvalue which corresponds to scaling behavior possessed by $\varepsilon_n$ with the property that $f(\theta) + \varepsilon_n$ has a periodic orbit containing the inflection point of $f$ and has rotation number $\phi_{n-1}/\phi_n$. The scaling ratio $\delta$ is found to be approximately $-2.83$. In terms of the specific map (6.8.12) one has

$$\delta^{-1} = \lim_{n \to \infty} \frac{\beta_{n+1} - \beta_n}{\beta_n - \beta_{n-1}}, \tag{6.8.21}$$

where $\beta_n$ is the parameter value for which $f_{\beta, 1}$ ($\gamma = 1$) has rotation number $\phi_{n-1}/\phi_n$. There is also a curve which describes the universal structure on small scales of the nonsmooth invariant curve of a planar mapping with rotation number $\omega$.

# CHAPTER 7
# Local Codimension Two Bifurcations of Flows

In this chapter we discuss bifurcations from equilibria which have a multiple degeneracies. We start with the analogues of the saddle-node and Hopf bifurcations with the same linear part, but additional degeneracy in the *nonlinear* terms of the Taylor expansion expressed in normal form. The theory here is complete, at least for the first few cases, and is essentially obtained by unfolding degenerate singularities of functions, since in each case we can reduce to a one-dimensional flow.

We then go on to consider cases in which the *linear* part of the vector field is doubly degenerate. As we saw in Section 3.1, there are three basic cases in which the reduced system on the center manifold is respectively two, three, and four dimensional, and the linear part takes the forms:

$$\begin{bmatrix} 0 & 1 \\ 0 & 0 \end{bmatrix}; \quad \begin{bmatrix} 0 & -\omega & 0 \\ \omega & 0 & 0 \\ 0 & 0 & 0 \end{bmatrix}; \quad \left[\begin{array}{cc:cc} 0 & -\omega_1 & \multicolumn{2}{c}{} \\ \omega_1 & 0 & \multicolumn{2}{c}{\text{\Large 0}} \\ \hdashline \multicolumn{2}{c:}{\text{\Large 0}} & 0 & -\omega_2 \\ \multicolumn{2}{c:}{} & \omega_2 & 0 \end{array}\right]. \quad (7.0.1)$$

The classification and unfolding of the first (nilpotent) type was done simultaneously (and independently) by Takens [1974a, b] and Bogdanov, cf. Arnold [1972]. The remaining cases have only been considered recently and the results are incomplete. We outline a program for obtaining additional information, but we shall see that the goal of obtaining structurally stable unfoldings appears unattainable in some cases.

We end this chapter, and the book, with some physical problems in which codimension two bifurcations of these types occur.

There are other types of codimension two bifurcations than those that involve equilibria, but they are not considered in this book. Indeed, the theory of such bifurcations is fragmentary and awaits further development. One area of substantial recent progress are theories devoted to understanding how quasiperiodic motions and invariant tori of flows disappear (Aronson et al. [1982], Feigenbaum et al. [1982], Rand et al. [1982], Mather [1982]). Other relevant work includes the analysis of Hopf bifurcations for periodic orbits at strong resonances (Arnold [1977]) and work on multiple bifurcation of periodic orbits (Jost and Zehnder [1972]). One motivation for the particular importance of studying multiple bifurcations of equilibria is that they provide an entrée for analytically finding complicated dynamics in large systems of equations and even in partial differential equations. Examples are discussed in Section 7.6.

Another topic which we do not discuss in detail in this book is that of bifurcation with symmetry. While we do give some examples of vector fields invariant under various symmetry groups, we do not attempt a systematic treatment. We refer the interested reader to the papers and forthcoming book of Golubitsky and Schaeffer [1979a, b, 1983] for an extensive treatment of the theory.

## 7.1. Degeneracy in Higher-Order Terms

Here we consider the unfolding of the saddle-node and Hopf bifurcations with an additional higher-order degeneracy of the simplest possible type. Thus our linear parts at bifurcation will still be $Ax = 0$ and $A\begin{pmatrix} x \\ y \end{pmatrix} = \begin{pmatrix} -\omega y \\ \omega x \end{pmatrix}$, respectively but the leading nonlinear coefficient in the normal form will be zero.

For the saddle-node, the truncated system takes the form

$$\dot{x} = a_3 x^3, \tag{7.1.1}$$

since $a_2 = f''(0) = 0$, by hypothesis. A little thought shows that there are small perturbations of the function $a_3 x^3$ which possess either one or three hyperbolic fixed points near $x = 0$, and that certain "unusual" perturbations possess two fixed points, one of which is nonhyperbolic. It is not possible to introduce more than three fixed points locally. All these behaviors can be captured by the addition of the lower-order terms $\mu_1 + \mu_2 x$, so that an unfolding is represented by

$$\dot{x} = \mu_1 + \mu_2 x + a_3 x^3. \tag{7.1.2}$$

The dynamics of such a one-dimensional vector field are determined up to topological equivalence by its fixed points and their stability types. In general, *singularity theory* provides the tools for a systematic study of the zeros of (families of) mappings $f: \mathbb{R}^n \to \mathbb{R}^m$, of which the right-hand side of (7.1.2) provides an example. Arnold [1972, 1981] provides an exposition of results of this kind. As we shall see, the multidimensional case still presents some problems, but the one-dimensional case has been solved completely. We will now indicate the main features of the solution.

Consider the polynomial

$$P_a(x) = x^n + a_{n-1}x^{n-1} + a_{n-2}x^{n-2} + \cdots + a_0. \qquad (7.1.3)$$

Setting $x_0 = -a_{n-1}/n$, we obtain the equivalent expression

$$P_\mu(x) = (x - x_0)^n + \mu_{n-2}(x - x_0)^{n-2} + \cdots + \mu_0; \qquad (7.1.4)$$

thus a translation of the x-axis produces a polynomial with no term of degree $(n - 1)$. We shall regard the coefficients $\mu_0, \ldots, \mu_{n-2}$ as *parameters* which perturb the polynomial $x^n$. Up to (small) translations of the x-axis, all polynomials of degree $n$ with small coefficients are represented among the $P_\mu$. If we regard $P_\mu(x)$ as a function on $\mathbb{R} \times \mathbb{R}^{n-1}$ $((x, \mu)$ space), then the gradient of $P_\mu(x)$ is nonzero at the origin. Therefore, the implicit function theorem implies that $Z = \{(x, \mu) | P_\mu(x) = 0\}$ is a submanifold of $\mathbb{R} \times \mathbb{R}^{n-1}$ near 0. This property is clearly preserved by perturbation of the family $P_\mu(x)$.

We can say more about the family $P_\mu(x)$ and its zeros. The polynomial $P_\mu(x)$ has a degenerate zero at $(x_0, \mu_0)$ as a function of x if $(d/dx)(P_\mu(x)) = 0$. This is equivalent to the statement that $Z$ has a direction of tangency with the plane $x = x_0$ in $\mathbb{R} \times \mathbb{R}^{n-1}$. Singularity theory implies that, if the family of polynomials $P_\mu(x)$ is perturbed (with a $C^\infty$ perturbation), then there are smooth changes of coordinates on $\mathbb{R} \times \mathbb{R}^{n-1}$ and $\mathbb{R}$ (the image space) restoring the perturbed family to its original form. In particular, the structure of the set of $\mu$ for which $P_\mu(x)$ has a degenerate zero does not change (cf. Golubitsky and Guillemin [1973]). This justifies that statement that (7.1.2) is the unfolding of (7.1.1).

For the family of one-dimensional vector fields

$$\dot{x} = P_\mu(x), \qquad (7.1.5)$$

the bifurcation set is given by the set of $\mu$ for which $P_\mu(x)$ has a degenerate zero. Let us calculate this set for the family (7.1.2). Differentiating $\mu_1 + \mu_2 x + a_3 x^3$ with respect to x gives $\mu_2 + 3a_3 x^2$. Equating both of these expressions to zero and eliminating x gives the bifurcation set $B$:

$$4\mu_2^3 + 27a_3\mu_1^2 = 0. \qquad (7.1.6)$$

Figure 7.1.1 shows the bifurcation set (7.1.6) of (7.1.2) and the corresponding phase portraits for $a_3 < 0$. Similar pictures (with reversal of time) are obtained for $a_3 > 0$. Note that $B$ consists of two *codimension one* open curves on which (codimension one) saddle-node bifurcations occur, and the *codimension two* point $(\mu_1, \mu_2) = (0, 0)$, at which we have the doubly degenerate codimension two singularity $a_3 x^3$. The form of the bifurcation set (7.1.6) provides the name sometimes attached to this unfolding: the cusp.

Throughout this chapter we shall use the device exemplified in Figure 7.1.1 of plotting the two-parameter bifurcation set and the corresponding phase portraits as a single figure.

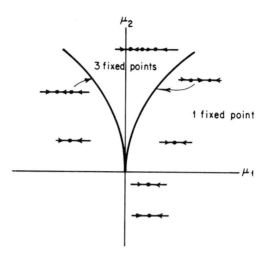

Figure 7.1.1. The bifurcation set and phase portraits for equation (7.1.2); $a_3 < 0$.

EXERCISE 7.1.1. Show that the universal unfolding of (7.1.2) is a family of gradient vector fields with potential function

$$V_{\mu_1,\mu_2}(x) = -\left(\mu_1 x + \mu_2 \frac{x^2}{2} + a_3 \frac{x^4}{4}\right).$$

Investigate the behavior of the critical points of $V_{\mu_1,\mu_2}$ and relate it to the bifurcation set of Figure 7.1.1.

EXERCISE 7.1.2. Compute the bifurcation set and associated phase portraits for the family of vector fields $\dot{x} = \mu_1 + \mu_2 x + \mu_3 x^2 + a_4 x^4$. Show that this bifurcation set in $\mathbb{R}^3$ contains cusps and folds. (Hint: Compare this problem with the "swallow tail" catastrophe of Thom [1975] (cf. Poston and Stewart [1978]).)

Readers familiar with Thom's catastrophe theory may wonder at this point whether there is a relationship between the catastrophe theory unfolding of a potential function and the unfolding of the corresponding vector field. Things are straightforward in one dimension, because all vector fields are gradients, but the example of the elliptic umbilic illustrates that in two or more dimensions the relationship is not in general a simple one. The elliptic umbilic comes from the universal unfolding of the function

$$V_0 = \frac{1}{2}\left(\frac{y^3}{3} - x^2 y\right). \tag{7.1.7}$$

The corresponding gradient vector field is

$$\dot{x} = xy$$
$$\dot{y} = \tfrac{1}{2}(x^2 - y^2). \tag{7.1.8}$$

Arnold [1972] shows that there are perturbations of the elliptic umbilic potential function which have closed orbits (cf. equations (1.8.5)–(1.8.7)). These closed orbits are not captured in the catastrophe theory discussion of the elliptic umbilic, since all of the vector fields considered there are gradient vector fields (cf. equation (1.8.13), but also see Exercise 7.1.3, below).

EXERCISE 7.1.3. Sketch the phase portraits of the gradient vector fields in $\mathbb{R}^2$ having potential functions given by the universal unfolding of the elliptic umbilic:

$$V_\mu(x, y) = \frac{1}{2}\left(\frac{y^3}{3} - x^2 y\right) + \mu_1 x + \mu_2 y + \mu_3(x^2 + y^2).$$

Show that, even in this case, *gradient* perturbations can exhibit bifurcations of hetero-clinic orbits which are not immediately derivable from the potential function. (Cf. Poston and Stewart [1978, Chapters 11 and 12] and Section 6.1 of the present book.)

EXERCISE 7.1.4. Show that cusps occur in the bifurcation sets of the following systems:

(a) the averaged van der Pol equation (2.1.14), in $(\sigma, \gamma)$ space;
(b) the averaged Duffing equation with positive linear stiffness (4.2.13)–(4.2.14), in $(\Omega, \gamma)$ space, with $\alpha > 0$ constant.

(These bifurcations each involve up to three fixed points of the averaged equations, and hence up to three periodic orbits of the full systems.)

The averaged van der Pol and Duffing equations of the preceding exercise are not gradient systems, but since in the neighborhood of the cusp point we can reduce to a family of systems with one-dimensional center manifolds, the bifurcations are effectively determined by an equivalent gradient field. This is *not* true near the degenerate point $(\sigma, \gamma) = (\frac{1}{2}, \frac{1}{2})$ for the van der Pol equation. The unfolding near this point is considered in Section 7.3, below.

The example of equation (1.8.5) illustrates another aspect of unfoldings. The degenerate vector field ($\lambda = \zeta = 0$) in polar coordinates, is

$$\dot{r} = \frac{r^2}{2} \sin 3\theta,$$

$$\dot{\theta} = \frac{r}{2} \cos 3\theta,$$

(7.1.9)

and is clearly invariant under rotations through $2\pi/3$, as is the special unfolding of equation (1.8.5):

$$\dot{r} = -\zeta r - \frac{r^2}{2} \sin 3\theta,$$

$$\dot{\theta} = \lambda - \frac{r}{2} \cos 3\theta.$$

(7.1.10)

Unfoldings do not generally preserve this symmetry (as Exercise 7.1.3 demonstrates, there are perturbations with two equilibrium points).

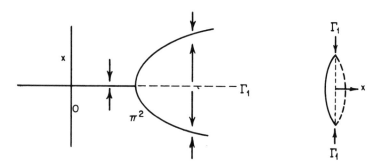

Figure 7.1.2. The symmetric buckling problem.

However, physical systems often do possess symmetries which their mathematical models necessarily inherit. In this example the invariance of the Poincaré map under rotations of $2\pi/3$ associated with a subharmonic of order three gives rise to the symmetry, cf. Section 4.2.

A classic example of the rôle of symmetry in a bifurcation problem is the buckling of a pinned column, first studied by Euler [1744]. Taking only a single mode, we obtain a second-order equation

$$\ddot{x} + \alpha\dot{x} + \pi^2(\pi^2 - \Gamma_1)x + \beta x^3 = 0; \qquad \alpha, \beta > 0, \qquad (7.1.11)$$

for the nondimensional displacement $(x)$ and velocity $(\dot{x})$ of that mode. The bifurcation diagram is easily calculated. If the axial load $\Gamma_1$ is less than $\pi^2$, then we have a single globally stable equilibrium state, while if $\Gamma_1$ exceeds the (first) buckling load, $\pi^2$, there are two additional symmetric fixed points at $x = \pm\pi(\sqrt{\Gamma_1 - \pi^2})/\beta$; Figure 7.1.2. If a side load, $\Gamma_0$, is now applied to the column, then the fixed points lie on the zero set of

$$\beta x^3 + \pi^2(\pi^2 - \Gamma_1)x - \Gamma_0 = 0, \qquad (7.1.12)$$

and we have the cusp of Figure 7.1.1. The degenerate symmetric bifurcation diagram of Figure 7.1.2 breaks, so that, as we follow the one-parameter path of increasing $\Gamma_1$, we encounter only a codimension one saddle-node. See Figure 7.1.3. Thus the question of whether or not perturbations are allowed which destroy a symmetry present in a normal form at a bifurcation must be addressed in bifurcation problems.

In the case of the Hopf bifurcation, the normal form (3.4.8)–(3.4.9) exhibits symmetry, when expressed in polar coordinates:

$$\begin{aligned}
\dot{r} &= \mu_1 r + \mu_2 r^3 + a_5 r^5 + \mathcal{O}(r^7), \\
\dot{\theta} &= \omega + b_1 r^2 + b_2 r^4 + \mathcal{O}(r^6).
\end{aligned} \qquad (7.1.13)$$

Note that these equations are invariant under the symmetry $(r, \theta) \to (-r, \theta)$. We have carried the normal form up to the fifth-order coefficient $a_5$, and have replaced the third-order coefficient $(a)$ by a variable parameter, $\mu_2$. This is appropriate in a problem for which the coefficient of $r^3$ can change

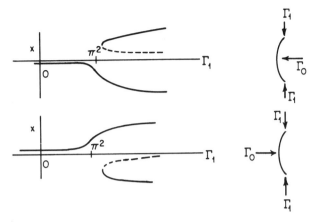

Figure 7.1.3. The asymmetric problem.

sign. We can then study the *generalized* Hopf bifurcation singularity with radial part

$$\dot{r} = a_5 r^5 + \mathcal{O}(r^7), \qquad (7.1.14)$$

by independently varying the coefficients $\mu_1$ and $\mu_2$ of (7.1.13). Moreover, for $|r| \ll 1$, the azimuthal component of (7.1.13) is dominated by $\omega$ ($\gg b_1 r^2 + b_2 r^4$) and the behavior of the vector field is thus captured solely by the radial component. It is easy to verify that, for $\mu_2 \neq 0$, the line $\mu_1 = 0$ is a bifurcation value at which a standard Hopf bifurcation occurs, subcritical if $\mu_2 > 0$, and supercritical if $\mu_2 < 0$. A second bifurcation set is the semi-parabola

$$\mu_2^2 = 4a_2 \mu_1; \qquad \frac{\mu_2}{a_2} < 0, \qquad (7.1.15)$$

on which a pair of closed orbits, one an attractor and the other a repellor, coalesce and vanish. We show the bifurcation set and associated phase portraits in Figure 7.1.4, below. For more information see Takens [1973b], Arnold [1972], and Golubitsky and Langford [1981].

EXERCISE 7.1.5. Compute the bifurcation set for the seventh-order generalized Hopf bifurcation

$$\dot{r} = r(\mu_1 + \mu_2 r^2 + \mu_3 r^4 \pm r^6), \qquad \dot{\theta} = \omega + \mathcal{O}(r^2),$$

and describe the associated phase portraits.

EXERCISE 7.1.6. Compute bifurcation sets and diagrams for the following generalized Hopf bifurcations with "degenerate" parameters $\mu_i$ (cf. Golubitsky and Langford [1981]).

(a) $\dot{r} = (\mu_1^2 + \mu_2) r \pm r^3$;
(b) $\dot{r} = (\mu_1^3 + \mu_1 \mu_2 + \mu_3) r \pm r^3$.

(In Golubitsky and Langford's treatment the rôle of a "distinguished bifurcation parameter" is stressed: one parameter ($\mu_1$) is singled out and the others regarded as unfolding or perturbation parameters, cf. Golubitsky and Schaeffer [1979a,b].)

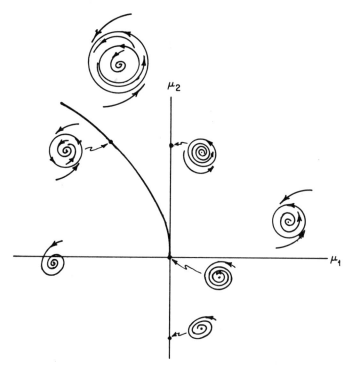

Figure 7.1.4. The generalized Hopf bifurcation, $a_5 < 0$ case.

## 7.2. A Note on $k$-Jets and Determinacy

Faced with a vector field $f(x)$ whose linearization at some fixed point $\bar{x}$ is hyperbolic, we can use Hartman's theorem to determine the local phase portrait. In such a case we say that the 1-jet $Df(\bar{x}) \cdot x$ *determines* the system locally. However, when the point $\bar{x}$ is nonhyperbolic, Hartman's theorem does not apply and we must include higher-order terms. In Chapter 3 we have already seen how the number of such terms can be reduced by using the normal form theorem on a Taylor series expansion. The question we now address is, how far need such an expansion go to determine the local vector field up to homeomorphism?

Thus far we have implicitly assumed that the lowest-order nonvanishing term(s) in the Taylor series *do* determine the local phase portrait and hence the stability type. For the Hopf, generalized Hopf, and saddle-node and cuspoid bifurcations discussed in Chapter 3 and in Section 7.1 above, this is not difficult to verify, since the normal forms show that the local behavior is governed by the one-dimensional systems

$$\dot{x} = \sum_{j=2}^{k} a_j x^j + \mathcal{O}(x^{k+1}), \qquad (7.2.1)$$

and

$$\dot{r} = \sum_{j=1}^{k} a_j r^{2j+1} + \mathcal{O}(r^{2k+3}), \qquad (7.2.2)$$

respectively. In these cases it is clear that, for sufficiently small $|x|$ or $r$, the neglect of higher-order terms can have no bad effects.

However, in cases for which the dimension of the center manifold is greater than one, the situation is more subtle and each case must be checked individually. Takens [1974a] performed the requisite computations for the three cases given in (7.0.1) (in the third case, he assumed that certain non-resonance conditions on the frequencies $\omega_1$ and $\omega_2$, described in Section 7.5, were met). Here we outline his treatment of the double zero case

$$Df(\bar{x}) = \begin{bmatrix} 0 & 1 \\ 0 & 0 \end{bmatrix} \overset{\text{def}}{=} A. \qquad (7.2.3)$$

To start with, we compute the normal form. A short calculation shows that, the range $B_2$ of the operator $\text{ad } A = [\cdot, A]$ described in the normal form theorem (Theorem 3.3.1) is spanned by the six vectors:

$$\left\{ \begin{pmatrix} 2xy \\ 0 \end{pmatrix}, \begin{pmatrix} y^2 \\ 0 \end{pmatrix}, \begin{pmatrix} 0 \\ 0 \end{pmatrix}, \begin{pmatrix} x^2 \\ -2xy \end{pmatrix}, \begin{pmatrix} xy \\ -y^2 \end{pmatrix}, \begin{pmatrix} y^2 \\ 0 \end{pmatrix} \right\}, \qquad (7.2.4)$$

(cf. Exercise 3.3.1). These vectors are, respectively, the Lie brackets of the linear part $\begin{pmatrix} y \\ 0 \end{pmatrix}$ with the six standard basis vectors for $H_2$:

$$\left\{ \begin{pmatrix} x^2 \\ 0 \end{pmatrix}, \begin{pmatrix} xy \\ 0 \end{pmatrix}, \begin{pmatrix} y^2 \\ 0 \end{pmatrix}, \begin{pmatrix} 0 \\ x^2 \end{pmatrix}, \begin{pmatrix} 0 \\ xy \end{pmatrix}, \begin{pmatrix} 0 \\ y^2 \end{pmatrix} \right\}. \qquad (7.2.5)$$

Thus quadratic terms of the forms $\begin{pmatrix} xy \\ 0 \end{pmatrix}, \begin{pmatrix} y^2 \\ 0 \end{pmatrix}, \begin{pmatrix} 0 \\ y^2 \end{pmatrix}$, and $\begin{pmatrix} x^2 \\ -2xy \end{pmatrix}$ can immediately be removed, and either

$$\text{span}\left\{ \begin{pmatrix} x^2 \\ 0 \end{pmatrix}, \begin{pmatrix} 0 \\ x^2 \end{pmatrix} \right\}, \qquad (7.2.6)$$

or

$$\text{span}\left\{ \begin{pmatrix} 0 \\ x^2 \end{pmatrix}, \begin{pmatrix} 0 \\ xy \end{pmatrix} \right\}, \qquad (7.2.7)$$

forms a complementary subspace to $B_2$ in $H_2$. Hence the 2-jet of the normal form can be conveniently written either as

$$\begin{aligned} \dot{x} &= y + a_2 x^2, \\ \dot{y} &= b_2 x^2, \end{aligned} \qquad (7.2.8)$$

or

$$\begin{aligned} \dot{x} &= y, \\ \dot{y} &= a_2 x^2 + b_2 xy. \end{aligned} \qquad (7.2.9)$$

EXERCISE 7.2.1. Show that

$$H_3 = \text{span}\{B_3\} + \text{span}\left\{\begin{pmatrix} x^3 \\ 0 \end{pmatrix}, \begin{pmatrix} 0 \\ x^3 \end{pmatrix}\right\} \quad \text{or} \quad \text{span}\{B_3\} + \text{span}\left\{\begin{pmatrix} 0 \\ x^3 \end{pmatrix}, \begin{pmatrix} 0 \\ x^2 y \end{pmatrix}\right\},$$

and, more generally that

$$H_k = \text{span}\{B_k\} + \text{span}\left\{\begin{pmatrix} x^k \\ 0 \end{pmatrix}, \begin{pmatrix} 0 \\ x^k \end{pmatrix}\right\} \quad \text{or} \quad \text{span}\{B_k\} + \text{span}\left\{\begin{pmatrix} 0 \\ x^k \end{pmatrix}, \begin{pmatrix} 0 \\ x^{k-1} y \end{pmatrix}\right\}.$$

We now outline Takens' [1974a] proof that the 2-jet (7.2.8) determines the local topological type of any vector field

$$\dot{x} = y + a_2 x^2 + \mathcal{O}(|x, y|^3),$$
$$\dot{y} = b_2 x^2 + \mathcal{O}(|x, y|^3), \tag{7.2.10}$$

provided that $b_2 \neq 0$. (A similar conclusion is necessarily true for (7.2.9), since the flows arising from these two vector fields are topologically equivalent.) The main tool is a technique called *blowing-up*. Singular changes of coordinates are introduced which expand degenerate fixed points into circles containing a finite number of fixed points. If these are hyperbolic after the first blow-up, then the local flow near the circle, and hence near the original fixed point, is stable with respect to higher-order terms. In the example (7.2.10), three blow-ups and an additional trick are necessary before the transformed vector field is stable.

Here we describe the first blow-up in detail and sketch the remaining ones, to illustrate the method. Letting $x = r \cos \theta$, $y = r \sin \theta$, we start with a transformation to polar coordinates, so that (7.2.10) becomes

$$\dot{r} = r \cos \theta (\sin \theta + ar \cos^2 \theta) + br^2 \cos^2 \theta \sin \theta + \mathcal{O}(r^3),$$
$$\dot{\theta} = br \cos^3 \theta - (\sin^2 \theta + ar \sin \theta \cos \theta) + \mathcal{O}(r^2); \tag{7.2.11}$$

where we have dropped the subscripts, writing $a_2 = a$, $b_2 = b$. We now expand the singularily $r = 0$ into a circle by regarding $(r, \theta)$ as coordinates on the surface of a cylinder, Figure 7.2.1. Looking down from the top, the upper half of the cylinder becomes the original phase plane $\mathbb{R}^2$ minus the origin ($r > 0$), while the circle $r = 0$ becomes the origin itself (the lower half

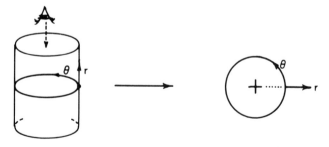

Figure 7.2.1. The singular blow-up of Takens.

of the cylinder does not correspond to any points on the original $(x, y)$ plane).

Equation (7.2.11) has equilibria at $\theta = 0, \pi$ on the circle $r = 0$, both of which are degenerate and have similar linear parts. Letting $\theta \approx 0, r \approx 0$ we expand (7.2.11) in Taylor series near $(r, \theta) = (0, 0)$ to obtain

$$\dot{r} = r\theta + ar^2 + \mathcal{O}(|r, \theta|^3),$$
$$\dot{\theta} = br - \theta^2 - ar\theta + \mathcal{O}(r^2) + \mathcal{O}(|r, \theta|^3).$$
(7.2.12)

A similar form is obtained near $(r, \theta) = (0, \pi)$. Since this vector field is still degenerate, we must blow-up once more, but first we can apply the normal form theorem to (7.2.12) to remove some terms of second order to obtain

$$\dot{r} = r\theta,$$
$$\dot{\theta} = br - \theta^2 + \mathcal{O}(|r, \theta|^3).$$
(7.2.13)

After two further blow-ups $(r, \theta) \to (\rho, \phi)$ defined by $\theta = \rho \cos \phi$, $r = \rho \sin \phi$ and $(\rho, \phi) \to (\eta, \psi)$ defined by $\phi = \eta \cos \psi$, $\rho = \eta \sin \psi$, the energetic reader can check that we obtain the vector field

$$\dot{\eta} = \eta^2(-b \cos^3 \psi + 2 \cos^2 \psi \sin \psi - \sin^3 \psi + b \cos \psi \sin^2 \psi + \cdots),$$
$$\dot{\psi} = \eta(b \cos^3 \psi \sin \psi - 3 \cos \psi \sin^2 \psi + b \cos^2 \psi \sin \psi + \cdots).$$
(7.2.14)

Now the phase portrait of (7.2.14) is unaffected (except possibly at $\eta = 0$) by division of the vector field by $\eta$, and, since the common factor $\eta$ occurs in both components, we may consider the "divided out" vector field

$$\dot{\eta} = \eta(-b \cos^3 \psi + \cdots),$$
$$\dot{\psi} = (b \cos^3 \psi \sin \psi + \cdots),$$
(7.2.15)

which has six *hyperbolic* fixed points at $\eta = 0$, $\psi = 0$, $\pi/2$, $\pi$, $3\pi/2$, and $\psi = \arctan(2b/3)$. Hence the flow of (7.2.15) is stable to small (higher-order) perturbations and consequently, the flow of (7.2.14) near $\eta = 0$ is similarly stable. We now "blow-down" three times $(\eta, \psi) \to (\rho, \phi) \to (r, \theta) \to (x, y)$ to conclude that the flow of (7.2.8) near the degenerate fixed point $(x, y) = (0, 0)$ is indeed stable with respect to the addition of (small) higher-order terms provided that $b \neq 0$. Note that $a$ is unimportant, since it vanished along the way. However, $a$ *will* play an important role in determining the *unfolding*, as we shall see in the next section. In general the conditions sufficient for determinancy of a degenerate vector field may not be sufficient for determinancy of the unfolding. In Figure 7.2.2 we illustrate the blow-up process geometrically.

Similar analyses can be carried out for the other codimension two degenerate singularities of Sections 7.4–7.5, see Takens [1974a]. For the three- and four-dimensional cases one has to check that certain invariant

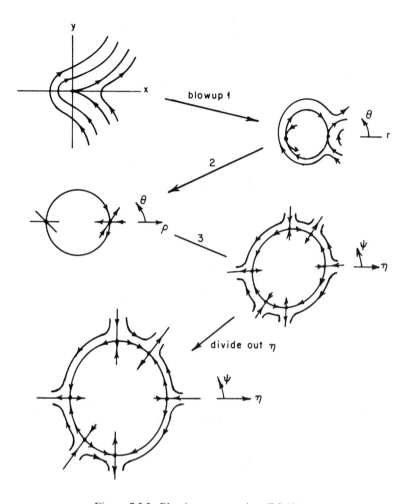

Figure 7.2.2. Blowing-up equation (7.2.10).

cones in the truncated symmetric systems are normally hyperbolic and hence persist under the addition of higher-order terms. We shall not consider this more complicated analysis here.

## 7.3. The Double Zero Eigenvalue

We now turn to a study of the degenerate 2- and 3-jets associated with the linear part $\begin{bmatrix} 0 & 1 \\ 0 & 0 \end{bmatrix}$, and a development of their universal unfoldings. As we saw in the preceding section, the $k$-jet of the normal form for this problem can be conveniently written in either of two ways:

$$
\begin{aligned}
\dot{x} &= y + \sum_{j=2}^{k} a_j x^j \\
\dot{y} &= \sum_{j=2}^{k} b_j x^j
\end{aligned}
\quad + \mathcal{O}(|x, y|^{k+1}),
\tag{7.3.1}
$$

or

$$
\begin{aligned}
\dot{x} &= y \\
\dot{y} &= \sum_{j=2}^{k} (a_j x^j + b_j x^{j-1} y)
\end{aligned}
\quad + \mathcal{O}(|x, y|^{k+1}).
\tag{7.3.2}
$$

Takens [1974a] uses the first form while Bogdanov [1975] and Arnold [1972] take the second. Here, in contrast to our outlines of Takens' results in Section 7.2, we shall follow Bogdanov's choice. Also see Kopell and Howard [1975] for another treatment.

We start with the assumption that the quadratic coefficients $a_2, b_2$ do not vanish and initially neglect terms of order 3 and higher. As in Section 7.2, we drop the subscripts; thus we wish to unfold the degenerate vector field

$$
\begin{aligned}
\dot{x} &= y, \\
\dot{y} &= ax^2 + bxy.
\end{aligned}
\tag{7.3.3}
$$

We shall see below that the signs of both $a$ and $b$ are important in a topological classification, and that *both* coefficients must be nonzero for the unfolding to be fully determined. (Applying the methods of Section 7.2 to this normal form we find that we only require $a \neq 0$ for determinacy of the degenerate field.)

A universal unfolding of (7.3.3) must provide a family of vector fields whose local flows contain all possible small perturbations of the degenerate flow of (7.3.3). In contrast to the nice situation in singularity theory (cf. Golubitsky and Schaeffer [1983]), here there is no general recipe available for constructing such a family. Each case must be considered individually. In the present case, since the equilibrium point has a zero eigenvalue, it must certainly be allowed to disappear or to split into at least two structurally stable equilibria, but other nonwandering sets such as periodic orbits may also be present in perturbations. The following two-parameter family provides a universal unfolding of (7.3.3):

$$
\begin{aligned}
\dot{x} &= y, \\
\dot{y} &= \mu_1 + \mu_2 y + ax^2 + bxy.
\end{aligned}
\tag{7.3.4}
$$

We will sketch the justification of this assertion later. We note that our form (7.3.4) differs from Bogdanov's [1975], but that we recover the same topological types in our unfolding as in his and in Takens' [1974b].

In our analysis, we fix $a$ and $b$ for simplicity. It should be clear that, with suitable rescaling and letting $(x, y) \to (-x, -y)$, for any $a, b \neq 0$ the possible cases can be reduced to two: $a = 1$ and $b = \pm 1$. We shall consequently take $a = b = 1$ and leave the other case as an exercise.

It is easy to find bifurcation curves in which (7.3.4) undergoes saddle-node and Hopf bifurcations. We first seek fixed points, which are given by

$$(x, y) = (\pm \sqrt{-\mu_1}, 0) \overset{\text{def}}{=} (x_\pm, 0), \tag{7.3.5}$$

and hence only exist for $\mu_1 \leq 0$. Linearizing at these points, we find

$$Df(x_\pm, 0) = \begin{bmatrix} 0 & 1 \\ \pm 2\sqrt{-\mu_1} & \mu_2 \pm \sqrt{-\mu_1} \end{bmatrix}. \tag{7.3.6}$$

Therefore $(x_+, 0)$ is a saddle for $\mu_1 < 0$ and all $\mu_2$, while $(x_-, 0)$ is a source for $\{\mu_2 > \sqrt{-\mu_1}, \mu_1 < 0\}$ and a sink for $\{\mu_2 < \sqrt{-\mu_1}, \mu_1 < 0\}$. Checking the conditions, we find that a Hopf bifurcation occurs on the curve $\mu_2 = \sqrt{-\mu_1}$ while saddle-node bifurcations occur on $\mu_1 = 0, \mu_2 \neq 0$.

EXERCISE 7.3.1. Verify the statements above.

To study the stability of the Hopf bifurcation, we change coordinates twice, first to bring the point $(x_-, 0)$ to the origin and then to put the vector field into standard form. Letting $\bar{x} = x - x_-, \bar{y} = y$, we obtain

$$\begin{pmatrix} \dot{\bar{x}} \\ \dot{\bar{y}} \end{pmatrix} = \begin{bmatrix} 0 & 1 \\ 2x_- & 0 \end{bmatrix} \begin{pmatrix} \bar{x} \\ \bar{y} \end{pmatrix} + \begin{pmatrix} 0 \\ \bar{x}\bar{y} + \bar{x}^2 \end{pmatrix}. \tag{7.3.7}$$

Then, using the linear transformation

$$\begin{pmatrix} \bar{x} \\ \bar{y} \end{pmatrix} = T \begin{pmatrix} u \\ v \end{pmatrix}, \tag{7.3.8}$$

where

$$T = \begin{bmatrix} 0 & 1 \\ \sqrt{-2x_-} & 0 \end{bmatrix}$$

is the matrix of real and imaginary parts of the eigenvectors of the eigenvalues $\lambda = \pm i\sqrt{-2x_-}$, we obtain the system with linear part in standard form:

$$\begin{pmatrix} \dot{u} \\ \dot{v} \end{pmatrix} = \begin{bmatrix} 0 & -\sqrt{-2x_-} \\ \sqrt{-2x_-} & 0 \end{bmatrix} \begin{pmatrix} u \\ v \end{pmatrix} + \begin{pmatrix} uv + \dfrac{1}{\sqrt{-2x_-}} v^2 \\ 0 \end{pmatrix}. \tag{7.3.9}$$

We can now use the stability algorithm (3.4.11). In this case, the only non-vanishing terms are $f_{uv} = 1$, $f_{vv} = 2/\sqrt{-2x_-}$ and we obtain the leading coefficient in the normal form of the Hopf bifurcation as

$$a = \frac{1}{16\sqrt{-2x_-}} \cdot 1 \cdot \frac{2}{\sqrt{-2x_-}} = \frac{1}{-16x_-} = \frac{1}{+16\sqrt{-\mu_1}} > 0. \quad (7.3.10)$$

The bifurcation is therefore *subcritical* and we have a family of *unstable* periodic orbits surrounding the sink for $\mu_2 < \sqrt{-\mu_1}$ and close to $\sqrt{-\mu_1}$. We can summarize our knowledge so far as in Figure 7.3.1, which shows the (partial) bifurcation set and associated phase portraits. The reader should check that the vector fields, and especially those of the degenerate saddle-nodes on $\mu_1 = 0$, are as shown in this figure.

EXERCISE 7.3.2. Show that there are no periodic orbits in the regions $\mu_1 > 0$ and $\mu_2 < -\sqrt{-\mu_1}$, $\mu_1 < 0$, and verify that the phase portraits of Figure 7.3.1 are correct. In particular, use center manifold theory near $\mu_1 = 0$, $\mu_2 \neq 0$ to show that the sink-saddle and source-saddle connections sketched above actually exist.

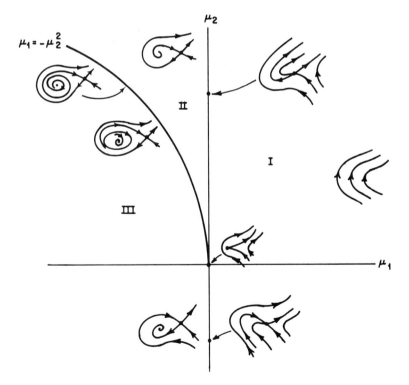

Figure 7.3.1. Unfolding equation (7.3.3), the partial bifurcation set and phase portraits of equation (7.3.4).

EXERCISE 7.3.3. Include the coefficient $b$ in equation (7.3.4) explicitly in your calculations, and show that a *supercritical* Hopf bifurcation occurs if $b < 0$.

We now note that the phase portraits in region III near $\mu_1 = 0$, $\mu_2 < 0$ and $\mu_2 = \sqrt{-\mu_1} > 0$ are not homeomorphic, for the latter possess limit cycles while the former do not. Hence there must be additional bifurcation points in region III. The saddle and the sink do not change topological type in this region, and thus a *global* bifurcation must take place, perhaps a saddle loop (cf. Section 6.1) at which the limit cycle vanishes and the stable and unstable manifolds of the saddle point "cross over." To study this, we use a rescaling transformation which differs somewhat from the blowing-up process described above (cf. Takens [1974b] and Carr [1981]). We set

$$x = \varepsilon^2 u, \qquad y = \varepsilon^3 v, \qquad \mu_1 = \varepsilon^4 v_1, \qquad \mu_2 = \varepsilon^2 v_2, \qquad \varepsilon \geq 0, \quad (7.3.11)$$

and rescale time $t \to \varepsilon t$, so that (with $a = b = 1$), (7.3.4) becomes

$$\begin{aligned} \dot{u} &= v, \\ \dot{v} &= v_1 + \varepsilon v_2 v + \varepsilon u v + u^2. \end{aligned} \qquad (7.3.12)$$

The analysis of the unfolding now becomes the analysis of the *three*-parameter problem (7.3.12) with $v_1$, $v_2$ of $\mathcal{O}(1)$ and $\varepsilon$ small. At first sight our transformation has merely introduced another parameter and enlarged the problem, but we note that we are only concerned with the case $v_1 < 0 \, (\mu_1 < 0)$, since there are no fixed points for $v_1 > 0$; and, more significantly, if we let $\varepsilon \to 0$ with $v_1 \neq 0$ fixed, (7.3.12) becomes an integrable Hamiltonian system

$$\begin{aligned} \dot{u} &= v, \\ \dot{v} &= v_1 + u^2, \end{aligned} \qquad (7.3.13)$$

with Hamiltonian

$$H(u, v) = \frac{v^2}{2} - v_1 u - \frac{u^3}{3}. \qquad (7.3.14)$$

Using our singular transformation (7.3.11), we see that $\varepsilon = 0$ implies that $\mu_1 = \mu_2 = 0$, and our degenerate fixed point has blown-up into a Hamiltonian system. Moreover, the transformation keeps the fixed points a finite distance apart as the parameter approaches the degenerate bifurcation point $\mu_1 = \mu_2 = 0$.

The motivation for the rescaling is now apparent, since we can perturb the *global* solution curves of (7.3.13) for small $\varepsilon$, and hence reveal the behavior of (7.3.4) for $\mu_1$, $\mu_2$ close to zero. We might say that the Hamiltonian vector field of (7.3.13), with $v_1$ fixed (and we take $v_1 = -1$, corresponding to $\mu_1 \leq 0$) contains all the behavior of the unfolding "in embryo." In particular, in Figure 7.3.2, note the closed orbits and the saddle connection $\Gamma_0$ corresponding to the level curve $H(u, v) = \frac{2}{3}$.

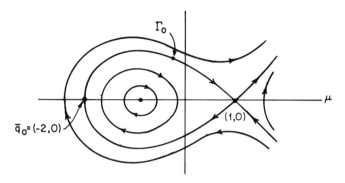

Figure 7.3.2. The phase portrait of (7.3.13), $v_1 = -1$.

The search for saddle loops is now a search for values of $v_2$, $\varepsilon \neq 0$ for which the saddle connection is maintained, a problem which we can solve by the Melnikov method (Sections 4.5–4.6). The solution on $\Gamma_0$ based at the point $\bar{q}_0 = (-2, 0)$ is given by

$$(u_0(t), v_0(t)) = \left(1 - 3 \operatorname{sech}^2\left(\frac{t}{\sqrt{2}}\right), 3\sqrt{2} \operatorname{sech}^2\left(\frac{t}{\sqrt{2}}\right) \tanh\left(\frac{t}{\sqrt{2}}\right)\right). \quad (7.3.15)$$

In this case the Melnikov function $M(t_0)$ is not time-dependent, since the perturbation is the constant vector field $\varepsilon\begin{pmatrix} 0 \\ v_2 v + uv \end{pmatrix}$, and we have

$$M(v_2) = \int_{-\infty}^{\infty} v_0(t)(v_2 v_0(t) + u_0(t)v_0(t))\, dt$$

$$= \frac{1}{\sqrt{2}}\left[ v_2 \int_{-\infty}^{\infty} 18 \operatorname{sech}^4 \tanh^2 \tau\, d\tau \right. \qquad (7.3.16)$$

$$\left. + \int_{\infty}^{\infty} (1 - 3 \operatorname{sech}^2 \tau)18 \operatorname{sech}^4 \tau \tanh^2 \tau\, d\tau \right],$$

where $\tau = t/\sqrt{2}$. The bifurcation situation, when the saddle connection is preserved, is given by $M \equiv 0$, or, for $\varepsilon$ small:

$$v_2 \approx -\frac{\int_{-\infty}^{\infty} (1 - 3 \operatorname{sech}^2 \tau) \operatorname{sech}^4 \tau \tanh^2 \tau\, d\tau}{\int_{-\infty}^{\infty} \operatorname{sech}^4 \tau \tanh^2 \tau\, d\tau}.$$

Noting that $\operatorname{sech}^2 \tau = 1 - \tanh^2 \tau$ and that

$$\int_{-\infty}^{\infty} \operatorname{sech}^2 \tau \tanh^k \tau\, d\tau = \frac{\tanh^{k+1}(\tau)}{k+1}\Big|_{-\infty}^{\infty} = \frac{2}{k+1},$$

we find that

$$v_2 = \frac{5}{7}, \qquad (7.3.17)$$

Finally, recalling that $v_1 = -1$ and using (7.3.11) ($\mu_1 = -\varepsilon^4$, $\mu_2 = \varepsilon^2 v_2$), we obtain the approximate bifurcation curve

$$\mu_1 = -(\tfrac{49}{25})\mu_2^2, \; \mu_2 \geq 0. \tag{7.3.18}$$

The true bifurcation curve will be tangent to this semi-parabola at $\mu_1 = \mu_2 = 0$. Comparing this with the equation for the Hopf bifurcation set $B_h$:

$$\mu_1 = -\mu_2^2; \qquad \mu_2 > 0, \tag{7.3.19}$$

we see that there is, indeed, a second bifurcation curve $B_{sc}$ lying to the left of (7.3.19) and tangent to it (and to $\mu_1 = 0$) at $(\mu_1, \mu_2) = (0, 0)$. On $B_{sc}$ the phase portrait has a saddle loop. The sign taken by the Melnikov function $M$ for $\mu_1 >$ (resp. $<$) $- (\tfrac{49}{25})\mu_2^2$ gives the relative position of the stable and unstable manifolds (separatrices of the saddle), and to conclude, we note that the trace of the "saddle quantity" (cf. Section 6.1) is positive on the curve (7.3.18):

$$\mathrm{tr}\, Df(+\sqrt{-\mu_1}, 0) = \mu_2 + \sqrt{-\mu_1} = \tfrac{12}{5}\mu_2 > 0, \tag{7.3.20}$$

and thus that the homoclinic orbit is an $\alpha$-limit set for nearby points. We illustrate this in Figure 7.3.3.

It remains to check that, throughout region IIIa, the system has a unique repelling limit cycle for each pair of parameter values $(\mu_1, \mu_2)$. Letting $\gamma = (u_\alpha(t), v_\alpha(t))$ denote one of the closed orbits within $\Gamma_0$ with Hamiltonian $H(u_\alpha, v_\alpha) = \alpha$ and period $T_\alpha$, it is sufficient by Melnikov theory to verify that

$$\begin{aligned} M^\alpha(v_2) &= \int_0^{T_\alpha} v_\alpha(t)[v_2 v_\alpha(t) + u_\alpha(t)v_\alpha(t)]\, dt \\ &= \int_{\gamma^\alpha} [v_2 v + uv]\, du \end{aligned} \tag{7.3.21}$$

is zero for just one parameter value $v_2(\varepsilon, \alpha)$ for each choice of $\varepsilon$ and $\alpha$. This can be done either by direct evaluation of the integrals, using elliptic functions, or by arguments of the type used by Carr [1981]. We do not give the details here, but leave the computation as an exercise:

EXERCISE 7.3.4. Convert (7.3.16) into a contour integral around the homoclinic orbit and evaluate it. Evaluate (7.3.21) using elliptic functions (substitute $v = \sqrt{2\alpha + 2v_1 u + \tfrac{2}{3}u^2}$ from (7.3.14)). You might find Byrd and Friedman [1971] useful.

EXERCISE 7.3.5. Find the bifurcation set and phase portraits for the unfoldings of (7.3.4) with $a = +1, b = -1$; hence show that the choice $a = b = 1$ essentially captures all the cases up to reversal of time. What can you conclude when $b = 0$?

Note that the bifurcation set of Figure 7.3.3 is a set of codimension one *curves* meeting at the point, $\mu_1 = \mu_2 = 0$, at which the vector field has the codimension two singularity. On each curve a codimension one bifurcation takes place, saddle-nodes on $\mu_1 = 0$; $\mu_2 \neq 0$, Hopf on $\mu_1 = -\mu_2^2$; $\mu_2 > 0$

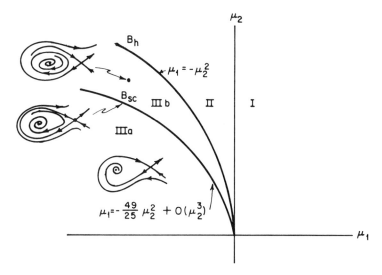

Figure 7.3.3. The global saddle connection bifurcation.

and a saddle connection or homoclinic bifurcation on $\mu_1 \approx -(\frac{49}{25})\mu_2^2$; $\mu_2 > 0$. The latter is an example of a global bifurcation which we have met before in Section 6.1. However, in this unfolding we have seen how the presence of such a *global* bifurcation can be detected by means of *local* analysis. This feature will occur repeatedly in this chapter.

We will not address the problem of proving that (7.3.4) is a universal unfolding of (7.3.3) here. Indeed we have not even clearly defined universal unfoldings, for there are technical issues at work here beyond the scope of our treatment (cf. Newhouse *et. al.* [1976]). Interested readers should consult Arnold [1972] and Bogdanov [1975] for more information on this particular example.

EXERCISE 7.3.6. Show that the codimension two bifurcation just described occurs at the point $(\sigma, \gamma) = (\frac{1}{2}, \frac{1}{2})$ in the averaged van der Pol equation (2.1.14) (cf. Section 2.1 and Holmes and Rand [1978]).

In the same [1974b] paper, Takens also studies unfoldings which preserve certain rotational symmetries. An important case in the study of nonlinear oscillations is that of "cubic" symmetry or symmetry under rotation through $\pi$. Here the degenerate field contains cubic terms at lowest order and the unfolding is provided by the two-parameter family

$$\dot{x} = y,$$
$$\dot{y} = \mu_1 x + \mu_2 y + a_3 x^3 + b_3 x^2 y. \qquad (7.3.22)$$

Allowing reversal and linear rescaling of time, we can set $b_3 = -1$ without loss of generality, but there are two distinct cases $a_3 = \pm 1$ to be considered.

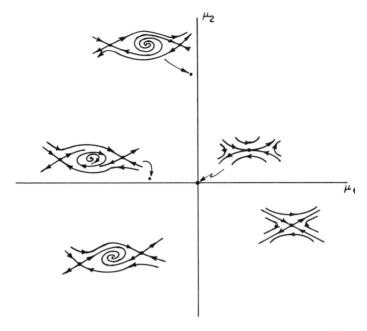

Figure 7.3.4. Partial bifurcation set and phase portraits of equation (7.3.22): $a_3 = +1$, $b_3 = -1$.

Taking $a_3 = +1$ first, local analyses yield the (partial) bifurcation set and phase portraits of Figure 7.3.4. At the origin we have a degenerate saddle point and strictly we should check that the three jet $\left( \begin{array}{c} y \\ \pm x^3 - x^2 y \end{array} \right)$ is determined, using the methods of Section 7.2.2, before studying the unfolding. We refer the reader to Takens [1974a, b] for details of these calculations.

EXERCISE 7.3.7. Verify that Figure 7.3.4 is correct.

In this case we suspect the existence of global bifurcations in the upper left-hand quadrant. To check this, we rescale as follows, letting

$$x = \varepsilon u, \qquad y = \varepsilon^2 v, \qquad \mu_1 = \varepsilon^2 v_1, \qquad \mu_2 = \varepsilon^2 v_2, \qquad t \to \varepsilon t, \quad (7.3.23)$$

to obtain

$$\dot{u} = v,$$
$$\dot{v} = v_1 u + \varepsilon v_2 v + u^3 - \varepsilon u^2 v. \quad (7.3.24)$$

Setting $v_1 = -1$ (so that $\mu_1 \leq 0$) we find that (7.3.24) is Hamiltonian with Hamiltonian

$$H(u, v) = \frac{v^2}{2} + \frac{u^2}{2} - \frac{u^4}{4}, \quad (7.3.25)$$

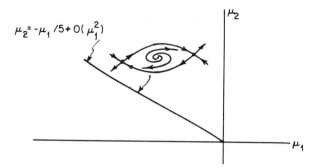

$$\mu_2 = -\mu_1/5 + O(\mu_1^2)$$

Figure 7.3.5. The completion of Figure 7.3.4.

for $\varepsilon = 0$. The phase portrait of this Hamiltonian system has a pair of symmetric heteroclinic orbits connecting the saddle points at $(u, v) = (\pm 1, 0)$ on the level curve $H(u, v) = \frac{1}{4}$, and a similar analysis to that above yields the bifurcation set of Figure 7.3.5, completing the analysis.

The case $a_3 = -1$ is a little more complicated, since the global bifurcations here are more varied, involving the coalescence of closed orbits as well as saddle connections. Rescaling with (7.3.23) and setting $v_1 = +1$ in this case (to obtain nontrivial fixed points at $(x, y) = (\pm\sqrt{\mu_1}, 0)$ or $(u, v) = (\pm 1, 0)$), we obtain the system

$$\dot{u} = v,$$
$$\dot{v} = u + \varepsilon v_2 v - u^3 - \varepsilon u^2 v. \tag{7.3.26}$$

When $\varepsilon = 0$, the resulting integrable system with Hamiltonian

$$H(u, v) = \frac{v^2}{2} - \frac{u^2}{2} + \frac{u^4}{4} \tag{7.3.27}$$

has the phase portrait of Figure 7.3.6 (a picture we have already met!).

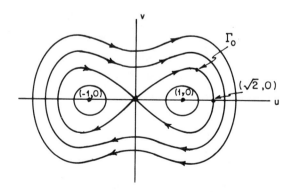

Figure 7.3.6. The phase portrait of (7.3.26), $\varepsilon = 0$.

Perturbing the (double) homoclinic orbit $\Gamma_0$ given by

$$(u_0(t), v_0(t)) = (\pm\sqrt{2} \text{ sech } t, \mp\sqrt{2} \text{ sech } t \tanh t), \qquad (7.3.28)$$

it is not hard to see that a (double) saddle connection occurs on

$$v_2 = \tfrac{4}{5}, \qquad (7.3.29)$$

or, in terms of the original unfolding parameters, on a curve tangent to

$$\mu_2 = \frac{4\mu_1}{5} \qquad (7.3.30)$$

at $(\mu_1, \mu_2) = (0, 0)$. This computation, and conventional analyses by linearization and Hopf bifurcation calculations, provide the partial bifurcation set and phase portraits of Figure 7.3.7.

EXERCISE 7.3.8. Verify that Figure 7.3.7 is correct (cf. Exercise 4.6.4).

To complete the analysis, we first note that, for $\mu_2 < 0$, no closed orbits exist, since in that case trace $Df = \mu_2 - x^2 < 0$ and we can apply Bendixon's criterion. This suggests that the closed orbit encircling all three fixed points

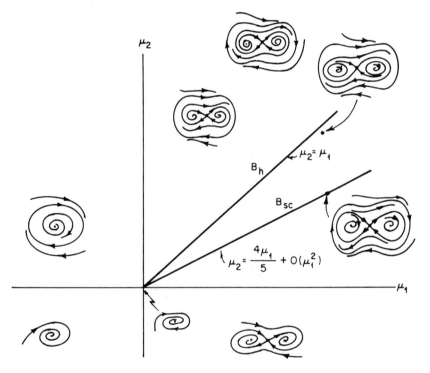

Figure 7.3.7. Partial bifurcation set for equation (7.6.22), $a_3 = b_3 = -1$, with associated phase portraits.

must somehow vanish on a bifurcation curve in the positive quadrant below $\mu_2 = 4\mu_1/5$ (the homoclinic bifurcation) but above $\mu_2 = 0$. To study this, we examine the transformed Hamiltonian field

$$\dot{u} = v,$$
$$\dot{v} = u - u^3, \tag{7.3.31}$$

and its perturbation (7.3.26), more closely. In addition to asking about perturbations of the homoclinic orbit $\Gamma_0$ we must also consider perturbations of the closed level curves of $H_0$ lying within and outside $\Gamma_0$. This involves lengthy calculations with elliptic functions, the results of which were reported (partially) by Takens [1974b] and Holmes and Rand [1980] and were repeated and completed by Carr [1981], and Knobloch and Proctor [1981].

Picking one of these level curves $\gamma^\alpha = H_0^{-1}(\alpha)$ and denoting the solution $(u_\alpha(t), v_\alpha(t))$ as in (7.3.21), we have the Melnikov function

$$M^\alpha(v_2) = \int_0^{T_\alpha} v_\alpha(t)[v_2 v_\alpha(t) - u_\alpha^2(t)v_\alpha(t)] \, dt$$
$$= v_2 \int_{\gamma^\alpha} v \, du - \int_{\gamma^\alpha} u^2 v \, du, \tag{7.3.32}$$

where we have used $v = du/dt$ to convert the time integral into a contour integral on the closed orbit $\gamma^\alpha$. To evaluate (7.3.32) we express $v$ as a function of $u$ and $\alpha$ via the Hamiltonian (7.3.25). For a given closed orbit to remain after perturbation, we require $M^\alpha = 0$, or

$$v_2 = \frac{\int_{\gamma^\alpha} u^2 v \, du}{\int_{\gamma^\alpha} v \, du} \overset{\text{def}}{=} R(\alpha). \tag{7.3.33}$$

Using certain properties of the integrals, Carr [1981] proves that $R(\alpha)$ has the form shown in Figure 7.3.8, having a unique minimum $c \approx 0.752$ at some finite $\alpha > 0$ and then increasing monotonically towards infinity as $\alpha$ increases. Thus, for $v_2 \in (1, \infty)$ only one closed orbit near one of the level

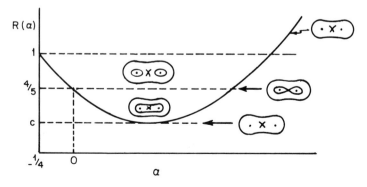

Figure 7.3.8. The graph of $R(\alpha)$, with associated preserved level curves.

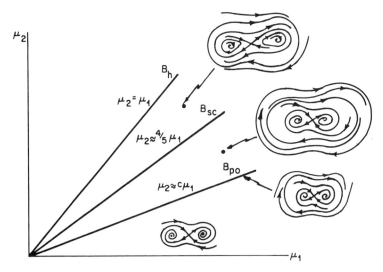

Figure 7.3.9. The completion of Figure 7.3.7.

curves, for some $\alpha > 0$, is preserved, while for $v_2 \in (\frac{4}{5}, 1)$ three orbits, two for $\alpha \in (-\frac{1}{4}, 0)$ and one for $\alpha > 0$, are maintained. At $v_2 = \frac{4}{5}$ we have the homoclinic saddle connection already described, coexisting with a closed orbit lying outside the connection, while for $v_2 \in (c, \frac{4}{5})$ we have two closed orbits, an attractor encircling a repellor, both enclosing all three fixed points. These orbits coalesce and vanish as $v_2$ passes through $c$. We can now complete the unfolding of Figure 7.3.7 as in Figure 7.3.9, adding the bifurcation curve $B_{po}$ on which the periodic orbits coalesce.

This concludes our analysis of unfoldings of vector fields with degenerate linear part $\begin{bmatrix} 0 & 1 \\ 0 & 0 \end{bmatrix}$. In Section 7.6 we shall show that this symmetric normal form occurs in interesting problems in solid and fluid mechanics.

## 7.4. A Pure Imaginary Pair and a Simple Zero Eigenvalue

In this and the next section we provide a partial analysis of the two remaining local bifurcations of codimension two. The results available are relatively incomplete due to the presence of homoclinic orbits in the unfoldings. In each case, the vector fields have imaginary eigenvalues. The normal forms of the vector fields are symmetric (equivariant) with respect to rotations in directions associated with the imaginary eigenvalues. If the original vector fields possess these symmetries, then the homoclinic and heteroclinic orbits which occur are necessarily degenerate and the phase portraits can be described completely. However, when the full vector fields are not symmetric, the homoclinic phenomena can be analyzed only in terms of three- or four-dimensional flows, respectively. Thus, in the general case, all the complexity of the Silnikov and Newhouse examples of Sections 6.5–6.7 can be expected to occur. A complete analysis of the present local problems will evidently have to await a fuller understanding of these global phenomena.

We note that partial results on the present problems using perturbation methods and Liapunov–Schmidt techniques were obtained by Keener [1976, 1981], Langford [1979], and Iooss and Langford [1980]. Their methods are essentially restricted to the analysis of a planar vector field and yield little or no indication of the global chaotic dynamics which can arise in specific cases. Here, after rederiving their results using normal form theory, we shall go on to consider the implications for the full three- and four-dimensional flows, drawing heavily on the global bifurcation results of Chapter 6. For earlier work on these global phenomena in this context, see Holmes [1980d] and Guckenheimer [1981].

Using the methods of Section 3.3, the reader can verify that the normal form of the degenerate $k$-jet with linear part

$$\begin{bmatrix} 0 & -\omega & 0 \\ \omega & 0 & 0 \\ 0 & 0 & 0 \end{bmatrix} \begin{pmatrix} x \\ y \\ z \end{pmatrix}, \tag{7.4.1}$$

can be conveniently written in cylindrical polar coordinates as

$$\begin{aligned} \dot{r} &= a_1 rz + a_2 r^3 + a_3 rz^2 + \mathcal{O}(|r, z|^4), \\ \dot{z} &= b_1 r^2 + b_2 z^2 + b_3 r^2 z + b_4 z^3 + \mathcal{O}(|r, z|^4), \\ \dot{\theta} &= \omega + \mathcal{O}(|r, z|^2). \end{aligned} \tag{7.4.2}$$

As in the simple Hopf bifurcation discussed in Section 3.4, this $k$-jet contains no $\theta$-dependent terms for arbitrarily high $k$, since the normal form can be chosen to be equivariant with respect to rotations about the $z$-axis. Thus the azimuthal component of (7.4.2) can be decoupled and we can work, at least initially, with a degenerate two-dimensional vector field with the special symmetry expressed by the fact that the first component is odd in $r$ and the second even. We will, however, ultimately have to restore the non-$S^1$-symmetric terms present in the "tail" of the Taylor series expansion when we return to the full three-dimensional problem.

EXERCISE 7.4.1. Compute the normal form for systems with linear part (7.4.1) up to terms of order 3 and verify that the polar coordinate version of equation (7.4.2) is correct.

We now begin our classification and unfolding of this degenerate singularity. Truncating (7.4.2) at $\mathcal{O}(|r, z|^2)$ and removing the azimuthal term, we obtain the planar system

$$\begin{aligned} \dot{r} &= a_1 rz, \\ \dot{z} &= b_1 r^2 + b_2 z^2, \end{aligned} \tag{7.4.3}$$

which was shown by Takens [1974a] to be 2-determined provided that $a_1, b_1, b_2 \neq 0$ and $b_2 - a_1 \neq 0.$* In this case we can rescale, with a possible reversal of time, to remove two of the coefficients. Letting $\bar{r} = \alpha r$, $\bar{z} = \beta z$, the equation becomes

$$\frac{d}{dt} \bar{r} = \alpha \left[ a_1 \frac{\bar{r}\bar{z}}{\alpha\beta} \right],$$

$$\frac{d}{dt} \bar{z} = \beta \left[ b_1 \frac{\bar{r}^2}{\alpha^2} + b^2 \frac{\bar{z}^2}{\beta^2} \right]. \tag{7.4.4}$$

Setting $\beta = -b_2$ and $\alpha = -\sqrt{|b_1 b_2|}$ and dropping the bars, (7.4.4) then yields

$$\dot{r} = arz,$$

$$\dot{z} = br^2 - z^2; \qquad b = \pm 1, \tag{7.4.5}$$

where $a = -a_1/b_2$ can be positive or negative, but will be assumed to be nonzero.

Takens [1974a] lists five distinct nondegenerate topological types for this normal form. There are in fact six, as given in Figure 7.4.1, but we note that types IIa and IIb ($b = +1$, $a < 0$) and IVa and IVb ($b = -1$, $a < 0$) might each be paired together, since their unfoldings are essentially identical, as we shall see.

A key idea in obtaining this classification is the determination of invariant lines $z = kr$ for the vector field. Substituting into (7.4.5), we find that the slopes, $k$, of such lines must satisfy

$$\frac{dz}{dr} = k = \frac{br^2 - (kr)^2}{ar(kr)} = \frac{b - k^2}{ak},$$

or

$$k = \pm\sqrt{\frac{b}{a+1}}. \tag{7.4.6}$$

We note that the $z$-axis ($r = 0$) is always invariant, and that other invariant lines $z = kr$ exist if $b/(a + 1) > 0$. The direction of the flow on (and near) such invariant lines is easily computed from the sign of the inner product of the vector field (7.4.5) with a radial vector field $(r, z)$:

$$s = (arz, br^2 - z^2) \cdot (r, z)|_{z=kr}$$

$$= azr^2(a + b - k^2) = \frac{a}{a+1} zr^2(1 + a + b). \tag{7.4.7}$$

---

* Actually $a_1 \neq 0$ is only required if $b_2 - a_1 < 0$, see Takens [1974a].

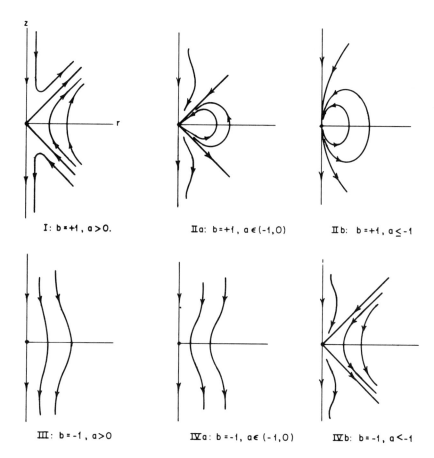

I: $b = +1$, $a > 0$.    IIa: $b = +1$, $a \in (-1, 0)$    IIb: $b = +1$, $a \leq -1$

III: $b = -1$, $a > 0$    IVa: $b = -1$, $a \in (-1, 0)$    IVb: $b = -1$, $a < -1$

Figure 7.4.1. Phase portraits in the $(r, z)$ half plane ($r \geq 0$) for the normal form (7.4.5).

If $s > 0$, then the flow is radially outward, if $s < 0$, the flow is inward. Other standard two-dimensional techniques (cf. Section 1.8) can be used to complete the classification of Figure 7.4.1. We leave the details to the reader.

EXERCISE 7.4.2. Verify that the topological classification of Figure 7.4.1 is correct. (Hint: In verifying that the closed solution curves in IIa and IIb are correct, use the symmetry of the vector field under reflection about the $r$-axis, or consider the integrals $F$ and $G$ given below in equations (7.4.18)–(7.4.19).)

In our attempt to obtain a universal unfolding of the normal form (7.4.2), we first add the two-parameter linear part

$$\begin{bmatrix} \mu_1 x \\ \mu_1 y \\ \mu_2 \end{bmatrix},$$

(7.4.8)

which provides all possible stable perturbations of fixed points, including their removal. In terms of the reduced planar system (7.4.5), we now have

$$\dot{r} = \mu_1 r + arz,$$
$$\dot{z} = \mu_2 + br^2 - z^2. \tag{7.4.9}$$

We easily find that (7.4.9) has fixed points at

$$(r, z) = (0, \pm\sqrt{\mu_2}) \quad \text{for } \mu_2 \geq 0,$$

and

$$(r, z) = \left(\sqrt{\frac{\mu_1^2}{a^2} - \mu_2}, -\frac{\mu_1}{a}\right) \quad \text{for } \mu_1^2 \geq a^2\mu_2, \, b = +1;$$
$$(r, z) = \left(\sqrt{\mu_2 - \frac{\mu_1^2}{a^2}}, -\frac{\mu_1}{a}\right) \quad \text{for } \mu_1^2 \leq a^2\mu_2, \, b = -1, \tag{7.4.10}$$

and we note that the linearized system has the matrix

$$\begin{bmatrix} \mu_1 + az & ar \\ 2br & -2z \end{bmatrix}. \tag{7.4.11}$$

We will now sketch the analysis of the cases I and IIa–IIb in some detail, leaving the derivation of the remaining results, which we quote below, to the reader. For case I we set $b = +1$, $a > 0$. The matrix (7.4.11) is diagonal, being

$$\begin{bmatrix} \mu_1 \pm a\sqrt{\mu_2} & 0 \\ 0 & \mp 2\sqrt{\mu_2} \end{bmatrix}, \tag{7.4.12}$$

at the fixed points $(r, z) = (0, \pm\sqrt{\mu_2})$, respectively, and it is easy to see that for $\mu_2 > 0$, these points are as follows:

| $(r, z)$ | $(0, +\sqrt{\mu_2})$ | $(0, -\sqrt{\mu_2})$ |
|---|---|---|
| $\mu_1 > a\sqrt{\mu_2}$ | saddle | source |
| $a\sqrt{\mu_2} > \mu_1 - a\sqrt{\mu_2}$ | saddle | saddle |
| $-a\sqrt{\mu_2} > \mu_1$ | sink | saddle |

Crossing the line $\mu_2 = 0$ for $\mu_1 \neq 0$, we can verify that saddle-node bifurcations occur in the one-dimensional center manifold given by the invariant line $r = 0$. Similarly, crossing $\mu_2 = \mu_1^2/a^2$ with $\mu_2$ decreasing for $\mu_1 \neq 0$, symmetric pitchfork bifurcations occur from the fixed point $(r, z) = (0, +\sqrt{\mu_2})$ ($\mu_1 < 0$) and $(0, -\sqrt{\mu_2})$ ($\mu_1 > 0$), respectively (only the bifurcating equilibrium with $r > 0$ is of interest here). To see this, we set

$\mu_1 \neq 0$ constant and let $\sqrt{\mu_2} = |\mu_1|/a - \varepsilon$. Note that $\varepsilon$ increasing corresponds to the transversal crossing of $\mu_2 = \mu_1^2/a^2$ with $\mu_2$ decreasing. We consider $\mu_1 > 0$ and the corresponding bifurcation from $(r, z) = (0, -\sqrt{\mu_2})$. First letting $z = -\sqrt{\mu_2} + \zeta$ we translate the degenerate fixed point to the origin, so that (7.4.9) becomes

$$\dot{r} = \varepsilon a r + a r \zeta,$$

$$\dot{\zeta} = 2\left(\frac{|\mu_1|}{a} - \varepsilon\right)\zeta + r^2 - \zeta^2, \tag{7.4.13}$$

or, as a suspended system,

$$\dot{r} = \varepsilon a r + a r \zeta,$$

$$\dot{\varepsilon} = 0, \tag{7.4.14}$$

$$\dot{\zeta} = 2\left(\frac{|\mu_1|}{a} - \varepsilon\right)\zeta + r^2 - \zeta^2.$$

We now apply the center manifold methods of Chapter 3. We note that the tangent space approximation $h = 0$ to the center manifold $\zeta = h(r, \varepsilon)$ yields no stability information and we therefore let $\zeta = \alpha r^2 + \beta r \varepsilon + \gamma \varepsilon^2$, to obtain the leading coefficients

$$2\left(\frac{|\mu_1|}{a}\right)(\alpha r^2 + \beta r \varepsilon + \gamma \varepsilon^2) + r^2 = \mathcal{O}(3)$$

or

$$\alpha = \frac{-a}{2|\mu_1|} \overset{\text{def}}{=} -\delta < 0; \qquad \beta = \gamma = 0, \tag{7.4.15}$$

for small $\varepsilon$. Substituting $\zeta = -\delta r^2$ into the first component of (7.4.13), we obtain the reduced system

$$\dot{r} = \varepsilon a r - a \delta r^3 + \cdots, \tag{7.4.16}$$

showing that a supercritical pitchfork bifurcation occurs as $\varepsilon$ passes through zero, increasing. Since $|\mu_1|/a > 0$, in this case the center manifold repels nearby solutions, and we obtain the behavior illustrated in Figure 7.4.2.

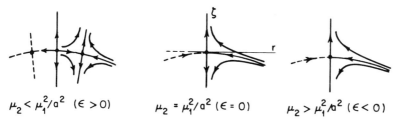

$\mu_2 < \mu_1^2/a^2 \ (\varepsilon > 0)$          $\mu_2 = \mu_1^2/a^2 \ (\varepsilon = 0)$          $\mu_2 > \mu_1^2/a^2 \ (\varepsilon < 0)$

Figure 7.4.2. A pitchfork bifurcation at $\mu_2 = \mu_1^2/a^2$, $\mu_1 > 0$.

Finally, we note that the matrix (7.4.11) for the fixed point at $(r, z) = (\sqrt{\mu_1^2/a^2 - \mu_2}, - \mu_1/a)$ when $b = +1$ is

$$\begin{bmatrix} 0 & \sqrt{\mu_1^2 - a^2\mu_2} \\ \dfrac{2}{a}\sqrt{\mu_1^2 - a^2\mu_2} & \dfrac{2\mu_1}{a} \end{bmatrix}, \qquad (7.4.17)$$

and that this point is therefore a saddle for all $\mu_2 < \mu_1^2/a^2$. Moreover, the fact that this fixed point is a saddle with (Poincaré) index $- 1$ implies that, in this case, there can be no periodic orbits for the vector field, since any such orbit must contain fixed points whose indices sum to $+1$ (cf. Section 1.8). Clearly no periodic orbit can intersect the invariant $z$-axis, and hence no such orbits exist, since there are no further fixed points with $r \neq 0$.

We are now in a position to give the complete unfolding of the planar vector field in case I. In Figure 7.4.3 we show the bifurcation set in $(\mu_1, \mu_2)$ space and the associated phase portraits.

EXERCISE 7.4.3. Perform any additional computations you feel necessary to convince yourself that the unfolded phase portraits of Figure 7.4.3 are correct.

We now turn to cases IIa–IIb, with $b = +1, a < 0$. Computations similar to those above show that the fixed points $(r, z) = (0, \pm\sqrt{\mu_2})$ are as given below for $\mu_2 > 0$. Moreover, saddle-node and pitchfork bifurcations occur on $\mu_2 = \mu_1^2/a^2$ and $\mu_2 = 0$ somewhat as in the previous case:

| $(r, z)$ | $(0, +\sqrt{\mu_2})$ | $(0, -\sqrt{\mu_2})$ |
|---|---|---|
| $\mu_1 > -a\sqrt{\mu_2}$ | saddle | source |
| $-a\sqrt{\mu_2} > \mu_1 > a\sqrt{\mu_2}$ | sink | source |
| $a\sqrt{\mu_2} > \mu_1$ | sink | saddle |

EXERCISE 7.4.4. Investigate the bifurcations occurring on $\mu_2 = 0, \mu_1 \neq 0$ and $\mu_2 = \mu_1^2/a^2, \mu_1 \neq 0$ for this case. In particular, show that, as $\mu_2$ passes through $\mu_1^2/a^2$ with $\mu_1 > 0$ (resp. $\mu_1 < 0$) the upper (resp. lower) saddle undergoes a pitchfork bifurcation.

However, the behavior of third fixed point $(r, z) = \sqrt{\mu_1^2/a^2 - \mu_2}, - \mu_1/a)$ is rather different in this case. Here the linearized matrix (7.4.17) has eigenvalues

$$\lambda_{1, 2} = \frac{\mu_1}{a} \pm \sqrt{\frac{\mu_1^2}{a^2} + \frac{2}{a}(\mu_1^2 - a^2\mu_2)}, \qquad (7.4.18)$$

and is therefore a sink for $\mu_1 > 0$, since $\mu_1/a < 0$, and a source for $\mu_1 < 0$ (the eigenvalues are complex conjugate for $\mu_2 < \mu_1^2(2 + 1/a)/2a^2$). Passing

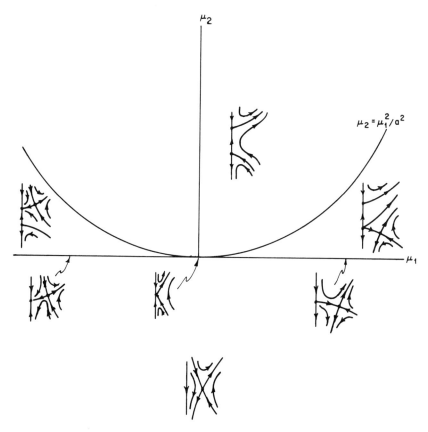

Figure 7.4.3. The unfolding of case I, $b = +1$, $a > 0$. Bifurcation set and phase portraits. Note that on the curve $\mu_2 = \mu_1^2/a^2$, the phase portrait is homeomorphic to that above $\mu_2 = \mu_1^2/a^2$, but the lower saddle is degenerate if $\mu_1 > 0$ and the upper saddle is degenerate if $\mu_1 < 0$ (cf. Figure 7.4.2).

transversely through $\mu_1 = 0$ for $\mu_2 < 0$ the reader can verify that a Hopf bifurcation occurs at this fixed point. However, computation of the stability coefficient by the methods of Section 3.4 yields $a_3 \equiv 0$, showing that terms of at least cubic order must be included in the normal form to determine the dynamics of this secondary bifurcation. In fact, for $\mu_1 = 0$ the system

$$\dot{r} = arz,$$
$$\dot{z} = \mu_2 + r^2 - z^2, \tag{7.4.19}$$

is completely integrable, since the function

$$F(r, z) = \frac{a}{2} r^{2/a} \left[ \mu_2 + \frac{r^2}{1 + a} - z^2 \right], \tag{7.4.20}$$

is constant on solution curves.

EXERCISE 7.4.5. Verify the statement above, and check that the Hopf stability coefficient for (7.4.19) is identically zero, as claimed.

Assembling the information obtained above we can now sketch the bifurcation set and phase portraits for case IIa–IIb as in Figure 7.4.4. We note that the partial unfolding obtained at this stage does not differ for the subcases $a \in [-1, 0)$ and $a < -1$.

We now give, without derivation, bifurcation sets and phase portraits for the two remaining cases III and IVa–IVb for which $b = -1$. We remark that case IVa–IVb is quite straightforward, and the results are comparable to case I, in that no secondary Hopf bifurcations occur from the fixed point $(r, z) = (\sqrt{\mu_2 - \mu_1^2/a^2}, -\mu_1/a)$, which remains a saddle point throughout the region $\mu_2 > \mu_1^2/a^2$ in which it exists. However, when $a < 0$, this point undergoes a Hopf bifurcation on $\mu_1 = 0$, $\mu_2 > 0$, the stability of which is

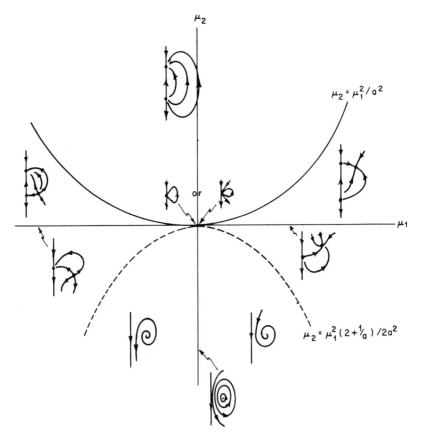

Figure 7.4.4. Partial bifurcation sets and phase portraits for case IIa–IIb; $b = +1$, $a < 0$. The phase portraits on $\mu_2 = \mu_1^2/a^2$, $\mu_1 \neq 0$ are topologically equivalent to those above the two branches of this curve (cf. Figure 7.4.3).

again undetermined by the 2-jet. Once more in this case we have a first integral:

$$G(r, z) = \frac{a}{2} r^{2/a} \left( \mu_2 - \frac{1}{1 + a} r^2 - z^2 \right),$$
(7.4.21)

and the reader can check that $dG/dt = (\partial G/\partial r)\dot{r} + (\partial G/\partial z)\dot{z} \equiv 0$ on the solution curves of

$$\dot{r} = arz,$$
$$\dot{z} = \mu_2 - r^2 - z^2.$$
(7.4.22)

This permits us to conclude that there is a one-parameter family of periodic orbits limiting on a homoclinic cycle connecting the saddle points at $(r, z) = (0, \pm\sqrt{\mu_2})$ for $\mu_1 = 0$, $\mu_2 > 0$ in case III. We refer the reader to Figures 7.4.5 and 7.4.6 for bifurcation sets and phase portraits.

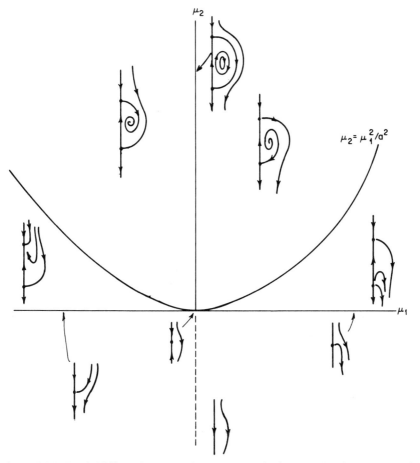

Figure 7.4.5. Partial bifurcation set and phase portraits for case III; $b = -1$, $a > 0$. The phase portraits on $\mu_2 = \mu_1^2/a^2$ are topologically equivalent to those below the two branches of this curve.

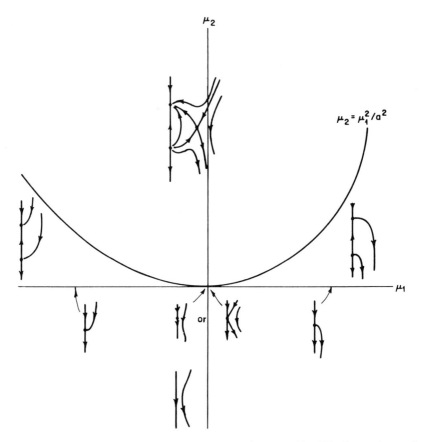

Figure 7.4.6. Bifurcation set and phase portraits for case IVa–IVb; $b = -1$, $a < 0$. The phase portraits on $\mu_2 = \mu_1^2/a^2$ are topologically equivalent to those below the two branches of this curve.

Exercise 7.4.6. Carry out the stability and bifurcation computations necessary to verify the correctness of Figures 7.4.5 and 7.4.6.

We have marked Figures 7.4.4 and 7.4.5 as *partial* bifurcation sets and phase portraits, since the quadratic terms included in the 2-jet (7.4.9) do not suffice to determine the type of the Hopf bifurcation in these cases. We will return to this point in a moment. First, however, we consider the implications of our unfolding of cases I and IVa–IVb for the full three-dimensional vector field (7.4.1)–(7.4.2) with which we started.

We remark that, while fixed points on the $z$-axis ($r = 0$) correspond to fixed points in the full system, fixed points $(r_0, z_0)$ with $r_0 > 0$ correspond to periodic orbits. To see this, consider a two-dimensional Poincaré cross section

$$\Sigma = \{(r, \theta, z)|\theta = 0; r > 0; r \text{ and } |z| \text{ sufficiently small}\},$$

for the three-dimensional flow of the unfolded system,

$$\dot{r} = \mu_1 + arz + \mathcal{O}(3),$$
$$\dot{\theta} = \omega + \mathcal{O}(r^2, z^2),$$
$$\dot{z} = \mu_2 + br^2 - z^2 + \mathcal{O}(3),$$

(7.4.23)

and suppose that the planar system (7.4.9) has a hyperbolic fixed point $(r_0, z_0)$, $r_0 > 0$, with $r_0$, $|z_0| \ll 1$. Clearly, the circle $\gamma = \{(r, \theta, z) | r = r_0$, $z = z_0$, $\theta \in [0, 2\pi)\}$ is a hyperbolic closed orbit for the $S^1$-symmetric truncated system, since $\dot{\theta} = \omega + \mathcal{O}(|r|^2, |z|^2) > 0$ for small $r$, $|z|$. Moreover, since such a limit set is hyperbolic, it will persist for small perturbations such as the addition of higher-order *and* non-$S^1$-symmetric terms, just as hyperbolic fixed points on the $z$-axis will persist (although they may leave the $z$-axis for nonsymmetric perturbations). We sketch the full three-dimensional phase portraits and the associated Poincaré maps for two of the structurally stable unfoldings in Figure 7.4.7.

We can also conclude that the "nondegenerate" codimension one bifurcations occurring in cases I and IVa–IVb carry over to "nondegenerate" codimension one bifurcations of the full three-dimensional flow in a natural way, much as the results of averaging carry over to the Poincaré map in forced oscillations (cf. Sections 4.1–4.3). Specifically, the saddle-node bifurcations occurring on $\mu_2 = 0$ remain saddle-nodes, but the symmetric

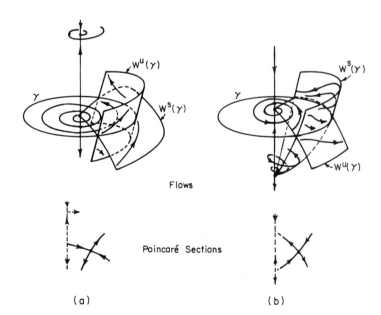

Figure 7.4.7. Two of the three-dimensional flows corresponding to unfoldings of the planar vector field. (a) Case I, $\mu_1 > 0$, $\mu_2 \in (0, \mu_1^2/a^2)$; (b) Case IVa–IVb, $\mu_2 > \mu_1^2/a^2$.

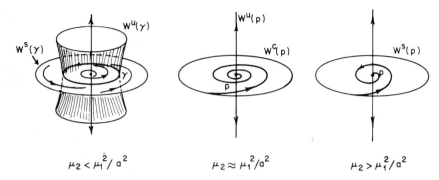

$$\mu_2 < \mu_1^2/a^2 \qquad \mu_2 \approx \mu_1^2/a^2 \qquad \mu_2 > \mu_1^2/a^2$$

Figure 7.4.8. A Hopf bifurcation for the three-dimensional flow, case I (cf. Figure 7.4.2).

pitchfork bifurcations on $\mu_2 = \mu_1^2/a^2$ correspond to Hopf bifurcations in the three-dimensional flow, cf. Figure 7.4.8.

We conclude that, in cases I and IVa–IVb, coupling a Hopf and a saddle-node bifurcation in a doubly degenerate fixed point with eigenvalues $(\pm i\omega, 0)$ yields nothing unexpected: we merely obtain up to two fixed points and one closed orbit, as in the individual codimension one bifurcations. The unfolding of these cases is now essentially complete, and we shall not consider them further.

The other two cases, IIa–IIb and III, are considerably more delicate, since here the planar vector field can admit closed orbits. Before considering what the presence of such orbits implies for the three-dimensional flow, we first address the effect of higher-order (cubic) terms in the planar system, and show how their presence typically determines the stability type of the Hopf bifurcation and the number of closed orbits that can appear. We start by restoring the truncated cubic terms to (7.4.9):

$$\dot{r} = \mu_1 r + arz + (cr^3 + drz^2),$$
$$\dot{z} = \mu_2 + br^2 - z^2 + (er^2z + fz^3). \tag{7.4.24}$$

Since the linear part of (7.4.24) is zero when $\mu_1 = \mu_2 = 0$, the normal form theorem (3.3.1) does not provide information about how coordinate changes can be used to remove higher-order nonlinear terms from (7.4.24). Nonetheless, coordinate changes of the form (identity and higher-order terms) as well as reparametrization of the vector field can be used to alter the cubic coefficients of (7.4.24). We do this in the most general way compatible with the symmetry of (7.4.24) as follows. First we introduce new coordinates

$$s = r(1 + gz),$$
$$w = z + hr^2 + iz^2, \tag{7.4.25}$$
$$\tau = (1 + jz)^{-1}t,$$

and compute the effect of transforming (7.4.24) to these coordinates:

$$\frac{ds}{d\tau} = \mu_1 s + asw + (c + bg - ah)s^3 + (d - g - ai + aj)sw^2 + R_s(s, w, \mu_1, \mu_2),$$

$$\frac{dw}{d\tau} = \mu_2 + bs^2 - w^2 + (e - 2bg + 2(a + 1)h + 2bi + bj)s^2 w \quad (7.4.26)$$

$$+ (f - j)w^3 + R_w(s, w, \mu_1, \mu_2).$$

The "remainder" terms have order at least four in $(s, w)$ except for some cubic terms divisible by $\mu_i$.

Ignoring these higher-order terms, we now choose $(g, h, i, j)$ to make (7.4.26) as simple as possible. The "new" cubic coefficients introduced by the coordinate change depend linearly on $(g, h, i, j)$ as described by the matrix:

$$\mathbf{M} = \begin{pmatrix} b & -a & 0 & 0 \\ -1 & 0 & -a & a \\ -2b & 2a + 2 & 2b & b \\ 0 & 0 & 0 & -1 \end{pmatrix}; \quad (7.4.27)$$

i.e. if $\mathbf{v}$ is the vector with entries $(g, h, i, j)$ then the vector $\mathbf{Mv}$ has the components $((bg - ah), \ldots, -j)$ which have been added to the terms $s^3$, $sw^2$, $s^2w$, and $w^3$, respectively of (7.4.26). This matrix has rank three with kernel spanned by $(a, b, -1, 0)$. Consequently, we may choose $(g, h, i, j)$ so that (7.4.26) has only one nonzero coefficient. Thus, in the remainder of this section, we assume $c = d = e = 0$ in (7.4.24), and consider the cubic perturbation $(0, fz^3)$.

The analysis now proceeds in a manner parallel to the analysis of the double zero eigenvalue case of Section 7.3 by blowing up the degenerate equilibrium at $(r, z) = (0, 0)$ occurring for $\mu_1 = u_2 = 0$ to reveal homoclinic orbits "in embryo." We therefore pick rescaled variables in which the periodic orbits of (7.4.19) (and (7.4.22)) retain their sizes as $\mu_1, \mu_2 \to 0$ along rays. Specifically, we let

$$r = \varepsilon u, \qquad z = \varepsilon v, \qquad \mu_1 = \varepsilon^2 v_1, \qquad \mu_2 = \varepsilon^2 v_2, \quad (7.4.28)$$

and rescale time $t \to \varepsilon t$, so that (7.4.24) becomes

$$\dot{u} = auv + \varepsilon v_1 u$$
$$\dot{v} = v_2 + bu^2 - v^2 + \varepsilon f v^3 \quad (7.4.29)$$

For $\varepsilon = 0$ we recover the degenerate 2-jet at the bifurcation point $\mu_1 = \mu_2 = 0$, and, as in Section 7.3, we now have to study small perturbations of an integrable system

$$\dot{u} = auv,$$
$$\dot{v} = v_2 + bu^2 - v^2, \quad (7.4.30)$$

with integral (for $a \neq -1$):

$$F(u, v) = \frac{a}{2} u^{2/a} \left[ v_2 + \frac{b}{1+a} u^2 - v^2 \right], \qquad (7.4.31)$$

(cf. equations (7.4.20)–7.4.21). Here, however, the system is not Hamiltonian (unless $a = 2$), and the presence of fractional powers in the integral makes the computations analytically intractible. We will outline a program for their completion without carrying it through.

We recall that the cases of interest which exhibit continuous families of periodic orbits are

IIa–IIb:                    $b = +1,$      $a < 0,$      $v_2 = \dfrac{\mu_2}{\varepsilon^2} = -1 < 0,$

and

III:                        $b = -1,$      $a > 0,$      $v_2 = +1 > 0.$

(Note that, as in Section 7.3, we can set $v_2 = \pm 1$ without loss of generality, since the variation in the original parameter $\mu_2$ is obtained as $\varepsilon$ varies.) In the first case we have a family of unbounded periodic orbits encircling the center at $(u, v) = (1, 0)$ and in the second a family of periodic orbits encircling the center at $(u, v) = (1, 0)$ and limiting on the homoclinic loop $F(u, v) = 0$ which connects the saddles at $(u, v) = (0, \pm 1)$.

It is more convenient to work with the system (7.4.29) modified by multiplication by the integrating factor $u^{(2/a)-1}$, giving (with $v_2 = \mp 1$ when $b = \pm 1$)

$$\begin{aligned}
\dot{u} &= au^\alpha v + \varepsilon v_1 u^\alpha, \\
\dot{v} &= -bu^{\alpha-1} + bu^{\alpha+1} - u^{\alpha-1} v^2 + \varepsilon f u^{\alpha-1} v^3, \qquad (7.4.32) \\
&\overset{\text{def}}{=} k(x) + \varepsilon l(x, \mu),
\end{aligned}$$

where $\alpha = 2/a$. Since for $u > 0$ (7.4.32) is merely a dilated version of the vector field (7.4.29), its solution curves are topologically equivalent to those of the latter system. Moreover, for $\varepsilon = 0$, (7.4.32) is now Hamiltonian with $F(u, v)$ of (7.4.31) playing the rôle of the Hamiltonian energy function. Thus, if we can show that, for $0 < \varepsilon \ll 1$ and suitable choices of the linear and cubic perturbation coefficients $(v_1, f)$, isolated closed level curves are preserved, then we will have proved that isolated periodic orbits occur in the original system (7.4.29) for the corresponding values of $\mu_1, \mu_2$, etc. Moreover, if these closed curves appear and disappear in a stable fashion as the coefficients vary, then the behavior will persist under other (smaller) perturbations to the planar vector field, such as the inclusion of higher-order terms. However, not all of this behavior will carry over to the three-dimensional flow, as we shall see.

Applying the Melnikov theory of Sections 4.5–4.6 to this problem and using Green's theorem, we deduce that a given level curve $\Gamma$ of $F$ will be preserved near the parameter value $\mu_0$ if the integral

$$\int_{\text{int } \Gamma} \text{trace } Dl(x, \mu) \, dx \qquad (7.4.33)$$

has a simple zero with respect to $\mu$ at $\mu = \mu_0$. In the present case the divergence calculation gives

$$\text{trace } Dl = \alpha v_1 u^{\alpha - 1} + 3 f u^{\alpha - 1} v^2$$
$$= (\alpha v_1 + 3 f v^2) u^{\alpha - 1}; \quad \alpha = 2/a. \qquad (7.4.34)$$

(We automatically have trace $Dk \equiv 0$, since $k$ is Hamiltonian.) We therefore must find values $K$ in the admissible range for which $F^{-1}(K)$ is a closed curve $\Gamma_K$ and

$$\alpha \iint_{\text{int } \Gamma_K} \left[ v_1 + \frac{3af}{2} v^2 \right] u^{2/a - 1} \, du \, dv = 0. \qquad (7.4.35)$$

In general, these integrals can only be evaluated numerically, although Keener [1981] has considered the special case $K = 0$ corresponding to the homoclinic loop in case III, in which the integrals reduce to gamma functions. However, Zholondek [1984] has estimated the integrals in the general case and proved that the zeros of (7.4.35) define $v_1$ as a monotone function of $K$ when $f \neq 0$. Consequently, there is a range of parameter values for which (7.4.24) has a unique limit cycle. Here we shall carry through only the simplest of the calculations by establishing where homoclinic orbits and Hopf bifurcations occur when $a = 2$ ($\alpha = 1$), for in that case the Melnikov calculations give elliptic integrals and the boundary values of $K$ corresponding to homoclinic orbits and Hopf bifurcations reduce to elementary integrals.

Equation (7.4.35) becomes (cancelling $\alpha$)

$$\iint_{\text{int } K} (v_1 + 3 f v^2) \, du \, dv = 0, \qquad (7.4.36)$$

and the level curves of $K$ are given by

$$u \left( 1 - \frac{u^2}{3} - v^2 \right) = K; \qquad (7.4.37)$$

cf. Figure 7.4.9.

The level curve $K = 0$ contains the homoclinic loop with the two saddle points. Integrating (7.4.35) and with $K = 0$ yields the following integrals

$$v_1 \int_0^{\sqrt{3}} \sqrt{1 - \frac{u^2}{3}} \, du + f \int_0^{\sqrt{3}} \left( 1 - \frac{u^2}{3} \right)^{3/2} \, du, \qquad (7.4.38)$$

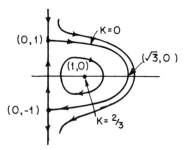

Figure 7.4.9. The unperturbed level curves of (7.4.34) (case III, $a = 2$).

and evaluating these integrals gives

$$\frac{\sqrt{3}\pi}{16}(4v_1 + 3f) = 0. \tag{7.4.39}$$

This is the equation which determines the location of homoclinic loops (to order $\varepsilon$), and so in terms of the variables used before scaling, we obtain the homoclinic bifurcation curve;

$$\mu_1 = \frac{-3f\mu_2}{4} + \mathcal{O}(\varepsilon). \tag{7.4.40}$$

(Recall that for case III, in which the homoclinic orbit exists, we have $v_2 = +1$, $b = -1$.)

To determine the stability of the homoclinic loop, we note that the two saddles are located at $(u, v) = (0, \pm 1 + \varepsilon f/2) + \mathcal{O}(\varepsilon^2)$. The sum of the logarithms of the eigenvalues at the two saddle points is

$$\varepsilon(v_1 + 3f) + \mathcal{O}(\varepsilon^2) = \varepsilon\left(\frac{9f}{4}\right) + \mathcal{O}(\varepsilon^2), \tag{7.4.41}$$

where we use $v_1 = -3f/4$ from (7.4.39). (See Exercise 6.1.4.) Thus the homoclinic orbit is stable if $f < 0$ and unstable if $f > 0$.

EXERCISE 7.4.7. Compute the integrals (7.4.36) for $K \in (0, \frac{2}{3})$ as complete elliptic integrals. Show that they give $v_1$ as a monotone function of $K$ (Byrd and Friedman [1971] will be useful).

EXERCISE 7.4.8. Differentiate (7.4.36) with respect to $K$ to obtain $dv_1/dK$ and show that the resulting function has no zero. (Hint: Since integrals of exact differentials are zero, the integrals can be simplified, cf. Sanders and Cushman [1984a, b].)

Next, we consider the effect of the cubic term on the Hopf bifurcation. The Hopf bifurcation occurs still at $v_1 = 0$ in the presence of the single cubic term of (7.4.29). The stability of the Hopf bifurcation is easily determined by computing the divergence of the right-hand side of this equation as $3\varepsilon f z^2$

(recall that $a = 2$ for this analysis). Consequently, the Hopf bifurcation point is a weak attractor if $f < 0$ and a weak repellor if $f > 0$.

The results of these calculations and those of Exercises 7.4.7–8 imply that if $f < 0$ (resp. $f > 0$) there is a unique attracting (resp. repelling) limit cycle for (7.4.29) for each $\mu_1$, $\mu_2$ parameter value in the wedge bounded by $\mu_1 = 0$ and $\mu_1 = -3f\mu_2/4 + \mathcal{O}(\varepsilon)$. See Figure 7.4.10.

EXERCISE 7.4.9. Let $X$ be a planar vector field of the form $f + g$ with $f$ Hamiltonian. Assume (a) $f$ has an equilibrium with imaginary eigenvalues at 0, (b) $g(0) = 0$, and (c) the divergence of $g$ has a nondegenerate maximum at 0 with value 0. Show that the Hopf coefficient in the normal form of $X$ at 0 is negative.

Before going on to consider the implications for the three-dimensional flow, we remark that, since the family of periodic orbits for case IIa–IIb becomes unbounded as $K \to 0$, perturbation calculations such as those above cannot provide complete information. For example, while the Hopf bifurcation might be "stabilized" by the addition of higher-order terms, and small attracting or repelling closed orbits found near this bifurcation, as $\mu_1$ and $\mu_2$ vary, these orbits can grow without bound and hence leave any fixed (small) neighborhood of the origin. We cannot deduce the fate of such orbits from perturbation calculations which are only valid locally.

EXERCISE 7.4.10. Investigate the bifurcations for case IIa–IIb, with $a = -2$. (In this case the integral becomes $F(u, v) = (1/u)(1 + u^2 + v^2)$.)

We end this section by considering the implications for the three-dimensional flow of the closed orbits and homoclinic loops found above. Note that, if the planar system has a hyperbolic (attracting or repelling) closed orbit, then the corresponding three-dimensional flow has a hyperbolic invariant torus of the same stability type. This follows from the fact that the Poincaré

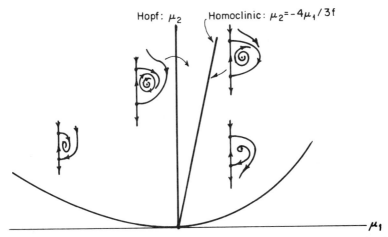

Figure 7.4.10. The completion of the case III bifurcation diagram when $f < 0$.

map of the truncated $S^1$-symmetric system has a smooth hyperbolic invariant curve, which will be preserved under small perturbations. The analysis is a little delicate, since the hyperbolicity of the invariant curve is relatively weak, being governed by the size of the small parameters $\mu_1$, $\mu_2$. However, the perturbation due to higher-order terms is yet smaller (Iooss and Langford [1980]).

We note that the resulting doubly periodic flow on the torus has one "fast" frequency ($\approx \omega$) associated with the angular variable $\theta$, and a slow frequency ($\approx \sqrt{2a\mu_2}$, cf. (7.4.18)) associated with the secondary Hopf bifurcation of the planar system. One therefore expects to see a rapid oscillation with a slow modulation. We remark that, if the case shown in Figure 7.4.10(b) occurs, then we expect this slow modulation to grow yet slower "after" the bifurcation, as the period of the secondary oscillation goes to infinity in the homoclinic bifurcation. Some experimental measurements by Libchaber et al. [1983] on the convection in mercury in a magnetic field (an essentially two-parameter problem) show precisely this behavior. Moreover, as the slow modulation grows in period, Libchaber observes the onset of chaotic oscillations, which should be expected in perturbations of the homoclinic loop, as we now indicate.

Consider the $S^1$-symmetric three-dimensional flow corresponding to the homoclinic bifurcation; see Figure 7.4.11(a), (b). The homoclinic loop becomes an invariant sphere filled with heteroclinic orbits connecting the two saddles, together with the segment of the $z$-axis lying between them. The coincidence of 2 two-manifolds in three-space is clearly exceptional and they can be expected to split as indicated in Figure 7.4.11(c), yielding transverse intersections. To prove that such splitting, and the resulting complex dynamics with horseshoes, actually occurs, we can show that, as the planar vector field passes through the homoclinic bifurcation, parameter values must occur for which the non-$S^1$-symmetric three-dimensional flow has a homoclinic orbit to a spiral saddle, just as in Silnikov's example of Section 6.5. We leave the details to the reader:

EXERCISE 7.4.11. Consider the generic homoclinic bifurcations of Figures 7.4.10(a), (b) and show that the conditions of Silnikov's theorem are met for non-$S^1$-symmetric perturbations of the full three-dimensional flow. State explicitly the required non-degeneracy assumptions (cf. Guckenheimer [1981]).

While the smooth torus can be expected to persist away from the homoclinic bifurcation curve, the dynamics on it will exhibit the complicated intervals of phase locking and irrational flow discussed in Section 6.2, as the rotation number decreases and one approaches the homoclinic bifurcation. However, before the horseshoes and chaotic dynamics associated with the homoclinic bifurcation are created, the torus must lose differentiability and break-up. Precisely how such tori interact with nearby periodic motions and "explode" is still imperfectly understood, but see Aronson et al. [1982],

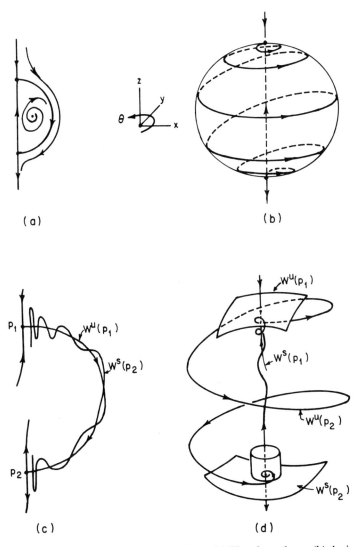

Figure 7.4.11. Homoclinic loops imply horeshoes. (a) The planar loop; (b) the invariant sphere for the $S^1$-invariant flow; (c) transversal heteroclinic orbits (the split sphere sectioned); (d) the analogue of Silnikov's construction.

Rand *et al.* [1982], Feigenbaum *et al.* [1982] and Greenspan and Holmes [1983] for some specific examples. Additional complications occur if there are saddle-node bifurcations of the periodic orbits in the planar system. Chenciner [1982] has systematically studied this situation.

We note that the fact that transverse homoclinic orbits and horseshoes exist near parameter values for the homoclinic bifurcation curves for the planar system, and that no such orbits exist away from these curves, implies that homoclinic tangencies must occur as the parameters vary. In turn, this implies that all phenomena discussed in Sections 6.6–6.7—wild hyperbolic sets and Newhouse sinks—will occur. It therefore seems unlikely that a complete stable unfolding of the full three-dimensional problem will be possible in a two-parameter family.

To summarize: we have found that the interaction of a saddle-node and a Hopf bifurcation viewed as a three-dimensional flow restricted to a center manifold can exhibit secondary bifurcation phenomena of surprising complexity, including doubly periodic flows, transverse homoclinic orbits and horseshoes. Such global phenomena arise naturally from the study of a local codimension two bifurcation, and are detected by a delicate analysis of a planar vector field.

Some obvious modifications to this example include the restriction that there always be a "trivial" fixed point at $z = 0$. Then the saddle-node becomes a transcritical bifurcation and the 2-jet is modified to:

$$\dot{r} = \mu_1 r + arz,$$
$$\dot{z} = \mu_2 z \pm r^2 - z^2 \qquad\qquad (7.4.42)$$

Alternatively, one can specify a $\mathbb{Z}_2$-symmetry in the $z$-direction so that the saddle-node becomes a pitchfork bifurcation and the $k$-jet starts with homogeneous cubic terms. This is essentially the problem considered in the next section and we will add some remarks on it later (cf. Iooss and Langford [1980]). The first problem was considered by Langford [1979] who did not include cubic terms, but noted the possible existence of invariant tori without determining their stability or number. We leave this case as a final exercise for the reader:

EXERCISE 7.4.12. Study the unfolding

$$\dot{r} = \mu_1 r + arz,$$
$$\dot{z} = \mu_2 z \pm r^2 - z^2,$$
$$(\dot{\theta} = \omega + \mathcal{O}(2)),$$

and classify the various cases, adding higher-order terms as necessary.

## 7.5. Two Pure Imaginary Pairs of Eigenvalues without Resonance

In this section we consider the four-dimensional flow obtained when there are two purely imaginary pairs of eigenvalues at an equilibrium. We assume that no low-order resonances* occur, so that the normal form of 3-jet takes

* Specifically $m\omega_1 + n\omega_2 \neq 0$, $|m| + |n| \leq 4$ (Takens [1974a]).

the form (cf. Takens [1974a]):

$$\dot{r}_1 = \mu_1 r_1 + a_{11} r_1^3 + a_{12} r_1 r_2^2 + \mathcal{O}(|r|^5),$$
$$\dot{r}_2 = \mu_2 r_2 + a_{21} r_1^2 r_2 + a_{22} r_2^3 + \mathcal{O}(|r|^5),$$
$$\dot{\theta}_1 = \omega_1 + \mathcal{O}(|r|^2),$$
$$\dot{\theta}_2 = \omega_2 + \mathcal{O}(|r|^2),$$

(7.5.1)

where $\mu_1$ and $\mu_2$ are the unfolding parameters. Once more, we can deduce much of the behavior of this system by considering the family of planar vector fields obtained by ignoring the azimuthal components.

As before, we first rescale to reduce the number of coefficients. Letting $\bar{r}_1 = r_1 \sqrt{|a_{11}|}$ and $\bar{r}_2 = r_2 \sqrt{|a_{22}|}$, dropping the bars and possibly rescaling time, we obtain the following

$$\dot{r}_1 = r_1(\mu_1 + r_1^2 + br_2^2),$$
$$\dot{r}_2 = r_2(\mu_2 + cr_1^2 + dr_2^2),$$

$$d = \pm 1,$$

(7.5.2)

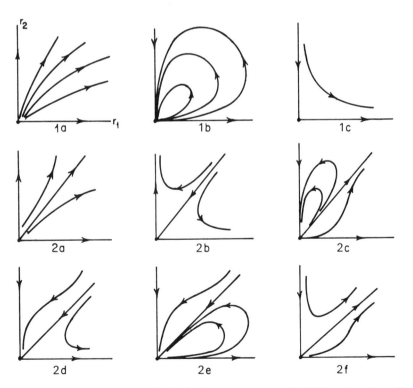

Figure 7.5.1. Nine phase portraits for the degenerate vector field. Note that 2c and 2e, 2d and 2f, respectively, are topologically equivalent if we allow reversal of time and reflection about the diagonal invariant line.

where $b = a_{12}/|a_{22}|$ and $c = a_{21}/|a_{11}|$. Takens [1974a] showed that the 3-jet of (7.5.1) with $\mu_1 = \mu_2 = 0$ is determined (with respect to suitably symmetric higher-order perturbations), provided that $a_{ij} \neq 0$ and $a_{11}a_{22} - a_{12}a_{21} \neq 0$, implying that $d - bc \neq 0$ in (7.5.2). Henceforth we shall assume this to be the case. Takens found nine topologically distinct equivalence classes, (some of) which we show in Figure 7.5.1 and Table 7.5.1.

Table 7.5.1. The Nine Degenerate Fixed Points.

Type 1: Two invariant lines; $(1 - c)/(d - b) < 0$.

1a: $d = +1$, all appropriate $b, c$ (Takens Ia, Id)
1b: $d = -1, c > 0 > b, -1 - bc > 0$ (Takens Ib)
1c: $d = -1$, all other $b, c$ (Takens Ic)

Type 2: Three invariant lines; $(1 - c)/(d - b) > 0$.

2a: $d = +1, (1 - bc)/(1 - b) > 0$, all appropriate $b, c$ (Takens II'a, II'd)
2b: $d = +1, (1 - bc)/(1 - b) < 0$, all appropriate $b, c$ (Takens II'e)
2c: $d = -1, (-1 - bc)/(-1 - b) > 0, b < -1, c < 1$ (Takens II'b)
2d: $d = -1, (-1 - bc)/(-1 - b) < 0, b < -1, c > 1$ (Takens II'c)
2e: $d = -1, (-1 - bc)/(-1 - b) < 0, b > -1, c < 1$ (Takens II'b)
2f: $d = -1, (-1 - bc)/(-1 - b) > 0, b > -1, c < 1$ (Takens II'c)

In particular, we note that invariant radial lines $r_2 = \sqrt{(1 - c)/(d - b)}\,r_1$ exist whenever $(1 - c)/(d - b) > 0$, but, in contrast to Takens, we do not distinguish between phase portraits which are equivalent up to reversal of time. For example, Takens' cases Ia and Id are lumped in our 1a. However, Takens allows interchange of the roles of $r_1$ and $r_2$, which, having fixed the coefficient of $r_1^3 = 1$, we cannot. In Figure 7.5.1 only the positive $(r_1, r_2)$ quadrant is shown, since the phase portraits are symmetric under reflection about both axes. Figure 7.5.1 should be read in conjunction with Table 7.5.1 below, in which we also give Takens' classification numbers for convenience.*

EXERCISE 7.5.1. Verify that the phase portraits of Figure 7.5.1 are correct; in particular, check the flow directions on the invariant lines and in the sectors between them. Show that no further topologically distinct types occur for $d = \pm 1$ and $b, c, d - bc \neq 0$ (cf. Section 1.8, equation (1.8.23) and Exercise 1.8.11).

The classification of degenerate fixed points given above is not the most natural when it comes to studying the unfoldings. Here there are twelve distinct cases, as set out in Table 7.5.2. (We note that in earlier work in which such unfoldings were studied (cf. Holmes [1980d], Iooss and Langford [1980]), a different numbering scheme was used.)

The present classification is based on a study of secondary pitchfork bifurcations from nontrivial equilibria for the planar vector field. We note

* In the published version of Takens' paper some figures are incorrectly labelled.

Table 7.5.2. The Twelve Unfoldings.

| Case | Ia | Ib | II | III | IVa | IVb | V | VIa | VIb | VIIa | VIIb | VIII |
|------|----|----|----|----|----|----|----|----|----|----|----|----|
| $d$ | $+1$ | $+1$ | $+1$ | $+1$ | $+1$ | $+1$ | $-1$ | $-1$ | $-1$ | $-1$ | $-1$ | $-1$ |
| $b$ | $+$ | $+$ | $+$ | $-$ | $-$ | $-$ | $+$ | $+$ | $+$ | $-$ | $-$ | $-$ |
| $c$ | $+$ | $+$ | $-$ | $+$ | $-$ | $-$ | $+$ | $-$ | $-$ | $+$ | $+$ | $-$ |
| $d - bc$ | $+$ | $-$ | $(+)$ | $(+)$ | $+$ | $-$ | $(-)$ | $+$ | $-$ | $+$ | $-$ | $(-)$ |

that $(r_1, r_2) = (0, 0)$ is always an equilibrium and that up to three other equilibria (in the positive quadrant) can appear, as follows:

$$(r_1, r_2) = (\sqrt{-\mu_1}, 0) \qquad \text{for } \mu_1 < 0,$$

$$(r_1, r_2) = (0, \sqrt{-\mu_2/d} \qquad \text{for } \mu_2 d < 0,$$

$$(r_1, r_2) = \left(\sqrt{\frac{b\mu_2 - d\mu_1}{A}}, \sqrt{\frac{c\mu_1 - \mu_2}{A}}\right) \quad \text{for } \frac{b\mu_2 - d\mu_1}{A}, \frac{c\mu_1 - \mu_2}{A} > 0,$$

$$(7.5.3)$$

where $A = d - bc$ and $d = \pm 1$.

EXERCISE 7.5.2. Show that pitchfork bifurcations occur from $(0, 0)$ on the lines $\mu_1 = 0$ and $\mu_2 = 0$ and also that pitchfork bifurcations occur from $(\sqrt{-\mu_1}, 0)$, on the line $\mu_2 = c\mu_1$, and from $(0, \sqrt{-\mu_2/d})$ on the line $\mu_2 = d\mu_1/b$. Investigate the stability types of these bifurcations.

As in the previous section, the behavior remains relatively simple as long as Hopf bifurcations do not occur from the fixed point $(\bar{r}_1, \bar{r}_2) = (\sqrt{(b\mu_2 - d\mu_1)/A}, \sqrt{(c\mu_1 - \mu_2)/A})$. To detect such bifurcations, we linearize at this fixed point to obtain the matrix

$$\begin{bmatrix} \mu_1 + 3\bar{r}_1^2 + b\bar{r}_2^2 & 2b\bar{r}_1\bar{r}_2 \\ 2c\bar{r}_1\bar{r}_2 & \mu_2 + c\bar{r}_1^2 + 3d\bar{r}_2^2 \end{bmatrix}, \qquad (7.5.4)$$

with trace

$$\frac{2}{A}[\mu_1 d(c - 1) + \mu_2(b - d)], \qquad (7.5.5)$$

and determinant

$$\frac{4}{A}[(b\mu_2 - d\mu_1)(c\mu_1 - \mu_2)]. \qquad (7.5.6)$$

Using the conditions for the existence of this fixed point from (7.5.3), we find that Hopf bifurcations can occur only on the line

$$\mu_2 = \frac{\mu_1 d(1 - c)}{b - d}, \qquad (7.5.7)$$

when

$$A = d - bc > 0. \tag{7.5.8}$$

We immediately see that Hopf bifurcations cannot occur in cases Ib, IVb, V, VIb, VIIb, and VIII. It is also easy to show that they cannot occur in cases Ia, II, III, and IVa, since for such bifurcations to occur, the slope $d(1 - c)/(b - d)$ of the line given by (7.5.7) must lie between the slopes of the pitchfork lines

$$\mu_2 = c\mu_1 \quad \text{and} \quad \mu_2 = \frac{d\mu_1}{b}, \tag{7.5.9}$$

and in the appropriate sector of the $(\mu_1, \mu_2)$ plane. In each of these four cases, simple computations reveal that this requirement contradicts the condition $A > 0$.

We will now consider some of these unfoldings, leaving the detailed computations and remaining cases to the reader. Since the computations necessary to check these bifurcation sets and phase portraits are relatively simple, if tedious, we merely sketch the results in Figures 7.5.2–7.5.4. As usual, we do not distinguish between nodes and spiral foci. We end with the two cases VIa and VIIa, in which Hopf bifurcations can occur. As in the previous section, we will close with a brief discussion of the implications of these results for the unfolding of the four-dimensional flow of (7.5.1) and for the three-dimensional flow of the $\mathbb{Z}_2$-symmetric pitchfork/Hopf problem:

$$\dot{r} = a_1 r^3 + a_2 r z^2,$$
$$\dot{\theta} = \omega + \mathcal{O}(|r, z|^2), \tag{7.5.10}$$
$$\dot{z} = b_1 r^2 z + b_2 z^3,$$

alluded to at the end of that section.

EXERCISE 7.5.3. Using the methods of Chapters 1 and 3, check that the bifurcation sets and phase portraits of Figures 7.5.2–7.5.4 are correct.

We note that, while the phase portraits of several of these unfoldings reduce to the same degenerate case as $\mu_1, \mu_2 \to 0$, the unfolded systems are genuinely distinct, since different numbers of fixed points, with different stability types, exist in corresponding sectors of their bifurcation sets. However, since the radial invariant lines of the type 2 degenerate singularities are not preserved, our unfoldings do not distinguish between degenerate types 1a and 2a, for example.

EXERCISE 7.5.4. Compute the bifurcation sets and phase portraits for the unfoldings of cases II, III, V, and VIII.

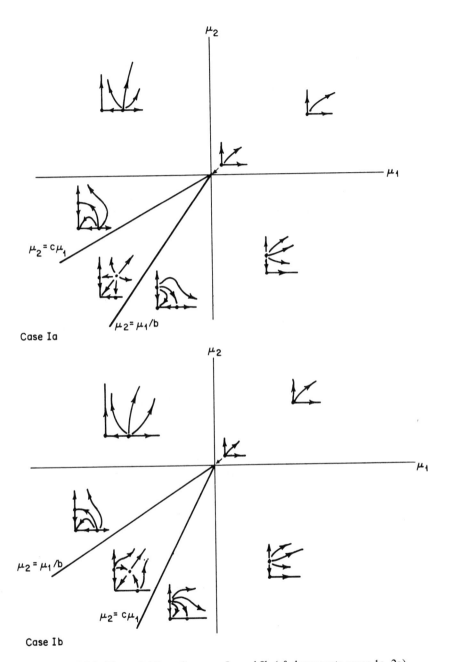

Figure 7.5.2. The unfoldings for cases Ia and Ib (cf. degenerate cases 1a, 2a).

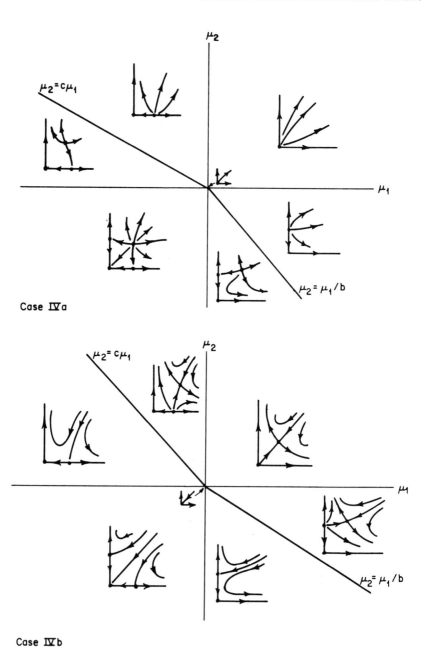

Figure 7.5.3. The unfoldings for cases IVa and IVb (cf. degenerate cases 1a, 2a, and 2b), respectively.

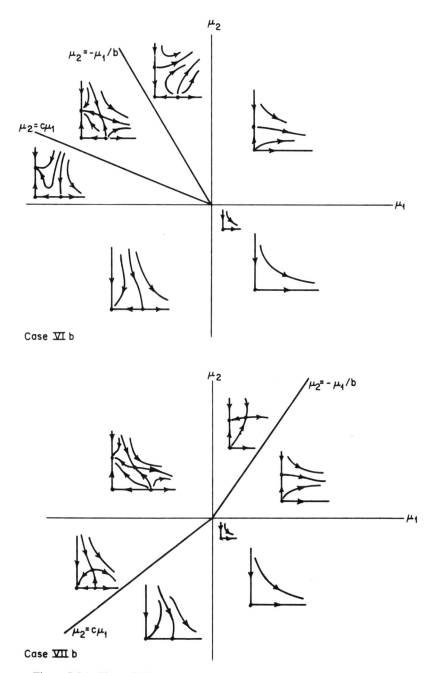

Figure 7.5.4. The unfoldings for cases VIb and VIIb (cf. degenerate case 1c).

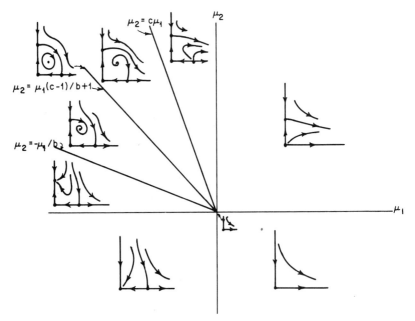

Figure 7.5.5. Partial bifurcation set and phase portraits for the unfolding of case VIa.

Now consider the case VIa, in which a Hopf bifurcation can occur. For this case, we obtain the partial bifurcation set and phase portraits of Figure 7.5.5. As in the previous section, on the Hopf bifurcation line given by equation (7.5.7) we find that the system

$$\dot{r}_1 = r_1(\mu_1 + r_1^2 + br_2^2),$$
$$\dot{r}_2 = r_2\left(\mu_1\left(\frac{c-1}{b+1}\right) + cr_1^2 - r_2^2\right),$$
(7.5.11)

is integrable, and that the function

$$F(r_1, r_2) = r_1^\alpha r_2^\beta(\mu_1 + (r_1^2 + \gamma r_2^2)),$$
(7.5.12)

where $\alpha = 2(1 - c)/A$, $\beta = 2(1 + b)/A$, and $\gamma = (1 + b)/(1 - c)$, is constant along solution curves. In case VIa, $b > 0 > c$, $A = -1 - bc > 0$, and $\mu_1 \overset{\text{def}}{=} -\mu < 0$, and the level curves of this function take the form shown in Figure 7.5.6.

EXERCISE 7.5.5. Check that the function $F(r_1, r_2)$ of equation (7.5.12) is constant on solution curves of (7.5.11), and verify that Figure 7.5.6 is correctly drawn.

Once more it is necessary to add higher-order terms to "stabilize" the degenerate Hopf bifurcation and to determine the topological type of this unfolding. In this case the terms are homogeneous of fifth order:

$$r_1(er_1^4 + fr_1^2r_2^2 + gr_2^4),$$
$$r_2(hr_1^4 + jr_1^2r_2^2 + kr_2^4).$$
(7.5.13)

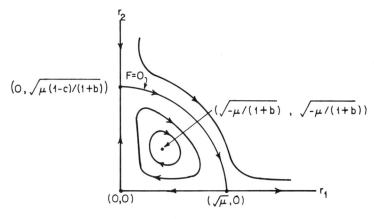

Figure 7.5.6. Level curves of $F(r_1, r_2)$ for case VIa ($b > 0 > c$).

However, we can choose a coordinate system in which all but one of these six coefficients $e, \ldots, k$ are zero. As in the case of one zero and a pair of imaginary eigenvalues, considered in Section 7.4, we compute the effect of coordinate changes of the form

$$
\begin{aligned}
s_1 &= r_1(1 + lr_1^2 + mr_2^2), \\
s_2 &= r_2(1 + nr_1^2 + or_2^2), \\
\tau &= (1 + pr_1^2 + qr_2^2)^{-1}t
\end{aligned}
\tag{7.5.14}
$$

on the fifth-degree terms. Again these are the most general degree three coordinate changes which preserve the appropriate symmetry. We obtain

$$
\frac{ds_1}{d\tau} = s_1^3 + bs_1s_2^2 + (e + p)s_1^5 + (f + 2bl + (2c - 2)m - 2bn + bp + q)s_1^3s_2
$$

$$
+ (g + 2dm - 2bo + bq)s_1s_2^4 + \mathcal{O}(5),
$$

$$
\frac{ds_2}{d\tau} = cs_1^2s_2 + ds_2^3 + (h - 2cl + 2n + cp)s_1^4s_2
\tag{7.5.15}
$$

$$
+ (j - 2cm + (2b - 2d)n + 2co + dp + cq)s_1^2s_2^3 + (k + dq)s_2^5 + \mathcal{O}(5).
$$

The effect of these coordinate changes on fifth-degree terms is again "linear" and is given by the matrix

$$
\begin{bmatrix}
0 & 0 & 0 & 0 & 1 & 0 \\
2b & 2c - 2 & -2b & 0 & b & 1 \\
0 & 2d & 0 & -2b & 0 & b \\
-2c & 0 & 2 & 0 & c & 0 \\
0 & -2c & 2b - 2d & 2c & d & c \\
0 & 0 & 0 & 0 & 0 & d
\end{bmatrix}.
\tag{7.5.16}
$$

This matrix has rank five with kernel spanned by the vector $(1, b, c, d, 0, 0)$. Accordingly, we may assume in (7.5.13) that all of the coefficients except one are zero. We shall take $e = f = g = h = j = 0$ and study the effects of the perturbation $(0, kr_2^5)$ on the 3-jet (7.5.2).

Adding this quintic term to our original equation (7.5.2) with $d = -1$, using the transformations

$$r_1 = \sqrt{\varepsilon} u, \quad r_2 = \sqrt{\varepsilon} v, \quad \mu_1 = \varepsilon v_1, \quad \mu_2 = \varepsilon v_1\left(\frac{c-1}{b+1}\right) + \varepsilon^2 v_2, \quad (7.5.17)$$

and rescaling time $t \to \varepsilon t$, we obtain:

$$\dot{u} = u(v_1 + u^2 + bv^2),$$

$$\dot{v} = v\left(v_1\left(\frac{c-1}{b+1}\right) + cu^2 - v^2\right) + \varepsilon v(v_2 + kv^4).$$

$$(7.5.18)$$

We again have to study a small perturbation of an integrable system. Multiplying (7.5.18) by the integrating factor $u^{\alpha-1}v^{\beta-1}$, we obtain the "equivalent" perturbed Hamiltonian system:

$$\dot{u} = u^\alpha v^{\beta-1}[v_1 + u^2 + bv^2],$$

$$\dot{v} = u^{\alpha-1}v^\beta\left[\left(v_1\left(\frac{c-1}{b+1}\right) + cu^2 - v^2\right) + \varepsilon(v_2 + kv^4)\right].$$

$$(7.5.19)$$

The reader can check that

$$F(u, v) = \frac{1}{\beta} u^\alpha v^\beta\left[v_1 + \left(u^2 + \left(\frac{1+b}{1-c}\right)v^2\right)\right]$$

$$(7.5.20)$$

is the Hamiltonian function for (7.5.19), with $\varepsilon = 0$, where

$$\alpha = \frac{2(1-c)}{A}, \quad \beta = \frac{2(1+b)}{A}, \quad A = -1 - bc,$$

as before. In studying case VIa, we can set $v_1 = -1$ without loss of generality.

Once again, we use the Melnikov theory of Sections 4.5–4.6 to analyze the dependence of periodic orbits on the parameter $v_2$ and the coefficient $K$. Denoting the closed level curve $F(u, v) = K$ by $\Gamma_K$, we compute the zeros of the function

$$\iint_{\text{int } \Gamma_K} [\beta v_2 + (\beta + 4)kv^4]u^{\alpha-1}v^{\beta-1} \, du \, dv \quad (7.5.21)$$

to find values of $v_2$ where periodic solutions occur. If (7.5.21) has a simple zero (as a function of $K$) then

$$v_2 = -\frac{(\beta + 4)k \iint_{\text{int } \Gamma_K} u^{\alpha-1}v^{\beta+3} \, du \, dv}{\beta \iint_{\text{int } \Gamma_K} u^{\alpha-1}v^{\beta-1} \, du \, dv} \quad (7.5.22)$$

is a value of the parameter for which there is a periodic solution near $\Gamma_K$. As in Section 7.4, these integrals can only be evaluated numerically in the general case, but Zholondek is reported (van Gils *et al.* [1985]) to have proved that (7.5.22) defines $v_2$ as a monotone function of $K$ when $k \neq 0$.

In the case $b = 3$, $c = -3$ of (7.5.2), $\alpha = \beta = 1$ and (7.5.2) is Hamiltonian when $\mu_2 = -\mu_1$. We illustrate the calculation of parameter values producing a homoclinic loop in this special case. The Hamiltonian is

$$F(u, v) = uv(-1 + u^2 + v^2) \qquad (7.5.23)$$

($v_1 = -1$ for this case) and the homoclinic loop lies in the level curve $F = 0$, and is a quadrant of the circle $u^2 + v^2 = 1$. Thus, we compute

$$\iint_{\text{int } \Gamma_0} [v_2 + 5kv^4] \, du \, dv = \int_0^1 (v_2\sqrt{1 - u^2} + k(1 - u^2)^{5/2}) \, du$$

$$= \frac{\pi v_2}{4} + \frac{5\pi k}{32}. \qquad (7.5.24)$$

We therefore find the condition

$$v_2 \approx \frac{-5k(\pi/32)}{\pi/4} = -\frac{5k}{8}. \qquad (7.5.25)$$

Using the transformation (7.5.17) with $\mu_2 = -\varepsilon v_1 + \varepsilon^2 v_2$ and $v_1 = -1$, $\varepsilon = \mu_1/v_1 = -\mu_1$, we find the bifurcation curve for homoclinic loops is approximated by

$$\mu_2 = -\mu_1 - \frac{5k}{8}\mu_1^2. \qquad (7.5.26)$$

In Figure 7.5.7 we show the bifurcation diagram in case $k < 0$. The reader can check the stability type of the Hopf and homoclinic bifurcations as in Section 7.4.

EXERCISE 7.5.6. Evaluate the integrals of (7.5.24) with $\Gamma_0$ replaced by $\Gamma_K$ for $K \in (-\frac{1}{8}, 0)$, using elliptic integrals in the latter case (at the center $(u, v) = (\frac{1}{2}, \frac{1}{2})$; $F(u, v) = -\frac{1}{8}$). Hence compute the periodic orbit bifurcation curves in the $(\mu_1, \mu_2)$ parameter plane.

EXERCISE 7.5.7. Carry out Hopf bifurcation stability computations for this problem analogous to those of Section 7.4. Thus obtain specific criteria for sub- and supercritical bifurcations involving the fifth-order coefficient $k$ of (7.5.13) (for $e, \ldots, j = 0$).

The remaining case, VIIa, is even more complicated than the one we have just outlined, since it really contains three distinct subcases, depending upon the relative sizes of the coefficients $b$ and $c$. These determine whether the Hopf bifurcation line lies in the first, third, or fourth quadrants. The three cases are shown in Figure 7.5.8, and represent unfoldings of the degenerate cases 2e,

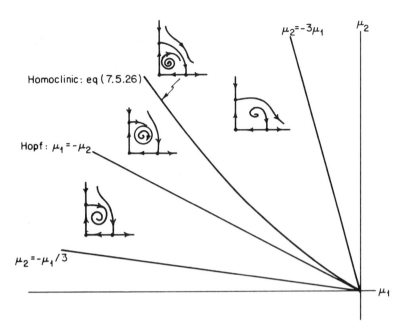

Figure 7.5.7. A completion of the unfolding of case VIa ($b = -c = 3$), with $k < 0$, showing the existence of a family of attracting closed orbits.

1b, and 2c, respectively. The first integral $F(r_1, r_2)$ of (7.5.12) is invariant on solution curves in all these cases when

$$\mu_2 = \mu_1 \left( \frac{c - 1}{b + 1} \right) \tag{7.5.27}$$

(i.e., on the Hopf bifurcation line). However, we note that in these cases at least one of the indices $\alpha$, $\beta$ is always negative and thus that the integral is singular at $(r_1, r_2) = (0, 0)$. This fact leads to the bunching of level curves consistent with the existence of a sink or a source at $(0, 0)$.

We will not discuss the completion of these bifurcation sets and phase portraits by the addition of fifth-order terms, other than to note, as in Section 7.4, that here the closed orbits occurring in the Hopf bifurcation can become unbounded, and hence that the local analysis cannot be expected to provide complete information.

In the discussions above, we have merely sketched the kind of analysis necessary to complete the classification of these two-dimensional unfoldings. A complete study of the general effects of fifth-order terms would be a valuable undertaking, but we must admit that we shrink from the task ourselves. Detailed studies will probably have to await the added stimulus of the detection of specific cases in physical applications. At least one such

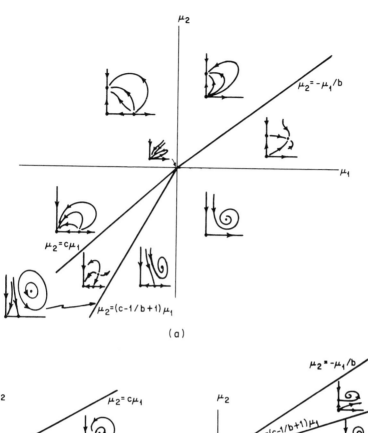

(a)

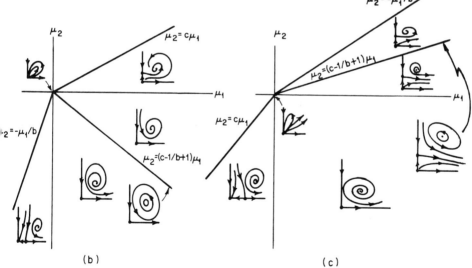

(b)　　　　　　　　　　　　　　　(c)

Figure 7.5.8. The three subcases for case VIIa (cf. degenerate cases 2e, 1b, and 2c, respectively). (a) $b > -1$, $c > 1$; (b) $b > -1$, $c < 1$ or $b < -1$, $c > 1$; (c) $b < -1$, $c < 1$.

has already been discovered (Di Prima *et al.* [1982, 1983]), as we mention in the next section.

Finally, we turn to a discussion of the implications of these results for the full four-dimensional flow of equation (7.5.1). Here there are *two* rotations $\dot{\theta}_i \approx \omega_i$ which must be restored to complete the analysis, and it is easy to see that the following correspondences exist:

| Planar system | Four-dimensional flow |
|---|---|
| Fixed point $(r_1, r_2) = (0, 0)$ | $\Rightarrow$ trivial fixed point; |
| $(r_1, r_2) = (\bar{r}_1, 0)$ | $\Rightarrow$ periodic orbit, period $\approx 2\pi/\omega_1$; |
| $(r_1, r_2) = (0, \bar{r}_2)$ | $\Rightarrow$ periodic orbit, period $\approx 2\pi/\omega_2$; |
| $(r_1, r_2) = (\bar{r}_1, \bar{r}_2)$ | $\Rightarrow$ invariant two torus, periods $\approx 2\pi/\omega_1$, $2\pi/\omega_2$; |
| Periodic orbit | $\Rightarrow$ invariant three torus, periods $\approx 2\pi/\omega_1$, $2\pi/\omega_2$, and $\mathcal{O}(1/\mu_i)$. |

The last (long) period on the three-torus is, of course, associated with the Hopf bifurcation of the planar system. Once more, delicate resonance phenomena can be expected to occur, and the four-dimensional analogue of heteroclinic orbits such as those shown in Figure 7.5.7 will include transversal intersections of the three- and two-dimensional invariant manifolds of a pair of closed orbits. Thus, as in Section 7.4, we expect to find a thin "wedge" around the Hopf bifurcation line in which chaotic dynamics occurs, with transversal homoclinic orbits and horseshoes. Moreover, it is known that small perturbations of triply periodic flows on three-dimensional tori can yield strange attractors (cf. Newhouse *et al.* [1978], and Section 5.5), so we can expect this case to be even more complicated than that of Section 7.4. Almost nothing is known about how such three-tori, or the flows on them, can be expected to bifurcate, or about how they can lose differentiability and break up.

We end this incomplete story by briefly noting that the planar unfoldings which we have constructed here apply to the case of a pure imaginary pair and a simple zero eigenvalue with $\mathbb{Z}_2$-symmetry, for which the degenerate 3-jet takes the form (7.5.10). Holmes [1980d] and Iooss and Langford [1980] considered some aspects of this case and we leave it to the reader to carry out the details of the correspondence. We also end by listing the normal forms of certain strong resonance cases, in which the two frequencies $\omega_1$ and $\omega_2$ are related by small integers. The analyses of these cases is still largely open, although some special subcases have been partially studied by, for example, Steen and Davis [1982] and Bajaj and Sethna [1982]:

Strong resonance: $\omega_2 = \omega_1$:

$$\dot{x}_1 = -y_1 + x_2,$$

$$\dot{y}_1 = x_1 + y_2,$$

$$\dot{x}_2 = -y_2 - (b_1 r_1^2 + b_2 r_2^2)y_2 + (b_3 r_1^2 + b_4 r_2^2)x_2 - (c_1 r_1^2 + c_2 r_2^2)y_1$$
$$+ (c_3 r_1^2 + c_4 r_2^2)x_1 + c_5[(x_1^2 - y_1^2)x_2 + 2x_1 y_1 y_2]$$
$$+ c_6[2x_1 y_1 x_2 - y_2(x_1^2 - y_1^2)] + c_7[x_1(x_2^2 - y_2^2) + 2y_1 x_2 y_2]$$
$$+ c_8[2x_1 x_2 y_2 - y_1(x_1^2 - y_1^2)],$$
$$\dot{y}_2 = x_2 + (b_1 r_1^2 + b_2 r_2^2)x_2 + (b_3 r_1^2 + b_4 r_2^2)y_2 + (c_1 r_1^2 + c_2 r_2^2)x_1$$
$$+ (c_3 r_1^2 + c_4 r_2^2)y_1 + c_5[2x_1 y_1 x_2 - (x_1^2 - y_1^2)y_2]$$
$$- c_6[(x_1^2 - y_1^2)x_2 + 2x_1 y_1 y_2] + c_7[2x_1 x_2 y_2 - y_1(x_2^2 - y_2^2)]$$
$$- c_8[x_1(x_2^2 - y_2^2) + 2y_1 x_2 y_2].$$

$$(7.5.28)$$

Here the $c_i[...]$ are additional terms due to the resonance and $r_i^2 = x_i^2 + y_i^2$. In terms of the complex variables $z_i = x_i + iy_i$, we have

$$\dot{z}_1 = iz_1 + z_2,$$
$$\dot{z}_2 = iz_2 + a_1 z_1^2 \bar{z}_1 + a_2 z_1^2 \bar{z}_2 + a_3 z_1 z_2 \bar{z}_1 + a_4 z_1 z_2 \bar{z}_2 + a_5 z_2^2 \bar{z}_1$$
$$+ a_6 z_2^2 \bar{z}_2.$$

$$(7.5.29)$$

Strong resonance: $\omega_2 = 2\omega_1 = 2\omega$

$$\left.\begin{array}{l} \dot{x}_1 = -\omega y_1 + a(x_1 x_2 + y_1 y_2) + b(x_2 y_1 - x_1 y_2), \\ \dot{y}_1 = \omega x_1 + a(x_1 y_2 - x_2 y_1) + b(x_1 x_2 + y_1 y_2), \\ \dot{x}_2 = -2\omega y_1 + c(x_1^2 - y_1^2) - 2d(x_1 y_1), \\ \dot{y}_2 = 2\omega x_1 + 2c(x_1 y_1) - d(x_1^2 - y_1^2). \end{array}\right\} \qquad (7.5.30)$$

## 7.6. Applications to Large Systems

In this final section, we briefly discuss some applications of the theory developed in this chapter to problems involving partial differential equations and an infinite number of degrees of freedom. When dealing with such large systems, it is often difficult to apply the techniques developed in this book, since they typically require a good geometric visualization of how trajectories lie in phase space. However, the need to integrate the equations directly can be circumvented in small parameter regions near a multiple bifurcation in many cases. The "solution" of such a problem involves substantial amounts of routine algebraic calculation, but all of the analysis of dynamics takes place in low-dimensional spaces. Each of the examples we describe has been discussed in much more detail in the references cited and here we merely sketch the main features of the analysis.

We would like to view systems with an infinite number of degrees of freedom as smooth flows on infinite-dimensional function spaces. The functional analysis required to formulate this idea rigorously introduces a new level of complexity to the analysis. This complexity disappears to a large extent with a suitable version of the center manifold theorem (Theorem 3.2.1), but the interested reader should consult Marsden and McCracken [1976], Holmes and Marsden [1978], Henry [1981], or Carr [1981] for further discussion of these issues. Here we shall assume that the center manifold theorem (Marsden–McCracken) is valid for the examples we describe (which is true, but must be proven in each case) and proceed with formal calculations on this basis.

Given a system, we want to find an equilibrium point at which a multiple bifurcation occurs. We are especially interested in equilibria for which the spectrum lies entirely in the left half plane, except for a few eigenvalues on the imaginary axis. This is the situation which occurs when a stable equilibrium loses stability, and here we have some hope of finding new asymptotic states with complicated dynamics determined by a normal form. Finding such a point of multiple bifurcation is often a formidable task and can seldom be carried through completely in realistic problems. However, in typical examples, some aspect of the problem, such as a symmetry, often serves to simplify the calculation.

We now introduce two examples which will serve as illustrations of the general features of this approach.

EXAMPLE I. The first example which we describe is thermohaline convection in a fluid layer. A horizontal layer of incompressible fluid of depth $d$ has fixed temperatures and salt concentrations on its upper and lower boundaries. Both the temperature and salt concentrations affect the buoyancy of the fluid, and it will be assumed that this dependence is linear. We then have the following systems of equations:

$$\mathbf{v}_t + \mathbf{v} \cdot \nabla \mathbf{v} = \frac{1}{\rho} \nabla p + \mathbf{g}(\alpha T - \beta S) + \nu \nabla^2 \mathbf{v},$$

$$\operatorname{div} \mathbf{v} = 0,$$

$$\frac{\partial T}{\partial t} + \mathbf{v} \cdot \nabla T - w \frac{\Delta T}{d} = k \nabla^2 T, \tag{7.6.1}$$

$$\frac{\partial S}{\partial t} + \mathbf{v} \cdot \nabla S - w \frac{\Delta S}{d} = k_s \nabla^2 S,$$

where

$\mathbf{v}$ = fluid velocity;
$w$ = vertical component of velocity;
$T$ = departure of temperature from linear function;
$S$ = departure of salt concentration from linear function;

$\Delta T$ = temperature difference across fluid layer;
$\Delta S$ = salinity difference across fluid layer;
$\rho$ = density;
$p$ = pressure;
$\mathbf{g}$ = gravitational vector;
$k$ = thermal diffusivity;
$k_s$ = salt diffusivity;
$v$ = kinematic viscosity;
$\alpha$ = buoyancy dependence on temperature;
$\beta$ = buoyancy dependence on salt.

Details of the derivation of these equations can be found in Huppert and Moore [1976]. For background information on convective and buoyancy driven flows, see Turner [1973]. The analysis sketched here was stimulated by the studies of Knobloch and co-workers (Knobloch and Proctor [1981], DaCosta et al. [1981], Knobloch and Weiss [1981]) in which amplitude expansions were used to study truncated models which are essentially the same as the normal form outlined below. Siegmann and Rubenfeld [1975] derived the same equations for a model of thermohaline convection in a loop. Subsequently Guckenheimer and Knobloch [1983] applied center manifold methods to the mathematically similar problem of convection in a rotating layer and compared the results explicitly with amplitude expansions. Also, Carr and Muncaster [1982] considered the loop problem from the viewpoint of center manifold theory.

By assuming that we study motions whose velocities never have a $y$-component, we can introduce a stream function $\psi$ for the velocity and eliminate the pressure term from (7.6.1) by taking the curl of the first equation. After rescaling, (7.6.1) is transformed to the following system:

$$\sigma^{-1}\nabla^2\psi_t - \sigma^{-1}J(\psi, \nabla^2\psi) = -R_T\partial_x T + R_S\partial_x S + \nabla^4\psi,$$

$$T_t + \psi_x - J(\psi, T) = \nabla^2 T, \qquad (7.6.2)$$

$$S_t + \psi_x - J(\psi, S) = \tau\nabla^2 S,$$

where $J(f, g)$ is defined to be $f_x g_z - f_z g_x$, and $\sigma = v/k$, $\tau = k_s/k$ and $R_T = g\alpha\Delta T d^3/kv$, $R_S = g\beta\Delta S d^3/kv$ are Rayleigh numbers.

To facilitate the calculation of the spectrum of the linearization of (7.6.2) at its trivial solution $\psi = T = S = 0$, we choose the boundary conditions $\psi = \psi_{zz} = T = S = 0$ when $z = 0$ or 1. We note that these are not the boundary conditions appropriate to most experiments. However, they have the advantage that the eigenfunctions of the linearized system for the pure conductive solution have the simple forms

$$\begin{pmatrix} \psi \\ T \\ S \end{pmatrix}(x, z, t) = e^{\lambda t}\sin n\pi z \begin{pmatrix} \psi_0 \sin \pi\kappa x \\ T_0 \cos \pi\kappa x \\ S_0 \cos \pi\kappa x \end{pmatrix}$$

where $(\rho, \eta, \kappa)$ satisfy the equations

$$\lambda^3 + (\sigma + \tau + 1)k^2\lambda^2 + ((\sigma + \tau + \sigma\tau)k^4 - \pi^2\sigma\tau^2k^{-2}(R_T - R_S))\lambda$$
$$+ \sigma\tau k^6 + \pi^2\sigma\kappa^2(R_S - \tau R_T) = 0,$$

$$(7.6.3)$$

and

$$k^2 = \pi^2(n^2 + \kappa^2).$$

The eigenvalues with maximum real part occur with $\kappa^2 = \frac{1}{2}$ and $n = 1$. These can appear as a simple zero eigenvalue, a simple pair of pure imaginary eigenvalues, or as a zero eigenvalue of multiplicity two. The last possibility occurs when

$$(\sigma + \tau + \sigma\tau)k^4 - \pi^2\sigma\tau^2k^{-2}(R_T - R_S) = 0,$$
$$\kappa\tau k^6 + \pi^2\sigma\kappa^2(R_S - R_T) = 0.$$

$$(7.6.4)$$

EXERCISE 7.6.1. Check that (7.6.3) and (7 6.4) are correct.

Without horizontal boundary conditions, (7.6.2) has a continuous spectrum ($\kappa$ can vary continuously in (7.6.3)) and the center manifold theorem of Marsden and McCracken [1976] cannot be used. If one imposes either periodic boundary conditions with period $L$ or the conditions that $\psi = T_x = S_x = 0$ at $x = 0$ and $x = L$, then the eigenfunctions of the linearized equations have the same form, but $\kappa$ is restricted to discrete values. Apart from discrete values of $L$ at which a pair of allowed values $\kappa = m/L$ give solutions to (7.6.3) with the same maximum real part, the most degenerate bifurcation which can occur from the stable trivial equilibrium is the one with a zero eigenvalue of multiplicity two described above. We will discuss the computation of the normal form (the nonlinear terms) after introducing the second example.

EXAMPLE II. The second example comes from the study of motions of an elastic panel subject to an axial load and a fluid flow along its surface. For more complete descriptions, and for details of the functional analytic methods necessary to obtain existence, uniqueness, and smoothness of solutions in the infinite-dimensional phase space, see Holmes [1977, 1981a] and Holmes and Marsden [1978].

In Figure 7.6.1 we sketch the physical problem. A supersonic stream of fluid passes above a thin plate fixed at the edges $z = 0$ and $z = 1$ and unconstrained at $y = 0$ and $y = l$. The panel is simultaneously subjected to an in-plane mechanical load $\Gamma \cdot l$. The fluid velocity is characterized in terms of the dynamic pressure $\rho$. Using nondimensional quantities, and assuming that the panel bends in a cylindrical mode (so that $w(z, y, t) = w(z, t)$ is

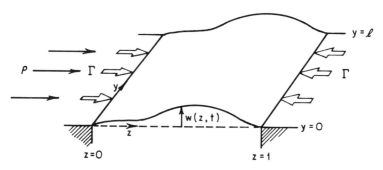

Figure 7.6.1. Panel flutter.

independent of $y$), we obtain the following nonlinear integro-differential equation, which is essentially a one-dimensional version of the von Karman equations for a thin plate:

$$w_{tt} + \alpha w_{tzzzz} + \sqrt{\rho \delta}\, w_t + \left\{ \Gamma - \kappa \int_0^1 w_\xi^2(\xi)\, d\xi - \sigma \int_0^1 w_\xi(\xi) w_{\xi t}(\xi)\, d\xi \right\} w_{zz}$$

$$+ w_{zzzz} + \rho w_z = 0. \tag{7.6.5}$$

Here $w = w(z, t)$ is the transverse displacement of the panel, $\alpha$, $\sigma \geq 0$ are (linear) viscoelastic damping parameters associated with the panel, $\delta > 0$ represents fluid damping and $\kappa > 0$ is a measure of the nonlinear axial (membrane) restoring forces generated in the panel due to transverse displacement. All these are regarded as fixed constants in a given problem, but the two parameters $\Gamma$ and $\rho$ (mechanical load and flow rate) are allowed to vary. From the general discussions of Chapter 3, we therefore expect to find both codimension one and two bifurcations in this problem. Along with (7.6.5) we shall assume simply supported (hinged) boundary conditions:

$$w(0, t) = w_{zz}(0, t) = w(1, t) = w_{zz}(1, t) = 0. \tag{7.6.6}$$

It does not seem possible to solve the linearized equations for (7.6.4) and (7.6.5) explicitly, as was possible for the thermohaline problem. However, we can use a *Galërkin* method to approximate the linearized problem by a sequence of finite-dimensional problems. In the Galërkin method, we view the system (7.6.5)–(7.6.6) as a flow defined on a Hilbert space $H$ and choose an orthonormal basis $\{\phi_j\}$ for $H$. Writing

$$w(z, t) = \sum_{j=1}^{\infty} a_j(t) \phi_j(t), \tag{7.6.7}$$

and substituting (7.6.7) into (7.6.5), formally produces an infinite system of second-order ordinary differential equations:

$$\ddot{a}_j = f_j(a_1, \ldots, a_n, \ldots, \dot{a}_1, \ldots, a_n, \ldots). \tag{7.6.8}$$

*Truncation* of this system by setting all of the modes $a_j, j > n$ to be zero and ignoring the equations for $\dot{a}_j, j > n$ produces a finite-dimensional system whose spectrum can be studied numerically.

In the panel flutter problem, we set $H = L^2(0, 1)$ and $\phi_j = \sin j\pi z$. The truncation with $n = 2$ is given by

$$\ddot{a}_1 + (\alpha\pi^4 + \sqrt{\rho\delta})\dot{a}_1 + \pi^2(\pi^2 - \Gamma)a_1 - (8\rho/3)a_2 + f_1(a_1, \dot{a}_1, a_2, \dot{a}_2) = 0,$$

$$(7.6.9)$$

$$\ddot{a}_2 + (16\alpha\pi^4 + \sqrt{\rho\delta})\dot{a}_2 + (8\rho/3)a_1 + 4\pi^2(4\pi^2 - \Gamma)a_2 + f_2(a_1, \dot{a}_1, a_2, \dot{a}_2) = 0;$$

where

$$f_1 = \frac{\pi^4}{2}\{\kappa(a_1^2 + 4a_2^2) + \sigma(a_1\dot{a}_1 + 4a_2\dot{a}_2)\}a_1,$$

$$(7.6.10)$$

$$f_2 = 2\pi^4\{\kappa(a_1^2 + 4a_2^2) + \sigma(a_1\dot{a}_1 + 4a_2\dot{a}_2)\}a_2.$$

Rewriting (7.6.9) as a set of four first-order equations, we have the system

$$\dot{x} = A_\mu x + f(x); \qquad x \in \mathbb{R}^4, \mu \in \mathbb{R}^2,$$   $$(7.6.11)$$

where

$$x = \begin{pmatrix} a_1 \\ a_2 \\ \dot{a}_1 \\ \dot{a}_2 \end{pmatrix},$$

$$A_\mu = \left[ \begin{array}{cc:cc} & & 1 & 0 \\ & 0 & & \\ & & 0 & 1 \\ \hdashline \pi^2(\Gamma - \pi^2) & 8\rho/3 & -(\alpha\pi^4 + \sqrt{\rho\delta}) & 0 \\ -8\rho/3 & 4\pi^2(\Gamma - 4\pi^2) & 0 & -(16\alpha\pi^4 + \sqrt{\rho\delta}) \end{array} \right],$$

$$f(x) = \begin{pmatrix} 0 \\ 0 \\ -f_1 \\ -f_2 \end{pmatrix},$$   $$(7.6.12)$$

and

$$\mu = \begin{pmatrix} \Gamma \\ \rho \end{pmatrix}$$

is our two-dimensional parameter.

We shall consider bifurcations from the trivial equilibrium position $w(z, t) \equiv 0$, or $x = 0$. The stability of this fixed point is determined by the eigenvalues of $A$, which are the roots of the fourth-order polynomial

$$\lambda^4 + \alpha_3 \lambda^3 + \alpha_2 \lambda^2 + \alpha_1 \lambda + \alpha_0 = 0,$$

where

$$\alpha_0 = 4\pi^4(\Gamma - \pi^2)(\Gamma - 4\pi^2) + 64\rho^2/9,$$

$$\alpha_1 = -\pi^2\{(\Gamma - \pi^2)(16\alpha\pi^4 + \sqrt{\rho}\delta) + 4(\Gamma - 4\pi^2)(\alpha\pi^4 + \sqrt{\rho}\delta)\}$$

$$\alpha_2 = (\alpha\pi^4 + \sqrt{\rho}\delta)(16\alpha\pi^4 + \sqrt{\pi}\delta) - \pi^2(5\Gamma - 17\pi^2), \qquad (7.6.13)$$

$$\alpha_3 = (17\alpha\pi^4 + 2\sqrt{\rho}\delta).$$

Since the trivial equilibrium is always a fixed point for (7.6.11) and the nonlinear terms (7.6.10) are cubic, we expect pitchfork bifurcations to occur from $x = 0$ at simple zero eigenvalues. From (7.6.13) it is easy to see $A_\mu$ has a simple zero eigenvalue when $\alpha_0 = 0$ and $\alpha_j \neq 0, j = 1, 2, 3$; that is, for

$$\rho = \tfrac{3}{4}\pi^2\sqrt{(\Gamma - \pi^2)(4\pi^2 - \Gamma)}. \qquad (7.6.14)$$

Such bifurcations can only occur for $\Gamma \in [\pi^2, 4\pi^2]$.

However, other "secondary" bifurcations can occur from the nontrivial fixed points which bifurcate from $x = 0$. These turn out to be fold bifurcations:

EXERCISE 7.6.2. Find the locus of equilibria of (7.6.11) directly and check that pitchfork bifurcations occur on the curve (7.6.14). Show that the fold bifurcations occur on the set $\{(\Gamma, \rho) | \rho = 9\pi^4/8, \Gamma > 5\pi^2/2\}$. Show that a bifurcation with a simple zero eigenvalue and higher-order degeneracy occurs at $(\Gamma, \rho) = (5\pi^2/2, 9\pi^4/8)$. Investigate this bifurcation.

A double zero eigenvalue occurs when $\alpha = .005$ and $\delta = .1$ at

$$\mu_0 = (\Gamma_0, \rho_0) \approx (2.23\pi^2, 1.11\pi^4), \qquad (7.6.15)$$

$(\alpha_0 = \alpha_1 = 0$ in (7.6.13)) and this is the most degenerate situation which can occur. One *hopes* that higher truncations of (7.6.8) will give multiple bifurcation points that converge to a multiple bifurcation point, but it is difficult to do the linear algebra required when there are many modes or to demonstrate convergence (but see Holmes and Marsden [1978]).

Having found points of multiple bifurcation in our examples, we next want to determine the normal forms associated with these bifurcations. This requires calculation of the center manifolds and the normal forms within these center manifolds to a sufficiently high order that results of the preceding sections can be applied. These calculations can be lengthy and require considerable stamina, but they follow a set algorithm, so there is the prospect that they can be automated. Since both of our examples have double zero eigenvalues, we expect the results of Section 7.3 to apply.

EXERCISE 7.6.3. Show that the thermohaline and panel flutter problems each have a $\mathbb{Z}_2$-symmetry.

The symmetries of both problems rule out a normal form with nonzero quadratic terms in a system of the form (7.3.3). Instead, cubic normal forms of type (7.3.22) occur and the Taylor series calculations must be carried through to at least third order.

For the two mode system (7.6.11), straightforward (but lengthy!) calculations shows that the cubic coefficients of the normal form (7.3.22) are approximately

$$a_3 \approx -0.303k\pi^4,$$

$$b_3 \approx -0.043k\pi^4. \tag{7.6.16}$$

These calculations are somewhat simplified by the fact that the tangent space approximation to the center manifold is sufficient. The bulk of the work is in putting the linear part of (7.6.11) into the appropriate block diagonal form of (3.2.13). For details, see Holmes [1981a] (but note that there are some numerical errors in the computations in that paper).

We therefore deduce that the bifurcation diagram in the $(\Gamma, \rho)$ plane near $\mu_0$ is described by Figures 7.3.7 and 7.3.9. Assuming that such a point of multiple bifurcation occurs for the full system (7.6.5), Holmes and Marsden [1978] argue that *forced* oscillations of the panel will be chaotic (cf. Section 4.5).

The normal form computations for the thermohaline problem can be carried through completely at its point of multiple bifurcation, but the ability to do so depends upon special features of the equations. In particular, because the nonlinearities of (7.6.2) are quadratic and because the product of two trigonometric functions has a finite Fourier series, only a finite number of modes play a rôle in determining the normal form on the center manifold. In this example, five modes need to be retained in order to compute the cubic terms in the normal forms. This calculation is carried through in a different format by Knobloch and Proctor [1981]. The relevant expansion of $\psi$, $T$, $S$ is the following

$$\psi = a(t) \sin(\pi\alpha x) \sin \pi z,$$

$$T = b(t) \cos(\pi\alpha x) \sin \pi z + c(t) \sin 2\pi z, \tag{7.6.17}$$

$$S = d(t) \cos(\pi\alpha x) \sin \pi z + e(t) \sin 2\pi z,$$

with the generalized eigenspace of zero in the three-dimensional subspace spanned by modes $a, b$, and $d$. Substitution into (7.6.2), followed by projection onto the space spanned by the five modes of (7.6.17), yields the system

$$\dot{a} = -\sigma k^2 a - \sigma \frac{\pi \alpha}{k^2} R_T b + \sigma \frac{\pi \alpha}{k^2} R_S d,$$

$$\dot{b} = -\pi \alpha a - k^2 b - \frac{\pi^2 \alpha}{2} ac,$$

$$\dot{c} = -4\pi^2 c + \frac{\pi^2 \alpha}{2} ab, \qquad (7.6.18)$$

$$\dot{d} = -\pi \alpha a - \tau k^2 d + \frac{\pi^2 \alpha}{2} ac,$$

$$\dot{e} = -4\pi^2 \tau e + \frac{\pi^2 \alpha}{2} ad.$$

In this example, the bifurcation set in the $(R_S, R_T)$ plane near the multiple bifurcation is given by Figures 7.3.4–5.

EXERCISE 7.6.4. Compute the normal form of (7.6.18) when $(R_S, R_T)$ have values which satisfy (7.6.4). (Hint: Though the zero eigenspace lies in the $(a, b, d)$ coordinate subspace, the projection of (7.6.17) onto this subspace is *linear*. To compute the cubic terms of the normal form, the center manifold must be computed to second order, with the result that $c \approx \alpha ab/8$ and $e \approx (\alpha/8\tau)ad$. These substitutions into the equations for the $(a, b, d)$ modes make the normal form computations much easier.)

EXERCISE 7.6.5. Show that the inclusion of more modes in the normal form computations for the thermohaline problem does not affect the cubic normal form equations.

The thermohaline and panel flutter problems described above seem to be typical applications of the codimension two bifurcation theory described in earlier sections. Two further examples deserve mention. The one-dimensional reaction–diffusion system (cf. Auchmuchty and Nicolis [1975, 1976]):

$$X_t = D_1 X_{xx} + X^2 Y - (B + 1)X + A,$$
$$Y_t = D_2 Y_{xx} - X^2 Y + BX, \qquad (7.6.19)$$

with $X(0) = X(\pi) = A$ and $Y(0) = Y(\pi) = B$, admits multiple bifurcations of several types from the stable trivial solution. Of particular note here is a multiple bifurcation with a zero eigenvalue and a pair of pure imaginary eigenvalues. The unfoldings of this bifurcation are described by the solution to the following:

EXERCISE 7.6.6. Show that there is a parameter region for which solutions with invariant two-dimensional tori occur. Apart from being unable to establish that these tori break up in *transversal* homoclinic trajectories, this example provides a system of partial differential equations for which one can prove that there are solutions which are chaotic with an irregular spatial dependence. See Guckenheimer [1981] for details concerning this example.

The second example arises in the classic fluid dynamic stability study of Couette flow between rotating cylinders. DiPrima *et al.* [1982, 1983] consider the case in which the cylinders can counter-rotate and there is a pair of parameters: the two rotation speeds. There are points in the parameter plane at which two modes simultaneously destabilize and at one such point they find a simple zero and a pure imaginary pair of eigenvalues. However, in this case, a $\mathbb{Z}_2$-symmetry inherent in the physical experiment leads to the normal form (7.5.10). One of the two cases considered leads to the unfolding of case VIIa (Figure 7.5.8(a)), and one therefore expect to find two-dimensional tori and chaotic solutions in this problem as well. DiPrima *et al.* have even computed the fifth-order terms necessary for a complete determination of the planar vector field in this case, although the computations still await completion.

The general approach of using multiple bifurcation theory to look for chaotic solutions of partial differential equations seems to be a fruitful one. Extrapolation from small amplitude solutions found near multiple bifurcations gives a procedure for guessing what the "routes to chaos" in specific problems should be and how these might vary with changes in secondary parameters. However, the chaos found thus far near codimension two bifurcations is mild and typically involves only a few spatial modes. Further exploitation of this idea should use the theory of bifurcations of higher codimension, a theory which is waiting for the energetic reader who has persevered this far to pursue as the next unfinished chapter of this subject. Good luck!

# Suggestions for Further Reading

In this book we have been highly selective in our choice of topics. Therefore, this appendix has been included as a signpost with pointers toward related areas which are likely to be of interest to many readers. Our bibliography has been limited to those works cited in the text. Extensive bibliographies for dynamical systems and catastrophe theory have been compiled recently by Shiraiwa [1981] and Zeeman [1981].

Our treatment of dynamical systems theory is formulated in the style of the "Smale school" which traces its origins to the work of Smale and his students during the 1960s. Smale's [1967] survey paper provides a clear statement of a program for the development of the subject. Smale [1980] includes a reprint and update of this important paper. There are several types of questions which have been pursued in this context. A rough separation gives three categories consisting of:

(1) structural stability questions;
(2) topological questions for Axiom A systems; and
(3) bifurcation problems.

To pursue these topics further, the reader has available lecture notes and texts, review articles, and conference proceedings. A series of proceedings from major conferences gives a good picture of the evolution of this body of work:

Berkeley 1968—Chern and Smale [1970].
Warwick 1969—Chillingworth [1971].
Bahia 1971—Peixoto [1973].
Warwick 1974—Manning [1975].
Rio de Janeiro 1976—Palis and de Carmo [1977].

Northwestern 1979—Nitecki and Robinson [1980].
Warwick 1980—Rand and Young [1981].
Rio de Janeiro 1981—to appear.

In addition to Smale's survey, there are Robbin [1972], Bowen [1978], and Franks [1982]. Lecture notes and texts in this area include those of Nitecki [1971], Bowen [1975], Shub [1978], Irwin [1980], Newhouse [1980], Kušnirenko et al. [1981], and Palis and de Melo [1982].

There are a number of closely related points of view concerning dynamical systems which parallel developments within the Smale school. Hale and his collaborators have based their work on invariant manifolds and their bifurcation analyses on alternative methods. Chow and Hale [1982] provides a systematic presentation of this work which includes discussion of codimension two bifurcations. Sattinger [1973], Iooss [1979], and Iooss and Joseph [1981] treat bifurcation theory from a similar viewpoint, these treatments being more analytical than the geometrical view presented in the present book. Arnold [1982] and Cornfeld et al. [1982] provide entrees to the extensive Soviet contributions to dynamical systems theory, some of which predate those of the Smale school.

The book by Cornfeld et al. [1982] presents a probabilistic approach to studying dynamical systems, which stems from Kolmogorov. The work of Sinai, Ruelle, and Bowen on the statistical mechanics of Axiom A systems is one manifestation of this point of view. Important recent developments using methods based in ergodic theory include theorems on the existence of invariant manifolds for general systems by Pesin [1977], Ruelle [1979], and Katok [1980].

The oldest traditions of studying dynamical systems are oriented toward Hamiltonian systems. The books of Arnold [1978] and Abraham and Marsden [1978] provide modern mathematical accounts of this theory. The monographs of Arnold and Avez [1968] and Moser [1973] and the book of Lichtenberg and Lieberman [1982] are more strongly oriented to problems of chaotic motion in Hamiltonian systems. We also note that the important earlier book of Birkhoff [1927] has recently been reprinted.

The introduction of renormalization and scaling methods from statistical physics has brought a new perspective into dynamical systems and bifurcation theory since 1975. This body of work is growing rapidly and has not yet been amalgamated into a unified treatment. A few of the important papers are Feigenbaum [1978], Collet et al. [1980], Campanino and Epstein [1981], Lanford [1982], Feigenbaum et al. [1982], and Rand et al. [1982].

Many bifurcation problems arise in the context of partial differential equations and physical systems with an infinite number of degrees of freedom. This brings complications of functional analysis to the fore when an attempt is made to apply results on ordinary differential equations. See Marsden and McCracken [1976] for an introduction to these issues and Marsden [1981] for a later review. Much of this work on partial differential equations is oriented toward the study of equilibria and their stability, with other dynamical properties in the background. Golubitsky and Schaeffer have pursued a recent program of studying such problems using methods of singularity theory. Their book, Golubitsky and Schaeffer [1985] provides an introduction. Thompson and Hunt [1973] examines problems of elastic buckling from an engineering point of view, while Antman [1984] provides elegant mathematical treatments of both classical and modern problems in the theory of beams and plates. Closely related to this work on bifurcation of equilibria is Thom's catastrophe theory (Thom [1975], Zeeman [1977], Poston and Stewart [1978]).

In this book we have chosen to illustrate the application of dynamical systems methods to specific problems in nonlinear oscillations. There is a considerable body of work in this area available in the engineering literature, although, with the exception of Andronov and co-workers [1966, 1971, 1973], it has been carried out largely in isolation from the mathematical developments outlined above. The textbook of Nayfeh and Mook [1979] provides an extensive listing of recent work on nonlinear oscillations from the engineering viewpoint, with a strong bias toward perturbation methods. The collections of articles edited by Lefschetz [1950, 1952, 1956, 1958, 1960] provides a mathematical viewpoint on earlier work in nonlinear oscillations.

There is a diverse and scattered literature on the dynamical systems analysis of specific problems in engineering and the applied sciences. As automated data collection techniques and awareness of aperiodic phenomena increase, this literature on applications is rapidly proliferating. The present book has been written with the hope of filling the needs of such "users," but we have devoted little space to the applications themselves. Conference proceedings from Santa Barbara (Jorna [1978]), Asilomar (Holmes [1980a]), the New York Academy of Sciences (Gurel and Rössler [1979], Helleman [1980]), and Los Alamos (Campbell and Rose [1983]) provide a taste of the diverse range of applications. We hope that the readers of this volume will be inspired to find and study others.

# Postscript Added at Second Printing

The wish expressed at the end of the Appendix to the first edition has been amply rewarded; perhaps too amply. Literally hundreds of papers have appeared in the last 3 years in which the methods (or at least the language) of dynamical systems theory are used to study differential equations and iterated mappings arising in virtually every field of the sciences. Several books and monographs have appeared. The collections of papers edited by Cvitanovic [1984] and Hao [1984] are useful and noteworthy in that they present physicists' selections of important papers, with a strong bias towards renormalization, scaling methods, and "universal" phenomena. Schuster [1984] also provides a physicists' introduction to some of these ideas. In contrast, Devaney [1985] gives a self-contained and rigorous development of the theory of iterated mappings starting at a somewhat more elementary level than assumed in the present book.

In a similar spirit, the book by Golubitsky and Schaeffer [1985] and the forthcoming second volume (Golubitsky and Stewart [1987]) provide complete treatments of bifurcation theory from singularity and group theoretic viewpoints, although they concentrate on equilibrium solutions and steady periodic behaviors.

There have been many national and international meetings on dynamical systems in recent years. Notable proceedings from three such conferences (Les Houches 1981, Los Alamos 1982, and Sitges 1982) were edited by Campbell and Rose [1983], Helleman *et al.* [1983], and Garrido [1983].

In what follows we indicate noteworthy developments in the topics treated in Chapters 3–7.

## Chapter 3

Since the theory of local, codimension one bifurcations of vector fields and maps was already relatively complete at the first printing, few results of substance have been added in this area since then. One interesting and potentially far-reaching development has, however, occurred in the computation of Taylor series approximations of center manifolds and of normal forms. Software packages have been developed which use symbolic manipulation languages such as MACSYMA and SMP. These programs make possible calculations which are effectively impossible by hand. See Rand [1984] for an introduction to the use of MACSYMA in a general applied mathematical context, with emphasis on perturbation theory.

## Chapter 4

Since the first printing of this book, the Melnikov method has been used to study homoclinic bifurcations in a number of planar and higher-dimensional systems. See, for example, Robinson [1983, 1985], Salam and Sastry [1985], Kopell and Washburn [1982], Sanders and Cushman [1984a], Koiller [1984], and Holmes [1986].

Also some interesting technical developments have taken place: Schecter [1985a, b] has studied the degenerate situation of homoclinic orbits to a saddle-node bifurcation point (cf. Figures 2.1.3 (at $S$), and 3.4.5, above) both in the purely planar and periodically forced context and Gruendler [1985] and Wiggins and Holmes [1985a, b, c] have generalized Melnikov methods to non-Hamiltonian systems with phase spaces of dimension $\geq 3$. Robinson [1985] provides a nice account, showing how various Melnikov functions occurring in different contexts are related.

## Chapter 5

The major area in which developments have occurred since the first printing of this chapter is that of dimension computations. Probably over 100 papers have appeared in which various algorithms are used for computation of capacity, "information dimension," "correlation dimension," etc. Unfortunately, there are still few rigorous mathematical results in this area. We therefore feel that it is inappropriate to provide a detailed account of all the work in this area at this stage. However, we give a very brief description of the major techniques.

The definition of dimension for an attractor which has been most amenable to numerical computation is the *pointwise dimension* of a measure $\mu$ (Farmer *et al.* [1983]). Given a measure $\mu$ (presumably an invariant measure on an attractor $\Lambda$) and a point $x$, consider the function $V_x(r) = \mu(B_x(r))$ where $B_x(r)$ is the ball of radius $r$ centered at $x$. If the limit $\lim_{r\to 0}(\log V_x(r)/\log r)$ exists and is independent of $x$, for $\mu$ almost all $x$, then this common limit is the *pointwise dimension* of $\mu$. Algorithms that give numerical estimates of the pointwise dimension of an asymptotic measure (see p. 283) are easy to implement. One assumes that the points in an observed or computed trajectory for a system are distributed in a fashion that is described by an invariant probability measure $\mu$. Selecting a point $x$ in the trajectory, $V_x(r)$ is estimated as the proportion of points in the trajectory that lie within distance $r$ of $x$. By computing the distances between $x$ and all the other points in the trajectory and then sorting the set of distances, one can easily compute this estimate of $V_x(r)$. If a log-log plot of this function is reasonably linear, its slope provides an estimate of the pointwise dimension of $\mu$.

There are a variety of difficulties associated with the procedure described above that are discussed by Guckenheimer [1984, 1986]. More robust statistics are obtained by choosing several "reference points" $x$ in the calculation described above and averaging. A slightly different definition, introduced by Grassberger and Procaccia [1983], has been frequently used in examining experimental and numerical data. Given a "typical" finite trajectory, define $N(r)$ to be the proportion of *pairs* of points in the trajectory whose distance from each other is less than $r$. Then $\lim_{r\to 0}(\log N(r)/\log r)$ is defined as the *correlation dimension* of the attractor containing the trajectory. This definition makes use of all the information contained in interpoint distances, but is harder to use theoretically because the distances are not statistically independent of one another.

There are a growing number of analyses of experimental data using these techniques. (Guckenheimer and Buzyna [1983] and Brandstater *et al.* [1983] were the first.) In many of these studies an embedding trick suggested by Ruelle and justified by Takens [1981] is used to create a picture of the attractor in its phase space from a single time series of observations. The idea is to introduce a delay time $\tau$ and use the observations $(x(t), x(t + \tau), \ldots, x(t + (n - 1)\tau)$ as an $n$-dimensional vector corresponding to time $t$. Takens [1981] and Mañé [1981] prove that, for generic systems, delay times $\tau$ and observables (the function being measured), if $n$ is larger than twice the dimension of the attractor, then the representation of the attractor in $R^n$ is a homeomorphism.

# Chapter 6

The elegant and deep work of Douady [1982/3], Douady and Hubbard [1982], Sullivan [1984], and others on iterated maps in the complex plane has illuminated various issues in the bifurcation theory of real analytic one-dimensional maps (cf. Sections 6.3 and 6.4). Devaney [1985] provides an excellent introduction to the theory of such discrete dynamical systems, including both one- and two-dimensional real mappings and iterated functions of a single complex variable. See also Blanchard [1984].

A number of papers have recently appeared on the topic of orbits with rational and irrational rotation numbers for maps of the circle and the annulus. In particular, the discovery of "Aubry-Mather" invariant Cantor sets for area-preserving maps of the annulus has sparked a revival of interest in the earlier work of Birkhoff [1932a, b] and others (cf. Herman [1983]). See Aubry [1983], Aubry and LeDaeron [1983], Chenciner [1983, 1984], Mather [1982a, b, 1984], Katok [1982, 1983], and Percival [1980]. These sets have the property that orbits within them possess the same order properties as orbits of rigid rotations of the circle $R_\alpha: S^1 \to S^1$ (cf. Section 6.2). The way in which such "well-ordered" orbits occur together with "badly ordered" orbits in both area-preserving and dissipative two-dimensional maps has been studied in several papers, e.g., Hall [1983], Boyland and Hall [1985], Casdagli [1985], Le Calvez [1985], and Hockett and Holmes [1986]. There has also been a considerable amount of work on noninvertible or "supercritical" maps of the circle, e.g. Ito [1981], Bernhardt [1982], Boyland [1983], and Newhouse et al. [1983]. In a remarkable series of papers, Chenciner [1982a, b, c, 1983a, 1985a, b] has studied the degenerate (codimension two) Hopf bifurcation for two-dimensional diffeomorphisms and found similar structures of ordered and nonordered orbits as those found in area-preserving maps. (This bifurcation is the map analogue of the codimension two Hopf bifurcation for vector fields discussed in Section 7.1.)

A third area in which an industry has developed somewhat akin to that of Melnikov functions is in "Šilnikov type" bifurcations for systems both with and without symmetries. See Sparrow's [1982] book on the Lorenz equations, Tresser [1983, 1984a, b], Glendinning [1985], Glendinning and Tresser [1985], and Glendinning and Sparrow [1983, 1985]. In particular, we draw attention to the "gluing" bifurcations studied by Coullet et al. [1983] and Gambaudo et al. [1984, 1985], and the occurrence of Farey sequences in their description. Also, we especially wish to draw attention to the recent Russian literature on the Lorenz and similar systems, see Afraimovich et al. [1983] and references therein and Belykh [1985] for an example involving continuation of homoclinic bifurcations.

## Chapter 7

In addition to the work on codimension two bifurcations with pure imaginary eigenvalues which we have incorporated in our revision of Sections 7.4 and 7.5, we note that a lot of work has been done on multiple bifurcations and unfolding degenerate singularities, especially with reference to the special cases of equilibrium and periodic solutions. As the analyses of Sections 7.4 and 7.5 show, the analysis of unfoldings of degenerate singularities in vector fields of dimension $\geq 3$ is beset with problems due to "global" homoclinic phenomena. However, analyses which ignore such features have been pushed quite far. The forthcoming proceedings from the AMS 1985 Summer Research Conference on Multiparameter Bifurcation Theory, to be published in the AMS Series Contemporary Mathematics, provides a good survey of recent work.

# Glossary

## General Mathematical Terms

*Boundary*: The boundary $\partial A$ of a closed set $A$ is the set of points of $A$ which are not in the interior of $A$.

*Cantor set*; *perfect set*: A Cantor set $\Gamma$ is a closed set with the properties that: (1) the largest connected subset of $\Gamma$ is a point; and (2) every point of $\Gamma$ is a limit point of $\Gamma$.

*$C^k$ function*: A function is $C^k$ if it is $k$-times differentiable.

*Closed set*: A set $A$ is closed if it contains all its limit points.

*Closure*: The closure $\bar{A}$ of a set $A$ is the union of $A$ and its set of limit points.

*Codimension*: The codimension of a $k$-dimensional submanifold of an $n$-dimensional manifold is $n - k$.

*Compact*: A set $A$ is compact if every open cover has a finite subcover. (Equivalently, for subsets of Euclidean space, $A$ is closed and bounded.)

*Compact support*: A function $f: \mathbb{R}^n \to \mathbb{R}$ has compact support if the set of points $x$ with $f(x) \neq 0$ is bounded.

*Connected*: A set $A$ is not connected if there exist two open sets $U_1$ and $U_2$ with: (1) $U_i \cap \bar{U}_j = \emptyset$ for $i \neq j$; (2) $U_i \cap A \neq \emptyset$ for $i = 1, 2$; and (3) $A \subset U_1 \cup U_2$.

*Diffeomorphism*: A $C^k$-diffeomorphism $f: M \to N$ is a mapping $f$ which is 1–1, onto, and has the property that both $f$ and $f^{-1}$ are $k$-times differentiable.

*Eigenvector*; *right, left eigenvector*; *generalized eigenvector*: An eigenvector (or right eigenvector) $v$ of an $n \times n$ matrix $A$ is a nonzero vector which satisfies $Av = \lambda v$ or $(A - \lambda I)v = 0$ for some $\lambda \in \mathbb{C}$; $\lambda$ is the *eigenvalue* of $v$. A left eigenvector satisfies $v^T(A - \lambda I) = 0$ ($T$ denotes transpose). A generalized (right) eigenvector is one which satisfies a polynomial equation $(A - \lambda I)^k v = 0$ for some $1 \leq k \leq n$.

*Global*: This term is applied to properties which cannot be analyzed in arbitrarily small neighborhoods of a single point.

*Homeomorphism*: A homeomorphism is a $C^0$ diffeomorphism, i.e. a continuous mapping $f: M \to N$ with a continuous inverse.

*Interior*: The interior of a set $A$, int $A$, is the largest open set contained in $A$.

*Jordan normal form*: The Jordan normal form of an $n \times n$- matrix $A$ is a matrix $B = PAP^{-1}$, $P$ an invertible $n \times n$ matrix, with the nonzero entries of $B$ in the form of diagonal blocks

$$\begin{pmatrix} \lambda & 1 & & 0 \\ & \ddots & \ddots & \\ & & \ddots & 1 \\ 0 & & & \lambda \end{pmatrix}$$

*Limit point*: The point $x$ is a limit point of a set $A$ if every neighborhood of $x$ contains a point of $A - \{x\}$.

*Local*: A property is local if it can be analyzed in an arbitrarily small neighborhood of a given point.

*Manifold*: An $n$-dimensional manifold $M \subset \mathbb{R}^N$ is a set for which each $x \in M$ has a neighborhood $U$ for which there is a smooth invertible mapping (diffeomorphism) $\phi: \mathbb{R}^n \to U$ ($n \leq N$).

*Measure*: A measure $\mu$ on a set $A$ is a countably additive, nonnegative function defined on a $\sigma$-algebra of subsets of $A$.

*Metric space*: A metric space $A$ is a set together with a distance function $d: A \times A \to \mathbb{R}$ which satisfies: (1) $d(x, y) \geq 0$ with equality if and only if $x = y$; (2) $d(x, y) = d(y, x)$; and (3) $d(x, y) + d(y, z) \geq d(x, z)$ (the triangle inequality).

*Neighborhood*: A neighborhood of a point $x$ is a set $U$ which contains $x$ in its interior.

*Open set*: A set $U$ in a metric space is open if for each $x \in U$ there is an $\varepsilon > 0$ such that $d(x, y) < \varepsilon$ implies $y \in U$.

*Probability measure*: A measure $\mu$ on $A$ is a probability measure if $\mu(A) = 1$.

*Submanifold*: A submanifold $M$ of a manifold $N$ is a subset of $N$ which is a manifold.

*Tangent bundle*: The tangent bundle $TM$ of an $n$-dimensional manifold $M$ is a $2n$-dimensional manifold, the disjoint union of the tangent spaces of $M$. If $U \subset M$ is an open set and $\phi: \mathbb{R}^n \to U$ is a parametrization of $U$, then $\Phi: \mathbb{R}^n \times \mathbb{R}^n \to T_U M$ defined by $\Phi(x, v) = (\phi(x), D\phi(x) \cdot v)$ is a parametrization of $T_U M$.

*Transverse intersection*: Two submanifolds $N$, $P$ of $M$ intersect transversely at $x$ if $T_x N$ and $T_x P$ span $T_x M$.

## Frequently Used Notation

| | |
|---|---|
| $\mathbf{a} = \{a_i\}_{i=-\infty}^{\infty}$ or $\mathbf{a} = \{a_i\}_{i=0}^{\infty}$ | symbol sequences |
| $\mathrm{ad}(L)$ | adjoint action of linear vector field $L$ |
| $C^k, C^\infty$ | $k(\infty)$ times differentiable |
| $C^r(M, N)$ | $r$ times differentiable maps of $M$ to $N$ |
| $\mathbb{C}; \mathbb{C}^n$ | complex numbers; complex $n$-space |
| $Df(p)$ | (total) derivative of $f$ evaluated at the point $p$ |
| $D_x f, f_x, \dfrac{\partial f}{\partial x}, \dfrac{\partial^2 f}{\partial x\,\partial\mu}$ | partial derivatives |
| $\det |A|$ or $\det(A)$ | determinant of an $n \times n$ matrix $A$ |
| $\mathrm{Diff}^r(M)$ | $r$-times differentiable diffeomorphisms of $M$ |
| $E^s, E^u, E^c$ | stable, unstable, center subspaces for linearized systems |
| $e^{tA}$ | exponential of matrix $A$ |
| $f, g, F, G$ | mappings or right-hand sides of differential equations |
| $f|_\Lambda$ | $f$ restricted to the set $\Lambda$ |
| $\bar{f}$ | time averaged value of a time-dependent function $f$ |
| $\mathscr{F}$ | strong stable foliation |
| $\gamma$ | curve |
| $\Gamma$ | chain recurrent set; Cantor set |
| $H, E$ | Hamiltonian functions |
| $h$ | homeomorphism; entropy (Section 5.8) |
| $HD(\mu)$ | Hausdorff dimension of measure $\mu$ |
| $\lambda$ | eigenvalue |
| $\Lambda$ | indecomposable invariant set |
| $L$ | linear vector field |
| $M(t_0)$ | Melnikov function |
| $\mu$ | parameters; measure (Section 5.8) |

| $|\ |$ | Euclidean norm |
|---|---|
| $n_s, n_u, n_c$ | stable, unstable, center manifold dimensions |
| $P(P_\varepsilon, P_\mu)$ | (parametrized) Poincaré map |
| $\phi_t(x), \phi(x, t)$ | flow |
| $R$ | rectangle for Markov partition |
| $\rho(f)$ | rotation number of $f: S^1 \to S^1$ |
| $\mathbb{R}; \mathbb{R}^n$ | real numbers; real $n$-space |
| $S(f)$ | Schwarzian derivative of $f$ |
| $\sigma$ | shift map |
| $\Sigma$ | symbol space of sequences; cross-section of periodic orbit |
| $T^n$ | $n$-dimensional torus |
| $T_\Lambda \mathbb{R}^n$ | union of tangent spaces for points of $\Lambda$ |
| $\tau(\Gamma)$ | thickness of Cantor set $\Gamma$ |
| $\theta, \phi$ | angular variables |
| $V(x)$ | potential function of gradient |
| $W^s, W^u, W^c,$ $W^s_\varepsilon, W^u_\varepsilon, W^c_\varepsilon,$ $W^s_{loc}, W^u_{loc}, W^c_{loc}$ | global and local invariant subspaces (submanifolds) for system |
| $\dot{x}, \dfrac{dx}{dt}$ | time derivatives of $x$ |
| $\mathcal{X}^r(M)$ | $C^r$ vector fields on $M$ |
| $\Omega$ | nonwandering set |
| $\wedge$ | wedge product in $\mathbb{R}^2$: $(v_1, v_2) \wedge (w_1, w_2) = v_1 w_2 - v_2 w_1$ |
| $[f, g]$ | Lie bracket (of vector fields) |
| $\{f, g\}$ | Poisson bracket (of functions) |
| $\cap$ | intersection |
| $\cup$ | union |
| $\subset$ | contained in |
| $\in$ | a member of |
| $\forall$ | for all |

## Analytic Expressions of Frequently Studied Systems

Lorenz equations:
$$\dot{x} = \sigma(y - x)$$
$$\dot{y} = \rho x - y - xz$$
$$\dot{z} = -\beta z + xy$$

Duffing equation: $\ddot{x} + \delta\dot{x} - x + x^3 = \gamma \cos \omega t$

van der Pol's equation: $\ddot{x} + \alpha(x^2 - 1)\dot{x} + x = \beta \cos \omega t$

Bouncing ball map:

$$\phi_{j+1} = \phi_j + v_j \qquad\qquad \text{or} \quad F(\phi, v) =$$

$$v_{j+1} = \alpha v_j - \gamma \cos(\phi_j + v_j) \qquad (\phi + v, \alpha v - \gamma \cos(\phi + v))$$

Henon map: $F(x, y) = (y, 1 + bx - ay^2)$

Quadratic map: $f(y) = a - y^2$ or $f(y) = ay(1 - y)$

# References

Abraham, R. H., and Marsden, J. E. [1978]. *Foundations of Mechanics*. Benjamin/Cummings: Reading, MA.

Abraham, R. H., and Robbin, J. [1967]. *Transversal Mappings and Flows*. Benjamin: Reading, MA.

Afraimovich, V. S., Bykov, V. V., and Silnikov, L. P. [1983]. On structurally unstable attracting limit sets of Lorenz attractor type. *Trans. Moscow. Math. Soc.*, **44** (2), 153–216.

Aizawa, Y., and Saitô, N. [1972]. On the stability of isolating integrals, I. *J. Phys. Soc. Jap.*, **32**, 1636–1640.

Allwright, D. J. [1977]. Harmonic balance and the Hopf bifurcation. *Math. Proc. Camb. Phil. Soc.*, **82**, 453–467.

Andronov, A. A., Leontovich, E. A., Gordon, I. I., and Maier, A. G. [1971]. *Theory of Bifurcations of Dynamic Systems on a Plane*. Israel Program of Scientific Translations, Jerusalem.

Andronov, A. A., Leontovich, E. A., Gordon, I. I., and Maier, A. G. [1973]. *Theory of Dynamic Systems on a Plane*. Israel Program of Scientific Translations, Jerusalem.

Andronov, A. A., and Pontryagin, L. [1937]. Systèmes Grossiers. *Dokl. Akad. Nauk. SSSR*, **14**, 247–251.

Andronov, A. A., Vitt, E. A., and Khaiken, S. E. [1966]. *Theory of Oscillators*. Pergamon Press: Oxford.

Antman, S. S. [1984]. *Nonlinear Problems of Elasticity* (Forthcoming book).

Arnéodo, A., Coullet, P., and Tresser, C. [1981]. Possible new strange attractors with spiral structure. *Comm. Math. Phys.*, **79**, 573–579.

Arnéodo, A., Coullet, P., and Tresser, C. [1982]. Oscillations with chaotic behavior: an illustration of a theorem by Šilnikov. *J. Stat. Phys.*, **27**, 171–182.

Arnold, V. I. [1963a]. Proof of A. N. Kolmogorov's theorem on the preservation of quasiperiodic motions under small perturbations of the Hamiltonian. *Russ. Math. Surv.*, **18** (5), 9–36.

Arnold, V. I. [1963b]. Small divisor problems in classical and celestial mechanics. *Russ. Math. Surv.*, **18** (6), 85–192.

Arnold, V. I. [1964]. Instability of dynamical systems with several degrees of freedom. *Sov. Math. Dokl.*, **5**, 581–585.

Arnold, V. I. [1965]. Small denominators, I: Mappings of the circumference onto itself. *AMS Transl. Ser. 2*, **46**, 213–284.

Arnold, V. I. [1972]. Lectures on bifurcations in versal families. *Russ. Math. Surv.*, **27**, 54–123.

Arnold, V. I. [1973]. *Ordinary Differential Equations*, M.I.T. Press: Cambridge, MA. (Russian original, Moscow, 1971.)

Arnold, V. I. [1977]. Loss of stability of self oscillations close to resonances and versal deformations of equivariant vector fields. *Funct. Anal. Appl.*, **11** (2), 1–10.

Arnold, V. I. [1978]. *Mathematical Methods of Classical Mechanics*. Springer-Verlag: New York, Heidelberg, Berlin. (Russian original, Moscow, 1974).

Arnold, V. I. [1981]. *Singularity Theory: Selected Papers*. London Mathematical Society Lecture Notes, Vol. 53. Cambridge University Press: Cambridge.

Arnold, V. I. [1982]. *Geometrical Methods in the Theory of Ordinary Differential Equations*. Springer-Verlag: New York, Heidelberg, Berlin. (Russian original, Moscow, 1977.)

Arnold, V. I., and Avez, A. [1968]. *Ergodic Problems of Classical Mechanics*. W. A. Benjamin: New York.

Aronson, D. G., Chory, M. A., Hall, G. R., and McGeehee, R. P. [1980]. A discrete dynamical system with subtly wild behavior. In *New Approaches to Nonlinear Problems in Dynamics*, P. Holmes (ed.), pp. 339–359. S.I.A.M. Publications: Philadelphia.

Aronson, D. G., Chory, M. A., Hall, G. R., and McGeehee, R. P. [1982]. Bifurcations from an invariant circle for two-parameter families of maps of the plane: a computer assisted study. *Comm. Math. Phys.*, **83**, 303–354.

Aubrey, S. [1983]. The twist map, the extended Frenkel-Kontorova model, and the Devil's staircase. *Physica*, **7D**, 240–258.

Aubrey, S., and LeDaeron, P. [1983]. The discrete Frenkel-Kontorova model and its extensions. *Physica*, **8D**, 381–422.

Auchmuchty, J. F. G., and Nicolis, G. [1975]. Bifurcation analysis of nonlinear reaction diffusion equations, I. *Bull. Math. Biol.*, **37**, 323–365.

Auchmuchty, J. F. G., and Nicolis, G. [1976]. Bifurcation analysis of nonlinear reaction diffusion equations, III. *Bull. Math. Biol.*, **38**, 325–349.

Bajaj, A. K., and Sethna, P. R. [1982]. Bifurcations in three dimensional motions of articulated tubes, I and II. *Trans. ASME J. Appl. Mech.* **49**, 606–611 and 612–618.

Bajaj, A. K., Sethna, P. R., and Lundgren, T. S. [1980]. Hopf bifurcation phenomena in tubes carrying a fluid. *SIAM J. Appl. Math.*, **39** (2), 213–230.

Beaman, J. J., and Hedrick, J. K. [1980]. Freight car harmonic response: a simplified nonlinear method. In *Nonlinear System Analysis and Synthesis*, Vol. 2, R. V. Rammath, J. K. Hedrick and H. M. Paynter (eds.), pp. 177–195. A.S.M.E. Publications: New York.

Bellman, R., and Kalaba, R. (eds.) [1964]. *Selected Papers on Mathematical Trends in Control Theory*. Dover: New York.

Belykh, V. N. [1985]. Bifurcation of separatrices of a saddle point of the Lorenz system. *Diff. Eq.*, **20** (10), 1184–1191.

Bennetin, G., Casartelli, M., Galani, L., Giorgilli, A., and Strelcyn, J. M. [1978]. On the reliability of numerical study of stochasticity, I: Existence of time averages. *Il. Nuovo Cimento*, **44B** (1), 183–195.

Bennetin, G., Casartelli, M., Galgani, L., Giorgilli, A., and Strelcyn, J. M. [1979]. On the reliability of numerical studies of stochastically, II: Identification of time averages. *Il. Nuovo Cimento*, **50B**, 211–232.

Bernhardt, C. [1982]. Rotation intervals for a class of endomorphisms of the circle. *Proc. Lond. Math. Soc.*, **45**, 258–280.

Bernoussou, J. [1977]. *Point Mapping Stability*. Pergamon: Oxford.

Birkhoff, G. D. [1927]. *Dynamical Systems*. A.M.S. Publications: Providence.

Birkhoff, G. D. [1932a]. Sur l'existence de régions d'instabilité en dynamique. *Ann. Inst. H. Poincaré*, **2**, 369–386.

Birkhoff, G. D. [1932b]. Sur quelques courbes fermées remarquables. *Bull. Soc. Math. France*, **60**, 1–26.

Blanchard, P. [1984]. Complex analytic dynamics in the Riemann sphere. *Bull. Amer. Math. Soc.*, **11**, 85–141.

Block, L., Guckenheimer, J. Misiurewicz, M., and Young, L.-S. [1979]. Periodic points of one-dimensional maps. Springer Lecture Notes in Mathematics, Vol. 819, 18–34, Springer-Verlag: New York, Heidelberg, Berlin.

Bogdanov, R. I. [1975]. Versal deformations of a singular point on the plane in the case of zero eigenvalues. *Functional Analysis and Its Applications*, **9** (2), 144–145.

Bowen, R. [1970]. Markov partitions for Axiom A diffeomorphisms. *Amer. J. Math.*, **92**, 725–747.

Bowen, R. [1978]. *On Axiom A Diffeomorphisms*. CBMS Regional Conference Series in Mathematics, Vol. 35. A.M.S. Publications: Providence.

Bowen, R., and Ruelle, D. [1975]. The ergodic theory of Axiom A flows. *Invent. Math.*, **79**, 181–202.

Boyland, P. L. [1983]. Bifurcations of circle maps, Arnold tongues, bistability, and rotation intervals. Boston University (preprint).

Boyland, P. L., and Hall, G. R. [1985]. Invariant circles and the order structure of periodic orbits in monotone twist maps. Boston University (preprint).

Brandstater, A., Swift, J., Swinney, H., Wolf, A., Farmer, J., Jen, E., and Crutchfield, J. [1983]. Low-dimensional chaos in a system with Avrogadro's number of degrees of freedom. *Phys. Rev. Lett.*, **51**, 1442–1445.

Braun, M. [1978]. *Differential Equations and Their Applications*. Springer-Verlag: New York, Heidelberg, Berlin.

Brin, M., and Katok, A. B. [1983]. *On Local Entropy*. Springer Lecture Notes in Mathematics, Vol. 1007, 30–38. Springer-Verlag: New York, Heidelberg, Berlin.

Byrd, P. B., and Friedman, M. D. [1971]. *Handbook of Elliptic Integrals for Scientists and Engineers*. Springer-Verlag: New York, Heidelberg, Berlin.

Campinino, M., and Epstein, H. [1981]. On the existence of Feigenbaum's fixed point. *Comm. Math. Phys.*, **79** (2), 261–302.

Campbell, D. K. and Rose, H. A. (eds.) [1983]. *Order in Chaos*. Proceedings of a conference held at Los Alamos National Lab., North-Holland: Amsterdam, and *Physica*, **7D**.

Carr, J. [1981]. *Applications of Center Manifold Theory*. Springer-Verlag: New York, Heidelberg, Berlin.

Carr, J., and Muncaster, R. [1982]. Centre manifolds and amplitude expansions, I: Ordinary differential equations. *J. Diff. Eqns.*, **50**, 280–288 [1983].

Cartwright, M. L. [1948]. Forced oscillations in nearly sinusoidal systems. *J. Inst. Elec. Eng.*, **95**, 88–94.

Cartwright, M. L., and Littlewood, J. E. [1945]. On nonlinear differential equations of the second order, I: The equation $\ddot{y} + k(1 - y^2)\dot{y} + y = b\lambda k \cos(\lambda t + a)$, $k$ large. *J. Lond. Math. Soc.*, **20**, 180–189.

Casdagli, M. [1985]. Periodic orbits for dissipative twist maps. University of Warwick (preprint).

Chenciner, A. [1982]. Bifurcations de difféomorphismes de $R^2$ au voisinage d'un point fixe elliptique. In *Chaotic behavior of deterministic systems*, pp. 273–348. North Holland: Amsterdam [1983].

Chenciner, A. [1982a]. Sur un énoncé dissipatif du théorème géométrique de Poincaré-Birkhoff. *C.R. Acad. Sci. Paris I*, **294**, 243–245.

Chenciner, A. [1982b]. Points homoclines au voisinage d'un bifurcation de Hopf dégénerée de difféomorphismes de $R^2$. *C.R. Acad. Sci. Paris I*, **294**, 269–272.

Chenciner, A. [1982c]. Points périodiques de longues périodes au voisinage d'une bifurcation de Hopf dégénérée de difféomorphismes de $R^2$. *C.R. Acad. Sci. Paris* I, **294**, 661–663.

Chenciner, A. [1983a]. Orbites périodiques et ensembles de Cantor invariantes d'Aubry-Mather au voisinage d'une bifurcation de Hopf dégénérée de difféomorphismes de $R^2$. *C.R. Acad. Sci. Paris* I, **297**, 465–467.

Chenciner, A. [1983b]. Bifurcations de difféomorphismes de $R^2$ au voisinage d'un point fixe elliptique. In *Les Houches Summer School Proceedings*. R. Helleman and G. Iooss (eds.) North Holland: Amsterdam.

Chenciner, A. [1985a]. Bifurcations de points fixes elliptiques, I: Courbes invariantes. *Publ. Math. IHES*, **61**, 67–147.

Chenciner, A. [1985b]. Bifurcations de points fixes elliptiques. II: Orbites périodiques et ensembles de Cantor invariantes. *Invent. Math.*, **80**, 81–106.

Chern, S. S., and Smale, S. (eds.) [1970]. *Global Analysis*. A.M.S. Proceedings of Symposium on Pure Mathematics, Vol. 14. University of California Press: Berkeley.

Chillingworth, D. R. J. (ed.) [1971]. *Symposium on Differential Equations and Dynamical Systems, University of Warwick, 1968–9*. Springer Lecture Notes in Mathematics, Vol. 206. Springer-Verlag: New York, Heidelberg, Berlin.

Chillingworth, D. R. J. [1976]. *Differentiable Topology with a View to Applications*. Pitman: London.

Chirikov, B. V. [1979]. A universal instability of many dimensional ocillator systems. *Phys. Rep.*, **52**, 263–379.

Choquet-Bruhat, Y., Dewitt-Morette, C., and Dillard-Bleick, M. [1977]. *Analysis, Manifolds and Physics*. North Holland: Amsterdam.

Chow, S. N., and Hale, J. K. [1982]. *Methods of Bifurcation Theory*. Springer-Verlag: New York, Heidelberg, Berlin.

Chow, S. N., Hale, J. K., and Mallet-Paret, J. [1980]. An example of bifurcation to homoclinic orbits. *J. Diff. Eqns.*, **37**, 351–373.

Chow, S. N., and Mallet-Paret, J. [1977]. Integral averaging and bifurcation. *J. Diff. Eqns.*, **26**, 112–159.

Churchill, R. C. [1982]. On proving the non-integrability of a Hamiltonian system. In *The Riemann Problem, Complete Integrability and Arithmetic Applications*. Springer Lecture Notes in Mathematics, Vol. 925, 103–122. Springer Verlag: New York, Heidelberg, Berlin.

Churchill, R. C., Kummer, M., and Rod, D. L. [1982]. On averaging, reduction and symmetry in Hamiltonian systems. *J. Diff. Eqns.*, **49**, 359–414 [1983].

Coddington, E. A., and Levinson, N. [1955]. *Theory of Ordinary Differential Equations*. McGraw-Hill: New York.

Cohen, A. H., Holmes, P. J., and Rand, R. H. [1982]. The nature of the coupling between segmental oscillators of the lamprey spinal generator for locomotion: a mathematical model. *J. Math. Biol.*, **13**, 345–369.

Cohen, D. S., and Neu, J. C. [1979]. Interacting oscillatory chemical reactors. In *Bifurcation Theory and Applications in Scientific Disciplines*, O. Gurel and O. S. Rössler (eds.), pp. 332–337. N.Y. Acad. Sci.: New York.

Collet, P., and Eckmann, J.-P. [1980]. *Iterated Maps on the Interval as Dynamical Systems*. Progress on Physics, Vol. I. Birkhäuser-Boston: Boston.

Collet, P., Eckmann, J.-P., and Koch, H. [1981]. Period doubling bifurcations for families of maps on $\mathbb{R}^n$. *J. Stat. Phys.*, **25** (1), 1–14.

Collet, P., Eckmann, J.-P., and Lanford, O. E. [1980]. Universal properties of maps on an interval. *Comm. Math. Phys.*, **76**, 211–254.

Conley, C. [1978]. *Isolated Invariant Sets and the Morse Index*. CBMS Regional Conferences in Mathematics, Vol. 38. A.M.S. Publications: Providence.

Cooperrider, N. K. [1980]. Nonlinear behavior in rail vehicle dynamics. In *New Approaches to Nonlinear Problems in Dynamics*, P. J. Holmes (ed.), pp. 173–194. S.I.A.M. Publications: Philadelphia.

Cornfeld, I. P., Fomin, S. V., and Sinai, Ya. G. [1982]. *Ergodic Theory*. Springer-Verlag: New York, Heidelberg, Berlin.

Coullet, P., Gambaudo, J. M., and Tresser, C. [1984]. Une nouvelle bifurcation de codimension 2: le collage de cycles. *C.R. Acad. Sci. Paris* I, **299**, 253–256.

Curry, J. [1978]. A generalized Lorenz system. *Comm. Math. Phys.*, **60**, 193–204.

Cushman, R., and Deprit, A. [1980]. Normal forms and representation theory. Preprint No. 180. Department of Mathematics, Rijksuniversiteit Utrecht.

Cushman, R., Deprit, A., and Mosak, R. [1983]. Normal forms and representation theory. (in preparation.)

Cushman, R., and Rod, D. L. [1982]. Reduction of the semisimple 1:1 resonance. *Physica*, **6D**, 105–112.

Cushman, R., and Sanders, J. A. [1983]. A codimension two bifurcation with a third order Picard–Fuchs equation. Rapport No. 249, Subfaculteit Wiskunde en Informatica, Vrije Universiteit, Amsterdam.

Cvitanovic, P. (ed.) [1984]. *Universality in Chaos*. Adam Hilger: Bristol.

DaCosta, L. N., Knobloch, E., and Weiss, N. O. [1981]. Oscillations in double-diffusive convection. *J. Fluid Mech.*, **109**, 25–43.

Denjoy, A. [1932]. Sur les courbes définies par les équations différentielles à la surface du tore. *J. Math.*, **17** (IV), 333–375.

Deprit, A. [1982]. Méthodes nouvelles de simplification en mécanique céleste (preprint).

Devaney, R. L. [1985]. *An Introduction to Chaotic Dynamical Systems*. Benjamin/Cummings: Menlo Park, CA.

Devaney, R., and Nitecki, Z. [1979]. Shift automorphisms in the Hénon mapping. *Comm. Math. Phys.*, **67**, 137–148.

DiPrima, R. C., Eagles, P. M., and Sijbrand, J. [1982]. Interaction of axisymmetric and nonaxisymmetric disturbances in the flow between concentric counterrotating cylinders: bifurcations near multiple eigenvalues. *Proceedings of NATCAM, Ninth U.S. Congress on Applied Mechanics*. Ithaca, June 21–25, 1982 (abstract).

DiPrima, R. C., Eagles, P. M., and Sijbrand, J. [1983]. Interaction of axisymmetric and nonaxisymmetric disturbances in the flow between concentric counter-rotating cylinders. (in preparation).

Douady, A. [1982/3]. Systèmes dynamiques holomorphes. *Seminaire Bourbaki 35 année*, **599**.

Douady, A., and Hubbard, J. H. [1982]. Itération du polynômes quadratiques complexes. *C.R. Acad. Sci. Paris* I, **294**, 123–126.

Dowell, E. H. [1966]. Nonlinear oscillations of a fluttering plate. *AIAA J.*, **4**, 1267–1275.

Dowell, E. H. [1975]. *Aeroelasticity of Plates and Shells*. Noordhoff International Publishing: Leyden.

Dowell, E. H. [1980]. Nonlinear aeroelasticity. In *New Approaches to Nonlinear Problems in Dynamics*, P. J. Holmes (ed.), pp. 147–172b. S.I.A.M. Publications: Philadelphia.

Dowell, E. H. [1982]. Flutter of a buckled plate as an example of chaotic motion of a deterministic autonomous system. *J. Sound Vib.*, **85**, 333–344.

Duffing, G. [1918]. *Erzwungene Schwingungen bei Veränderlicher Eigenfrequenz*. F. Vieweg u. Sohn: Braunschweig.

Euler, L. [1744]. Additamentum I de Curvis Elasticis, Methodus Inveniendi Lineas Curvas Maximi Minimivi Proprietate Gandentes. In [1960] *Opera Ommia I*, Vol. 24, pp. 231–297. Füssli: Zurich.

Evans, J. W., Fenichel, N., and Feroe, J. A. [1982]. Double impulse solutions in nerve axon equations. *SIAM J. Appl. Math.*, **42** (2), 219–234.

Farmer, J. D., Ott, E., and Yorke, J. A. [1983]. The dimension of chaotic attractors. *Physica*, **7D**, 153–180.

Feigenbaum, M. J. [1978]. Quantitative universality for a class of nonlinear transformations. *J. Stat. Phys.*, **19**, 25–52.

Feigenbaum, M. J. [1979]. The onset spectrum of turbulence, *Phys. Lett.*, **74A**, 375–378.

Feigenbaum, M. J. [1980]. Universal behavior in nonlinear systems. *Los Alamos Sci.*, **1**, 4–27.

Feigenbaum, M. J., Kadanoff, L. P., and Shenker, S. J. [1982]. Quasiperiodicity in dissipative systems: a renormalization group. *Physica*, **5D**, 370–386.

Feroe, J. A. [1982]. Existence and stability of multiple impulse solutions of a nerve equation. *SIAM J. Appl. Math.*, **42** (2), 235–246.

Fiala, V. [1976]. *Solution of Nonlinear Vibration Systems by Means of Analogue Computers.* Monograph 19, National Research Institute for Machine Design, Bechovice, Prague.

Flaherty, J. E., and Hoppensteadt, F. C. [1978]. Frequency entrainment of a forced van der Pol oscillator. *Stud. Appl. Math.*, **18**, 5–15.

Franceschini, V. [1982]. Bifurcations of tori and phase locking in a dissipative system of differential equations. *Physica*, **6D**, 285–304 [1983].

Franks, J. M. [1970]. Anosov diffeomorphisms. In *Global Analysis*, S. S. Chern and S. Smale (eds.). University of California Press: Berkeley.

Franks, J. M. [1974]. Time dependent stable diffeomorphisms. *Invent. Math.*, **24**, 163–172.

Franks, J. M. [1982]. *Homology and Dynamical Systems.* CBMS Regional Conference Series in Mathematics, Vol. 49, A.M.S. Publications: Providence.

Fredrickson, P., Kaplan, J. L., Yorke, E. D., and Yorke, J. A. [1982]. The Liapunov dimension of strange attractors. University of Maryland, College Park, MD (preprint).

Gambaudo, J. M., Glendinning, P., and Tresser, C. [1984]. Collage de cycles et suites de Farey. *C.R. Acad. Sci. Paris* I, **299**, 711–714.

Gambaudo, J. M., Glendinning, P., and Tresser, C. [1985]. The gluing bifurcation, I: symbolic dynamics of the closed curves. Université de Nice (preprint).

Garrido, L. (ed.) [1983]. *Proceedings, Sitges 1982*, Springer-Verlag: Berlin, Heidelberg, New York.

Gavrilov, N. K., and Šilnikov, L. P. [1972]. On three-dimensional dynamical systems close to systems with a structurally unstable homoclinic curve, I. *Math. USSR Sb.*, **88** (4), 467–485.

Gavrilov, N. K., and Šilnikov, L. P. [1973]. On three-dimensional dynamical systems close to systems with a structurally unstable homoclinic curve, II. *Math. USSR Sb.*, **90** (1), 139–156.

Gillies, A. W. [1954]. On the transformation of singularities and limit cycles of the variational equations of van der Pol. *Quart. J. Mech. Appl. Math.*, **7**, 152–167.

Gilmore, R. [1981]. *Catastrophe Theory for Scientists and Engineers.* Wiley: New York.

Glass, L. and Perez, R. [1982]. The fine structure of phase locking. *Phys. Rev. Lett.*, **48**, 1772–1775.

Glendinning, P. [1985]. Bifurcations near homoclinic orbits and symmetry. *Phys. Lett.* A (to appear).

Glendinning, P., and Sparrow, C. [1985]. T-points: a codimension two heteroclinic bifurcation. D.A.M.T.P., Cambridge, (preprint).

Glendinning, P., and Tresser, C. [1985]. Heteroclinic loops leading to hyperchaos. *J. Phys., Lett., Paris*, **46** (8), 347–352.

Goldstein, H. [1980]. *Classical Mechanics*, 2nd ed. Addison-Wesley, Reading.

Gollub, J. P., and Benson, S. V. [1980]. Many routes to turbulent convection. *J. Fluid Mech.*, **100**, 449–470.

Golubitsky, M., and Guillemin, V. [1973]. *Stable Mappings and Their Singularities.* Springer-Verlag: New York, Heidelberg, Berlin.

Golubitsky, M., and Langford, W. F. [1981]. Classification and unfoldings of degenerate Hopf bifurcations. *J. Diff. Eqns.*, **41**, 375–415.

Golubitsky, M., and D. Schaeffer, D. [1979a]. A theory for imperfect bifurcation via singularity theory. *Comm. Pure Appl. Math.*, **32**, 21–98.

Golubitsky, M., and Schaeffer, D. [1979b]. Imperfect bifurcation in the presence of symmetry. *Comm. Math. Phys.*, **67**, 205–232.

Golubitsky, M., and Schaeffer, D. [1985]. *Singularities and Groups in Bifurcation Theory*, I. Springer Verlag: New York, Heidelberg, Berlin.

Golubitsky, M., and Stewart, I. [1987]. *Singularities and Groups in Bifurcation Theory*, *II* (forthcoming book).

Grassberger, P., and Procaccia, I. [1983]. Characterization of strange attractors. *Phys. Rev. Lett.*, **50**, 346–349.

Greene, J. M. [1980]. Method for determining a stochastic transition. *J. Math. Phys.*, **20** (6), 1183–1201.

Greenspan, B. D. [1981]. Bifurcations in periodically forced oscillations: subharmonics and homoclinic orbits. Ph.D. thesis, Center for Applied Mathematics, Cornell University.

Greenspan, B. D., and Holmes, P. J. [1982]. Homoclinic orbits, subharmonics and global bifurcations in forced oscillations. In *Nonlinear Dynamics and Turbulence*, G. Barenblatt, G. Iooss, and D. D. Joseph (eds.), Pitman: London.

Greenspan, B. D., and Holmes, P. J. [1983]. Repeated resonance and homoclinic bifurcations in a periodically forced family of oscillators, *SIAM J. Math. Anal.* **15**, 69–97 [1984].

Gruendler, J. [1982]. A generalization of the method of Melnikov to arbitrary dimension. Ph.D. thesis, University of North Carolina, Chapel Hill.

Gruendler, J. [1985]. The existence of homoclinic orbits and the method of Melnikov for systems in $R^n$. *SIAM J. Math. Anal.*, **16**, 907–931.

Guckenheimer, J. [1973]. Bifurcation and catastrophe. In *Dynamical Systems*, M. M. Peixoto (ed.), Academic Press: New York.

Guckenheimer, J. [1976]. A strange strange attractor. In *The Hopf Bifurcation and its Applications*, J. E. Marsden and M. McCracken (eds.), pp. 368–381. Springer-Verlag: New York, Heidelberg, Berlin.

Guckenheimer, J. [1977]. On the bifurcation of maps of the interval. *Invent. Math.*, **39**, 165–178.

Guckenheimer, J. [1979]. Sensitive dependence on initial conditions for one-dimensional maps. *Comm. Math. Phys.*, **70**, 133–160.

Guckenheimer, J. [1980a]. Symbolic dynamics and relaxation oscillations. *Physica*, **1D**, 227–235.

Guckenheimer, J. [1980b]. Bifurcations of dynamical systems. In *Dynamical Systems*, CIME Lectures, Bressanone, Italy, June 1978, pp. 115–231. Progress in Mathematics, No. 8. Birkhäuser-Boston: Boston.

Guckenheimer, J. [1981]. On a codimension two bifurcation. In *Dynamical Systems and Turbulence*, D. A. Rand and L. S. Young (eds.), pp. 99–142. Springer Lecture Notes in Mathematics, Vol. 898. Springer-Verlag: New York, Heidelberg, Berlin.

Guckenheimer, J. [1984]. Dimension estimates for attractors. *Contemp. Math.*, **28**, 357–367.

Guckenheimer. J. [1986]. Strange attractors in fluids: another view. *Ann. Rev. Fluid Mech.* **18**, 15–31.

Guckenheimer, J., and Buzyna, G. [1983]. Dimension measurements for geostrophic turbulence. *Phys. Rev. Lett.*, **51**, 1438–1441.

Guckenheimer, J., and Knobloch, E. [1983]. Nonlinear convection in a rotating layer: amplitude expansions and center manifolds. *Geophys. and Astrophys. Fluid Dyn.*, **23**, 247–272.

Guckenheimer, J., and Williams, R. F. [1979]. Structural stability of Lorenz attractors. *Publ. Math. IHES*, **50**, 59–72.

Gumowski, I., and Mira, C. [1980]. *Recurrences and Discrete Dynamical Systems*. Springer Lecture Notes in Mathematics, Vol. 809. Springer-Verlag: New York, Heidelberg, Berlin.

Gurel, O., and Rössler, O. E. (eds.) [1979]. *Bifurcation Theory and Applications in Scientific Disciplines*. Annals of the New York Academy of Sciences, Vol. 316. New York Academy of Sciences: New York.

Hale, J. K. [1963]. *Oscillations in Nonlinear Systems*. McGraw-Hill: New York.

Hale, J. K. [1969]. *Ordinary Differential Equations*. Wiley: New York.

Hall, G. R. [1983]. A topological version of a theorem of Mather on twist maps. MRC Technical Report, University of Wisconsin, Madison, WI.

Hamilton, R. S. [1982]. The inverse function theorem of Nash and Moser. *Bull. Amer. Math. Soc.*, **7** (1), 65–222.

Hao, B.-L. (ed.) [1984]. *Chaos*. World Scientific: Singapore.

Hartlan, R. T., and Currie, I. G. [1970]. Lift oscillator model of a vortex-induced vibration. *Proc. ASCE*, **EM5**, 577–591.

Hartman, P. [1964]. *Ordinary Differential Equations*. Wiley: New York.

Hassard, B. D., Kazarinoff, N. D., and Wan, Y.-H. [1980]. *Theory and Applications of the Hopf Bifurcation*. Cambridge University Press: Cambridge.

Hassard, B. D., and Wan. Y.-H. [1978]. Bifurcation formulae derived from center manifold theory. *J. Math. Anal. Appl.*, **63** (1), 297–312.

Hastings, S. P. [1982]. Single and multiple pulse waves for the Fitzhugh–Nagumo equations. *SIAM J. Appl. Math.*, **42** (2), 247–260.

Hayashi, C. [1964]. *Nonlinear Oscillations in Physical Systems*. McGraw-Hill: New York.

Helleman, R. H. G. (ed.) [1980]. *Nonlinear Dynamics*. Annals of the New York Academy of Sciences.Vol. 357. New York Academy of Sciences: New York.

Helleman, R. H. G., Iooss, G., and Stora (eds.) [1983]. *Chaotic Behavior of Deterministic Systems*. Proceedings of the Cours a l'Ecole des Houches, July 1981. North Holland: Amsterdam.

Hénon, M. [1976]. A two-dimensional mapping with a strange attractor. *Comm. Math. Phys.*, **50**, 69–77.

Hénon, M., and Heiles, C. [1964]. The applicability of the third integral of motion: some numerical experiments. *Astron. J.*, **69**, 73.

Henry, D. [1981]. *Geometric Theory of Semilinear Parabolic Equations*. Springer Lectures Notes in Mathematics, Vol. 840. Springer-Verlag: New York, Heidelberg, Berlin.

Herman, M. R. [1976]. Sur la conjugaison Différentiable des Diffeomorphismes du Cercle à des Rotations. Thesis, Université de Paris, Orsay and *Publ. Math. IHES* **49**.

Herman, M. R. [1977]. Mesure de Lesbesque et Nombre de Rotation. In *Geometry and Topology*, J. Palis and M. de Carmo (eds.), pp. 271–293. Lecture Notes in Mathematics, Vol. 597. Springer-Verlag: New York, Heidelberg, Berlin.

Herman, M. R. [1983]. Sur les Courbes Invariantes par les difféomorphismes de l'Anneau, I, *Astérisque*, 103–104.

Hirsch, M. W. [1976]. *Differential Topology*. Springer-Verlag: New York, Heidelberg, Berlin.

Hirsch, M. W., and Pugh, C. C. [1970]. Stable manifolds and hyperbolic sets. *Proc. Symp. Pure. Math.*, **14**, 133–163.

Hirsch, M. W., Pugh, C. C., and Shub, M. [1977]. *Invariant Manifolds*. Springer Lectures Notes in Mathematics, Vol. 583. Springer-Verlag: New York, Heidelberg, Berlin.

Hirsch, M. W., and Smale, S. [1974]. *Differential Equations, Dynamical Systems and Linear Algebra*. Academic Press: New York.

Hockett, K., and Holmes, P. J. [1986]. Josephson's junction. annulus maps, Birkhoff attractors, horseshoes and rotation sets. *Ergodic Theory Dyn. Syst.* (in press).

Hocking, J. G., and Young, G. S. [1961]. *Topology*. Addison-Wesley: Reading, MA.

Holmes, C., and Holmes, P. J. [1981]. Second order averaging and bifurcations to subharmonics in Duffing's equation. *J. Sound Vib.*, **78**, 161–174.

Holmes, P. J. [1977]. Bifurcations to divergence and flutter in flow-induced oscilla-tions: a finite-dimensional analysis. *J. Sound Vib.*, **53** (4), 471–503.

Holmes, P. J. [1979a]. A nonlinear oscillator with a strange attractor. *Phil. Trans. Roy. Soc. A*, **292**, 419–448.

Holmes, P. J. [1979b]. Domains of stability in a wind induced oscillation problem. *Trans. ASME. J. Appl. Mech.*, **46**, 672–676.

Holmes, P. J. (ed.) [1980a]. *New Approaches to Nonlinear Problems in Dynamics.* S.I.A.M. Publications: Philadelphia.

Holmes, P. J. [1980b]. Averaging and chaotic motions in forced oscillations. *SIAM J. Appl. Math.*, **38**, 65–80; Errata and addenda. *SIAM J. Appl. Math.*, **40**, 167–168.

Holmes, P. J. [1980c]. A strange family of three-dimensional vector fields near a degenerate singularity. *J. Diff. Eqns.*, **37**, 382–404.

Holmes, P. J. [1980d]. Unfolding a degenerate nonlinear oscillator: a codimension two bifurcation. In *Nonlinear Dynamics*, R. H. G. Helleman (ed.), pp. 473–488. New York Academy of Sciences: New York.

Holmes, P. J. [1981a]. Center manifolds, normal forms and bifurcations of vector fields with application to coupling between periodic and steady motions. *Physica*, **2D**, 449–481.

Holmes, P. J. [1981b]. Space- and time-periodic perturbations of the sine-Gordon equation. In *Dynamical Systems and Turbulence*, D. A. Rand and L.-S. Young (eds.), pp. 164–191. Springer Lecture Notes in Mathematics, Vol. 898. Springer-Verlag: New York, Heidelberg, Berlin.

Holmes, P. J. [1982a]. The dynamics of repeated impacts with a sinusoidally vibrating table. *J. Sound Vib.*, **84**, 173–189.

Holmes, P. J. [1982b]. Proof of nonintegrability for the Henon–Heiles Hamiltonian near an exceptional integrable case. *Physica*, **5D**, 335–347.

Holmes, P. J. [1986]. Chaotic motions in a weakly nonlinear model for surface waves. *J. Fluid Mech.* (in press).

Holmes, P. J., and Marsden, J. E. [1978]. Bifurcations to divergence and flutter in flow-induced oscillations: an infinite-dimensional analysis. *Automatica*, **14** (4), 367–384.

Holmes, P. J., and Marsden, J. E. [1981]. A partial differential equation with infinitely many periodic orbits: chaotic oscillations of a forced beam. *Arch. Ration. Mech. Anal.* **76**, 135–166.

Holmes, P. J., and Marsden, J. E. [1982a]. Horseshoes in perturbations of Hamiltonians with two degrees of freedom. *Comm. Math. Phys.*, **82**, 523–544.

Holmes, P. J., and Marsden, J. E. [1982b]. Melnikov's method and Arnold diffusion for perturbations of integrable Hamiltonian systems. *J. Math. Phys.*, **23** (4), 669–675.

Holmes, P. J., and Marsden, J. E. [1983a]. Horseshoes and Arnold diffusion for Hamiltonian systems on Lie groups. *Indiana Univ. Math. J.* **32**, 273–310.

Holmes, P. J., Marsden, J. E., and Scheurle, J. [1986]. On averaging and exponentially small Melnikov functions (in preparation).

Holmes, P. J., and Rand, D. A. [1976]. The bifurcations of Duffings' equation: an application of catastrophe theory. *J. Sound Vib.*, **44** (2), 237–253.

Holmes, P. J., and Rand, D. A. [1978]. Bifurcations of the forced van der Pol oscillator. *Quart. Appl. Math.*, **35**, 495–509.

Holmes, P. J., and Rand, D. A. [1980]. Phase portraits and bifurcations of the non-linear oscillator $\ddot{x} + (\alpha + \gamma x^2)\dot{x} + \beta x + \delta x^3 = 0$. *Int. J. Nonlinear Mech.*, **15**, 449–458.

Holmes, P. J., and Whitley, D. C. [1983a]. On the attracting set for Duffing's equation, I: Analytical methods for small force and damping. In *Partial Differential Equations and Dynamical Systems*, W. E. Fitzgibbon III (ed.), pp. 211–240. Pitman: London [1984].

Holmes, P. J., and Whitley, D. C. [1983b]. On the attracting set for Duffing's equation, II: A geometrical model for moderate force and damping. *Physica*, **7D**, 111–123.

Hopf, E. [1942]. Abzweigung einer periodischen Lösung von einer stationären Lösung eines differential-systems. *Ber. Math.-Phys. Kl. Sächs Acad. Wiss. Leipzig*, **94**, 1–22 and *Ber. Verh. Sachs. Acad. Wiss. Leipzig Math.-Nat. Kl.*, **95** (1), 3–22. (A translation of this paper, and comments on it appear as Section 5 of Marsden–McCracken [1976].)

Hsu, C. S. [1977]. On nonlinear parametric excitation problems. *Adv. Appl. Mech.*, **17**, 245–301.

Huberman, B., and Crutchfield, J. P. [1979]. Chaotic states of anharmonic systems in periodic fields. *Phys. Res. Lett.*, **43** (23), 1743–1747.

Huppert, H. E., and Moore, D. R. [1976]. Nonlinear double-diffusive convection. *J. Fluid. Mech.*, **78**, 821–854.

Iooss, G. [1979]. *Bifurcation of Maps and Applications*. Mathematical Studies, Vol. 36. North Holland: Amsterdam.

Iooss, G., and Joseph, D. D. [1981]. *Elementary Stability and Bifurcation Theory*. Springer-Verlag: New York, Heidelberg, Berlin.

Iooss, G., and Langford, W. F. [1980]. Conjectures on the routes to turbulence via bifurcation. In *Nonlinear Dynamics*, R. H. G. Helleman (ed.), pp. 489–505. New York Academy of Sciences: New York.

Irwin, M. C. [1980]. *Smooth Dynamical Systems*. Academic Press: New York.

Ito, R. [1981]. Rotation sets are closed. *Math. Proc. Camb. Phil. Soc.*, **89**, 107–111.

Jakobson, M. V. [1978]. Topological and metric properties of one-dimensional endomorphisms. *Sov. Math. Dokl.*, **19**, 1452–1456.

Jakobson, M. V. [1981]. Absolutely continuous invariant measures for one-parameter families of one-dimensional maps. *Comm. Math. Phys.*, **81**, 39–88.

Jorna, S. (ed.) [1978]. *Topics in Nonlinear Mechanics: A Tribute to Sir Edward Bullard*. A.I.P. Conference Proceedings, Vol. 46, A.I.P.: New York.

Jost, R., and Zehnder, E. [1972]. A generalization of the Hopf bifurcation theorem. *Helv. Phys. Acta*, **45**, 258–276.

Kaplan, J. L., and Yorke, J. A. [1979a]. Chaotic behavior of multidimensional difference equations. In *Functional Differential Equations and Approximation of Fixed Points*, H.O. Peitgen and H. O. Walther (eds.), pp. 228–237. Springer Lecture Notes in Mathematics, Vol. 730, Springer-Verlag: New York, Heidelberg, Berlin.

Kaplan J. L., and Yorke, J. A. [1979b]. Preturbulence, a regime observed in a fluid flow model of Lorenz. *Comm. Math. Phys.*, **67**, 93–108.

Katok, A. B. [1980]. Lyapunov exponents, entropy and periodic points for diffeomorphisms. *Publ. Math. IHES*, **51**, 137–173.

Katok, A. B. [1981]. Dynamical Systems with a hyperbolic structure. In *Three Papers on Dynamical Systems*, A. G. Kušnirenko, A. B. Katok, and V. M. Alekseev (eds.), pp. 43–95. A.M.S.: Providence.

Katok, A. B. [1982]. Some remarks on Birkhoff and Mather twist map theorems. *Ergodic Theory Dyn. Syst.*, **2**, 185–194.

Katok, A. B. [1983]. Periodic and quasi-periodic orbits for twist maps. In *Proceedings, Sitges 1982*, L. Garrido (ed.). Springer-Verlag: Berlin, Heidelberg, New York.

Keener, J. P. [1976]. Secondary bifurcation in nonlinear diffusion reaction equations. *Stud. Appl. Math.*, **55**, 187–211.

Keener, J. P. [1981]. Infinite period bifurcation and global bifurcation branches. *SIAM J. Appl. Math.*, **41** (1), 127–144.

Kelley, A. [1967]. The stable, center stable, center, center unstable and unstable manifolds. *J. Diff. Eqns.*, **3**, 546–570.

Knobloch, E., and Proctor, M. R. E. [1981]. Nonlinear periodic convection in double-diffusive systems. *J. Fluid Mech.*, **108**, 291–316.

Knobloch, E., and Weiss, N. O. [1981]. Bifurcations in a model of double-diffusive convection. *Phys. Lett.*, **85A** (3), 127–130.

Koiller, J. [1984]. A mechanical system with a "wild" horseshoe. *J. Math. Phys.*, **25** (5), 1599–1604.

Kolmogorov, A. N. [1954]. On conservation of conditionally periodic motions under small perturbations of the Hamiltonian. *Dokl. Akad. Nauk. SSSR*, **98**, 527–530.

Kolmogorov, A. N. [1957]. General theory of dynamical systems and classical mechanics. *Proceedings of the 1954 International Congress of Mathematics*, pp. 315–333. North Holland: Amsterdam. (Translated as an Appendix in Abraham and Marsden [1978]).

Kopell, N., and Howard, L. N. [1975]. Bifurcations and trajectories connecting critical points. *Adv. Math.*, **18** (3), 306–358.

Kopell, N., and Washburn, R. B. [1982]. Chaotic motions in the two degrees of freedom swing equations. *IEEE Trans. Circuits Syst.*, CAS **29**, 738–746.

Krylov, N. M., and Bogoliubov, N. N. [1934]. *New Methods of Nonlinear Mechanics in their Application to the Investigation of the Operation of Electronic Generators, I*. United Scientific and Technical Press: Moscow. (Also see Krylov and Bogoliubov [1947].)

Krylov, N. M., and Bogoliubov, N. N. [1947]. *Introduction to Nonlinear Mechanics*. Princeton University Press: Princeton. (Russian original, Moscow, 1937.)

Kušnirenko, A. G., Katok, A. B., and Alekseev, V. M. [1981]. *Three Papers on Dynamical Systems*. A.M.S. Translations Series 2, Vol. 116. A.M.S. Publications: Providence.

Lanford, O. E. [1977]. Computer pictures of the Lorenz attractor. Appendix to Williams [1977].

Lanford, O. E. [1982]. A computer assisted proof of the Feigenbaum conjecture. *Bull. Amer. Math. Soc.*, **6** (3), 427–434.

Langford, W. F. [1979]. Periodic and steady mode interactions lead to tori. *SIAM J. Appl. Math.*, **37** (1), 22–48.

LaSalle, J. P., and Lefschetz, S. [1961]. *Stability by Liapunov's Direct Method with Applications*. Academic Press: New York.

LeCalvez, P. [1985]. Existence d'orbites quasi-périodiques dans les attracteurs de Birkhoff. University of Paris, Orsay (preprint).

Lefschetz, S. (ed.) [1950]. *Contributions to the Theory of Nonlinear Oscillations, I*. Annals of Mathematical Studies, Vol. 20. Princeton University Press: Princeton.

Lefschetz, S. (ed.) [1952]. *Contributions to the Theory of Nonlinear Oscillations, II*. Annals of Mathematical Studies, Vol. 29. Princeton University Press: Princeton.

Lefschetz, S. (ed.) [1956]. *Contributions to the Theory of Nonlinear Oscillations, III*. Annals of Mathematical Studies, Vol. 36. Princeton University Press: Princeton.

Lefschetz, S. [1957]. *Ordinary Differential Equations: Geometric Theory*. Interscience Publishers: New York (Reissued by Dover New York, 1977.)

Lefschetz, S. (ed.) [1956]. *Contributions to the Theory of Nonlinear Oscillations, IV*. Annals of Mathematical Studies, Vol. 41, Princeton University Press: Princeton.

Lefschetz, S., Cesari, L., and LaSalle, J. P. (eds.) [1960]. *Contributions to the Theory of Nonlinear Oscillations, V*. Annals of Mathematical Studies, Vol. 45. Princeton University Press: Princeton.

Levi, M. [1978]. Qualitative analysis of the periodically forced relaxation oscillations. Ph.D. thesis, New York University.

Levi, M. [1981]. Qualitative analysis of the periodically forced relaxation oscillations. *Mem. AMS*, **214**, 1–147.

Levi, M., Hoppensteadt, F., and Miranker, W. [1978]. Dynamics of the Josephson junction. *Quart. Appl. Math.*, **35**, 167–198.

Levinson, N. [1949]. A second-order differential equation with singular solutions. *Ann. Math.*, **50**, 127–153.

Li, T. Y., and Yorke, J. A. [1975]. Period three implies chaos. *Amer. Math. Monthly*, **82**, 985–992.

Liapunov, A. M. [1949]. *Problème Général de la Stabilité du Mouvement*. Annals of Mathematical Studies, Vol. 17. Princeton University Press: Princeton.

Libchaber, A., Faure, S., and Laroche, C. [1983]. Two-parameter study of routes to chaos. *Physica*, **7D**, 73–84.

Libchaber, A., and Maurer, J. [1982]. A Rayleigh-Bénard experiment: helium in a small box. In *Nonlinear Phenomena at Phase Transitions and Instabilities*. T. Riste (ed.), pp. 259–286. Plenum Publication Corp.: New York.

Lichtenberg, A. J., and Lieberman, M. A. [1982]. *Regular and Stochastic Motion*. Springer-Verlag: New York, Heidelberg, Berlin.

Lieberman, M. A. [1980]. Arnold diffusion in Hamiltonian systems with three degrees of freedom. In *Nonlinear Dynamics*, R. H. G. Helleman (ed.), pp. 119–142. New York Academy of Sciences, New York.

Lorenz, E. N. [1963]. Deterministic non-periodic flow. *J. Atmos. Sci.*, **20**, 130–141.

Lorenz, E. N. [1964]. The problem of deducing the climate from the governing equations. *Tellus*, **16** (1), 1–11.

Lozi, R. [1978]. Un attracteur étrange? du type attracteur de Hénon. *J. Phys.* (Paris), **39** (C5), 9–10.

Mañé, R. [1981]. *On the Dimension of the Compact Invariant Sets of Certain Nonlinear Maps*. Springer Lecture Notes in Mathematics, Vol. 898, 230–242. Springer-Verlag: New York, Heidelberg, Berlin.

Manning, A. (ed.) [1975]. *Dynamical Systems, University of Warwick 1974*. Springer Lecture Notes in Mathematics, Vol. 468. Springer-Verlag: New York, Heidelberg, Berlin.

Marcus, P. S. [1981]. Effects of truncation in modal representations of thermal convection. *J. Fluid Mech.*, **103**, 241–256.

Marion, J. B. [1970]. *Classical Dynamics of Particles and Systems*. Academic Press: New York.

Markus, L. [1971]. *Lectures in Differentiable Dynamics*. A.M.S. Publications: Providence.

Marsden, J. E. [1981]. *Lectures on Geometric Methods in Mathematical Physics*. S.I.A.M. Publications: Philadelphia.

Marsden, J. E., and McCracken, M. [1976]. *The Hopf Bifurcation and Its Applications*. Springer-Verlag: New York, Heidelberg, Berlin.

Mather, J. N. [1982a]. Existence of quasi-periodic orbits for twist homeomorphisms of the annulus. *Topology*, **21**, 457–467.

Mather, J. N. [1982b]. Non-uniqueness of solutions of Percival's Euler–Lagrange equation. *Comm. Math. Phys.*, **86**, 465–476.

Mather, J. N. [1984]. More Denjoy minimal sets for area preserving diffeomorphisms. Princeton University (preprint).

Maynard-Smith, J. [1971]. *Mathematical Ideas in Biology*. Cambridge University Press: Cambridge.

Mees, A. I. [1981]. *Dynamics of Feedback Systems*. Wiley: New York.

Melnikov, V. K. [1963]. On the stability of the center for time periodic perturbations. *Trans. Moscow Math. Soc.*, **12**, 1–57.

Meriam, J. L. [1975]. *Dynamics*, 2nd ed. Wiley: New York.

Metropolis, N., Stein, M. L., and Stein, P. R. [1973]. On finite limit sets for transformations on the unit interval. *J. Combin. Theor.*, **A15**, 25–44.

Milnor, J., and Thurston, R. [1977]. On iterated maps of the interval I and II. Unpublished notes, Princeton University Press: Princeton.

Minorsky, N. [1962]. *Nonlinear Oscillations*. Van Nostrand: New York.

Misiurewicz, M. [1980]. The Lozi mapping has a strange attractor. In *Nonlinear*

*Dynamics*, R. H. G. Helleman (ed.), pp. 348–358. New York Academy of Sciences: New York.

Misiurewicz, M. [1981]. The structure of mapping of an interval with zero entropy. *Publ. Math. IHES*, **53**, 5–16.

Moon, F. C., and Holmes, P. J. [1979]. A magnetoelastic strange attractor. *J. Sound Vib.*, **65** (2), 285–296.

Moon, F. C., and Holmes, P. J. [1980]. Addendum: a magnetoelastic strange attractor. *J. Sound Vib.*, **69** (2), 339.

Morosov, A. D. [1973]. Approach to a complete qualitative study of Duffing's equation. *USSR J. Comp. Math. and Math. Phys.*, **13**, 1134–1152.

Moser, J. [1962]. On invariant curves of area-preserving mappings of an annulus. *Nachr. Akad. Wiss. Göttingen Math. Phys. Kl.*, **2**, 1–20.

Moser, J. [1973]. *Stable and Random Motions in Dynamical Systems*. Princeton University Press: Princeton.

Murdock, J., and Robinson, C. [1980]. Qualitative dynamics from asymptotic expansions: local theory. *J. Diff. Eqns.*, **36**, 425–441.

Nauenberg, M. and Rudnick, J. [1981]. Universality and power spectrum at the onset of chaos. *Phys. Rev.*, **B27**, 493–498.

Nayfeh, A. H., and Mook, D. T. [1979]. *Nonlinear Oscillations*. Wiley: New York.

Neishtadt, A. I. [1984]. The separation of motions in systems with rapidly rotating phase. *Prikl. Matem. Mekhan.* **48** (2), 197–204.

Nemytskii, V. V., and Stepanov, V. V. [1960]. *Qualitative Theory of Differential Equations*. Princeton University Press: Princeton. (Russian original, Moscow, 1949.)

Neu, J. C. [1979]. Coupled chemical oscillators. *SIAM J. Appl. Math.*, **37** (2), 307–315.

Newhouse, S. E. [1970]. Nondensity of Axiom A (a) on $S^2$. *Proc. Symp. Pure Math.*, **14**, 191–202.

Newhouse, S. E. [1974]. Diffeomorphisms with infinitely many sinks. *Topology*, **13**, 9–18.

Newhouse, S. E. [1979]. The abundance of wild hyperbolic sets and non-smooth stable sets for diffeomorphisms. *Publ. Math. IHES*, **50**, 101–151.

Newhouse, S. E. [1980]. Lectures on dynamical systems. In *Dynamical Systems*, C.I.M.E. Lectures Bressanone, Italy, June 1978, pp. 1–114. Progress in Mathematics, No. 8, Birkhäuser-Boston: Boston.

Newhouse, S. E., Palis, J., and Takens, F. [1976]. Stable arcs of diffeomorphisms. *Bull. Amer. Math. Soc.*, **82**, 491–502.

Newhouse, S. E., Palis, J., and Takens, F. [1983]. Bifurcations and stability of families of diffeomorphisms. *Publ. Math. IHES*, **57**, 5–72.

Newhouse, S. E., Ruelle, D., and Takens, F. [1978]. Occurrence of strange axiom A attractors near quasiperiodic flows on $T^m$, $m \geq 3$. *Comm. Math. Phys.*, **64**, 35–40.

Nitecki, Z. [1971]. *Differentiable Dynamics*. M.I.T. Press.: Cambridge.

Nitecki, Z., and Robinson, C. (eds.) [1980]. *Global Theory of Dynamical Systems*. Springer Lecture Notes in Mathematics, Vol. 819. Springer-Verlag: New York, Heidelberg, Berlin.

Novak, M., and Davenport, A. G. [1970]. Aeroelastic instability of prisms in turbulent flow. *Proc. ASCE J. Eng. Mech. Div. EMI*, **96**, 17–39.

Oseledec, V. I. [1968]. A multiplicative ergodic theorem: Liapunov characteristic numbers for dynamical systems. *Trans. Moscow Math. Soc.*, **19**, 197–231.

Palis, J. [1969]. On Morse–Smale dynamical systems. *Topology*, **8**, 385–405.

Palis, J., and de Carmo, M. (eds.) [1977]. *Geometry and Topology, Rio de Janeiro, 1978*. Springer Lectures Notes in Mathematics, Vol. 597, Springer-Verlag: New York, Heidelberg, Berlin.

Palis, J., and de Melo, W. [1982]. *Geometric Theory of Dynamical Systems: An Introduction*. Springer-Verlag: New York, Heidelberg, Berlin.

Palis, J., and Smale, S. [1970]. *Structural Stability Theorems*. A.M.S. Proceedings of Symposium on Pure Mathematics, Vol. 14, 223–232. A.M.S. Publications: Providence.

Parkinson, G. V. [1974]. Mathematical models of flow induced vibrations of bluff bodies. In *Proceedings of the International Association for Hydraulics Research Symposium on Flow-Induced Structural Vibrations*, Karlsruhe, Germany, pp. 81–127.

Peixoto, M. M. [1962]. Structural stability on two-dimensional manifolds. *Topology*, **1**, 101–120.

Peixoto, M. C., and Peixoto, M. M. [1959]. Structural stability in the plane with enlarged boundary conditions. *Ann. Acad. Brasil. Ciencias.*, **31**, 135–160.

Peixoto, M. M. (ed.) [1973]. *Dynamical Systems*. Academic Press: New York.

Percival, I. C. [1980]. Variational principles for invariant tori and cantori. In *Symposium on Nonlinear Dynamics and Beam–Beam Interactions*, M. Month and J. C. Herrara (eds.), A.I.P. Conference Proceedings, Vol. 57, 310–320. A.I.P.: New York.

Pesin, J. B. [1977]. Characteristic Lyapunov exponents and smooth ergodic theory. *Russ. Math. Surv.*, **32**, 55–114.

Piangiani, G., and Yorke, J. A. [1979]. Expanding maps on sets which are almost invariant: decay and chaos. *Trans. Amer. Math. Soc.*, **252**, 351–366.

Plykin, R. [1974]. Sources and sinks for *A*-diffeomorphisms. *USSR Math. Sb.*, **23**, 233–253.

Poincaré, H. [1880–1890]. *Mémoire sur les córbes définies par les équations différentielles I-VI*, Oeuvre I. Gauthier-Villar: Paris.

Poincaré, H. [1890]. Sur les équations de la dynamique et le problème des trois corps. *Acta Math.*, **13**, 1–270.

Poincaré, H. [1899]. *Les Méthodes Nouvelles de la Mécanique Céleste*, 3 Vols. Gauthier-Villars: Paris.

Pomeau, Y. and Manneville, P. [1980]. Intermittent transition to turbulence in dissipative dynamical systems. *Commun. Math. Phys.*, **74**, 189–197.

Poston, T., and Stewart, I. [1978]. *Catastrophe Theory and Its Applications*. Pitman: London.

Pounder, J. R., and Rogers, T. D. [1980]. The geometry of chaos: dynamics of a nonlinear second-order difference equation. *Bull. Math. Biol.*, **42** (4), 551–597.

Pugh, C. C. [1967a]. The closing lemma. *Amer. J. Math.*, **89**, 956–1009.

Pugh, C. C. [1967b]. An improved closing lemma and a general density theorem. *Amer. J. Math.*, **89**, 1010–1022.

Pugh, C. C. [1969]. On a theorem of P. Hartman. *Amer. J. Math.*, **91**, 363–367.

Pugh, C. C., and Shub, M. [1970]. Linearization of normally hyperbolic diffeomorphisms and flows. *Invent. Math.*, **10**, 187–198.

Pustylnikov, L. D. [1978]. Stable and oscillating motions in non-autonomous dynamical systems. *Trans. Moscow Math. Soc.*, **14**, 1–101.

Rand, D. A. [1978]. The topological classification of Lorenz attractors. *Math. Proc. Camb. Phil. Soc.*, **83**, 451–460.

Rand, D. A., Ostlund, S., Sethna, J., and Siggia, E. D. [1982]. A universal transition from quasi-periodicity to chaos in dissipative systems. *Physica*, **8D**, 303–342 [1983].

Rand, D. A., and Young, L. S. (eds.) [1981]. *Dynamical Systems and Turbulence*. Springer Lecture Notes in Mathematics, Vol. 898. Springer-Verlag: New York, Heidelberg, Berlin.

Rand, R. H. [1984]. *Computer Algebra in Applied Mathematics: An Introduction to MACSYMA*. Pitman: Boston, London, Melbourne.

Rand, R. H., and Holmes, P. J. [1980]. Bifurcation of periodic motions in two weakly coupled van der Pol oscillators. *Int. J. Nonlinear Mech.*, **15**, 387–399.

Reyn, J. W. [1979]. A stability criterion for separatrix polygons in the phase plane. *Niew Arch. Voor Wiskunde*, **38** (3), 238–254.

Robbin, J. [1972]. Topological conjugacy and structural stability for discrete dynamical systems. *Bull. Amer. Math. Soc.*, **78**, 923–952.

Robbins, K. A. [1979]. Periodic solutions and bifurcation structure at high $R$ in the Lorenz model. *SIAM J. Appl. Math.*, **36** (3), 457–472.

Robinson, C. [1981a]. Differentiability of the stable foliation of the model Lorenz equations. In *Dynamical Systems and Turbulence*, D. A. Rand and L.-S. Young (eds.), pp. 302–315. Springer Lecture Notes in Mathematics, Vol. 898. Springer-Verlag: New York, Heidelberg, Berlin.

Robinson, C. [1981b]. Stability of periodic solutions from asymptotic expansions. In *Classical Mechanics and Dynamical Systems*, R. Devaney and Z. Nitecki (eds.), pp. 173–185. Marcel Dekker: New York, Basel.

Robinson, C. [1982]. Bifurcation to infinitely many sinks. *Comm. Math. Phys.*, **90**, 433–459 [1983].

Robinson, C. [1983]. Sustained resonance for a nonlinear system with slowly varying coefficients. *SIAM J. Math. Anal.*, **14**, 847–860.

Robinson, C. [1985]. Horseshoes for autonomous systems using the Melnikov integral. Northwestern University (preprint).

Ruelle, D. [1977]. Applications conservant une mesure absolument continue par rapport à dx sur [0, 1]. *Comm. Math. Phys.*, **55**, 47–52.

Ruelle, D. [1979]. Sensitive dependence on initial condition and turbulent behavior of dynamical systems. In *Bifurcation Theory and Applications in Scientific Disciplines*, O. Gurel and O. E. Rössler (eds.), pp. 408–446. New York Academy of Sciences: New York.

Ruelle, D. [1981]. Small random perturbations of dynamical systems and the definition of attractors. *Comm. Math. Phys.*, **82**, 137–151.

Ruelle, D., and Takens, F. [1971]. On the nature of turbulence. *Comm. Math. Phys.*, **20**, 167–192; **23**, 343–344.

Rüssman, H. [1970]. Über invariante Kurven differenzierbarer Abbildungen eines Kreisringes. *Nachr. Akad. Wiss. Göttingen II Math. Phys. Kl.*, 67–105.

Salam, F. M. A., and Sastry, S. [1985]. Dynamics of the forced Josephson junction circuit: the regions of chaos. *IEEE Trans. Circuits Syst.* (to appear).

Salzmann, B. [1962]. Finite amplitude free convection as an initial value problem. *J. Atmos. Sci.*, **19**, 239–341.

Sanders, J. A. [1980]. A note on the validity of Melnikov's method. Report No. 139, Wiskundig Seminarium, Vrije Universiteit, Amsterdam.

Sanders, J. A. [1982]. Melnikov's method and averaging. *Celestial Mechanics*, **28**, 171–181.

Sanders, J. A., and Chow, S.-N. [1984]. On the number of critical points of the period. Rapport No. 262, Subfaculteit Wiskunde en informatica, Vrije Universiteit, Amsterdam.

Sanders, J. A., and Cushman, R. [1984b]. Abelian integrals and global Hopf bifurcations. Rapport No. 265, Subfaculteit Wiskunde en informatica, Vrije Universiteit, Amsterdam.

Sanders, J. A., and Chow, S.-N. [1984]. On the number of critical points of the period. Rapport No. 262, Subfaculteit Wiskunde en informatica, Vrije Universiteit, Amsterdam.

Sanders, J. A., and Verhulst, F. [1982]. *The Theory of Averaging* (in preparation).

Šarkovskii, A. N. [1964]. Coexistence of cycles of a continuous map of a line into itself. *Ukr. Math. Z.*, **16**, 61–71.

Sattinger, D. H. [1973]. *Topics in Stability and Bifurcation Theory*. Springer Lectures Notes in Mathematics, Vol. 309. Springer-Verlag: New York, Heidelberg, Berlin.

Schecter, S. [1985a], The saddle-node separatrix loop bifurcation. North Carolina State University (preprint).

Schecter, S. [1985b]. Melnikov's method at a saddle-node and the dynamics of the forced Josephson junction. North Carolina State University (preprint).

Schuster, H. G. [1984]. *Deterministic Chaos, An Introduction*. Physik-Verlag: Weinheim.

Seigmann, W. L., and Rubenfeld, L. A. [1975]. A nonlinear model for double-diffusive convection. *SIAM J. Appl. Math.*, **29**, 540–557.

Shenker, S. J. [1982]. Scaling behavior in a map of a circle onto itself: empirical results. *Physica*, **5D**, 405–411.

Shiraiwa, K. [1981]. *Bibliography for Dynamical Systems*. Department of Mathematics, Nagoya University.

Shiraiwa, K. [1985]. *Bibliography for Dynamical Systems, March 1985*. Preprint Series, No. 1. Department of Mathematics, Nagoya University.

Shub, M. [1978]. Stabilité globale des systèmes dynamiques. *Astérisque*, **56.**

Shub, M. [1982]. Personal communication.

Siegel, C. L. [1952]. Über die normal form analytischer Differential-Gleichungen in der Nähe einer Gleichgewichtslösung. *Nach. der Akad. Wiss. Göttingen*, 21–30.

Sijbrand, J. [1981]. Studies in nonlinear stability and bifurcation theory, Ph.D. thesis, Rijksuniversiteit Utrecht.

Šilnikov, L. P. [1965]. A case of the existence of a denumerable set of periodic motions. *Sov. Math. Dokl.*, **6**, 163–166.

Šilnikov, L. P. [1967]. The existence of a denumerable set of periodic motions in four-dimensional space in an extended neighborhood of a saddle-focus. *Sov. Math. Dokl.*, **8** (1), 54–58.

Šilnikov, L. P. [1970]. A contribution to the problem of the structure of an extended neighborhood of a rough equilibrium state of saddle-focus type. *Math. USSR Sb.*, **10** (1), 91–102.

Sinai, J. G. [1976]. *Introduction to Ergodic Theory*. Princeton University Press: Princeton.

Sinai, J. G., and Vul, E. [1981]. Hyperbolicity conditions for the Lorenz model. *Physica*, **2D**, 3–7.

Singer, D. [1978]. Stable orbits and bifurcations of maps of the interval. *SIAM J. Appl. Math.*, **35**, 260–267.

Smale, S. [1961]. On gradient dynamical systems. *Ann. Math.*, **74**, 199–206.

Smale, S. [1963]. Diffeomorphisms with many periodic points. In *Differential and Combinatorial Topology*, S. S. Cairns (ed.), pp. 63–80. Princeton University Press: Princeton.

Smale, S. [1967]. Differentiable dynamical systems. *Bull. Amer. Math. Soc.*, **73**, 747–817.

Smale, S. [1980]. *The Mathematics of Time: Essays on Dynamical Systems, Economic Processes and Related Topics*. Springer-Verlag: New York, Heidelberg, Berlin.

Sotomayor, J. [1973]. Generic bifurcations of dynamical systems. In *Dynamical Systems*, M. M. Peixoto (ed.), pp. 549–560. Academic Press: New York.

Sparrow, C. [1982]. *The Lorenz Equations*, Springer-Verlag: New York, Heidelberg, Berlin.

Steen, P., and Davis, S. H. [1982]. Quasi-periodic bifurcation in nonlinearly coupled oscillators near a point of strong resonance. *SIAM J. Appl. Math.*, **42** (6) 1345–1368.

Stefan, P. [1977]. A theorem of Sarkovskii on the existence of periodic orbits of continuous endomorphisms of the real line. *Comm. Math. Phys.*, **54**, 237–248.

Sternberg, S. [1958]. On the structure of local homeomorphisms of Euclidean $n$-space, II. *Amer. J. Math.*, **80**, 623–631.

Stoker, J. J. [1950]. *Nonlinear Vibrations*. Wiley: New York.

Stoker, J. J. [1980]. Periodic forced vibrations of systems of relaxation oscillators. *Comm. Pure Appl. Math.*, **33**, 215–240.

Sullivan, D. [1984]. Quasi-conformal homeomorphisms and dynamics, I–III. City College, New York (preprints).

Takens, F. [1973a]. Normal forms for certain singularities of vector fields. *Ann. Inst. Fourier*, **23**, 163–195.

Takens, F. [1973b]. Unfoldings of certain singularities of vector fields: generalized Hopf bifurcations. *J. Diff. Eqns.*, **14**, 476–493.

Takens, F. [1973c]. Introduction to global analysis. *Comm. Math. Inst., Rijksuniversiteit Utrecht*, **2**, 1–111.

Takens, F. [1974a]. Singularities of vector fields. *Publ. Math. IHES*, **43**, 47–100.

Takens, F. [1974b]. Forced oscillations and bifurcations. *Comm. Math. Inst., Rijkuniversiteit Utrecht*, **3**, 1–59.

Takens, F. [1980]. Detecting strange attractors in turbulence. In *Dynamical Systems and Turbulence*, D. A. Rand and L.-S. Young (eds.), pp. 366–381. Springer Lecture Notes in Mathematics, Vol. 898, Springer-Verlag: New York, Heidelberg, Berlin.

Taylor, D. L. [1980]. Nonlinear stability and response of car-trailer combinations. In *S.A.E. Transactions 8-00152*, pp. 944–957.

Thirring, W. [1978]. *A Course in Mathematical Physics, I: Classical Dynamics*. Springer-Verlag: New York, Heidelberg, Berlin.

Thom, R. [1975]. *Structural Stability and Morphogenesis*. W. A. Benjamin: Reading, MA. (Original edition, Paris, 1972.)

Thompson, J. M. T., and Hunt, G. W. [1973]. *A General Theory of Elastic Stability*. Wiley: New York.

Tresser, C. [1982]. On some theorems of L. P. Silnikov and some applications. Université de Nice: Nice. (preprint).

Tresser, C. [1983]. Un théorème de Silnikov en $C^{1.1}$. *C.R. Acad. Sci. Paris* I, **296**, 545–548.

Tresser, C. [1984a]. About some theorems of L. P. Silnikov. *Ann. Inst. H. Poincaré*, **40**, 441–461.

Tresser, C. [1984b]. Homoclinic orbits for flows in $R^3$. *J. Phys. Lett. Paris*, **45** (5), 837–841.

Turner, J. S. [1973]. *Buoyancy Effects in Fluids*. Cambridge University Press: Cambridge.

Ueda, Y. [1981a]. Personal communication.

Ueda, Y. [1981b]. Explosion of strange attractors exhibited by Duffing's equation. In *Nonlinear Dynamics*, R. H. G. Helleman (ed.), pp. 422–434. New York Academy of Sciences: New York.

Ulam, S. M., and von Neumann, J. [1947]. On combinations of stochastic and deterministic processes. *Bull. Amer. Math. Soc.* **53**, 1120.

Uppal, A., Ray, W. H., and Poore, A. [1974]. On the dynamic behavior of continuous stirred tank reactors. *Chem. Eng. Sci.*, **29**, 967–985.

van der Pol, B. [1927]. Forced oscillations in a circuit with nonlinear resistance (receptance with reactive triode). *London, Edinburgh and Dublin Phil. Mag.*, **3**, 65–80. (Reprinted in Bellman and Kalaba [1964].)

van der Pol, B., and van der Mark, J. [1927]. Frequency demultiplication. *Nature*, **120**, 363–364.

van Gils, S. A., Carr, J., and Sanders, J. A. [1985]. Nonresonant bifurcations with symmetry. Rapport No. 287, Subfaculteit Wiskunde en Informatica, Vrije Universiteit, Amsterdam.

van Strien, S. J. [1979]. Center manifolds are not $C^\infty$. *Math. Z.*, **166**, 143–145.

van Strien, S. J. [1981]. On the bifurcations creating horseshoes. In *Dynamical Systems and Turbulence*, D. A. Rand and L. S. Young (eds.), pp. 316–351. Springer Lecture Notes in Mathematics, Vol. 898. Springer-Verlag: New York, Heidelberg, Berlin.

Wan, Y. H. [1978]. Computations of the stability condition for the Hopf bifurcation of diffeomorphisms on $\mathbb{R}^2$. *SIAM J. Appl. Math.*, **34**, 167–175.

Whitley, D. C. [1982]. The bifurcations and dynamics of certain quadratic maps of the plane. Ph.D. thesis, Southampton University, Southampton.

Whittaker, E. T. [1959]. *A Treatise on the Analytical Dynamics of Particles and Rigid Bodies*, 4th edn. Cambridge University Press: Cambridge.

Wiggins, S. W., and Holmes, P. J. [1985a]. Periodic orbits in slowly varying oscillators. Cornell University (preprint).

Wiggins, S. W., and Holmes, P. J. [1985b]. Periodic orbits in slowly varying oscillators with time-dependent excitation. Cornell University (preprint).

Wiggins, S. W., and Holmes, P. J. [1985c]. Homoclinic orbits in slowly varying oscillators. Cornell University (preprint).

Willard, S. [1970]. *General Topology*. Addison-Wesley: Reading, MA.

Williams, R. F. [1967]. One-dimensional nonwandering sets. *Topology*, **6**, 473–487.

Williams, R. F. [1974]. Expanding attractors. *Publ. Math. IHES*, **43**, 169–203.

Williams, R. F. [1976]. The structure of Lorenz attractors. Northwestern University Press: Evanston, IL (preprint).

Williams, R. F. [1977]. The structure of Lorenz attractors. In *Turbulence Seminar Berkeley 1976/77*, P. Bernard and T. Ratiu (eds.), pp. 94–112. Springer-Verlag: New York, Heidelberg, Berlin.

Williams, R. F. [1979]. The structure of Lorenz attractors. *Publ. Math. IHES*, **50**, 101–152.

Wilson, K. G. [1971a]. The renormalization group and critical phenomena I: Renormalization and the Kadanoff scaling picture. *Phys. Rev.*, **B4**, 3174–3183.

Wilson, K. G. [1971b]. The renormalization group and critical phenomena II: Phase space cell analysis of critical behavior. *Phys. Rev.*, **B4**, 3184–3205.

Wood, L. A., and Byrne, K. P. [1981]. Analysis of a random repeated impact process. *J. Sound Vib.*, **82**, 329–345.

Young, L.-S. [1982]. Dimension, entropy, and lyapunov exponents. *Ergodic Theory and Dynamical Systems* **2**, 109–124.

Zeeman, E. C. [1977]. *Catastrophe Theory: Selected Papers 1972–1977*. Addison-Wesley: Reading, MA.

Zeeman, E. C. [1981]. *1981 Bibliography on Catastrophe Theory*. Mathematics Institute, University of Warwick: Coventry.

Zehnder, E. [1973]. Homoclinic points near elliptic fixed points. *Comm. Pure Appl. Math.*, **26**, 131–182.

Zholondek, K. [1984]. On the versality of a family of symmetric vector fields in the plane. *Math. USSR Sbornik*, **48**, 463–492.

Ziglin, S. L. [1982]. Self-intersection of the complex separatrices and the nonexistence of the integrals in the Hamiltonian systems with one-and-half degrees of freedom. *J. Appl. Math. Mech. (PMM)*, **45** (3), 411–413.

# Index

Primary references are *italicized*

α-limit point, α-limit set   *34, 235, 236*
(Definition 5.2.1)
Address   *307*
Anosov diffeomorphisms   20–21, *261–262*
Area preserving maps   *216–226*
Arnold diffusion   220
Asymptotic measure   *283*
Asymptotically stable (fixed point)   *3–4*
Attracting set   *34, 75, 91, 92, 256–259*
Attractor   *36, 256–257* (Definition 5.4.1),
*259–267*
Autonomous (differential equation,
dynamical system)   2
Autonomous averaged system. *See* Averaged
system
Averaged system   *167*
Averaging (method of)   68, 153 (footnote),
*166–184* (Sections 4.1–4.4), 206
Averaging theorem   *168* (Theorem 4.1.1)


Badly approximated (irrational numbers)
*303*
Band mergings   348
Basin of attraction. *See* Domain of
attraction
Bendixson's criterion   *44* (Theorem 1.8.2)
Bifurcation diagram   105–106, *118–120*
Bifurcation set   71, *119*
Bifurcation value (bifurcation point)   71,
105, 117, *119* (Definition 3.1.1)
Bi-infinite (symbol) sequence   *233*

Blowing up (a degenerate singularity)
*362–364*, 368
Bouncing ball (model for the dynamics of)
*102–116* (Section 2.4)
hyperbolic invariant set   *242–245*
Boundary layer   69
Branch of equilibria   *118–119*
Branched manifold   262
Bridge (of a Cantor set)   *333*


$C^k$-conjugacy   *38* (Definitions 1.7.2–1.7.3)
$C^k$ eigenvalence   *38* (Definitions 1.7.2–1.7.3)
$C^k$ perturbation of size ε   *38* (Definition
1.7.1)
Cantor set   88, 110, *229*, 232, 258, 266,
285–286, *332–334*
Cantor book   *279*
Capacity   *285* (Definition 5.8.3)
Catastrophe theory   356–357
Center   4
Center manifold   *123–138* (Section 3.2)
definition of   *124*
approximation of   *130–138* (Theorems
3.2.2–3.2.3)
nonuniqueness of   124–125
loss of smoothness of   124–127, 380–381
Center manifold theorem   *127* (Theorem
3.2.1)
Center subspace (eigenspace)
for flows   *10–12*
for maps   *17*

# Applied Mathematical Sciences

*(continued from page ii)*